T0391205

Biologically-Inspired Systems

Volume 4

More information about this series at http://www.springer.com/series/8430

Hermann Ehrlich

Biological Materials of Marine Origin

Vertebrates

Hermann Ehrlich
Institute of Experimental Physics
TU Bergakademie Freiberg
Freiberg, Sachsen, Germany

ISSN 2211-0593 ISSN 2211-0607 (electronic)
ISBN 978-94-007-5729-5 ISBN 978-94-007-5730-1 (eBook)
DOI 10.1007/978-94-007-5730-1
Springer Dordrecht Heidelberg New York London

Library of Congress Control Number: 2013934350

Printed on acid-free paper

Springer is part of Springer Science+Business Media (www.springer.com)

Preface

The higher chordate subgroup includes all the vertebrates: fish, amphibians, reptiles, birds, and mammals. All of them are found in marine environments and coastal regions. Probably the animal that more closely defines human thoughts of life in the sea is a fish. In fact, fish are an ancient group of animals whose origins date back more than 500 million years. They are the most common and diverse group of animals with backbones in the ocean and in the world today.

These animals are the real goldmine for material scientists because of their astonishing variety of shapes and sizes, as well as the diversity of biological materials that compose their organs and structures. Herein are only a few examples. Fish possess structures as barbels, claspers, denticles, scales, egg-cases, oral and pharyngeal teeth, bones, otoliths, cartilage, swim bladders, sucking disks, epidermal brushes, fins, pelvic spines and girdle, gills and bony operculums, unculi and breeding tubercles, and even wings in the case of flying fish. All of the listed structures are hierarchically organised from nano to micro and macro scales. They possess very specific biopolymers like collagens, elastoidines, elastins, keratins, and other cross-linked structural macromolecules. Moreover, we can also find such unique biocomposites of fish origin with exotic names as hyaloine, ganoine, or cosmine. Did you know that terms as enameloid, adameloid, coronoin, acrodin, and prelomin are related to fish scales? Or the recent research detailing differences between orthodentine and osteodentine, durodentine and vasodentine, plicidentine and mesodentine, semidentine and petrodentine, or elasmoidine, as forms of dentine in different fish species? If no, I hope you are now intrigued by this book, which was announced in my first monograph entitled *Biological Materials of Marine Origin: Invertebrates* published by Springer in 2010.

In addition to fish, I also analyse biological materials from marine turtles, iguanas, snakes, and crocodiles as well as sea birds. Special attention is paid to whales and dolphins, as representatives of marine mammals. In terms of species number, marine mammals are a relatively small taxonomic group; yet given their biomass and position in the food web, they represent an ecologically important part of marine biodiversity. Furthermore they are of significant conservation concern, with 23 % of species currently threatened by extinction. Therefore, marine mammals often feature prominently in marine conservation planning and protected area design.

Both non-mineralized and biomineral-containing structures have been described and discussed. Thus, bone, teeth, otoconia and otoliths, egg shells, biomagnetite, and silica-based minerals are analyzed as biominerals. A separate chapter is dedicated to pathological biomineralization. Furthermore, in this book, I take the liberty to introduce the term "Biohalite" for the biomineralized excretion produced by the salt glands of marine fish, reptiles, and birds. Further chapters are dedicated to material design principles, tissue engineering, material engineering, and robotics. Marine structural proteins are discussed from the biomedical point of view.

Altogether, the recent book consist of four parts: 14 chapters, including Introduction, addendums, an epilogue, and addendums to each chapter including more than 2,000 references. Many of the photos are shown here for the first time. I have also paid much attention to the historic factors, as it is my opinion that the names of the discoverers of unique biological structures should not be forgotten. As this is highly interdisciplinary research, fully satisfying the curiosity of expert readers is difficult to do in this rather short survey of a very broad field. However, I hope it will provoke thought and inspire further work in both applied and basic research areas.

There are so many institutions and individuals to whom I am indebted for the gift or loan of material for study that to mention them all would add pages to this monograph. It may be sufficient to say that without their cooperation, this work could hardly have been attempted. First of all, I am very grateful to Prof. Kurt Biedenkopf and his wife Mrs. Ingrid Biedenkopf as well as to the German Research Foundation (DFG, Project EH 394/3-1) for financial support. I also thank Prof. Catherine Skinner, Prof. Edmund Bäeuerlein, Prof. Victor Smetacek, Prof. Dan Morse, Prof. Peter Fratzl, Prof. Matthias Epple, Prof. George Mayer, Prof. Christine Ortiz, Prof. Marcus Buehler, Prof. Andrew Knoll, Prof. Adam Summers, Prof. Stanislav N. Gorb, Prof. Arthur Veis, Prof. Gert Wörheide, Prof. Alexander Ereskovsky, Prof. Hartmut Worch, and Prof. Dirk-Carl Meyer for their support and permanent interest in my research. Especially I would like to thank Prof. Bernd Meyer and Dr. Andreas Handschuh for the excellent scientific atmosphere at TU Bergakademie Freiberg where I enjoyed the time to prepare this work. I am grateful to Prof. Joseph L. Kirschvink, Dr. Martin T. Nweeia, and Dr. Regina Campbell-Malone for their helpful discussions of some chapters, and to Dr. Vasilii V. Bazhenov, Marcin Wysokowski, Dr. Andrey Bublichenko, Dr. Yuri Yakovlev, Alexey Rusakov, and Andre Ehrlich for their technical assistance. To Dr. Allison L. Stelling, I am thankful for taking excellent care of manuscripts and proofs. To my parents, my wife, and my children, I am under deep obligation for their patience and support during the years.

Freiberg, Germany Hermann Ehrlich

Structure and function of biological systems as inspiration for technical developments

Throughout evolution, organisms have evolved an immense variety of materials, structures, and systems. This book series deals with topics related to structure-function relationships in diverse biological systems and shows how knowledge from biology can be used for technical developments (bio-inspiration, biomimetics).

Contents

Part I
Biomaterials of Vertebrates Origin. An Overview

Chapter 1
Introduction

Abstract Marine vertebrates include fish, amphibians, reptiles, birds, and mammals. The Part I describes the classification of marine vertebrates. Included is information about the broad diversity seen in specific biological materials. These materials include mineralized tissues (cartilage, bones, teeth, dentin, egg shells), biominerals (otoliths and otoconia), and skeletal structures (carapaces, sucking disks, spines, scales, scutes, plates, denticles etc.). Elastomers (egg case) and structural proteins (collagen, keratins) are also mentioned. Special attention is payed to the biomimetic applications of biomaterials originating from marine vertebrates.

1.1 Species Richness and Diversity of Marine Vertebrates

The diversity of life forms on Earth is one of the most intriguing aspects for human community. Therefore, knowing how many species inhabit the planet is one of the most fundamental questions in modern science (Mora et al. 2011). The taxonomic classification of animal species into higher taxonomic groups (from genera to phyla) follows a consistent pattern from which the total number of species in any taxonomic group can be predicted. Assessment of this pattern for all kingdoms of life on Earth predicts about 8.7 million species globally, of which ca. 2.2 million are marine (Poore and Wilson 1993; Briggs and Shelgrove 1999; Bouchet 2006; Appeltans et al. 2012). It suggests that some 86 % of the species on Earth, and 91 % in aquatic niches, still await description (Mora et al. 2011).

Vertebrates, as important players in nearly all marine food webs, occupy all marine habitats. The vertebrates in the ocean include fish, amphibians, reptiles, birds, and mammals. The fish are the most successful in terms of numbers of individuals as well as numbers of species (ca. 50 % of living vertebrates) (Berg 1940; Long 1995; Nelson 2006) and below, give an overview of classifications for marine vertebrates. I include additional information about common and specific biological materials like mineralized tissues, skeletal structures (spines, scales, denticles), elastomers, structural proteins etc.

Among the most structurally complex organisms, marine vertebrates are classified under the Kingdom Animalia, Phylum Chordata and Subphylum Vertebrata. The four main marine superclasses and classes, as well as one representative of marine amphibians in Vertebrata, are discussed below.

H. Ehrlich, *Biological Materials of Marine Origin*, Biologically-Inspired Systems 4,
DOI 10.1007/978-94-007-5730-1_1

1.2 Part I: Biomaterials of Vertebrate Origin. An Overview

1.2.1 Supraclass Agnatha (Jawless Fishes)

Fossil evidence indicates that the group of agnathan (jawless fishes) species was once highly successful and extremely varied. The oldest fossil remnants of agnathans were found in Cambrian rocks. Only two groups, the lampreys and the hagfish, with about 100 species in total, still survive today. The relationship between the two clades, however, has not been resolved. There are two competing views: the cyclostome (circular mouth) hypothesis and the vertebrate hypothesis. In the first, hagfish together with lampreys form a monophyletic group, the Cyclostomata. In the second, lampreys are sister to jawed fishes and all other jawed animals (Gnathostomata) and together form the clade Vertebrata. The hagfishes, which lack vertebrae, are the sister–group to the Vertebrata. The data in support of the cyclostome hypothesis are mostly molecular, whereas those in support of the vertebrate hypothesis are mostly morphological.

The problem of establishing homologies within and among the ingroups and outgroups remains a challenge. It is of interest to note that Linnaeus (1758) classified hagfishes in the class Vermes and the order Intestina (intestinal worms) and lampreys in the class Amphibia and the order Nantes (swimming amphibians), erroneous placements that nevertheless reflect their great divergence. Both, identifiable stomach or any appendages have not been identified in all living and most extinct Agnatha. These animals possess fertilization and development in external form without any parental care. The Agnatha are cold blooded (ectothermic), and have a cartilaginous skeleton. Extensively developed bony plates of many extinct agnathans are localized directly under the skin. These served as protective armor and can be most often found in the region of the skull. The extant agnathan species possess no bony plates (see for review Xian–Guang et al. 2002; Janvier 2010; Renaud 2011).

1.2.1.1 Order Osteostraci

The Osteostraci are jawless and represent the sister taxon to jawed vertebrates. Principally, they are integral to understanding the evolution of gnathostomes from a jawless ancestor (Sansom 2009). The Osteostraci as a relatively compact group, first appeared in the Late Silurian, flourished in the Early Devonian, but were represented by only a few survivors by the Middle and Late Devonian (Robertson 1935). The surface of the exoskeleton of these animals was smooth, however possess some small dorsal tubercles. Also the pores and grooves of the sensory canal system and the related lateral lines were to detect. The thin layer of enamel, underlain by a much thicker layer of dentine-like tissue, was located externally. Together, the two formed the superficial layer. The dentine was perforated with tubules that arose from a network of small vascular canals at the very top of the middle layer (Denison 1947).

1.2.1.2 Order Anaspida

The anaspida were a now extinct order of fish-like vertebrates. In contrast to the Osteostraci, their body, was less flattened and covered with dermal scales including the head region. These animals possess hypocercal (tilting downwards) tail. Usually, Anaspids were up to 15 cm in length. The Anaspids, which ranged from the Late Silurian to the Late Devonian, included: Jaymoytius, Pharyngolepis, and Pterygolepis (Allaby and Allaby 1999).

1.2.1.3 Order Heterostraci

The fossil group of heterostracans (Heterostraci) represents a large clade of the Pteraspidomorphi. These armored, but jawless, vertebrates lived about 430–370 million years ago (from the Early Silurian to the Late Devonian). Their armored head was generally fusiform and a tail fan-shaped. Both, the large dorsal and a large ventral shield shields were formed by two plates. The exact morphological traits with respect to scales are differing from one group of Heterostraci to another. For example, such primitive forms as Lepidaspis, possess dorsal and ventral shields which are composed of a mosaic of tiny scales.

From histological point of view, the scales observed in representatives of this group, are distinct from other vertebrates. Their scales have dentine- and aspidine-based layers as well as an acellular bony tissue that is known to be unique to this class (Halstead Tarlo 1963).

The so called "cancella" represents the honeycombed middle layer. In pteraspidomorphs, the unique biological material aspidin is present both in the attachment of bone associated with the superficial dentine–enameloid tubercles, and in the dermoskeleton which contains characteristic "spongy" and basal "lamellar" layers. Aspidin is dominated by a collagenous fibrillar organic matrix and is acellular (Donoghue and Sansom 2002).

There are about 300 species related to Heterostracans. They habituates were sandy lagoons or deltas including marine environments with exception of some fresh water species. They are known to occur in Europe, Siberia as well as in North America. It is suggested that these animals were poor swimmers and probably bottom-dwellers, and fed by scraping the bottom with their fan-shaped oral plates that armed their lower lip. The representatives of Psammosteidae, developed steer-like branchial plates and could grow up to 1.5 m in length and, however, most heterostracans were relatively small (5–30 cm in total length). Their internal anatomy is only known from the impressions of the internal organs on the internal surface of the dermal armor because heterostracans have no calcified endoskeleton. One may trace the impressions of two distinct vertical semicircular canals of the labyrinth as well as of the gills, eyeballs, paired olfactory organ, and brain. Similar to extant hagfish, their olfactory organs seem to have opened ventrally into a large, median inhalant duct (see for review Halstead 1973; Janvier and Blieck 1979; Janvier 1997a).

1.2.1.4 Order Coelolepida

Thelodonts (from Greek: "nipple teeth", formerly coelolepids) are an ensemble of fossil jawless vertebrates, distinguished from other jawless vertebrate groups by the structure of their minute scales-based exoskeleton (Wilson and Caldwell 1993). These scales superficially resemble the placoid scales of sharks. Thelodonts lived in shallow-water marine environments from the Lowermost Silurian (and possibly the Late Ordovician) to the Late Devonian (430–370 million years ago) (Turner 1992). It is suggested that some thelodonts migrated into fresh water, perhaps to spawn (Van der Brugghen and Janvier 1993).

Probably, thelodonts were closely related and morphologically very similar to fish of the taxa Heterostraci and Anaspida, differing mainly in their covering of distinctive small, spiny scales. The small size and resilience of these scales makes them the most common vertebrate fossil of their time.

The bony scales of the thelodont group were formed and shed throughout the organisms' lifetimes, and quickly separated after their death. Correspondingly, they are the most abundant form of fossil as most resistant materials to the process of fossilisation and thus most useful for analysis because of exceptional preservation of internal details (Piepenbrink 1989). The scales contain an aspidine base and comprise a non-growing "crown" composed of dentine, with a specifically ornamented enameloid upper surface. The cell-free bone is the main element of its growing base, including anchorage structures which fix it to the side of the fish.

It is established that five types of bone-growth, which may represent five natural groupings within the thelodonts exist. Moreover, each of these scale morphs appears to resemble the scales of more derived groupings of fish. Therefore, it can be hypothesized that thelodont groups may have been stem groups to succeeding clades of fish.

The taxa of thelodonts have traditionally been defined on the basis of histological and morphological investigations of their scales. However, because a wide range of scale morphologies can occur in the same individual, some recent studies on articulated thelodonts show that scale morphology can be also misleading. On the basis of their scale morphology and histology, thelodonts are currently classified into following groups: Achanolepida, Loganiida, Turiniida, and Katoporida (see for review Janvier 1997a).

1.2.1.5 Order Cyclostomata

Class Myxini (Myxinoidea)

Hagfishes or Hyperotreti. A taxon of ocean-dwelling fish, which are small and jawless, as well as scavenging their food from both invertebrates and dead and dying fish. They habituate in cold ocean waters of both hemispheres. *Myxinikela siroka* is the only fossil hagfish, that remnants are localized in the Francis Creek Shale of northeastern Illinois (Bardack 1991). Fragments of the head and jaws, paired tentacles, internal organs were found within an iron carbonate (siderite) concretion. Because the similarity to modern hagfishes is striking, it was suggested that little evolutionary change in Myxini has been over the last 300 million years. According to modern point of view (Kuratani and Ota 2008), these animals are unique among living chordates. For example, they have a partial skull, but no vertebrae, and so they are not truly vertebrates. The skeleton lacks bone and is composed of cartilage. Hagfish are almost blind, have no cerebrum or cerebellum, no jaws or stomach, but three accessory hearts. They have four pairs of sensing tentacles arranged around their mouth and also have well developed senses of touch and smell. Interestingly, they can "sneeze" when their nostrils clog with their own slime. Being jawless, a hagfish is equipped with two pairs of tooth-like structures, the rasps, which are located on the top of a tongue-like projection. The pairs of rasps pinch together after this tongue is pulled back into the hagfish's mouth. This bite is used in catching and eating marine invertebrates like polychaete worms, or to tear into the flesh of dying and dead fish which have sunk to the muddy ocean bottom. Principally, their metabolism is very slow, therefore hagfish may go for up to 7 months without eating any food. These vertebrates are slimy and capable of tying their body into a knot. Furthermore, hagfish are known for producing large amounts of slime when stressed (see also Sect. 11.2 in this work). The production of the slime is believed to be some kind of defence mechanism against gill-breathing predators. It was reported that the slime can reduce water flow over the gills of fish. Slime thread skeins and mucin vesicles are two interacting components of hagfish slime. Both are released from glands along the ventrolateral length of these primitive vertebrates (see for review Downing et al. 1981; Lim et al. 2006).

Class Cephalaspidomorphi (Petromyzontida)

Lampreys or Hyperoartii are another group of primitive (Ruud 1954) and jawless fishes. Non-parasitic species are able to eat only in their larval form, dying as adults soon after reproducing. Parasitic species, however, latching onto the bodies of freshwater fish (Renaud and Economidis 2010). Lampreys have no jaws but possess an annular cartilage that supports the supraoral and infraoral laminae. Their body is naked and elongated. They possess seven branchial openings (or pores) on either side of the body. The seven pairs of gill pouches are supported by a surrounding branchial basket consisting of an elaborate network of fused cartilaginous elements. Lamprey cartilage is unique to the group and contains hydrophobic protein lamprin (Robson et al. 2000). The teeth on the oral disc and tongue-like piston of the adult lamprey are made of keratin (Fig. 1.1). They possess a hollow core allowing for a number of replacement teeth to occur one on top of the other. It has been estimated that over the course of 2 years, an adult Sea Lamprey, *Petromyzon marinus*, may replace its teeth about 30 times (Renaud 2011). The skeleton contains no bone, only cartilage, although this cartilage may be calcified. The main axial support for the body is the notochord, which is persistent throughout the life of the animal. Rudimentary vertebral elements termed arcualia are arranged two per myomere on either side along the dorsal nerve cord.

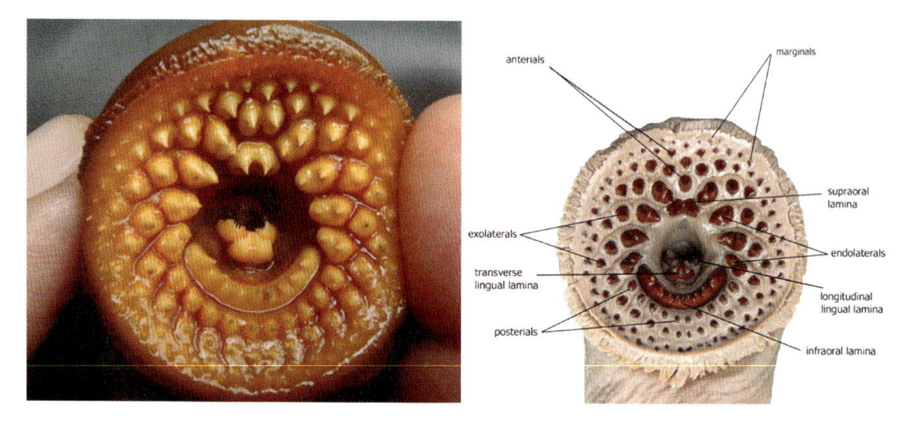

Fig. 1.1 Sea Lamprey teeth are located on the oral disc of lampreys (Photograph by Brian W. Coad, image manipulation by Noel Alfonso. Canadian Museum of Nature www.briancoad.com. Reprinted with permission.)

1.2.2 Gnathostomes

As reviewed by Janvier (1997b), the Gnathostomata, or gnathostomes, comprise the majority of vertebrates from the Middle Devonian (about 380 million years ago) to Recent. These animals differ morphologically from all other vertebrates by having a vertically biting device, the jaws, which consist of and a variety of exoskeletal grasping, crushing, or shearing organs, and endoskeletal mandibular arch. Gnathostomes possess the teeth, and jaw bones (Mallatt 1984). This group of recent vertebrates include chimaeras, sharks, rays, ray-finned and lobe-finned fishes and terrestrial vertebrates (see for review Schultze 2010).

The Chondrichthyes and Osteichthyes are two major clades of extant gnathostomes. In addition, the Placodermi (Early Silurian-Late Devonian) and the Acanthodii (Latest Ordovician or Earliest Silurian – Early Permian) are two extinct major gnathostome clades. Outside of the taxa listed above, there may be other fossil gnathostome groups, too.

The Chondrichthyes are characterized by the prismatic calcified cartilage, a special type of hard tissue lining the cartilages of their endoskeleton. The pelvic clasper is another chondrichthyan characteristic. This special copulatory organ is derived from the posterior part of the pelvic fin (metapterygium). However, a pelvic clasper may be present also in the fossil Placodermi. The Elasmobranchii and the Holocephali represent two major extant clades of Chondrichthyans. Several fossil clades like Cladoselachidae, Symmoriida, Xenacanthiformes, Iniopterygia, Eugeneodontida may fall outside these two clades.

The Osteichthyes possess such specific characteristics in the endoskeleton as endochondral ("spongy") bone, dermal fin rays made up by modified, tile-shaped scales (lepidotrichiae), and three pairs of tooth-bearing dermal bones lining the jaws (premaxillary, maxillary and dentary). The Actinopterygii and the Sarcopterygii are two major clades of the Osteichthyes.

The representatives of Placodermi possess a dermal armor consisting of head armor and a thoracic armor. The thoracic armor is characterized by the foremost dermal plates which form a complete "ring" around the body. It always includes at least one median dorsal plate.

The Acanthodii differ from other clades by dermal spines inserted in front of all fins but the caudal one. These animals also possess minute, growing scales with a special onion-like structure, i.e. the crown consists of overlying dentine or mesodentine-based layers.

1.2.2.1 Superclass Gnathostomata

Jawed fishes (99.7 % of all living fishes) and tetrapods are related to the Superclass Gnathostomata. These vertebrates possess jaws and usually a set of paired appendages. The superclass includes all the tetrapods: amphibians, reptiles, birds, and mammals.

Class Chondrichthyes

Cartilaginous fishes (about 1,000 species) possess primitive characters like cartilaginous endoskeleton, single nostrils, and absence of the gas bladder. Modern studies suggest that these vertebrates have a terminal position in the piscine tree (Rasmussen and Arnason 1999a, b; Botella et al. 2009). They first appear in Upper Silurian, and some fossil record starts in Lower Devonian. Representatives of this class possess placoid scales, bony teeth of ectodermal and mesodermal origin in jaws as well as teeth arranged in replacement whorls. Their endoskeleton is partially covered with prismatically patterned perichondral bone, and their gill septum extending to lamellar margin. The mosaic calcium carbonate-based granule pattern is unique to these fishes. It is known that these mineralized structures on the outside of the cartilage add strength. Both eggs and embryos of Chondrichthyans are large. Similar to ray-finned fishes, representatives of this class have also large adults (from 20 cm and 15 g to 12 m and 12,000 kg).

From biological materials point of view, these fish possess numerous structures with high biomimetic potential because of their unique physical, chemical and material properties (Oeffner and Lauder 2012).

Below are listed some of these formations in alphabetical order:

Barbels – From morphological point of view, barbels are long conical paired dermal lobes on the snouts of sharks. Their function is to locate prey. In contrast to most sharks which have barbels associated with the nostrils, Sawsharks have barbels in front of the nostrils.

Claspers – These paired copulatory organs are located on the pelvic fins of male cartilaginous fishes. Animals use them for internal fertilization of eggs.

Dermal denticle – (or placoid scale) is an example of small tooth-like scale that is unique to cartilaginous fish.

Egg-case – Flexible, horn-like protein-based envelope that surrounds the eggs of cartilaginous fishes. In egg-laying species this is robust and necessary for protection of the egg. However, in live-bearers it is often membranous and soft, and disintegrates while the fetuses are developing.

Rostrum – This cartilaginous structure is necessary to support the snout.

Saw or saw-snout – This elongated snout in sawsharks and sawfish possess numerous side teeth formed from enlarged denticles. Usually is used to kill or dig the prey (Fig. 1.2).

Spin-brush complex – unique spin-based structure described for *Akmonistion zangerli*, one of the widely known representatives of Paleozoic chondrichthyans. The spine of this fish (Fig. 1.3) consists of osteonal dentine surrounded by acellular bone. It lacks any enamel-like surface tissue. The non-prismatic globular

Fig. 1.2 The saw- snout is formed from enlarged denticles. This is an ideal example of bioinspiration for engineers and materials scientists (Image courtesy of Mason Dean)

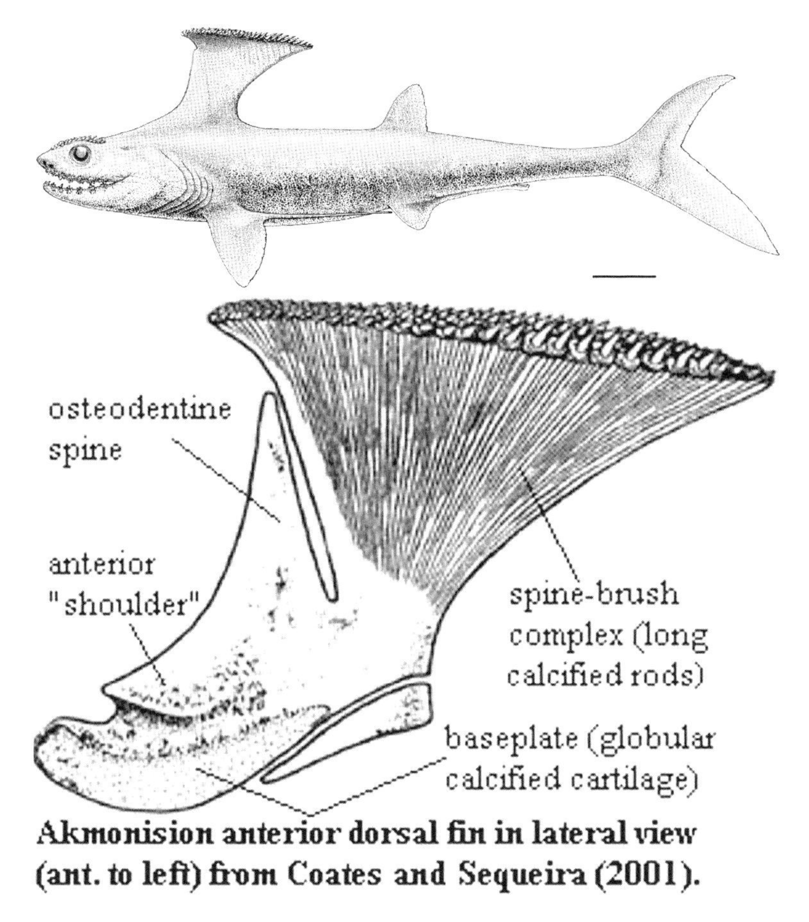

Fig. 1.3 *Akmonistion zangerli* possesses unique Spin-brush complex (Coates and Sequeira (2001), reprinted by permission of Taylor & Francis Ltd, (http://www.tandf.co.uk/journals))

calcified cartilage is the main structural component of the brush and basal plate. The peripheral regions of this specific matrix include a meshwork of crystal fibre bundles. The leading edge and base of the brush are coated with a thin, acellular bone layer of variable thickness (see for details Coates and Sequeira 2001).

Sting – The upper surfaces of the tails of most members of the stingray group (Myliobatoidei) possess this large, flattened spine-like structure with several side barbs.

Subclass Elasmobranchii

Two superorders, Batoidea (rays and their relatives) and Selachii (sharks), are typical representatives of Elasmobranchs. These animals possess cylindrical or flattened bodies covered by placoid scales. They have five to seven pairs of gill slits, their upper jaw not fused to the cranium.

Batoidea include skates, electric rays, stingrays, guitarfish, and sawfish (see for review Daniel 1922). They have five or six pairs of ventral gill slits and are characterized by a dorsoventrally flattened body. Their pectoral fins are fused to the head. There are four orders and ca. 470 species of batoids.

Selachians include all species of sharks. These animals are characterized by a fusiform body and five to seven pairs of lateral gill slits. Today, 8 orders and about 355 species of selachians are described. The diversity of elasmobranchs is well described in the literature (see for review Thiel et al. 2009; Neumann 2006; Howe and Springer 1993; Pollerspöck 2012).

Order Cladoselachiformes

Cladoselachidae (frilled sharks) is an extinct family of cartilaginous fish and among the earliest predecessors of recent sharks. As the only members of the order Cladoselachiformes, these animals were characterized by having an elongated body with a spine in each of the two dorsal fins. According to Martin (2012), *Cladoselache* specimens exhibited a combination of ancestral and derived characteristics. Their jaw joints are weaker in comparison to modern-day sharks. However, they possess very strong jaw-closing muscles. Its teeth were smooth-edged and multi-cusped, making them ideal for grasping, but not tearing or chewing. It is suggested that *Cladoselache* only seized prey by the tail and swallowed it whole. Unlike all modern and most ancient sharks, frilled sharks swam the seas virtually naked. *Cladoselache*'s skin seems to have been almost devoid of the tooth-like scales with exception of small, multi-cusped structures along the edges of their fins, in the mouth cavity, and around the eye.

Cladoselache's scales serve as more than simple armor against injury. These animals strengthen the skin to provide firmer attachments for their swimming muscles. Their fin spines were odd, too.

Being short and blade-like, composed of a porous bony material, and located some distance anterior to the origin of each dorsal fin, their spine fins were very unusual. In contrast to other sharks with denser, more spike-like fin spines, that of *Cladoselache* may have been lighter and sturdier. These structures may have reduced swimming effort yet provided solid discouragement to would-be predators.

Order Xenacanthiformes

Xenacanthiforms are known from the Lower Carboniferous up to the Triassic (Ginter 2004; Fischer et al. 2011). Their characteristics include articulated skeletons and diplodont teeth, i.e. teeth with two lateral cusps evidently larger than the median ones. For example, the recently discovered taxon *Reginaselache morrisi* is identified by its teeth. These robust teeth possess multicristate cusps, as well as prominent rounded coronal button, and a horseshoe-shaped labial boss. This shark was about 1 m long, and probably fed on smaller paleoniscoid, invertebrates and fish (Turner and Burrow 2011).

There is also paleontological evidence that xenacanthids, predominantly adapted to freshwater, have also lived in marine environments (Hampe and Ivanov 2007; Ginter et al. 2010).

Order Selachii (Typical Sharks)

The order Selachii and the class Chondrichthyes (sometimes also called Selachii) include different species of sharks, which majority can be traced back to around 100 million years ago (Martin 2006). These carnivorous animals use of food items ranging from plankton and fish to marine mammals and garbage. The young are born alive as well as hatch within the female. About 350 living species of sharks are estimated today. Some of them are small like pygmy ribbontail catfish shark (less than 30 cm). Others are huge in size like the whale shark with up to 12 m length. These magnificent creatures are able to live in every marine niche from the Arctic to the tropics, and play a crucial role in keeping aquatic wildlife in balance (see for review Compagno 2001). These animals keep prey populations in check and also function as apex predators eating the weakest individuals (Stelbrink et al. 2010).

Living sharks are divided into eight suborders (see for review http://saltwaterlife.co.uk/ws/sharkiologist/articles/shark-evolution-and-classification/):

1. "*Squatiniformes* (Angelsharks) have been around since the Triassic period (200–250 mya). They are comprised of 19 species. They are found mainly in mud and sand from cool temperate continental shelves, intertidal and continental slopes, and in deeper water in the tropics. They are identified by their broad flattened body, short snout, large pectoral and pelvic fins, two dorsal fins towards the end of their tail, no anal fin and five gill slits. They look similar to rays superficially. However, the gill openings are on the sides of the head, not beneath as in rays; and the large pectoral fins are clearly defined and separate from their heads. They have large wide mouths at the front of their head perfectly designed for ambushing their prey as they swim by the often sand covered shark. They are ovoviviparous (they produce live young from eggs which hatch within the body) with litter sizes between 1 and 25.
2. *Pristiophoriformes* (Sawsharks) first evolved during the Jurassic period (160–200 mya). The suborder contains 9 species normally found on the continental and insular shelves, in shallow water in temperate regions and deeper in the tropics. They are probably the most distinct order amongst the shark groups, easily identified by their flattened heads and long, flat, saw-like snout (the rostrum) complete with barbells in front of the nostrils. Lateral and ventral teeth are used to capture and kill prey, and possibly for courtship, competition and defence. The lateral teeth erupt as the young are developing but lie flat along the rostrum until after they are born. The eyes are located on the side of the head and they have large spiracles, two dorsal fins and no anal fins. They are bottom dwelling predators. Although data is lacking for most of the sawshark species it is known that the Sixgill Sawshark *Pliotrema warreni* (the remaining eight species all have five gill slits) is ovoviviparous and produces 5–7 pups per litter.

3. *Squaliformes* (Dogfish) have been around since the Jurassic period (160–200 mya). This is a large and varied order containing 106 identified species in seven families; bramble sharks (*Echinorhinidae*), dogfish sharks (*Squalidae*), gulper sharks (*Centrophoridae*), lantern sharks, (*Etmopteridae*), sleeper sharks (*Somniosidae*), roughsharks (*Oxynotidae*), and kitefin sharks (*Dalatiidae*). The dogfish sharks' habitat is wide ranging, with species found in marine estuarine environments world-wide. Currently they are the only known sharks to be found at high latitudes close to the poles. Their greatest diversity occurs in deepwater. They have a cylindrical/torpedo shaped body with the eyes on the side of the head, 5 gill slits, 2 dorsal fins (in some species these are spined), and no anal fins. They are ovoviviparous, with some species having a low fertility of just 1 pup per litter (e.g.: the gulper shark *Centrophorus granulosus*) and other species having a very long gestation period of 18–24 months (e.g.: the spiny dogfish *Squalus acanthias*). Some species are thought to be solitary; others form schools that range long distances during seasonal and annual migrations.
4. *Hexanchiformes* (Frilled and Cow sharks) are considered to be amongst the most "ancient" forms of living sharks dating back to the Permian period (260–300 mya). They are comprised of 2 families; cow sharks (Hexanchidae of which there are 2 species) and frilled sharks (Chlamydoselachidae of which there are 4 species). The frilled sharks are eel-like with distinctive spaced out teeth, the cow sharks are the more conventional cylindrical shape. Both families have 6 or seven gill slits, 1 dorsal fin, and anal fins are present. Most of the species are found worldwide, predominantly in the deep cold water of the tropics. They are ovoviviparous with frilled sharks producing 6–12 pups per litter and cow sharks ranging from 6 to 108 pups depending on the species.
5. *Carcharhiniformes* (Ground Sharks) are the largest, most diverse and widespread order of sharks. Dating from the Jurassic period (160–200 mya) the ground sharks are comprised of around 247 species in 8 families; catsharks (*Scyliorhinidae*), finback catsharks (*Proscylliidae*), false catsharks (*Pseudotriakidae*), barbeled houndsharks (*Leptochariidae*), houndsharks (*Triakidae*), weasel sharks (*Hemigaleidae*), requiem sharks (*Carcharhinidae*), and hammerhead sharks (*Sphyrnidae*).

   These shark species inhabit cold to tropical seas, intertidal to deep water and pelagic Open Ocean. Their physical appearances can be quite different; from the Daggernose shark (*Isogomphodon oxyrhynchus*) to the Great Hammerhead (*Sphyrna mokarran*), however all have 5 gill slits, 2 dorsal fins (with one exception; the aptly named "Onefin catshark" *Pentanchus profundicolus*) and anal fins.

   Reproductive strategies are also quite varied. Both the catsharks and the finback catsharks are oviparous (egg laying) and ovoviviparous with typically 1–2 eggs or live pups per litter, false catsharks are ovoviviparous (2–4 pups per litter), oophagous (meaning "egg eating", pups feed off eggs produced by the ovary whilst inside the uterus, 2 pups per litter), viviparous (pups nurtured via a placental connection, 7 pups per litter). The barbeled houndsharks reproductive method is unknown. Houndsharks are ovoviviparous (2–52 pups per litter) and viviparous (2–20 pups per litter), the exact method in weasel sharks is unknown but they do bare 1–4 live young, requiem sharks are ovoviviparous (10–80 pups per litter) and viviparous (1–135 pups per litter) and hammerhead sharks are viviparous (30–55 pups per litter).

6. *Lamniformes* (Mackerel sharks) date from the Jurassic Period (160–200 mya), and consist of 15 species in 7 families: thresher sharks (*Alopiidae*), *Cetorhinidae*, *Lamnidae*, *Megachasmidae*, *Mitsukurinidae*, *Odontaspididae*, *Pseudocarchariidae* (no common names exist for the latter 6 families). These species are predominantly large active pelagic sharks with cylindrical/torpedo shaped bodies with 5 gill slits, 2 dorsal fins and anal fins. They are found worldwide from the intertidal zone to deeper water and the open ocean. Reproduction is ovoviviparous (2–25 pups per litter depending on species). In one species, the sandtiger shark (*Carcharias taurus*), cannibalism occurs as the dominant pup will consume the other embryos.
7. *Orectolobiformes* (Carpet sharks) originate from the Jurassic period (160–200 mya). This order consists of 33 species in 7 families: collared carpetsharks (*Parascyllidae*), blind sharks (*Brachaeluridae)*, wobbegongs (*Orectolobidae*), longtail carpetsharks (*Hemiscyllidae*), nurse sharks (*Ginglymostomatidae*), zebra shark (*Stegostomatidae*), whale shark (*Rhincodontidae*). They can be found worldwide in warm temperate and tropical seas, from the intertidal zone to deep water. With the exception of the whale shark (*Rhincodon typus*) all are bottom dwelling, they have 5 gill slits, 2 dorsal fins and anal fins. They utilise a variety of reproductive strategies including oviparity (6–8 egg cases), ovoviviparous (20–30 pups per litter), viviparous (up to 300 pups per litter) and oophagy (litter size unknown).
8. *Heterodontiformes* (Bullhead sharks) date back to the Triassic period (200–260 mya). Comprised of 9 species they are bottom dwelling stout-bodied sharks, with 2 spined dorsal fins and anal fins. Their habitat varies between species and ranges from the intertidal zone to continental and insular shelves. They are nocturnal, preferring to rest in rocky crevices and caves during daylight. Reproduction is oviparous, laying very distinctive egg cases which are screw shaped; the number of eggs laid is unknown." (Smith 2013: «Shark Evolution and Classification», published online: http://saltwaterlife.co.uk/ws/sharkiologist/articles/shark-evolution-and-classification/)

Because of diversity and special properties of their skeletal structures like teeth, denticles, fins, eggs capsules, cartilage as well as biomechanics of their skin and body shape, sharks seem to be a veritable gold mine for material scientists as well as experts in bionics and biomimetics (Fig. 1.4)

Order Batoidea

Skates, stingrays, sawfishes, and guitarfish are representatives of cartilaginous batoid fishes (Claeson 2011), which possess dorsoventrally compressed bodies ranging in shape from circular to rhomboidal (Compagno 1977; Aschliman et al. 2012). In addition to marine habitats, they inhabit freshwater niches on five continents

Fig. 1.4 The "tooth-whorl" of ancient *Helicorpion* sharkstrongly reminiscent of a circular saw (*above*; published on-line http://en.wikipedia.org/wiki/Helicoprion#mediaviewer/File:Spirale_dentaire_d%27helicoprion.jpg © Citron / CC-BY-SA-3.0) Rough sketch of *Helicoprion* made by Todd Marshall (*below*; Reprinted by permission Todd Marshall. Copyright © of Todd Marshall)

(Compagno 2001). Their wing-like structures are greatly enlarged and fused to the cranium, forming large pectoral fins. These specifically modified pectoral fins function as the primary locomotor propulsors and inspired numerous studies to learn about their unique bionics, biomechanics and biodynamics (Blake 1983; Rosenberger 2001; Lauder 2000; Lauder and Drucker 2002; Lauder et al. 2002, 2003; Vogel 2003; Wilga and Lauder 2004; Schaefer and Summers 2005). Most of batoids use their pectoral fins to swim, and fall on a continuum from undulatory to oscillatory locomotion. However, such representatives as electric rays or sawfish, do not use these structures to swim. Instead, similar to their shark relatives, these species rely on the plesiomorphic caudal fin-based locomotor mode. As reviewed by Schaefer and Summers (2005), the skeleton of batoids is cartilaginous, however mineralized to varying degrees. This usually takes the form of a thin layer of tiles, tesserae, arranged on the surface of an unmineralized core. Serially repeating cartilaginous elements are the base elements of the wing skeleton upon which the locomotor waves are propagated.

Subclass Holocephali

Of all groups of living fishes, the Holocephali are the least well known (de Beer and Moy–Thomas 1935). Holocephali (Holocephalomorphi) ("*complete heads*") is a subclass of a mostly extinct species of cartilaginous fish with big heads and a long tails, which relationships to other taxa are considered controversial (see for review Kriwet and Gadzdicki 2003). These animals possess only one single gill opening in each side. Deep-sea species have large eyes. Extant species habituate mostly deep–water environments. They have continuously growing tooth plates in the upper and lower jaws. Typical fossilized remnants of holocephalian chimeroids are exceptionally well preserved tooth plates, fin spines, and egg cases. The oldest record of chimeroid fish was found in Early Jurassic deposits of Europe (Ward and Duffin 1989).

Order Chimaeriformes

Chimaeras split off from the rest of the groups earliest and therefore are known as the most primitive representatives of the cartilaginous fish (Patterson 1965). They have no difficulties to crunch very hard food including mollusc's shells. For this, they use permanent bony plates like nutcrackers Echinochimeroidei, Squalorajoidei, Myriacanthoidei, and Chimeroidei are four suborders of Chimaeriformes. Following extant taxa – *Callorhinchus*, *Chimaera*, *Hydrolagus*, *Rhinochimera*, *Hariotta*, *Neohariotta* (Stahl 1999), – are known. These taxa are arranged in three families, the Callorhynchidae (*Callorhinchus*), Rhinochimaeridae (*Rhinochimaera, Neoharriotta, Hariotta*), and Chimaeridae (*Chimaera, Hydrolagus*).

The dorsal spine of some recent *Chimaera* species is positioned anterior to the first dorsal fin. The spine is believed to function as a defensive device, particularly in juveniles and sub-adults when it is capable of inflicting a painful and venomous wound (Evans 1923; Patterson 1965). Interestingly, the dorsal spine is also considered to reduce turbulence and aid the hydrodynamics of the first dorsal fin. In addition, it is equipped with a smooth anterior keel, which runs longitudinally along the centre of the anterior margin of the spine, and is thought to reduce spine erosion (Maisey 1979).

The structure of the holocephalian spines is of interest for materials scientist. The dorsal spine of *Chimaera monstrosa* is composed of an outer and an inner layer of dentine, which collectively form the trunk dentine (Maisey 1979). A distinct boundary termed the trunk primordium is present between the two concentric dentine layers. This consists of collagen fibres that run longitudinally through the spine. A single layer of odontoblasts, located on the external side of the trunk primordium, centrifugally deposits the outer dentine. A similar layer occurs between the inner dentine and the spine lumen, and centripetally deposits the inner dentine. The odontoblasts produce an intracellular matrix into which they secrete amorphous cement materials through their dendritic processes. These processes leave anastomosing dendritic odontoblast canaliculi in their wake as the dentine increases in density (Calis et al. 2005). Moreover, growth increments, apparent as rings generally within the inner dentine layer, are the consequence of variability in the growth rate between summer and winter. The rings are associated with metabolic variability, which is effected by fluctuating water temperatures and food availability. Slow winter growth leads to the accumulation of continually deposited dentine in the inner dentine, visible as dark or opaque zones of deposition, whereas fast summer growth is manifested by light, translucent zones (Calis et al. 2005).

Chimeroids have internal fertilization. Their eggs are usually large and enclosed in brownish colored horny capsules. There is broad variety of their forms in different species: from spindle-shaped or tadpole-shaped elliptical. Chimeroids eggcases are called mermaid's purses or devil's purses, and vary in size and shape. Some look much like a mermaid's purse might: a pouch with four long tendrils on each corner to anchor the egg somewhere safe. These biopolymer-containing structures are the reason for bioinspiration by biological materials scientists (see Chap. 12).

Class Acanthodii

Acanthodii is derived from the Greek root acantha (Ακανθα), which refers to a spine (Owen 1846). The reference is to the spines at the anterior of each of the body fins. This group include mostly shark-like, small, early jawed fishes. With exception of the tail fin, they possess sharp spines along the leading edges of all of their fins. It is established that the earliest acanthodians were marine. Freshwater species became predominant only during the Devonian (see for review Janvier 1996). Most of them had large eyes, which suggests that they lived at great depth. Both, the evolutionary significance of acanthodians and their relationships to modern jawed vertebrates have been poorly understood until now. For example, it was thought that the Acanthodii and Osteichthyes (bony fish) share a common ancestor separately from other groups like sharks. The clade based on this common ancestor has been called Teleostomi (see for review Zhu et al. 1999). However, the acanthodian fish *Ptomacanthus anglicus* was reported as an earlier and more basal form than such representatives as *Acanthodes* (Brazeau 2009).

The braincase (the internal head skeleton) of *P. anglicus* more closely resembles that of early shark-like fish, and shares very few features in common with *Acanthodes* and the bony vertebrates. Therefore, it is established that *Ptomacanthus* species were either a very early relative of sharks, or close to the common ancestry of all modern jawed vertebrates.

By and large, the acanthodians were lightly armored with scales of bone and dentine. The scales covered the body of the animal and showed evidence of concentric growth rings. Also had growing scales that resembled an onion-like structure, as well as streamlined bodies, which enabled them to be fast swimmers (Janvier 1996). Following orders of Acanthodii have been described: Climatiiformes, Ischnacanthiformes and Acanthodiformes. Climatiiforma had shoulder armor and many small sharp spines, while Ischnacanthiforma had teeth fused to the jaw. The Acanthodiforma possess long gill rakers and were filter feeders, with no teeth in the jaw. It is suggested that jaws of acanthodians evolved from the first gill arch of some ancestral jawless fishes with a cartilagenous gill skeleton. These fish, although not the first vertebrate in history, are the earliest whole vertebrates to be represented in the fossil record (Wilson 2010).

Class Osteichthyes (Higher Bony Fishes)

Fossil representatives of the higher bony fishes are known from the latest Silurian period, 418 Ma ago, to the present (Zhu et al. 1999, 2009; Botella et al. 2007). According to the different attachment of their fins to the body, this class split into two main lineages, the Actinopterygii and the Sarcopterygii (lobe-fins) (Zhu et al. 2006). In Sarcopterygians fins can be moved freely in different directions because they are connected to the body via a single radial bone (Janvier 1996; Zhu and Schultze 1997). Nowadays, the lobe-fins are represented only by six species of lungfishes (*Lepidosiren paradoxa*, *Neoceratodus forsteri*), and four species of *Protopterus* and the famous coelacanth (*Latimeria chalumnae*). These animals were mostly widespread during the Paleozoic Era. Lobe-fins exhibited a greater diversity than the ray-fins during the Devonian and Carboniferous Periods. As active predators they occupied many of the marine and freshwater habitats. The actinopterygian lineage (with 26,981 living species) is related to osteichthyes and includes sturgeons, gars, teleosts and their relatives. The sarcopterygian lineage includes 26,742 living species (Eschmeyer 1990; Zhu et al. 2001; Zhu and Yu 2002; Zhu and Ahlberg 2004; Yu et al. 2010).

There is also a great variety of structures known within extant (see for review Maisey 1996) and living bony fish, which are of great interest for biological material scientists, including:

- oral (Fraser et al. 2006) and pharyngeal (Carr et al. 2006) teeth,
- jaw apparatus (Lauder 1983),
- sucking disks (Ritter 2002),
- otoliths (Brothers 1984),
- epidermal brushes (Geerinck et al. 2007),
- dermal denticles (odontodes) (Sire et al. 1998),
- armored skin (Lin et al. 2011),
- fins (Fujita 1990),
- wings of flying fish (Fish 1990),
- pelvic spines (Mok and Chang 1986) and girdle (Stiassny and Moore 1992),
- bones (Patterson and Johnson 1995) and interarcual cartilage (Travers 1981),
- eyes and tapetum lucidum (Arnott et al. 1970),
- gills and bony operculums (Hughes 1972),
- swim bladders (Davenport 2005),
- unculi (Roberts 1982),
- breeding tubercles (Wiley and Collette 1970; Kratt and Smith 1978)
- and hierarchically organized scales made of numerous and unique enamel-like substances like cosmine and ganoine (see for review Sudo et al. 2002; Ortiz and Boyce 2008; Bruet et al. 2008; Song et al. 2011) (Fig. 1.5)

Fig. 1.5 Bony fishes represent wide variety of biomaterial-based structures (Photograph by David Wrobel, SeaPics, Reprinted with permission)

Subclass Actinopterygii (Ray-Finned Fishes)

These fishes form the other lineage, are probably the most successful of vertebrates and certainly the most "successful" fish, which are the dominant aquatic vertebrates today. Actinopterygians appeared in the fossil record during the Devonian period, between 400 and 350 (Ma). During the Carboniferous period (360 Ma) they were dominant in freshwaters and started to invade the seas (Lund 2000; Basden and Young 2001). From the Carboniferous to the Triassic, generally small fish, palaeoniscoids were the dominant ray-finned fishes. They possess scales covered by a biological material known as ganoine. Their morphological features were quite diverse from elongated eel-like forms to compressed forms resembling living angelfish. It is believed that all palaeoniscoids were extinct by the end of the Mesozoic, leaving only a few distant, primitive species alive today- the paddlefish, sturgeons, and bichirs (see for review Lauder and Liem 1983). At present, approximately 42 orders and 431 families are recognized within this subclass. There are about 23,000 of Teleosts of the 24,000 species within the actinopterygians, and 96 % of all living fish species.

The anatomical features of ray-finned fish are as follow: pharyngeal slits, body wall muscles arranged in segmented block, a nerve cord, as well as the lateral line (Romer and Parsons 1986). This is a long canal running down each side of the fish body. This organ possesses specialized sensors that can detect water movements and currents.

The swim bladder is the next important organ seen in ray-finned fish (also occurring in Sarcopterygii) but not in cartilaginous fish. The swim bladder is a sac containing gas that originally develops as a pouch budding off the embryonic digestive tract. Because of the presence of this organ, the fish is able to adjust its buoyancy and thus its position in the water column by adjusting the amount of gas in the swim bladder. It retains an open connection to the esophagus in such fish as sturgeons, gars and eels. However, in most bony fish, the swim bladder is completely closed off. This organ is homologous to the lungs of tetrapods because both develop in the same way. In some fish species, especially those with an open swim bladder, it may be used as a breathing organ too (see for details Harder 1976).

The prehistoric looking fish "bichir", or Polypterus, is the most primitive representative of Actinopterygians of living today (Allis 1922). It resembles early actinopterygians from the Devonian as it has a covering of thick ganoin-based rhomboidal scales, which are non-overlapping and instead are connected by fibres. The skeleton of bichirs is mostly cartilagenous. Eleven species of bichirs habituate shallow floodwater areas in tropical African rivers. Usually, they feed on worms, as well as on larvae, and imagoes of insects.

Intriguingly, the paired lung-like swim bladders of Polypterus are connected to the esophagus and correspondingly are used for respiration. The animal is able to survive for hours out of water.

Order Beloniformes

Some representatives of the order Beloniformes demonstrate how marine bioinspiration can find applications in aerodynamics and engineering. The flying fish is a unique marine flying vertebrate. These animals are utilizing the advantages of moving in two different media, i.e. flying in air, as well as swimming in water. The hypertrophied fins and cylindrical body with a ventrally flattened surface both are good examples of aerodynamic designs and ideally useful for proficient gliding flight (Park and Choi 2010). The abrupt transition from predominantly swimming locomotion directly to flight has evolved, for example, in representatives of Exocoetidae. Because of the exceptional wing design and scaling with regard to flight performance, flying fish were the objectives of numerous scientific studies, which started on the end of nineteenth century (Ahlborn 1897) and were continued at the beginning of twentieth century (Gill 1905; Adams 1906; Durnford 1906; Shadbolt 1908; Crossland 1911; Hubbs 1918) and continue to this day (Fish 1990; Davenport 1994, 2003; Kutschera 2005; Park and Choi 2010). The structural specialization of flying fish from an aerodynamics standpoint, as well as the source of propulsive power used by these animals have been of particular interest in previous works (Shoulejkin 1929; Breder 1930; Forbes 1936; Loeb 1936; Mills 1936a, b).

Interestingly, the aerodynamic performance of various forms of bird wings is comparable to those of flying fish. Moreover, some morphological characteristics observed in flying fish are similar with aerodynamically designed modern aircrafts (Park and Choi 2010). Crucial role play abnormally large pectoral fins of the fish that act as airfoils and provide lift when the animal launches itself out of the water surface. Their maximum flight speed is about of 10–20 m/s. The animal can rich a total distance of as much as 400 m in 30 s (Fish 1990).

It was found that the role of the pelvic fins is to increase the lift-to-drag ratio as well as the lift force. This is confirmed by the jet-like flow structure that exists between the pelvic and pectoral fins. According to Park and Choi (2010), "with both the pectoral and pelvic fins spread, the longitudinal static stability is also more enhanced than that with the pelvic fins folded. Although the flying fish uses its hypertrophied pectoral and pelvic fins while gliding, the fins are usually folded against the body when the fish swims in the sea," (Park and Choi 2010). The flying fish will surely be the subject of detailed future studies for the development of special robots. A compact robot that could both swim underwater and glide in the air above water has many potential applications in ocean exploration, mapping, surveillance, and forecasting.

Infraclass Chondrostei

Broad variety of Chondrosteans lived during the late Palaeozoic. The sturgeons (family Acipenseridae) of Europe, Asia and Canada, as well as the paddlefish (family Polyodontidae) of Canada and China, are only the lineages that survive today (Bemis et al. 1997; Birstein et al. 1997).

Both lineages have secondarily lost following morphological features that are characteristic to actinopterygians. There are no on most of the body. They possess cartilaginous skeleton as well as shark-like, heterocercal tail. In some species, the rostrum extending past the mouth, which forms the paddle as in representatives of the Polyodontidae family.

Infraclass Holostei

The freshwater bowfin *Amia* of eastern Canada as well as some species of garpikes (*Lepisosteus* and *Atractosteus*) from North and Central America, are representatives of holosteans which survive today. In contrast to modern species, Mesozoic holosteans invaded and diversified extensively in marine habituates. It is established that these animals arose in the Permian (Cavin and Giner 2012). Typical representatives are *Lepidotes* (Triassic—Cretaceous), and *Dapedius* (Jurassic). Because of the development of a shorter and more mobile jaw, holosteans have more advanced jaws than the chondrosteans. It helps, instead of snapping the mouth shut, to produce a powerful suction. The fish possess scales that tend to lose their shiny ganoid covering, developed symmetrical tail and used the air sac to control buoyancy.

Infraclass Teleostei

Teleostei with 96 % of living fish species are the most number-rich taxon. It was in the Late Triassic that the earliest teleosts first appeared. However, as most dominant fish in the world's waters, they were distributed by the beginning of the Cenozoic.

After their radiation in the Cretaceous, teleosts were most successful fish group. Representatives of Teleostei have been found to survive in extreme aquatic niches like hot springs (up to 44 °C), alkaline lakes, or acid streams (Gash and Bass 1973) as well as in freezing Antarctic waters (Kellermann 1990), in the deep sea and in shallow rivers.

Both fully movable maxilla and premaxilla, which form the biting surface of the upper jaw, are characteristic features of the teleosts. Furthermore, the movable upper jaw makes it possible for these animals to protrude their jaws when opening the mouth. Teleosts possess fully symmetrical tails (see for detail Diogo 2007, 2008).

This taxon include eels, catfish, tuna, tarpon, flounder, halibut, trout, salmon, cod, herring, and many other fishes (see for review on teleost classification Wiley and Johnson 2010).

Bony fish habituate in all marine zones and are of amazing scientific interest because of their diversity in shapes and sizes. They range in size from the pigmy species like 7 mm large stout infant fish (*Schindleria brevipinguis*) to the up to 3 m long bluefin tuna (*Thunnus orientalis*).

Most teleosts are able to regulate the temperature of their bodies. However, they are only slightly endothermic in comparison to mammals. Characteristic representatives of endothermic bony fish are about 122 species such as bonitos, cutlassfishes, hairtails, kingfishes, frostfishes, scabbardfishes, seerfishes, albacores, tuna, and wahoo. All of them belong to the Suborder Scombroidei.

Endothermy requires a lot of energy, however results in improved digestion, better nerve signals, and greater muscle control. However, there are representatives of bony fish that require psychrophilic conditions in order to survive. For example, species of ice fishes (family Channichthyidae) (see for review Eastman 2005; Kock 2005a, b). They habituate in colder waters that hold more dissolved oxygen. Correspondingly, their red blood cells became dispensable. Low temperatures reduce the metabolic rates of the fish, reducing their demand for oxygen.

Because of some evolutionary innovations, including occurrence of both antifreeze proteins and proteins which can work at cold temperatures, these animals ultimately dominated (see for review Cheng and Chen 1999; Maher 2009). According to Maher B (2009), "as competitors in the freezing Antarctic waters disappeared millions of years ago, some icefish began to explore niches above the sea floor, something for which they needed buoyancy," (Maher 2009). It is established that icefish had lost their swim bladders for a long time. Instead, their originally hard mineralized skeletons began to soften. Here, we can speak about some kind of adaptive osteoporosis in icefish species which are adapted "to living happily with extreme anaemia," (Maher 2009). This phenomenon mimics the detrimental human condition osteopenia. Detrich and co-workers (Albertson et al. 2009, 2010) proposed that these Antarctic fishes can be useful as model systems for better understanding of human diseases like osteoporosis. Especially unknown genes and gene/environment interactions in icefish as evolutionary mutant models are of crucial scientific interest.

Subclass Sarcopterygii (Lobe-Finned Fishes)

The fleshy pelvic and pectoral fins with well-developed muscles and bones are characteristic morphological features of lobe-fins. Their humerus joins to the shoulder or pectoral girdle, and femur to the pelvis. Sarcopterygians possess a hinged braincase and a corresponding intracranial joint in the skull roof. This determines corresponding flexibility within the head that provides additional bite strength (see for review Thomson 1969).

Fossil remnants of lobe-fins are reported in the Lower Devonian rocks. For example, such genera as *Powichthys* and *Youngolepis*, those show some affinity to the Dipnomorpha. The diversity of lobe-fins, however, remained high during the Upper Devonian, Carboniferous and Lower Permian. Most Sarcopterygians belong to one of three major groups: the Actinistia (coelacanths), the Dipnomorpha (lungfishes and Porolepiformes), and the Tetrapodomorpha (Rhizodontoformes, Osteolepiformes and Elpistostegalia).

Intriguingly, the Actinistians which first appear in the Middle Devonian were thought to have become extinct until the well-known discovery of *Latimeria chalumnae* in 1938 (see for review Fricke and Plante 1988).

Both Porolepiformes and the Dipnoi (lungfishes) are related to Dipnomorpha (Clément 2004). Porolepiforms with the size up to 2.5 m are known as predators in Middle and Late Devonian. These species habituate nearshore and in freshwater niches (Ahlberg 1989) and possess following morphological features: "a broad head, small eyes, in the cheek they had a prespiracular bone, and the large fangs had a special style of infolding of the enamel and dentine – called a dendrodont tooth structure," (Monroe MH "Australia: The Land Where Time Began"). The large whorl of stabbing teeth was located at the front of the jaw.

Lungfishes are among the first Sarcopterygians to appear in the fossil record with two diversity maxima in the Upper Devonian, and in the Triassic, respectively (see for review Graham 1997). Lungfishes and the Actinsitia are the only representatives of lobe-fins to have survived the Permian; both also have living species (Bemis and Northcutt 1992). New discoveries of the lungfish *Rhinodipterus* from marine limestones in Australia confirms that these animals habituate in marine environments. It is suggested that they developed specializations to breathe air about 375 Ma ago (Clement and Long 2010). It is believed that global decline in oxygen levels during the Middle Devonian together with higher metabolic costs is a more likely driver of air-breathing ability in lungfish. This happened in both marine and freshwater lungfish, and in tetrapodomorph fish.

Rhizodonts, osteolepiforms, elpistostegalids and tetrapods are typical representatives of Tetrapodomorpha, a clade of lobe-fins (see for review Ruta et al. 2003). Some of the Rhizodonts, which first appeared in the Middle Devonian, were up to 7 m large and habituate in freshwater habitats. Osteolepiformes, which first appeared in the late Lower Devonian, and reached a maximum diversity in the Late Devonian, is the group of Sarcopterygians that gave rise to the tetrapods. This order contains two monophyletic families, the Megalichthyidae and the Tristichopteridae (see for review Ahlberg and Johanson 1998). The Megalichthyids are known as the only osteolepiform group to survive past the end of the Devonian. Tristichopterids with *Eusthenopteron foordi*

(Arsenault 1982) as one of the best known of any fossil vertebrate (Jarvik 1980), are related to the tetrapods and the Elpistostegalids. E. foordi seems to be unique because of special characters of its vertebral development (Cote et al. 2002).

The Elpistostegelia (also known as the Panderichthyida) possess an enlarged and flattened head made rigid by the loss of the intracranial joint, infolded (labyrinthodont) teeth, an elongated humerus, and expanded ribs, and no anal and dorsal fins. The genera like *Elpistostege*, *Livonia*, *Panderichthys*, and *Tiktaalik* are typical representatives of this group (see for review Shubin et al. 2006).

Sargopterygii possess hierarchically constructed and nanostructured cosmoid scales (Zylberberg et al. 2010), which are found in several ancient lobe-finned species. Similar structures are known for the earliest lungfishes, and were probably derived from a fusion of placoid scales. Placoid scales are found in the cartilaginous fish. According to Meunier (1980), cosmoid scales possess complex multi-layered structure made by lamellar, dense bone tissue termed isopedine and of upper spongy bone layer that contains blood vessels. Correspondingly, cosmine (Thomson 1975) that represents a complex dentine layer covers the bone layers additionally to a superficial outer coating of vitrodentine. The lamellar bone layer is responsible for increase in size of cosmoid scales (Meunier 1980).

1.2.2.2 Class Placodermi

Placoderms (armor-skinned) arrived during the Silurian and Devonian time replacing the Ostracoderms (Gardiner 1984; Goujet 1984; Young 1986). The armor of the placoderm covered their entire head and some of their body. Some placoderms even had armor surrounding their eyes, and unlike the ostracoderms the placoderms, they had functional jaws (Fig. 1.6). Placoderms bore this heavy bony head and neck armor,

Fig. 1.6 Armored head of *Dunkleosteus* (Placodermi) the Devonian fish that probably had the most powerful bite of any fish. Similar to *Tyrannosaurus rex* and modern crocodiles, this fish was able to concentrate a force of up 3,628 kg per square inch at the tip of its mouth (Image courtesy of www.fossilmuseum.net)

often with a specific joint in the dorsal armor between the head and neck regions. Because of this special joint construct, the head of the animal can move upwards as the jaw dropped downwards, creating a larger gape. Placoderms body was mostly naked or, partially covered with small scales.

It is suggested that placoderms share a common ancestor with sharks. In both placoderms and sharks the males have external clasping organs for internal fertilisation (Ahlberg et al. 2009). But only the ptyctodontids, one group of placoderms, have these organs. Ptyctodontids are examples of, probably, oldest definite evidence for vertebrate copulation. According to modern point of view, placoderms had a remarkably advanced reproductive biology (Long et al. 2008) with respect to the first internal fertilization and viviparity in vertebrates (Long et al. 2009).

In contrast to other jawed vertebrates, placoderms did not descend from toothed ancestors, and, correspondingly, never had teeth. Razor-like, literally self-sharpening edges on bony plates associated with the jaws performed the function of teeth. Placoderms differ from all other jawed vertebrates because their nasal capsules were not fused to the rest of the braincase. These early vertebrates with more than 250 genera became dominant in most brackish and near-shore ecosystems by the start of the Devonian period (see for review Denison 1975). The Placodermi includes six clades, they are: the Acanthoraci, Rhenanida, Antiarchi, Petalichthyida, Ptychodontida and Arthrodira. Of these the Arthrodires are the largest group. Class Placoderm included rhenanids (flattened, stingray-like forms), petalichthyids (forms with long spines), and ptyctodontids (slender, streamlined forms with crushing tooth plates).

1.2.3 Tetrapoda

Terrestrial vertebrates are typical representatives of Tetrapods which, however, also include numerous aquatic, amphibious, and flying groups. These animals occupy the highest levels of the food chain on land and in both freshwater and marine aquatic environments (Laurin 2010). Under Superclass Tetrapoda, important marine classes include Reptilia (the reptiles) which contains at least 3,082 species, Aves (the birds) containing at least 9,842 species, and Mammalia (all mammals) with at least 4,835 species. These taxa have species common in both marine as well as terrestrial environments.

The fish-tetrapod transition is defined as "the greatest step in vertebrate history" (see for review Long and Gordon 2004). Nowadays, the transition between fish with fins and tetrapods with limbs and digits is in focus of many scientific groups, especially because of some intriguing finds of new material (Ahlberg 1991, 1993; Boisvert 2005; Boisvert et al. 2008; Clack 2009; Long et al. 2010). For example, discoveries of new tetrapod-like fish and very primitive tetrapods stimulated the resolution of questions with regard to sequence of acquisition of tetrapod characters during the time and geographic location.

It is suggested that "forelimbs and skulls became modified in advance of hind limbs, adapted for supporting the head and front of the body out of the water, probably in connection with air breathing," (Clack 2009). Prototetrapods and aquatic tetrapods habituate during Eifelian mostly in humid regions like coastal lagoon or estuary margin soils (Retallack 2011). According to the modern "woodland hypothesis of tetrapod evolution, limbs and necks were selected for by scavenging and hunting in shallow-flooded woodlands and oxbow lakes during a unique period in Earth history, after the evolution of flood-ponding trees and before effective terrestrial predator resistance," (Retallack 2011).

Limbed tetrapods originated, probably, between 385 and 380 Ma ago, in the northern continent of Laurussia (see for review Clack 2002; Laurin 2002). The diversity of extant tetrapods is recently reviewed by Benson and co-workers (Benson et al. 2010).

1.2.3.1 Class Amphibia

It is suggested that the common ancestors of modern fish, living fossil fishes and amphibians, are phylogenetically separated (Wang et al. 2012). In general, both saltwater and freshwater fish (including living fossil fish) and amphibians are clustered in different clades. Thus, "the ancestor of living amphibians probably arose from a type of primordial freshwater fish, rather than the coelacanth, lungfish, or modern saltwater fish. Modern freshwater fish and modern saltwater fish were probably separated from a common ancestor by a single event, caused by crustal movement," (Wang et al. 2012).

A recent hypothesis about this transition suggests that the diverse assemblages of marine amphibious fish that occur primarily in tropical, high intertidal zone habitats are analogs of early tetrapods. This suggests that the intertidal zone, not tropical freshwater lowlands, was the springboard habitat for the Devonian land transition by vertebrates (Graham and Lee 2004). The extant marine amphibious fish, which occur mainly on rocky shores or mudflats, have reached the limit of their niche expansion onto land and remain tied to water by respiratory structures that are less efficient in air and more vulnerable to desiccation than lungs.

Indeed, of the 6,500 recognised amphibians, only one species can enter the sea (Neill 1958). However, the early stegocephalians (fossil amphibians whose skullcap formed a continuous covering and whose trunk was frequently covered with bony scales) tolerated saltwater, even although they also lived in freshwater (Schultze 1999; Niedzwiedzki et al. 2010; Laurin and Soler–Gijón 2010). Thus, the crab eating frog, *Fejervarya cancrivora* – is considered to possess the highest salinity tolerance among amphibians (Fig. 1.7). In this species, 50 % of the larvae survived in up to 80 % sea water – equivalent to 0.5 % NaCl (see for review Gordon and Tucker 1968). Thus, the skin of "marine amphibians" seems to be a subject of interest for experts in biological materials science.

Fig. 1.7 The crab-eating frog (*Fejervarya cancrivora*) is considered to possess the highest salinity tolerance among amphibians. Size (snout to vent): Female 8 cm, Male 7 cm (Image courtesy of Nick Baker, www.ecologyasia.com)

1.2.3.2 Class Reptilia (Reptiles)

This class includes cold-blooded (ectothermic) animals, which use lungs to breathe, and have tough featherless or hairless skin. Reptiles cannot survive in extremely cold climates because they cannot regulate their internal body temperature. The evolution of marine reptiles started about 250 Ma in the Early Triassic. These vertebrates dominated Mesozoic seas until their demise by the end of the Cretaceous, 65 Ma (see for review McGowan and Motani 2003; Motani 2009; Benson et al. 2010). Typical representatives of marine reptiles in the Triassic are—the mollusk eating, armored placodonts as well as pachypleurosaurs and nothosaurs. Both related to the long-necked fish-eating eosauropterygians (Rieppel 1995). Also serpentine thalattosaurs, and the streamlined ichthyosaurs (Motani 2005)—are examples of faunal recovery in the oceans following the devastation of the end Permian mass extinction. However, most of these marine reptile species disappeared in the Late Triassic. During Jurassic period, predators like plesiosaurs, marine crocodilians and ichthyosaurs dominated in the oceans (Thorne et al. 2011).

The first tetrapods with a fish-shaped body profile were parvipelvian ichthyosaurs. These animals are characteristic examples of the secondary adaptation of reptiles to marine life. According to Bernard et al. (2010), "ichthyosaurs evolved from basal neodiapsid reptiles, with the most obvious aquatic adaptations: a dolphin-like streamlined body without a neck, paddles, and a fish-like tail," (Bernard et al. 2010). They possess morphological features of cruising forms, similar to a living tuna. Several species were deep divers. The cancellous bone is the common character observed in genera such as Caypullisaurus, Stenopterygius, Temnodontosaurus, and Ichthyosaurus. Because of this feature, the fish-shaped ichthyosaurs are suggested as fast and far cruisers (Talevi and Fernández 2012). "The evolution of 'thunniform' body shapes in several different groups of vertebrates, including ichthyosaurs, supports the view that physical and hydromechanical demands provided important selection pressures to optimize body design for locomotion during vertebrate evolution," (Donley et al. 2004). Paleontological records of Ichthyosaurs in open marine sediments confirm the hypothesis that they were very well adapted to the marine environment (Sander et al. 2011). These ancient marine species possess a streamlined body, greatly enlarged eyes, an elongated rostrum with numerous conical teeth. However, some of them, like *Shastasaurus*, were toothless. According to Sander and co-workers (2011), "*Shastasaurus* is interpreted as a specialized suction feeder that preyed on unshelled cephalopods and fish, suggesting a unique but widespread Late Triassic diversification of toothless, suction-feeding ichthyosaurs," (Sander et al. 2011). Some other unique morphological properties of ichthyosaurs are also known (see for review Sander 2000). For example, *Ophthalmosaurus* was roughly 4 m long with a mass of 930 kg, and possess the largest eyes with more than 220 mm in diameter. The species had also the largest sclerotic ring aperture, with a diameter of about 100 mm. The animal could dive to a depth of 600 m (Motani et al. 1999).

Plesiosaurs possess powerful paddle-like limbs as well as heavily reinforced limb girdles. This group belong to the Sauropterygia, the sister group of the Lepidosauria (lizards and snakes). The varanoid anguimorphs distributed Late Cretaceous are known as mosasaurs. They also were highly adapted to marine life because of elongated body, paddle-like limbs and deep tail (Bernard et al. 2010). The anatomical features of these predators could afford high cruising speeds similar to that of modern tunas. Probably, they were successful as apex predators of Mesozoic aquatic ecosystems because of their thermophysiological status with respect to thermoregulation.

Recently, Bernard and co-workers compared the oxygen isotope compositions of the tooth phosphate of ichthyosaurs, plesiosaurs, and mosasaurs to those of coexisting fish. Obtained results confirm that these marine predators were able to maintain a constant and high body temperature in oceanic environments that ranged from tropical, to cold or temperate. "Their estimated body temperatures, in the range from 35 to 39 °C, suggest high metabolic rates required for predation and fast swimming over large distances offshore," (Bernard et al. 2010).

Thus, extant marine reptiles are still under investigations. Furthermore, numerous recent studies on their biomechanics, locomotion, biophysics of swimming and diving, buoyancy, and fin kinematics (Riess 1986; Taylor 1987; Massare 1994; Fish 2000; Motani 2001, 2002a, b; Rayfield 2007;) as well as on their bone microstructure (Lopuchowycz and Massare 2002) and fibre-like structures within their skin (Lingham–Soliar 1999, 2001), suggests high interest in these fossils for the biological materials science, bionics and biomimetics scientific communities.

Today (see for review Andrews 1910, 1913), the Class Reptilia, includes following orders:

- Testudines (turtles, terrapins, and tortoises);
- Squamata (lizards, worm lizards, and snakes);
- Crocodilia (crocodiles, alligators, gavials, and caimans);
- Rhynchocephalia (two species of lizard-like tuataras).

As recently reviewed by Rasmussen and co-workers (Rasmussen et al. 2011a, b), only about 100 of the more than 12,000 species and subspecies of extant reptiles, have re-entered the sea. Among them are sea turtles (7 species) and sea snakes (about 80 species and subspecies). Also, various other snakes, the saltwater crocodile, and the marine iguana of the Galapagos Islands are occasionally or regularly found in brackish waters. The sea snakes occur in the tropical and subtropical waters of the Indian and Pacific Oceans and represent the species richest group of marine reptiles. "They inhabit shallow waters along coasts, around islands and coral reefs, river mouths and travel into rivers more than 150 km away from the open ocean. A single species has been found more than 1,000 km up rivers," (Rasmussen et al. 2011b).

Order Testudines

The robust protective shell surrounding their body is the structural feature of the more than 260 species of turtles and tortoises (Fig. 1.8). Their shells show broad diversity in shape, size, ornamentation and color. Thus, the leatherback sea turtle *Dermochelys coriacea* (whose shell consists of bones beneath thick skin), possess up to 2.4 m in length huge shell which can weigh as much as 907 kg. However, the smallest species is, probably, the bog turtle (*Glyptemys muhlenbergii*) that measures only 11.4 cm at most. In contrast to the terrestrial species, reptiles that live in water, including marine turtles have a flatter and lighter shell (see for review Rhodin 1985; Eckert 2002). Marine species arose about 100 million years ago from fresh-water or terrestrial turtles (Naro–Maciel et al. 2008). Marine turtles, move forward by simultaneous action of the front flippers and, in this way, may attain a speed of about

Fig. 1.8 The carapace of sea turtles is unique keratin-based hierarchically structured construct. Loggerhead Sea Turtle (*Caretta caretta*) – "Saddle" a rehab patient at The Turtle Hospital, Marathon, Florida swimming among sea grass in his tank. Saddle had lost three flippers to predators prior to being rescued (Photographer is Jack Eng. Image courtesy of http://www.turtlehospital.org., Reprinted with permission)

35 km per hour. Plankton feeding species like leatherback turtles (*D. coriacea*), spend entire years in the ocean. Being the paradigmatic, skillful oceanic navigators, they are known to return to specific places to breed every 2–3 years. Intriguingly, these turtles, finding their way back home after long periods in the oceanic environment (Sale and Luschi 2009).

Recently, it was proposed that "sea turtles imprint on the magnetic field of their natal areas and later use this information to direct natal homing," (Lohmann et al. 2008).

The majority of marine species are predominantly carnivorous. However, for example, the green sea turtle changes to a vegetarian diet at the end of the juvenile stage. Some marine turtles are omnivorous. The hawksbill (*Eretmochelys imbricata*), an endangered marine turtle distributed in the coral reef regions in Caribbean, feeds almost exclusively on sponges, even though they contain a huge mass of very sharp siliceous spicules (Meylan 1988). Since marine invertebrates and plants are generally similar in salt content and contain three times as much salt/kg body water as sea turtles (Holmes and McBean 1964), the salt burden for feeding turtles can be considerable. Therefore, sea turtles use their tear ducts to excrete salts. We can observe this "crying" when the animal is out of the water. The need for salt excretion in such reptiles as sea turtles and marine iguanas is determined by the fact that their kidneys are much less efficient than those of mammals (see for review Ellis and Abel 1964; Marshall 1989; Reina et al. 2002). Unlike the skin of amphibians, reptile skin is impermeable to NaCl. The transition to a tougher skin meant a loss in salt-releasing ability (see for details Peaker and Linzell 1975). Calculations suggest that the sea turtle salt gland has a high volume-handling capacity, equivalent to that of the mammalian kidney (Nicolson and Lutz 1989).

It is proposed that, the turtle "provides an ideal case study for understanding changes in the developmental program associated with the morphological evolution of vertebrates," (Kuratani et al. 2011). In contrast to armored constructs observed in other vertebrate groups, the turtle shell exhibits very special topography of musculo-skeletal elements as well as involves the developmental re-patterning of the axial skeleton (Gilbert et al. 2001). The carapace (dorsal part of shell) is joined at the sides to the plastron (ventral part). This is in turn notched at the front and rear ends where the limbs emerge from the shell. The scutes on a turtle's carapace and plastron are the equivalent of scales on other reptiles. However, the body plan of the turtle represents "an example of evolutionary novelty in the acquisition of the shell," (Kuratani et al. 2011). The epidermis that covers the body shell of turtles forms these structures. When new layers of dermis are deposited over the bone, old epidermal cells are pushed towards the outside. As these epidermal cells die, they are keratinized and form the scutes. A depression forms between old, keratinized layers; and new layers of epidermis are called scute annuli (Wilson et al. 2003). Counting the number of scute annuli on the carapace or plastron of a turtle is a common way of determining the age of turtles (see for review Germano and Bury 1998; Ergene et al. 2011).

Turtles lack teeth, and therefore their shearing surfaces are formed via a keratinized beak, the rhamphotheca. The rhamphotheci are the structural elements of both the upper and lower jaws of cheloniids. They cover the premaxillary, maxillary, and vomer bones of the upper jaw, and the dentary of the lower jaw. The upper and lower beak surfaces meet at a ridged surface that can be knife-sharp like in case of the marine loggerhead turtle (*Caretta caretta*). This animal is capable of slicing through prey as hard as mineralized molluscs or crabs shells (see for review Wyneken 2001).

From biomaterials point of view, the multiscale structure, material properties, and mechanical responses (Richmond 1964) of the turtles shells were recently intensively studied to better understand "naturally occurring biological penetrator-armor systems," (Rhee et al. 2009; additionally see for review Krauss et al. 2009; Balani et al. 2011; Damiens et al. 2012). Both, the chemical analysis and the structural observations revealed that the turtle shell carapace "comprises a multiphase sandwich composite structure of functionally graded material having exterior bone layers with a foam-like bony network of closed-cells between the two layers," (Rhee et al. 2009).

The eggshell in turtles is of interest for biomaterials science community. This structure is nothing else as a protective multilayered barrier between the external environment and the embryo (Decker 1967). Additionally to protection of the embryo from any mechanical injury, the eggshell is responsible for moisture, heat and mostly for gas exchanges. It also can function as a substrate for rotational movements of the embryo within the egg (Ewert 1979; Al–Bahry et al. 2009). Recently, numerous comparative studies on structural biology and chemistry of various aspects of eggshell formation in some sea turtles (Erben 1970; Schleich and Kastle 1988), including biomineralization features, have been reported (Al–Bahry et al. 2009).

Order Squamata

Sixty percent of all reptiles (5,500 species) are Lizards (Suborder Sauria), the very specious and diverse group of reptiles which includes marine animals, too (Uetzt 2011). With exception of worm lizard (Amphisbaenia) which usually is legless, lizards usually have four legs with claws. These mostly insectivores reptiles inhabit shrubs, rocks and trees. However, some species will eat plants or even other vertebrates. They range from a few centimeters to nearly 3 m long.

Iguanas are representatives of lizards. The only iguana of the 416 known species that ventures into the ocean is distributed on the Galapagos Islands, which arose through volcanic activity 960 km off the coast of Ecuador. It is suggested that "lizards and other animals probably reached the island on rafts of vegetation washed down the rivers of western South America" (Rasmussen et al. 2011b).

Thus, the Galapagos marine iguana (*Amblyrhynchus cristatus*) (Bell 1825) represents the diverse and species rich family of Iguanidae (see for review Carpenter 1966; Dawson et al. 1977). *Amblyrhynchus*, however, is unique because of the feeding specialization. The diet of these marine animals is based predominantly on soft-bodied macrophytic algae (Dawson et al. 1977). Some authors (Hobson 1965; Bartholomew et al. 1976) described their behavior as follow: "These lizards must swim through the surf zone, dive beneath the surface (lungs remain inflated during diving so that the animals must initially overcome their own buoyancy), hold their breath during submergence, surface, and then swim back to shore through the surf" (Dawson et al. 1977). Adult marine iguanas cruise at velocities averaging about

0.45 m/s (28 body lengths/min) (Dawson et al. 1977). They can arrived depths up to 12 m, and remain submerged for as long as 30 min (Hobson 1965) because their powerful claws help them to hold on to rocks in the heavy sea. Similar to other reptiles, marine iguanas also have reduced the number of heart beats per minute from about 43 on land to 7–9 while diving (Higgins 1978). The feeding of *Amblyrhynchus* take place under relatively cool water temperatures, (15–25 ° C) typical for the Galapagos Archipelago. This situation potentially hinder development of the intensity of activity required during feeding. Therefore, it was observed that these large reptiles bask in the sun and reach body temperatures in the vicinity of 35 °C (Bartholomew and Lasiewski 1965; Dawson et al. 1977).

To cope with marine high-salt diet, *Amblyrhynchus* successfully uses large cranial salt glands. These formations being located over the eye (Dunson 1969), are responsible for excretion of the sodium, potassium, and chloride ingested (see for review Hazard 2004) (see also Sect. 3.6 in this work). "Forceful expulsion of the secreted fluid is the cause of the dramatic snorting and sneezing observed in these animals," (Schmidt–Nielsen and Fange 1958). Probably, this periodically occurring ejection is also used by iguanas to warn of intruders. The white "wig" often seen on marine iguanas is the result of the salt spray that often shoots up into the air and then falls back on the head of the animal.

According to my personal opinion, this phenomenon is an example of very unusual "biominerals production" in the animal world. Unfortunately, the salt glands have not yet to be investigated as producers of biominerals.

Intriguingly for experts in biomaterials: "adult marine iguanas can switch repeatedly between growth and shrinkage during their lifetime, depending on environmental conditions," (Wikelski and Thom 2000). Note, that in humans, bone regrowth after osteoporosis is usually impossible (Dennison and Cooper 2011). Especially in postmenopausal women, osteoporosis leads to several health problems (Pinkerton et al. 2010).

Snakes (Suborder Serpentes, or Ophidia, with about 2,000 species) are basically legless reptiles, measuring between 10 cm and 7.6 m in length. It is suggested that they evolved from lizards. However, the taxonomic status of "sea snakes" is still under review (Rasmussen 1997). In contrast to other reptiles, representatives of Serpentes lack outside ears. They have eyes which are covered only with permanent transparent scales. Snakes are distributed in different environmental niches including terrestrial area, lakeshores and even salt waters.

Sea snakes (Hydrophiinae) (Smith 1926) usually habituate in the tropical and subtropical waters of the Indian and Pacific oceans. Diverse species are found in the Australian region, Indonesia, Indo-Malayan Archipelago as well as in China seas. However, two specimens of *Pelamis platurus* were reported from Namibia coastal waters (Branch 1998). Sea snakes were observed in lakes in Thailand, Cambodia, the Philippines, and Rennell Island. They habituate shallow waters, river mouths, and even can penetrate into rivers up to more than 100 miles from the ocean (see for review Dunson 1975; Rasmussen 2001).

Several anatomical features of the sea snake's body confirm numerous adaptations for life in the ocean (see for detailed review Graham et al. 1987;

Heatwole 1999; Ineich and Laboute 2002; Aubret and Shine 2008). For example, these adaptations were listed as follow: "Valved nostrils are located high on the snout for breathing while swimming or basking at the surface. The belly is tapered like the keel of a boat for stability in the water, and this snake lacks the flattened belly scales that give land snakes traction on the ground. The sea snake's scales are knob-like and fit against one another like bricks, rather than overlapping as in land snakes. The tail is flattened and broadened to form an effective paddle for swimming," ("Yellow-Bellied Sea Snake" http://www.waikikiaquarium.org/experience/animal-guide/reptiles/yellow-bellied-sea-snake/, last accepted 25.05.2014).

Sea snakes possess corresponding physiological adaptations because they may spend up to 90 % of their time underwater. These animals have no gill; however it is known that their skin can act in the same manner. Space between the scales is richly supplied with blood vessels, and, correspondingly, oxygen enters and waste carbon dioxide leaves across the skin surface (Heatwole 1999). Here, one example reported by Vogel (2003): "The sea snake *Pelamis platurus*, typically descends about 30 m with enough air in its lung to be neutrally buoyant at that depth. But it loses gas, mainly carbon dioxide and nitrogen, from its skin, whose permeability lets it act as a kind of gill and supplement its lungs. So as it swims along it gradually ascends, thus maintaining its neutral buoyancy as part of its requisite return to the surface" (Vogel 2003).

Similar to marine iguanas (see above), sea snakes are able to excess salt from the seawater and fish diet, however using special glands in the snake's mouth (Taub and Dunson 1967; Voris and Voris 1983). The presence of a salt gland in these reptiles can explain their ability to survive in a hyperosmotic medium (Dunson et al. 1971) in contrast to their fresh water and terrestrial relatives which lacking this gland and quickly die when placed in sea water. It was suggested (Dunson and Robinson 1976) about correlation between skin permeability, dermal respiration and differences in salt gland size among sea snakes. The frequent skin shedding of sea snakes is well known phenomenon and may be related to maintenance of low water permeability as well as to prevention the biofouling in marine environment (Dunson and Freda 1985).

Fish eggs or crustaceans, mollusks or bottom dwelling fish like eels, are the main components of the sea snake diet (Goldek and Voris 1982; Rasmussen et al. 2011a, b). However, sea snake bite is the cause of human fatalities as well. These animals are known to produce venom that contains neurotoxins (Carey and Wright 1960), some of them (erabutoxins) are similar to curare. These toxins can block the nicotinic acetylcholine receptors on the post synaptic membranes (Tamiya and Yagi 2011). Westhoff and co-workers (2005) described them as "predators that hunt during the day, at dawn or at night. Like land snakes, sea snakes have scale sensilla that may be mechanoreceptive, i.e. that may be useful for the detection of water motions produced by prey fish. In addition, inner ear hair cells of sea snakes may also be involved in the detection of hydrodynamic stimuli" (Westhoff et al. 2005).

Sea snakes provide an excellent model for studying the evolutionary transition between the terrestrial and marine environments; as sea snake species encompass a wide continuum in terms of the degree of their dependence on the ancestral terrestrial environment (see for review Brischoux and Bonnet 2009). Studies about the mechanical properties of sea snake skin (Jyane 1988) are of importance for bionics, as well as for biomaterials science. For example, different variations in the surface hydrophobic properties of the sea snake skin in seawater are currently the subject of ongoing investigations (Lillywhite et al. 2009). It was shown that the external skin surfaces of marine *Laticauda* snakes studied by Lillywhite and co-workers are hydrophobic. Thus, the skin can interact with a microfilm of air at the surface of the stratum corneum. Probably, "variation in the nature of this air film is correlated with species differences in the efflux of water when these snakes are aquatic" (Lillywhite et al. 2009).

Sea snakes are known to shed their skin with relative frequency. Probably due to hydration damage of corresponding layers of their skin, which seems inevitable over time (Tu et al. 2002)? Moreover effective hydrophobic properties of the skin surfaces can be changed due frequent shedding (Lillywhite et al. 2009). Also the presence of neuronal processes, which is almost certainly important for detecting vibrations in the aquatic environment cannot be excluded (Lillywhite et al. 2009).

The Crocodilians (Order Crocodylia) include crocodiles, gavials, caimans and alligators. Altogether – 23 species. None of them is truly distributed in marine environment. Only the saltwater or "estuary" crocodile (*Crocodylus porosus*) habituates in brackish waters of south-east Asia and Australia (Martin 2008). Species like *Crocodylus johnstoni* or *C. acutus* have been found in tidal waters (Rasmussen et al. 2011a, b). Crocodilians are usually distributed in tropical waters. Alligators in USA and China live, however, in temperate climates. Crocodylians are cold-blooded, amphibious, egg-laying reptiles with 4 legs, partially webbed feet, and a long powerful tail. Exemplars of saltwater crocodile *C. porosus* can be up to 6.2 m long (Montague 1983). The weight of a 6 m adult is 1,100 kg. Their external skin has scales, often strengthened by bone deposits (osteoderms), and they have a flattened, tooth-lined skull with broad or narrow snouts. Crocodilians' long flattened tails possess morphological features that enable them to swim efficiently through water. These reptiles breathe through nostrils located at the top of their head.

The presence of the lingual salt glands which help to secrete excess of salt ions is known in crocodiles (Grigg et al. 1980; Taplin and Grigg 1981; Taplin 1984, 1985; Franklin and Grigg 1993). These glands in crocodilians are stimulated by a salt load, and their morphology is conserved (Kirschner 1980; Mazzotti and Dunson 1984). From physiological point of view "these tissues are typified by their abundance of ion pumps, including Na^+/K^+-ATPase (NKA), a basolateral, transmembrane ion pump responsible for the maintenance of cellular electrochemical gradients through the movement of Na^+ and K^+ ions against their osmotic gradients," (Cramp et al. 2010).

The enzyme Na^+/K^+-ATPase plays an important role in salt regulation. Its abundance, distribution, activity, and expression in the salt glands of *C. porosus* increase following chronic saltwater acclimation (Cramp et al. 2010). It was demonstrated that the salt glands of this species are "phenotypically plastic, both morphologically and functionally and acclimate to changes in environmental salinity" (Cramp et al. 2008). Similar glands are known also for Nile crocodile; however they may be not as efficient as those in the *C. porosus* (Taplin and Loveridge 1988).

It can be suggested that such adaptations to saline environment are evidence for a marine evolutionary origin of crocodiles (Rasmussen et al. 2011a, b).

Interestingly, the estuarine crocodile *C. porosus* possesses one additional organ for salt regulation – the cloaca. It was demonstrated (Kuchel and Franklin 2000) that the function of the cloaca/lower intestine is highly plastic. "Plasticity of function may be reflected in the plasticity of morphology of the cloaca, as it was for the lingual salt glands in *C. porosus*" (Kuchel and Franklin 2000). Urodaeum is the part of the cloaca into which the ureters and genital ducts empty, and, probably the primary site for postrenal modification of urine in this species (Kuchel and Franklin 2000). Examination of its mucosal surface by scanning electron microscopy showed a plastic response to environmental salinity. Thus, a possible increase in surface area in *C. porosus* kept in hyperosmotic water compared with species from fresh water. However, the renal/cloacal complex is responsible only for 2 % of the total Na efflux in comparison with 55 % of the lingual salt glands account. The urinary system in crocodile appears to be the principal route for excretion also of potassium (Taplin 1985).

1.2.3.3 Class Aves (Birds)

Archaeopteryx lithographica, the feathered archaeornithine, arose in late Jurassic (145 Ma), and is the earliest known bird today (Ostrom 1976; Ruben 1991). "It possessed (at least) complete wing and tail plumage and a striking superficial skeletal similarity to some carnivorous dinosaurs" (Ruben and Jones 2000). Another ancient "bird-like" animal is the *Ichthyornis* (meaning "fish bird", after its fish-like vertebrae). The 95–85 Ma old fossil remains of these toothed seabirds from are known from the chalks of Alberta, Alabama, Kansas, New Mexico (see for review Clarke 2004). As proposed by Olson (1985), Ichthyornis has been scientifically important from the view of bird evolution. As the first discovered prehistoric bird with exceptionally well preserved teeth, this animal was noted by Charles Darwin because of its significance during the early years of the theory of evolution. Today, the fossilized remains of Ichthyornis are important as examples of the few birds of Mesozoic origin known from more than a few specimens (Olson 1985).

It is suggested that that the basal lineages of modern birds originated deep within the Cretaceous (Brown et al. 2008) as well as that avian endothermy is likely to have been fully developed by about Late Cretaceous–Early Tertiary Periods (Goedert 1989). The paleontological studies indicate that modern bird orders were well defined by about 60 Ma (Feduccia 1996).

Because the oceans cover more than 3/4 of the Earth's surface, many birds have adapted to life over open water or along the coasts. Numerous species of birds use aquatic niches for feeding. Especially seabirds gather all of their food at sea where they spend most of their time (Enticott and Tipling 1997). Seabirds as a group (Kennedy and Page 2002), are very successful (Ainley 1980). The populations of seabirds are numbering in the hundreds of millions, however only 3 % of the 8,600 known bird species are seabirds. Furthermore, seabirds are impressive examples of physiological plasticity. This makes seabirds among the most abundant flying vertebrates in the world (see for review Harrison 1983, 1997; Lofgren 1984; Lockley 1984; Gaston 2004).

The Pelecaniformes Order

About 50 species of the seabirds like tropic-birds, pelicans, boobies, cormorants and frigate birds are representatives of the order Pelecaniformes (see for review Johnsgard 1993; Smith 2010). Representatives of this group can be distinguished by their toes, all four of which are webbed and point forward. Their chicks hatched naked and are thus extremely dependent on their parents. Pelicanimormes are active hunter and divers. Correspondingly, their nostrils are protected as an adaptation for diving (Nelson 2005).

Order Sphenisciformes (Penguins)

Phylogenetic hypotheses place penguins (16–19 species) within Aves and Neornithes, but further relationships remain contentious. Penguins are among the oldest of extant avian clades because their fossil remains were found in the Early Palaeocene (61 Ma) rocks (Clarke et al. 2010; Ksepka and Clarke 2010). Representatives of this order are distributed from the tropics to Antarctica, especially in the regions of high ocean productivity with respect to their prey. For example, a small penguin (about 30 cm), the Galápagos penguin (*Spheniscus mendiculus*) is the only penguin to live north of the equator. However, this species is living in the surprisingly cool waters surrounding the Galápagos Archipelago (Boersma 1977). *S. mendiculus* are excellent swimmers. They quickly achieving speeds of up to 35 km per hour (almost 22 mph) (see for review Williams 1995).

"Penguins exhibit a remarkable range of adaptations to life in the marine environment, including flattened limb bones, feathers that are unique among birds, and solid bones" (Walsh et al. 2008; see for more information Davis and Renner 2003).

Order Procellariiformes (Tube-Nosed Birds)

The tube-nosed birds (Procellariiformes) possess hooked bill as well as long, tubular nostrils, which help smell food up to 30 km away (Warham 1990). The most of these birds spend the vast majority of their lives on the ocean, only setting foot on land in order to mate. Birds in the order Procellariiformes includes following representatives: diving-petrels (4 species), albatrosses (13 species), shearwaters and petrels (66 species), and storm-petrels (21 species). Procellariiformes are diverse in size and weight. For example, Wandering Albatross (*Diomedea exulans*) at 11 kg show the wingspan of about 3.6 m. The tiny Least Storm Petrel (*Halocyptena microsoma*) with the weight of 20 g has the wingspan of 32 cm (see for review Onley and Scofield 2007).

Order Charadriiformes

Diverse small to medium-large birds are members of Charadriiformes. The order includes about 350 species (see for review Ericson et al. 2003). The typical representatives are waders (or "Charadrii"), the gulls and their allies (or "Lari"), and the auks (or "Alcae"). Because different species of Charadriiformes gather in large numbers during breeding, migration, or wintering seasons, they are frequently attractive targets for hunters. Many species have been hunted for their meat, feathers, oil, and eggs. Hunting by humans is believed to have resulted in the extinction of at least two Charadriiformes, the Eskimo curlew and the great auk. Hunting continues to contribute to the threatened state of many other charadriiform species.

Most marine birds rest and sleep on the water surface, while others roost on land for a few hours a day. Principally, all of them still need to come to land to lay eggs and raise their young (see for review Ashmole 1971; Schreiber and Burger 2001).

Below, I would like to analyze several physical adaptations that help to make a successful seabird from the viewpoint of biological materials.

Eyes

Seabirds have excellent air as well as amphibian vision. The eyes of birds are small and located on either side of their head. They play crucial role in seabird life because eyes protect them from the rather bright light at sea surface and give them a wide field of view. So called nictitating membrane of diving seabirds is responsible to close the eye when the bird is underwater, but does not play a refractive role. Some species possess a clear lens in the centre of this membrane that enhances their vision. The cornea of the seabird eye has an important refractive role in air but does not usually work underwater because its refractive index is similar to that of water. However, seabirds' (similar to water birds) (Tyler and Ormerod 1994) eyes had enhanced powers of accommodation. The iris of the eye had a very well developed sphincter muscle. It is suggested (Goodge 1960), that pressure exerted by this muscle could change the curvature of the lens, enabling the greater powers of accommodation.

According to Sivak (1980), "the suggested mechanisms by which aquatic birds can compensate for the refractive loss of the cornea underwater include an exaggerated underwater accommodative increase in refractive power, the development of a flattened cornea which is non-refractive in either medium, and use of the nictitating membrane as an underwater refractive goggle. Recent research indicates that certain waterfowl are capable of altering the shape of the lens dramatically by means of the ciliary and iris muscles, and that penguins have flattened corneas," (Sivak 1980). It was reported (see for review Katzir and Howland 2003), that seabirds like the cormorants, possess specific an iris accommodative mechanism capable of producing dramatic lens changes in these animals. Such phenomena were not described for non-diving birds.

Respiratory Turbinates

Albatrosses, petrels, fulmars and shearwaters possess large external nostrils and are related to the group of seabirds called "tubenoses". These specifically structured nostrils help them to find good feeding areas, other birds, breeding areas and nest sites by smell (see for review Nunn and Stanley 1998; Brooke 2004; Onley and Scofield 2007). Intriguingly, some representatives of "tubenoses" can smell food up to 30 km away!

Thus, seabirds possess cartilaginous, epithelially covered projections within their nasal cavity known as conchae, or turbinates. Respiratory turbinates of seabirds are situated directly in the path of respiratory airflow, greatly increasing the surface area of the nasal epithelial mucosa and simultaneously reducing the effective distance of respiratory air from the mucosal surfaces. As inspired air passes through the nasal cavities and over the moist surfaces of the respiratory turbinates, heat and water are exchanged, warming and humidifying the air while simultaneously cooling the epithelium of the turbinates. The efficiency of this evaporative exchange is such that the temperature of the nasal surfaces may occasionally drop below the temperature of the ambient air (Geist 2000).

Physiology of the respiratory turbinates is a very interesting scientific field because of a strong functional association with endothermy (Scholander et al. 1950; Hillenius 1994; Ruben 1995). It is well established that high levels of oxygen consumption and concomitant elevated rates of lung ventilation are tightly linked to endothermy in both mammals and birds. As written by Ruben and Jones (2000), "respiratory turbinates, which occur in greater than 99 % of all extant birds and mammals, facilitate an intermittent countercurrent exchange of respiratory heat and water between respired air and the moist, epithelial linings of the turbinates. In doing so, they significantly reduce respiratory water and heat loss that would otherwise be linked to the high rates of lung ventilation associated with mammalian and avian endothermy" (Ruben and Jones 2000). Interestingly, some kind of vascular shunts exists between respiratory turbinates and the brain. It has long been suggested that respiratory turbinates may be utilized as brain "coolers" in birds and mammals (Baker 1982).

Salt Gland

Feeding in saline environments can be dangerous for animals which have no physiological mechanisms for removing of NaCl and which are able to prevent the blood from becoming too salty. Birds which habituate in marine areas ingest much more salt than their kidneys can process. Correspondingly, they use special salt or nasal glands (Marples 1932; Schmidt–Nielsen and Sladen 1958; McFarland 1959; Schmidt–Neilsen et al. 1970; Ellis and Goertemiller 1977). These structures are located above the eye and can excrete a strong salty liquid. The bird kidney is able to excrete salts at concentrations of about one-half that found in sea water!

Why is it principally possible that seabirds can drink seawater? Because their cephalic salt glands secrete a NaCl solution more concentrated than seawater!

According to the modern view (Hughes 2003), “salt gland secretion generates osmotically free water that sustains their other physiological processes. Acclimation to saline induces interstitial water and Na move into cells. When the bird drinks seawater, Na enters the plasma from the gut and plasma osmolality increases. This induces water to move out of cells expanding the extracellular fluid volume,” (Hughes 2003). Increases in plasma osmolarity and extracellular fluid volume stimulate secretion using salt gland. The augmented intracellular fluid content should allow more rapid expansion of extracellular fluid volume in response to elevated plasma osmolarity. Of course, finally intestinally absorbed sodium chloride must be reabsorbed by the kidneys.

The capillaries around salt glands are arranged so that the flow of blood is in the direction opposite to the flow of secretory fluid. This flow maintains a minimum concentration gradient between blood and the tubular lumen along the entire length of the tubule. The avian salt gland is a countercurrent system that concentrates the secreted salt solution.

Intriguingly, the salt gland can be an example for microenvironments of some halophilic microorganisms. For example, a suitable environment for the colonization of extreme halophilic prokaryotes like *Halococcus morrhuae* and *Hcc. Dombrowskii* have been reported within the nostrils of the Cory’s Shearwater (*Calonectris diomedea*), which are endowed with a salt-excreting gland (Brito–Echeverria et al. 2009). Because of migration routes of this seabird between Mediterranean as the South Atlantic, it is suggested that dispersal mechanisms of haloarchaea across the Earth’s surface can be determined by this phenomenon, too.

Uropygial (Preen) Glands

The preen gland, which is located above the base of the seabird tail, produce oil-based substance. Seabirds use this liquid to keep their feathers clean, flexible and waterproof by constantly preening (see for review Hou 1928a, b; Elder 1954; Jacob and Ziswiler 1982). The gland is diverse in size, shape and presence/absence of tufts of feathers. The nipple-like protuberance of the gland exudes oil containing phospholipids, glycolipids, neutral lipids as well as acidic mucins (see for review Salibian and Montalti 2009). The bird uses the bill or the sides and back of the head to spread the oil throughout the plumage when a bird preens. “Preen oil helps keep the plumage of waterfowl in good condition. The oil maintains the flexibility of feathers, and keeps feather barbules from breaking,” (Moyer et al. 2003). Also antifungal as well as antibacterial properties of the oil are known. This substance also preserves the feather keratin. Additionally, control of plumage hygiene, thermal insulation, defense against predators, and pheromone production are examples of the physiological roles associated with the preen gland secretion (see for review Lequette et al. 1989; Nevitt and Bonadonna 2005; Nevitt 2008).

Waterproof Feathers

Sebum-like lipids are synthesized and secreted in the form of droplets by sebokeratinocytes, which are unique structural components of the avian epidermis. This epidermis also elaborates, but rarely secretes, the multigranular bodies (lipid—enriched organelles) (Menon and Menon 2000). In order to make their feather waterproof, a bird spreads waxes and fats, which are originally secreted by the preening gland. Additionally, birds also have, special feathers made of keratin that break into small dust-like pieces and known as powder downs (Chandler 1916; Stettenheim 2000). This dust is spread throughout the feathers of seabirds and plays very important role in the phenomenon of waterproofing. The powder downs are an "elaborate lipid rich material, which can be classified as secretion" (Menon and Menon 2000).

Birds and mammals distinguish from reptiles, amphibians and fish because of their "warm-bloodedness", or endothermic homeothermy. The success of birds and mammals in aquatic and terrestrial environments may be largely determined by endothermy. "Elevated rates of lung ventilation, oxygen consumption, and internal heat production (via aerobic metabolism), which are the hallmarks of endothermy, enable birds and mammals to maintain thermal stability over a wide range of ambient temperatures" (Ruben and Jones 2000).

Feathers are associated with both the flight (Feduccia 1996) and the thermoregulation of seabirds. However, "it is likely that avian endothermy probably evolved long after the origin of avian flight, in association with selection for enhanced capacity for long-distance flight rather than for thermoregulatory purposes" (Ruben and Jones 2000).

Recently, a report was released (Grémillet et al. 2005) on the specific feather structure of the great cormorant (*Phalacrocorax carbo*) that allows partial plumage wettability in diving birds. It was observed that each body feather of this seabird species has a loose, instantaneously wet, outer section and a highly waterproof central portion. Because of this structural feature, the plumage of *P. carbo* is only partly wettable. Thus, the bird maintains a thin layer of air in their plumage. "These findings suggest an unusual morphological-functional adaptation to diving which balances the antagonist constraints of thermoregulation and buoyancy," (Grémillet et al. 2005).

Wings and Flight

The shape of the seabird's wing seems to be an ideal construct for soaring and gliding along the sea's surface (for review see Savile 1957; Videler 2005; Brewer and Hertel 2007). For example, shearwaters and albatross are well known superb gliders, often soaring over large distances without taking a wing beat. The wing planform, camber lane and thickness distribution in seabirds are species dependent. These structural features are crucial from aero dynamical point of view. Seabirds possess wings which are generally larger than that of terrestrial species including the length of the flight feathers (see for review Ashmole 1971). Additionally, they use their amazing wingspans to ride the ocean winds and "sometimes to glide for hours without rest or even a flap of their wings" (Pennyquick 1987). Some species also float on the sea's surface, though the position makes them vulnerable to sharks. Wielding their up to 3.5 m long wings, a parent albatross can cover a distance of about 16,000 km to deliver one meal to its nestlings (Safina 2007; see for review Harrop 1994; Tickell 2000; Brooke 2004). It is suggested that, for example, the 50-year-old albatross has flown, at the very least, a total of 6 million km (Fisher 1975).

Unfortunately, selected albatross species like the Laysan albatross (*Phoebastria immutabilis*) were intensively hunted for feathers that were used as in the manufacture of women's hats as well as of down. According to archaeological excavations, these birds were also important part of the human diet in the area of the settlement of Aleut and Eskimos (Brooke 2004).

Recently, the styles of the bird's hovering, intermittent flight and locomotion have great potential to inspire the future design of autonomous flying vehicles (Tobalske 2010). Therefore, studies on seabird wing design (Norberg 1995), loadings and shapes (Warham 1977), as well as their structural mechanics (Habib 2010), determine the current science in terms of bioinspiration and bionics.

Wings and Diving

Because the optimum design of the propulsive organs is different for the two media (air and water), which differ substantially in density and buoyancy, an intermediate adaptive stage probably would involve a loss of efficiency in each medium as the price of adequacy in the other (Raikow et al. 1988). Diving petrels, auks, penguins use their wings for propulsion under water. However, propulsion underwater can be also provided by feet as it was observed for fish-eating ducks, loons, cormorants, or grebes. These foot-propelled divers are generally slower than wing-propelled divers.

The most extreme underwater adaptations occur in penguins. The limb and its skeleton are flattened, and the wing is reduced in surface area by the loss of differentiated flight feathers, patagia, and the alula. The shoulder joint and the extrinsic muscles of the wing are functionally specialized, and the limb is relatively rigid as joint mobility is restricted. In all, the wing is converted to a "flipper" similar in external form to those of other aquatic tetrapods (Raikow et al. 1988). Maximum diving depths of some seabirds are amazing (Adams and Walter 1993; Prince et al. 1994). For example, predators of pelagic fish in southern African waters, the pursuit diving Cape Cormorant (*Phalacrocorax capeis*) and African Pengui (*Spheniscus demersus*) are capable of diving to depths of 92 m and 130 m, respectively (Burger 1991).

Studies of underwater propulsion indicate that alcids employ a different method from that of penguins (Watanuki et al. 2006). In alcids, the manus stays in the flexed position during the propulsive stroke, in which the wing moves down and backward in a rowing action. In penguins the manus is extended, and the wing rotated so the leading edge is lower than the trailing edge in the downstroke. This entails little caudal movement from the wing, and the upstroke can then be used to generate thrust as well (see for review Raikow et al. 1988).

Detailed analysis of the wing propulsion during deep diving by Brünnich's guillemots (*Uria lomvia*) was recently reported by Watanuki and co-authors. "At the start of descent, the birds produced frequent surges (3.2 Hz) against buoyancy during both the upstroke and the downstroke to attain a mean speed of 1.2–1.8 m/s; close to the expected optimal swim speed. As they descended deeper, the birds decreased the frequency of surges to 2.4 Hz, relaying only on the downstroke" (Watanuki et al. 2003). However, "during their ascent, they stopped stroking at 18 m depth, after which the swim speed increased to 2.3 m/s. This may be due to increasing buoyancy as their air volumes expanded. This smooth change of surge frequency was achieved while maintaining a constant stroke duration (0.4–0.5 s), presumably allowing efficient muscle contraction," (Watanuki et al. 2003).

Bills, or Beaks

The Rostrum is the Latin term for a bird's bill or beak. Bird's rostrum consists of an upper, chiefly premaxillary and maxillary, and of a lower, or mandibular, half (Bock and Kummer 1968). The bills of seabirds are typical examples of hierarchical and complex multi-layered biological constructs with specific mechanical properties (Soons et al. 2010) "consisting of a horny external layer with underlying bone and a (epi-) dermal layer in between" (Soons et al. 2012a). This keratinous covering is to a certain extent molded along the shape of the supporting bones.

The soft cutaneous fragment of the skin is commonly restricted to a thin layer between the periosteum and the stratum germinativum (also known as Malpighian layer) of the epiderm. It is rich on many blood- vessels and nerves, the latter occasionally penetrating the keratinous layer to terminate in tactile or sensory corpuscles. The outer surface of the bill, the rhamphotheca (Lönnberg 1904), is a continuously growing structure in most birds (Stettenheim 1972); so bill morphology is determined by rates of both growth and wear. The thickness of this keratinous layer is on the order of 3 mm. Inter-individual changes in bill morphology have been viewed as passive reflections of changes in dietary protein content, changes in abrasion, as an adaptive response to dietary shifts. Though strategic adjustment of growth rates it was proposed (Bonser and Witter 1993) that changes in bill coloration from different melanin contents may have a mechanical function. The deposition of melanin can result in increased bill hardness, and this may have important implications for the maintenance of bill shape and, thus, foraging behavior.

Keratin is a robust biological material, which can protect the bone using absorption of energy from impacts (see Chap. 11). Thus, keratin on the bird beak possesses unique properties as a shock absorber with regard to distribution of corresponding forces and to prevention of bone microfracture (Frenkel and Gillespie 1976). If the beak only contained of bone, or only contained of keratin, it would not be durable. It is the interaction between the two different biological materials makes this organ firm while remaining very light. These properties seems to be the driving force in biomimetic key way with respect to design and development of self-repairing keratin-like protective sheaths (Soons et al. 2012a, b).

Rhamphothecae sheaths show amazing diversity, and provide some of the most compelling and easily appreciated examples of morphological adaptation in vertebrates (Hieronymus and Witmer 2010). These include forceps for probing in sandpipers, filters in ducks and flamingos, "teeth" for gripping fish in mergansers and gannets, and nutcrackers in hawfinches. A widely distributed feature of rhamphothecae that appears to be unrelated to their adaptive roles in feeding and display can be seen in marine birds such as albatrosses, in which the skin of the rhamphotheca is separated into several plates (Soldaat et al. 2009) (Fig. 1.9). This condition, referred to as a "compound rhamphotheca" (Lönnberg 1904) contrasts with the continuous cornified

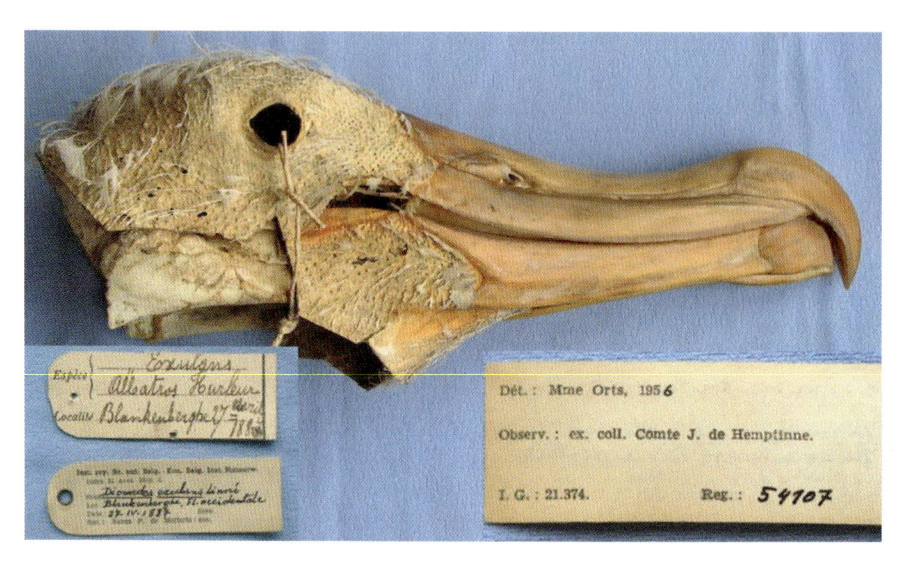

Fig. 1.9 The keratinous "compound rhamphotheca" of the complex structured bill of albatross (Adapted from Soldaat et al. (2009) with permission)

sheaths seen in birds such as American Crows that possess a simple rhamphotheca. Avian compound rhamphothecae have long been used as taxonomic characters (Hieronymus and Witmer 2010).

Egg Tooth

No birds have true teeth, just an appendage on the beak called an egg tooth (see for review Clarck 1961). The egg tooth plays the crucial role in the avian piping mechanism. The egg teeth consist of thin, apparently calcareous, sheets that cover the entire tips of the mandibles. The avian egg itself possess heavily mineralized calcareous eggshell. This shell, first of all, provides mechanical protection for the developing embryo from outside pressures. "Egg strength is highly species-specific, because it represents a compromise between different selective forces. Among these are the need for mechanical protection of an embryo, which favors great strength, and the necessity for the young to hatch without assistance from its parents, presumably favouring low strength of a shell," (Honza et al. 2001).

The main function of the egg teeth is to help hatchlings to break through the shell during hatching (Wetherbee and Barlett 1962). Additionally, egg teeth could have some function by increasing nestlings' visibility. Corresponding investigations with respect to the size, color, and persistence of egg teeth in woodpeckers, which nest in dark cavities has been carried out by Wiebe (2010). This author reported that "many species of woodpecker have two egg teeth, one each on the tip of the maxilla and of the mandible, which, together with the pale flanges, frame the open mouth when nestlings gape. Reflectance of flanges peaked in the ultraviolet, which is less visible to woodpeckers. Therefore, parent woodpeckers can probably see egg teeth better than flanges. Some burrow-nesting seabirds also retain egg teeth for a long time. This reinforces the idea that egg-tooth reflectance may have evolved independently in several phylogenetic groups in which parents must find nestlings in the dark," (Wiebe 2010).

Gular Pouches

It is well known that such seabirds as pelicans dive into water for fish from altitudes of 9–16 m (Allen 1923). They also attain speeds of up to 18 m/s before water impact (Johnsgard 1993). Pelican's gular pouches are highly distensible constructs. They can hold up to 11 L of water (Schreiber et al. 1975). "When the mandibles bow (from a resting position of about 5 cm apart to over 15 cm apart) the opening created is approximately 500 cm^2. This bowing, termed *streptognathism*, is well-developed in birds that are able to swallow large objects," (Meyers and Myers 2005). Another type of joint in which bones are joined by connective tissue has been observed in gulls and other species, including pelicans. This is known as "*syndesmosis*" (Judin 1961).

Recently, Meyers and Meyers (2005) investigated the bending of the Brown Pelican (*Pelecanus occidentalis*) lower jaw. They showed "that both mineral content and bone shape contribute to the high degree of bowing observed when this species dives into the water for food," (Meyers and Myers 2005). The bones have a mineral content of about 52 % and play crucial role within the syndesmotic joint. This joint permits these bones to move relative to one another, much like a leaf-spring functions (Buhler 1981).

Bones

Hierarchically structured bones of vertebrates distinguish in their details. Both ectothermy and endothermy play here very important role. Thus, so called lamellar-zonal bone of ectothermic extant amphibians and most reptiles "has a layered appearance, within which incremental growth lines are occasionally recognized; it is also poorly vascularized," (Ruben et al. 1998). Endothermic animals like birds, mammals, and dinosaurs, mostly possess well vascularized fibro-lamellar bone (Reid 1997; Margerie de 2002). As discussed by Ruben and co-authors: "fibro-lamellar bone is often held to be correlated with high growth rate that requires rapid deposition of calcium salts. Such rapid growth is supposedly possible only in systems with high metabolic rates associated with endothermy," (Ruben et al. 1998). Recently, it was proposed (Simons and O'Connor 2012) that laminarity is "an adaptation for resisting torsional loading. This may be explained by overall wing shape: while dynamic soaring birds have long slender wings, flappers and static soaring birds have broader wings with a larger wing chord that would necessarily impart a higher torsional moment on the feather-bearing bones," (Simons and O'Connor 2012).

We can analyses the bone as construct that represents a balance between hard tissue economy and flexural strength (see for review McGowan 1999; Currey 2003) (see also Sect. 3.1). From this point of view, the evolutionary expediency of development of hollow, tubular structures within tetrapod limb bones is not surprizing.

"Dense compact bone bears bending or torsional stresses on the periphery of the shaft. Less dense internal supportive tissues lend secondary support, or are lacking entirely, with a marrow-filled cavity lending haematopoietic function," (Bostwick et al. 2012).

Wings contain some of the longest and strongest bones. Flying animals appear to have denser bones than terrestrial counterparts of similar size. In bones, differences in density generally mean differences in mineral content. Higher mineral contents are linked to greater stiffness and strength. Dumont (2010) thinks that such material properties may be central to understanding why bone density is relatively high in birds. Along with flight, birds have also evolved small body size; and one way to design a small skeleton while preserving its strength is to make its bones denser. Investigating links between the material properties of bone and the evolution of flight and small body size in birds is going to require much more work – as we would need to obtain bone density measurements from transitional forms in the fossil record like *Archaeopteryx* (Hone et al. 2008).

The skeletons of birds are described in detail (see for review Shufeldt 1890; Baumel et al. 1993; Kaiser 2007; Hospitaleche et al. 2009). These constructs are examples of selection for minimizing the energy required for the bird's flight. "From a functional perspective, the weight (mass) of an animal relative to its lift-generating surfaces is a key determinant in the metabolic cost of flight. The evolution of birds has been characterized by many weight-saving adaptations that are reflected in bone shape, many of which strengthen and stiffen the skeleton" (Dumont 2010). According to this author "the round and thin-walled humeral shafts of birds are an optimal shape for resisting both torsion and bending, and fused skeletal elements have been interpreted as increasing force resistance (i.e. stiffness; Buhler 1992)," (Dumont 2010). It was observed that flightless and diving birds possess thicker-walled bones which may serve as reinforcement or ballast. In diving seabirds, the rib cages and bones must be stronger and heavier, respectively, to recover from the increased pressure during a dive. During this process, the rib cage collapses inwards but springs back when the dive is achieved (see for review Norberg 1985).

The evolution of flighted birds can be characterized by the expansion of pneumatized spaces within some bones as well as by gradual reduction and fusion of many skeletal elements. Correspondingly their bones are typically lighter, but denser than that in mammal skeletons. Also bird's skeletons must be stronger and stiffer because of their ability to withstand the mechanical forces encountered during flight.

It is little wonder that most seabirds have partially hollow or hollow bones (Fig. 1.10). The morphological properties of seabird bones that allow flight (Casinos and Cubo 2001; Simons 2010; Simons et al. 2011), as well as their biomechanics (Pennycuick 1967; Kirkpatrick 1994; Cubo and Casinos 1998, 2000; Mi et al. 2005; de Margerie et al. 2005, 2006), are still the subjects of numerous investigations. Aeronautical engineers mimic seabirds skeletal structures "by designing load-bearing structures with shapes that confer strength, and by using materials that have high strength-to-weight and stiffness-to-weight ratios," (Dumont 2010).

The important role of the shape in the mechanical properties of bones is well accepted in the literature. For example, "it is well known that given a tube and rod

Fig. 1.10 Frigatebird (*Fregata species*) wing bones with one sectioned to show hollow structure (Photo: Kim Taylor/Warren Photographic)

of similar material and equal mass, the hollow rod will resist bending more than a solid rod," (Meyers and Myers 2005). This property allowed seabird bones' has to be used as tools without much modification. For example, awls and points are the only tools made of bird bone found in the Stone Age in the Baltic Sea area. Albatross wing bones in particular are the premier material for some tools. For example, "*Koauau Ponga Ihu*" is the name of a small gourd with the top removed and two small holes. The flute-like sound can be achieved by blowing across the top with one nostril.

1.2.3.4 Marine Mammals (Class Mammalia)

In contrast to amphibians which never quite part with the aquatic environment, vertebrates like reptiles, birds, and mammals have numerous representatives that have returned to the both fresh water and marine aquatic niches for an amphibious lifestyle. Nowadays some marine mammals living a fully aquatic existence. According to (Uhen 2007), mammals in particular have returned to the ocean in at least seven separate lineages:

- Cetacea (whales, dolphins, and porpoises) (Thewissen and Williams 2002; McGowen et al. 2009),
- Sirenia (sea cows) (Domning 2001),

- Desmostylia (an extinct order of marine mammals) (Reinhart 1953, 1982),
- Pinnipedia (seals, sea lions, walrus) (Ronald et al. 1991),
- *Ursus maritimus* (polar bear),
- *Enhydra lutris* (sea otter),
- and *Thalassocnus* spp. (aquatic sloths)

"Some of these lineages have retained most of their terrestrial form while spending a great deal of time in the water, whereas others have changed their morphology dramatically and spend almost all, if not all, of their time in the water" (Uhen 2007; additionally see for review Rice 1998).

There is growing interest in marine mammals from the biological materials community arising from the numerous physical adaptations; as well as of specific biomaterials like baleen, teeth, tusks, bones, skin which possess unique structure and high biomimetic potential (Fig. 1.11). Some of these materials are analyzed in detail in this work (see Sects. 3.1.1, 3.1.2, 3.2.6.4, 3.2.6.5, 3.2.6.6, and 11.3).

Mechanical properties define how a material deforms in response to an applied force, and biomechanics goes a step further to examine how the response is related to function (Waite and Broomell 2012). There are several examples how marine mammals have modified their external shape with respect to effectiveness of corresponding propulsion mechanisms for locomotion under water. Thus, as reviewed by

Fig. 1.11 Marine mammals possess numerous biological materials like baleen fringes, giant porous bones, walrus tusk and narwhal tusk (Images courtesy of Dr. Andrey Bublichenko)

Reidenberg (2007), "seals use alternating horizontal sweeps of their hind flippers. Fur seals and sea lions 'fly' underwater by beating their fore flippers. Walruses sometimes use their tusks to grip the sea floor or ice and push their body forward with a downward nod of the head. Cetaceans have excelled in the attainment of streamlined form, and are thus the fastest swimmers," (Reidenberg 2007).

I definitively agree with Joy Reidenberg that knowledge "of marine mammals' unusual specializations will hopefully inspire us to copy nature in the development of new technologies. For example, continued investigations into flukes, flippers, axial movements, feeding mechanics, skin, and body shape may lead to development of more efficient hydrodynamic designs for water- and aircraft," (Reidenberg 2007).

From biomechanical point of view, the flukes of whale act like a pair of wings (Vogel 1994). In contrast to such static constructs as the wings of airplanes, the flukes are distinguished by their ability to generate a lift-derived thrust with high efficiency and are known as high-performance oscillatory propulsors. The mechanical features of this wing like structure are based also on biological materials properties they are designed during evolution. For examples, "the collagenous internal structure provides the framework forms a flexible hydrofoil with relatively high aspect ratio and moderate sweepback" (Fish 1998).

Dolphin and whale skin is also under investigations at nanolevel as a possible key to subaquatic speed (Baum et al. 2003; Pavlov 2006; Pavlov et al. 2012). As summarized by Reidenberg (2007):

> "further study of how aquatic mammals regulate buoyancy, control bone density, or manage dramatic changes in temperature and pressure as they rise and fall in the water column may lead to new treatments for osteoporosis, or the invention of protective gear for exposure to the extreme environmental changes of high and low altitude, space, or ocean depths. A more complete understanding of neural organization, underwater vision, or sound generation and sound reception mechanisms may lead to the creation of better artificial sensory systems," (Reidenberg 2007).

1.3 Conclusion

It was in the oceans that life first evolved and where complex animals have thrived for over 600 million years. There are marine representatives from every major animal phylum. Marine animals survive in environments as diverse as tropical coral reefs, polar ice-capped oceans, and the lightless abyssal depths. The diversity of habitats available in marine systems has led to a vast array of body designs, as well as physiological and behavioral mechanisms. These adaptations that evolved in animals are used to overcome the biotic and abiotic challenges in the ocean (Fish and Kosak 2011).

Investigating the relationships between organisms and their environments at many length and time scales is necessary for a mechanistic understanding of the limits of adaptation and the survival strategies of individuals and communities. With dimensions that range from molecular to global, the problem is a staggering one. Discovering solutions will require scientific and technical expertise, and creative thinking from many different disciplines (Waite and Broomell 2012). The biological, chemical and materials diversity of the ocean is amazing and therefore is an extraordinary resource for the discovery of new bioactive substances, biopolymers, bioadhesives, bioelastomers and hierarchically structured biocomposites. As reported in this chapter above, marine vertebrates possess amazing diversity of biological materials that determine and that serves to drive current and future progress in modern biomimetics.

Biomimetics can be defined as the use of natural principles and phenomena to inspire design and technology both at nano- and macrolevels. Biomimetics reinforces the need to protect ocean biodiversity, the patent library for the technologies of the future (Gorb 2011). Therefore, biomimetic materials and marine inspiration (Bellamkonda 2008) are the modern directions in materials science and engineering. It's not surprising that one of the special issues of the

Marine Technology Society Journal in August 2011 is entitled "*Biomimetics and Marine Technology*". This special issue brings together both biologists and engineers who are currently involved with the development of biomimetic devices for applications in the marine environment. The titles of the articles therefore were dedicated to, among others, applying biomimetics to underwater robotics, the bio-inspired propulsion mechanisms based on Manta ray locomotion (see for review Pennisi 2011), the use of a biomimetic skeleton for the flapping, propulsive tail of an aquatic robot, and additionally in the marine applications of a biomimetic humpback whale flipper. Thus, there are no doubts that marine biosystems are quite rich in capabilities that have and can further benefit humans from biomimicking (Bar Cohen 2011) (see Chaps. 5 and **7** in this work).

References

Adams LE (1906) The flight of flying fish. Zoologist 4:145–148
Adams NJ, Walter CB (1993) Maximum diving depths of cape gannets. Condor 95:136–138
Ahlberg PE (1989) Paired fin skeletons and relationships of the fossil group Porolepiformes (Osteichthyes: Sarcopterygii). Zool J Linnean Soc 96:119–166
Ahlberg PE (1991) Tetrapod or near tetrapod fossils from the Upper Devonian of Scotland. Nature 354:298–301
Ahlberg PE (1993) *Elginerpeton pancheni* and the earliest tetrapod clade. Nature 373:420–425
Ahlberg PE, Johanson Z (1998) Osteolepiforms and the ancestry of tetrapods. Nature 395(6704):792–794
Ahlberg P, Trinajstic K, Johanson Z, Long J (2009) Pelvic claspers confirm chondrichthyan–like internal fertilization in arthrodires. Nature 460:888–889
Ahlborn F (1897) Der flug der Fische. Zool Jb Abt Syst 9:329–338
Ainley DG (1980) Birds as marine organisms. CalCOFI Rep 21:48–53
Al–Bahry SN, Mahmoud IY, Al–Amri IS et al (2009) Ultrastructural features and elemental distribution in eggshell during pre and post hatching periods in the green turtle, *Chelonia mydas* at Ras Al–Hadd, Oman. Tissue Cell 41:214–221
Albertson RC, Cresko W, Detrich HW III, Postlethwait JH (2009) Evolutionary mutant models for human disease. Trends Genet 25:74–81

Albertson RC, Yan YL, Titus TA et al (2010) Molecular pedomorphism underlies craniofacial skeletal evolution in Antarctic notothenioid fishes. BMC Evol Biol 10:4
Allaby A, Allaby M (1999) Anaspida. In: A dictionary of earth sciences. Oxford University Press, Oxford
Allen WE (1923) Fishing activities of the California Brown Pelican. Condor 25:107–108
Allis EP (1922) The cranial anatomy of *Polypterus*, with special reference to *Polypterus bichir*. J Anat 56:180–294
Andrews CW (1910) A descriptive catalogue of the marine reptiles of the Oxford clay: Part 1. The British Museum (Natural History), London
Andrews CW (1913) A descriptive catalogue of the marine reptiles of the Oxford clay: Part 2. The British Museum (Natural History), London
Appeltans W, Bouchet P, Boxshall GA et al (2012) (eds) World register of marine species. Accessed at http://www.marinespecies.org on 29 Sept 2012
Arnott HJ, Maciolek NJ, Nicol JAC (1970) Retinal tapetum lucidum: a novel reflecting system in the eye of teleosts. Science 169:478–480
Arsenault M (1982) *Eusthenopteron foordi*, a predator on *Homalacanthus concinnus* from the Escuminac formation, Miguasha, Quebec. Can J Earth Sci 19:2214–2217
Aschliman NC, Claeson KM, McEachran JD (2012) Phylogeny of Batoidea. In: Carrier JC, Musick JA, Heithaus MR (eds) Biology of sharks and their relatives, 2nd edn. CRC Press, Boca Raton
Ashmole NP (1971) Seabird ecology and the marine environment. In: Farner DS, King JR (eds) Avian biology, vol 1. Academic, New York
Aubret F, Shine R (2008) The origin of evolutionary innovations: locomotor consequences of tail shape in aquatic snakes. Funct Ecol 22:317–322
Baker MA (1982) Brain cooling in endotherms in heat and exercise. Annu Rev Physiol 44:85–96
Balani K, Patel RR, Keshri AK et al (2011) Multi–scale hierarchy of *Chelydra serpentina*: microstructure and mechanical properties of turtle shell. J Mech Behav Biomed Mater 4(7):1440–1451
Bar Cohen Y (2011) Biomimicking marine mechanisms and organizational principles. Mar Technol Soc J 45:14–15
Bardack D (1991) First fossil hagfish (Myxinoidea): a record from the Pennsylvanian of Illinois. Science 254:701–703
Bartholomew GA, Lasiewski RC (1965) Heating and cooling rates, heart rate and simulated diving in the Galapagos marine iguana. Comp Biochem Physiol 16:573–582
Bartholomew GA, Bennett AF, Dawson WR (1976) Swimming, diving, and lactate production of the marine iguana, *Amblyrhynchus cristatus*. Copeia 1976:709–720
Basden AM, Young GC (2001) A primitive actinopterygian neurocranium from the Early Devonian of southeastern Australia. J Vertebr Paleontol 21:754–766
Baum C, Simon F, Meyer W et al (2003) Surface properties of the skin of the pilot whale *Globicephala melas*. Biofouling 19(Supplement):181–186
Baumel JJ, Witmer LM, Baumel JJ et al (1993) Osteologia. Handbook of avian anatomy: nomina anatomica avium. Publ Nuttall Ornithol Club 23:45–132
Bell T (1825) On a new genus of Iguanidae. Zool J 2:204–208
Bellamkonda RV (2008) Biomimetic materials: marine inspiration. Nat Mater 7:347–348
Bemis WE, Northcutt RG (1992) Skin and blood vessels of the snout of the Australian lungfish, *Neoceratodus forsteri*, and their significance for interpreting the cosmine of Devonian lungfishes. Acta Zool 73:115–139
Bemis WE, Findeis EK, Grande L (1997) An overview of Acipenseriformes. Environ Biol Fishes 48:25–71
Benson RBJ, Butler RJ, Lindgren J et al (2010) Mesozoic marine tetrapod diversity: mass extinctions and temporal heterogeneity in geological megabiases affecting vertebrates. Proc R Soc Lond B Biol Sci 277:829–834
Berg LS (1940) Classification of fishes, both recent and fossil. Travaux de l'Institute zoologique de l'Academie des Sciences de l'URSS 5(2):85–517

Bernard A et al (2010) From Bernard A, Lécuyer C, Vincent P et al (2010) Regulation of body temperature by some mesozoic marine reptiles. Science 328:1379–1382. Reprinted with permission from AAAS

Birstein VJ, Hanner R, DeSalle R (1997) Phylogeny of the Acipenseriformes: cytogenetic and molecular approaches. Environ Biol Fishes 48:127–155

Blake RW (1983) Fish locomotion. Cambridge University Press, Cambridge

Bock WJ, Kummer B (1968) The avian mandible as a structural girder. J Biomech 1:89–96

Boersma PD (1977) An ecological and behavioral study of the Galápagos penguin. Living Bird 15:43–93

Boisvert C (2005) The pelvic fin and girdle of *Panderichthys* and the origin of tetrapod locomotion. Nature 438:1145–1147

Boisvert CA, Mark–Kurik E, Ahlberg PE (2008) The pectoral fin of *Panderichthys* and the origin of digits. Nature 456:636–638

Bonser RHC, Witter MS (1993) Indentation hardness of the bill keratin of the European starling. Condor 95:136–138

Bostwick KS et al (2012) Reprinted from Bostwick KS, Riccio ML, Humphries JM (2012) Massive, solidified bone in the wing of a volant courting bird. Biol Lett (doi:10.1098/rsbl.2012) by permission of the Royal Society

Botella H, Blom H, Dorka M, Ahlberg PE et al (2007) Jaws and teeth of the earliest bony fishes. Nature 448:583–586

Botella H, Donoghue PCJ, Martínez–Pérez C (2009) Enameloid microstructure in the oldest known chondrichthyan teeth. Acta Zool 90:103–108

Bouchet P (2006) The magnitude of marine biodiversity. In: Duarte CM (ed) The exploration of marine biodiversity: scientific and technological challenges. Funbdacion BBVA, Bilbao

Branch B (1998) Field guide to the snakes and other reptiles of southern Africa. Ralph Curtis Books Publishing, Florida

Brazeau MD (2009) The braincase and jaws of a Devonian 'acanthodian' and the origin of modern gnathostomes. Nature 457:305–308

Breder CM Jr (1930) On the structural specialization of flying fishes from the standpoint of aerodynamics. Copeia 4:114–121

Brewer ML, Hertel F (2007) Wing morphology and flight behavior of pelecaniform seabirds. J Morphol 268:866–877

Briggs JC, Shelgrove P (1999) Marine species diversity. Bioscience 49:351–352

Brischoux F, Bonnet X (2009) Life history of sea kraits in New Caledonia. Zoologia Neocaledonica 7 Memoires du Museum National d'Histoire Naturelle 198:37–51

Brito–Echeverria J, Lopez–Lopez A, Yarza P et al (2009) Occurrence of Halococcus spp. in the nostrils salt glands of the seabird *Calonectris diomedea*. Extremophiles 13(3):557–565

Brooke M (2004) Albatrosses and petrels across the world. Oxford University Press, Oxford. Copyright © 2004, Oxford University Press. This material is used by the permission of Oxford University Press

Brothers EB (1984) Otolith studies. In: Moser HG, Richards WJ, Cohen MD et al (eds) Ontogeny and systematics of fishes, Spec. publ. no. 1. American Society of Ichthyologists and Herpetologists, Lawrence, pp 50–57

Brown JW, Rest JS, García–Moreno J et al (2008) Strong mitochondrial DNA support for a Cretaceous origin of modern avian lineages. BMC Biol 6:6

Bruet BJF, Song JH, Boyce MC et al (2008) Materials design principles of ancient fish armor. Nat Mater 7:748–756

Buhler P (1981) Functional anatomy of the avian jaw apparatus. In: King AS, McLelland J (eds) Form and function in birds, vol 2. Academic, London

Buhler P (1992) Light bones in birds. Los Angel Cty Mus Nat Hist Sci Ser 36:385–394

Burger AE (1991) Maximum diving depths and underwater foraging in alcids and penguins. In: Montevecchi WA, Gaston AJ (eds) Studies of high latitude seabirds. 1. Behavioral, energetic, and oceanographic aspects of seabird feeding ecology. Canadian Wildlife Service, Occasional paper no 68. Canadian Wildlife Service, Ottawa, pp 9–15

Calis E, Jackson EH, Nolan CP et al (2005) Preliminary age and growth estimates of the rabbitfish, *Chimaera monstrosa*, with implications for future resource management. J Northwest Atl Fish Sci 35:15–26
Carey JE, Wright EA (1960) Isolation of the neurotoxic component of the venom of the sea snake, *Enhydrina schistosa*. Nature 185:103–104
Carpenter CC (1966) The marine iguana of the Galapagos Islands, its behavior and physiology. Proc Calif Acad Sci 34:329–376
Carr A, Kemp AR, Tibbetts IR et al (2006) Microstructure of pharyngeal tooth enameloid in the Parrotfish *Scarus rivulatus* (Pisces: *Scaridae*). J Microsc 221:8–16
Casinos A, Cubo J (2001) Avian long bones, flight and bipedalism. Comp Biochem Physiol A 131:159–167
Cavin L, Giner S (2012) A large halecomorph fish (Actinopterygii: Holostei) from the Valanginian (Early Cretaceous) of southeast France. Cretac Res 37:201–208
Chandler AC (1916) A study of the structure of feathers, with reference to their taxonomic significance. Univ Calif Publ Zool 13:243–446
Cheng C–HC, Chen L (1999) Evolution of an antifreeze glycoprotein: a blood protein that keeps Antarctic fish from freezing arose from a digestive enzyme. Nature 401:443–444
Clack JA (2002) Gaining ground: the origin and early evolution of tetrapods. Indiana University Press, Bloomington
Clack JA (2009) With kind permission from Springer Science+Business Media: Clack JA (2009) The fish–tetrapod transition: new fossils and interpretations. Evo Edu Outreach 2:213–223. Copyright © 2009, Springer
Claeson KM (2011) The synarcual cartilage of batoids with emphasis on the synarcual of Rajidae. J Morphol 272:1444–1463
Clarck GA Jr (1961) Occurrence and timing of egg teeth in birds. Wilson Bull 73:268–278
Clarke JA (2004) Morphology, phylogenetic taxonomy, and systematics of *Ichthyornis* and *Apatornis* (Avialae: Ornithurae). Bull Am Mus Nat Hist 286:1–179
Clarke JA, Ksepka DT, Salas–Gismondi R et al (2010) Fossil evidence for evolution of the shape and color of penguin feathers. Science 330:954–957
Clément G (2004) Nouvelles données anatomiques et morphologie générale des "*Porolepididae*" (Dipnomorpha, Sarcopterygii). Rev Paléobiol 9:193–211
Clement A, Long JA (2010) Air–breathing adaptation in a marine Devonian lungfish. Biol Lett 6:509–512
Coates MI, Sequeira SEK (2001) A new stethacanthid chondrichthyan from the lower Carboniferous of Bearsden, Scotland. J Vertebr Paleontol 21:438–459
Compagno LJV (1977) Phyletic relationships of living sharks and rays. Am Zool 17:303–322
Compagno LJV (2001) Sharks of the world, an annotated and illustrated catalogue of shark species known to date – bullhead, mackerel & carpet sharks. FAO species catalogue for fishery purposes no 1, vol 2. FAO, Rome
Cote S, Carroll R, Cloutier R et al (2002) Vertebral development in the Devonian Sarcopterygian fish Eusthenopteron foordi and the polarity of vertebral evolution in non–amniote tetrapods. J Vertebr Paleontol 22:487–502
Cramp R et al (2008) Republished with permission of The Company of Biologists Ltd, from Cramp R, Meyer EA, Sparks N et al (2008) Functional and morphological plasticity of crocodile (*Crocodylus porosus*) salt glands. J Exp Biol 211:1482–1489. Copyright (2009); permission conveyed through Copyright Clearance Center, Inc
Cramp R et al (2010) Republished with permission of The Company of Biologists Ltd, from Cramp R, Hudson N, Franklin CE (2010) Activity, abundance, distribution and expression of Na^+/K^+–ATPase in the salt glands of *Crocodylus porosus* following chronic saltwater acclimation. J Exp Biol 213:1301–1308, Copyright (2010); permission conveyed through Copyright Clearance Center, Inc
Crossland C (1911) The flight of Exocoetus. Nature (London) 86:279–280
Cubo J, Casinos A (1998) Biomechanical significance of cross–sectional geometry of avian long bones. Eur J Morphol 36:19–28

Cubo J, Casinos A (2000) Mechanical properties and chemical composition of avian long bones. Eur J Morphol 38:112–121
Currey J (2003) The many adaptations of bones. J Biomech 36:1487–1495
Damiens R, Rhee H, Hwang Y et al (2012) Compressive behavior of a turtle's shell: experiment, modeling, and simulation. J Mech Behav Biomed Mater 6:106–112
Daniel JF (1922) The elasmobranch fishes. University of California Press, Berkeley
Davenport J (1994) How and why do flying fish fly? Rev Fish Biol Fish 40:184–214
Davenport J (2003) Allometric constraints on stability and maximum size in flying fishes: implications for their evolution. J Fish Biol 62:455–463
Davenport J (2005) Swimbladder volume and body density in an armored benthic fish, the streaked gurnard. J Fish Biol 55:527–534
Davis LS, Renner M (2003) Penguins. T&AD Poyser, London
Dawson WR, Bartholomew GA, Bennett AF (1977) A reappraisal of the aquatic specialization of the Galapagos marine iguana (*Amblyrhynchus cristatus*). Evolution 31:891–897. Republished with permission of Society for the Study of Evolution; permission conveyed through Copyright Clearance Center, Inc
de Beer GR, Moy–Thomas JA (1935) On the skull of Holocephali. Philos Trans R Soc Lond Ser B Biol Sci 224:287–312
de Margerie E (2002) Laminar bone as an adaptation torsional loads in flapping flight. J Anat 201:521–526
de Margerie E, Sanchez S, Cubo J et al (2005) Torsional resistance as a principal component of the structural design of long bones: comparative multivariate evidence in birds. Anat Rec A 282:49–66
de Margerie E, Tafforeau P, Rakotomanana L (2006) In silico evolution of functional morphology: a test on bone tissue biomechanics. J R Soc Interface 3:679–687
Decker JD (1967) Motility of the turtle embryo, *Chelydra serpentina* (Linné). Science 157:952–954
Denison RH (1947) The exoskeleton of *Tremataspis*. Am J Sci 245:337–365
Denison RH (1975) Evolution and classification of placodermi fishes. Breviora 432:1–24
Dennison E, Cyrus Cooper C (2011) Osteoporosis in 2010: Building bones and (safely) preventing breaks. Nat Rev Rheumatol 7:80–82
Diogo R (2007) On the origin and evolution of higher–clades: osteology, myology, phylogeny and macroevolution of bony fishes and the rise of tetrapods. Science Publishers, Enfield
Diogo R (2008) Comparative anatomy, homologies and evolution of the mandibular, hyoid and hypobranchial muscles of bony fish and tetrapods: a new insight. Anim Biol 58:123–172
Domning DP (2001) The earliest known fully quadrupedal sirenian. Nature 413:625–627
Donley JM et al (2004) Reprinted by permission from Macmillan Publishers Ltd., Nature (Donley JM, Sepulveda CA, Konstantinidis P et al (2004) Convergent evolution in mechanical design of lamnid sharks and tunas. Nature 429:61–65) copyright (2004)
Donoghue PCJ, Sansom IJ (2002) Origin and early evolution of vertebrate skeletonization. Microsc Res Tech 59:352–372
Downing SW, Salo WL, Spitzer RH et al (1981) The hagfish slime gland: a model system for studying the biology of mucus. Science 214:1143–1145
Dumont ER (2010) Reprinted from Dumont ER (2010) Bone density and the lightweight skeletons of birds. Proc R Soc B 277:2193–2198, by permission of the Royal Society
Dunson WA (1969) Electrolyte excretion by the salt gland of the Galapagos marine iguana. Am J Physiol 216:995–1002
Dunson WA (1975) The biology of the sea snakes. University Park Press, Baltimore/London/Tokyo
Dunson WA, Freda J (1985) Water permeability of the skin of the amphibious snakes, Agkistrodon piscivorus. J Herpetol 19:93–98
Dunson WA, Robinson GD (1976) Sea snake skin: permeable to water but not to sodium. J Comp Physiol I08:303–311
Dunson WA, Packer RK, Dunson MK (1971) Sea snakes: an unusual salt gland under the tongue. Science 173:437–441
Durnford CD (1906) Flying fish flight. Am Nut 40:1–11

Eastman JT (2005) The nature of the diversity of Antarctic fishes. Polar Biol 28:93–107
Eckert SA (2002) Swim speed and movement patterns of gravid leatherback sea turtles (*Dermochelys coriacea*) at St. Croix, US Virgin Islands. J Exp Biol 205:3689–3697
Elder WH (1954) The oil gland of birds. Wilson Bull 66:6–31
Ellis RA, Abel JH (1964) Intercellular channels in the salt–secreting glands of marine turtles. Science 144:1340–1342
Ellis RA, Goertemiller C (1977) Significance of extensive 'leaky' cell junctions in the avian salt gland. Nature 268:555–556
Enticott J, Tipling D (1997) Seabirds of the world. Stackpole Books, London
Erben HK (1970) Ultrastruckturen and mineralisation rezenter und fossiler eierschalen bei bogen und reptilien. Biomineralisation 1:1–66
Ergene S, Aymak C, Ucar AH (2011) Carapacial scute variation in green turtle (*Chelonia mydas*) and loggerhead turtle (*Caretta caretta*) hatchlings in Alata, Mersin, Turkey. Turk J Zool 35(3):343–356
Ericson PG, Envall I, Irestedt M et al (2003) Inter–familial relationships of the shorebirds (Aves: Charadriiformes) based on nuclear DNA sequence data. BMC Evol Biol 3:16
Eschmeyer WN (1990) Catalog of the genera of recent fishes. California Academy of Sciences, San Francisco. 697 pp
Evans HM (1923) The defensive spines of fishes, living and fossil, and the glandular structures in connection therewith, with observations on the nature of fish venoms. Philos Trans R Soc Lond 212:1–33
Ewert MA (1979) The embryo and its egg: development and natural history. In: Harless M, Morloch H (eds) Turtles: perspectives and research. Wiley, New York
Feduccia A (1996) The origin and evolution of birds. Yale University Press, New Haven
Fischer J, Voigt S, Schneider JW, Buchwitz M et al (2011) A Selachian freshwater fauna from the Triassic of Kyrgyzstan and its implication for Mesozoic Shark Nurseries. J Vertebr Paleontol 31(5):937–953
Fish FE (1990) Wing design and scaling of flying fish with regard to flight performance. J Zool 221(3):391–403
Fish FE (1998) Biomechanical perspective on the origin of cetacean flukes. In: Thewissen JGM (ed) The emergence of whales. Plenum Press, New York, pp 303–324
Fish FE (2000) Biomechanics and energetics in aquatic and semiaquatic mammals: platypus to whale. Physiol Biochem Zool 73:683–698
Fish FE, Kosak DM (2011) Biomimetics and marine technology: an introduction. Mar Technol Soc J 45:8–13
Fisher HI (1975) Longevity of the Laysan albatross, *Diomedea immutabilis*. Bird-Banding 46:1–100
Forbes A (1936) Flying fish. Science 83:261–262
Franklin CE, Grigg GC (1993) Increased vascularity of the lingual salt glands of the estuarine crocodile, *Crocodylus porosus*, kept in hyperosmotic salinity. J Morphol 218:143–151
Fraser GJ, Graham A, Smith MM (2006) Developmental and evolutionary origins of the vertebrate dentition: molecular controls for spatio–temporal organisation of tooth sites in osteichthyans. J Exp Zool B Mol Dev Evol 306:183–203
Frenkel MJ, Gillespie JM (1976) The proteins of the keratin component of bird's beaks. Aust J Biol Sci 29(5–6):467–479
Fricke H, Plante R (1988) Habitat requirements of the living Coelacanth *Latimeria chalumnae* at Grande Comore, Indian Ocean. Naturwissenschaften 75:149–151
Fujita K (1990) The caudal skeleton of Teleostean fishes. Tokai University Press, Tokyo
Gardiner BG (1984) The relationships of placoderms. J Vertebr Paleontol 4:379–395
Gash SL, Bass JC (1973) Age, growth and population structures of fishes from acid and alkaline Strip–Mine Lakes in Southeast Kansas. Trans Kans Acad Sci 76:39–50
Gaston AJ (2004) Seabirds: a natural history. Yale University Press, New Haven
Geerinck T, De Poorter J, Adriaens D (2007) Morphology and development of teeth and epidermal brushes in loricariid catfishes. J Morphol 268:805–814

Geist HR (2000) Nasal respiratory turbinate function in birds. Physiol Biochem Zool 73:581–589
Germano DJ, Bury RB (1998) Age determination in turtles: evidence of annual deposition of scute rings. Chel Conserv Biol 3(1):123–132
Gilbert SF, Loredo GA, Brukman A et al (2001) Morphogenesis of the turtle shell: the development of a novel structure in tetrapod evolution. Evol Dev 3:47–58
Gill T (1905) Flying fishes and their habits. A Rep Smithon Inst 1904:495–515
Ginter M (2004) Devonian sharks and the origin of Xenacanthiformes. In: Arratia G, Wilson MVH, Cloutier R (eds) Recent advances in the origin and early radiation of vertebrates. Verlag Dr. Friedrich Pfeil, München
Ginter M, Hampe O, Duffin C (2010) Paleozoic Elasmobranchii. In: Schultze H–P (ed) Handbook of paleoichthyology, vol 3D. Verlag Dr Friedrich Pfeil, München
Goedert J (1989) Giant Late Eocene marine birds (Pelecaniformes: Pelagornithidae) from Northwestern Oregon. J Paleontol 63(6):939–944
Goldek SG, Voris KH (1982) Marine snake diets, prey composition, diversity and overlap. Copeia 3:661–666
Goodge WR (1960) Adaptations for amphibious vision in the dipper (*Cinclus mexicanus*). J Morphol 107:79–91
Gorb SN (2011) Biomimetics: a million ideas from the ocean. In: Future Ocean, Kiel Marine Sciences (ed) The ocean is our future: Kiel marine scientists on a time trip to 2100. Cluster of Excellence "The Future Ocean", Kiel, pp 70–75
Gordon MS, Tucker VA (1968) Further observations on the physiology of salinity adaptation in the Crab–eating Frog (*Rana cancrivora*). J Exp Biol 49:185–193
Goujet D (1984) Placoderm interrelationships: a new interpretation, with a short review of placoderm classification. Proc Linnean Soc NSW 107:211–241
Graham JB (1997) Air–breathing fishes: evolution, diversity and adaptation. Academic, San Diego
Graham JB, Lee HJ (2004) Breathing air in air: in what ways might extant amphibious fish biology relate to prevailing concepts about early tetrapods, the evolution of vertebrate air breathing, and the vertebrate land transition? Physiol Biochem Zool 77(5):720–731
Graham JB, Gee JH, Motta J et al (1987) Subsurface buoyancy regulation by the sea snake *Pelamis platurus*. Physiol Zool 60:251–261
Grémillet D, Chauvin C, Wilson RP et al (2005) Unusual feather structure allows partial plumage wettability in diving great cormorants *Phalacrocorax carbo*. J Avian Biol 36(1):57–63. Copyright © 2005 Journal of Avian Biology. Reprinted with permission from John Wiley and Sons
Grigg GC, Taplin LE, Harlow P et al (1980) Survival and growth of hatchling *Crocodylus porosus* in salt water without access to fresh drinking water. Oecologia 47:264–266
Habib M (2010) The structural mechanics and evolution of aquaflying birds. Biol J Linn Soc 99:687–698
Halstead Tarlo LB (1963) Aspidin: the precursor of bone. Nature 199:46–48
Halstead Tarlo LB (1973) The heterostracan fishes. Biol Rev 48:279–332
Hampe O, Ivanov A (2007) First xenacanthid shark from the Pennsylvanian (Moscovian) of the northern Caucasus (Russia). Fossil Rec 10:179–189
Harder W (1976) Anatomy of fishes, 2nd edn. Science Publishers, Stuttgart
Harrison P (1983) Seabirds: an identification guide. A&C Black, London
Harrison P (1997) Seabirds of the world: a photographic guide. A&C Black, London
Harrop H (1994) Albatrosses in the Western Palearctic. Birding World 7:241–245
Hazard LC (2004) Sodium and potassium secretion by Iguana salt glands. University of California Press, Berkeley
Heatwole H (1999) Sea snakes. UNSW Press, Hong Kong
Hieronymus TL, Witmer LM (2010) Homology and evolution of avian compound rhamphothecae. Auk 127(3):590–604
Higgins PJ (1978) The Galapagos iguanas: models of reptilian differentiation. Bioscience 28:512–515

Hillenius WJ (1994) Turbinates in therapsids: evidence for Late Permian origins of mammalian endothermy. Evolution 48:207–229

Hobson ES (1965) Observations on diving in Galapagos marine iguana, *Amblyrhynchus cristatus* (Bell). Copeia 1965:249–250

Holmes WN, McBean RL (1964) Some aspects of electrolyte excretion in the green turtle, *Chelonia mydas mydas*. J Exp Biol 41:81–90

Hone DWE, Dyke GJ, Haden M et al (2008) Body size evolution in Mesozoic birds. J Evol Biol 21:618–624

Honza M, Picman J, Grim T et al (2001) How to hatch from an egg of great structural strength: a study of the common cuckoo. J Avian Biol 32:249–255. Copyright © 2001 Journal of Avian Biology. Reprinted with permission from John Wiley and Sons

Hospitaleche CA, Montalti D, Marti LJ (2009) Skeletal morphoanatomy of the Brown Skua Stercorarius antarcticus lonnbergi and the South Polar Skua *Stercorarius maccormicki*. Polar Biol 32:759–774

Hou HC (1928a) Studies on the glandula uropygialis of birds. Am J Physiol 85:380

Hou HC (1928b) Studies on the glandula uropygialis of birds. Chin J Phys 2:345–380

Howe JC, Springer VG (1993) Catalog of type specimens of recent fishes in the National Museum of Natural History, Smithsonian Institution, 5: Sharks (Chondrichthyes: Selachii). Smithson Contrib Zool 540:i–iii–1–19

Hubbs CL (1918) The flight of the California flying fish. Copeia 1918:85–88

Hughes GM (1972) Morphometrics of fish gills. Respir Physiol 14:1–26

Hughes MR (2003) Reprinted from Hughes MR (2003) Regulation of salt gland, gut and kidney interactions. Comp Biochem Physiol A Mol Integr Physiol 136(3):507–524. Copyright (2003) with permission from Elsevier

Ineich I, Laboute P (2002) Sea snakes of New Caledonia. IRD, Paris

Jacob J, Ziswiler V (1982) The uropygial gland. In: Farner DS, King JR, Parkes KC (eds) Avian biology, vol 6. Academic, New York, pp 199–314

Janvier P (1996) Early vertebrates. Oxford University Press, Oxford

Janvier P (1997a) Heterostraci. http://tolweb.org/Heterostraci/16904/1997.01.01. In: The tree of life web project. http://tolweb.org/. Accessed 20 Jan 2011

Janvier P (1997b) Gnathostomata. Jawed Vertebrates. http://tolweb.org/Gnathostomata/14843/1997.01.01. In: The tree of life web project. http://tolweb.org/. Accessed 20 Jan 2011

Janvier P (2010) MicroRNAs revive old views about jawless vertebrate divergence and evolution. Proc Natl Acad Sci U S A 107:19137–19138

Janvier P, Blieck A (1979) New data on the internal anatomy of the Heterostraci (Agnatha), with general remarks on the phylogeny of the Craniota. Zool Scr 8:287–296

Jarvik E (1980) Basic structure and evolution of vertebrates, vol 1. Academic, London

Johnsgard PA (1993) Cormorants, darters, and pelicans of the world. Smithsonian Institution Press, Washington, DC

Judin KA (1961) On mechanism of the jaw in Charadriformes, Procellariiformes, and some other birds. Tr Zool Inst Leningr 29:257–302

Jyane BC (1988) Mechanical behavior of snake skin. J Zool 214:125–140

Kaiser GW (2007) The inner bird: anatomy and evolution. UBC Press, Vancouver

Katzir G, Howland HC (2003) Corneal power and underwater accommodation in great cormorants (*Phalacrocorax carbo sinensis*). J Exp Biol 206:833–841

Kellermann A (1990) Catalogue of early life history stages of Antarctic notothenioid fishes. Berichte zur Polarforsch 67:45–136

Kennedy M, Page RDM (2002) Seabird supertrees: combining partial estimates of procellariiform phylogeny. Auk 119:88–108

Kirkpatrick SJ (1994) Scale effects on the stresses and safety factors in the wing bones of birds and bats. J Exp Biol 190:195–215

Kirschner LB (1980) Comparison of vertebrate salt–excreting organs. Am J Physiol 238:R219–R223

Kock K–H (2005a) Antarctic icefishes (*Channichthyidae*): a unique family of fishes. A review, part I. Polar Biol 28:862–895

Kock K–H (2005b) Antarctic icefishes (*Channichthyidae*): a unique family of fishes. A review, part II. Polar Biol 28:897–909
Kratt LF, Smith RJF (1978) Breeding tubercles occur on male and female Arctic Grayling (*Thymallus arcticus*). Copeia 1:185–188
Krauss S, Monsonego–Ornan E, Zelzer E et al (2009) Mechanical function of a complex three–dimensional suture joining the bony elements in the shell of the red–eared slider turtle. Adv Mater 21:407–412
Kriwet J, Gadzdicki A (2003) New Eocene Antarctic chimeroid fish (Holocephali, Chimaeriformes). Pol Polar Res 24(1):29–51
Ksepka DT, Clarke JA (2010) The basal penguin (Aves: Sphenisciformes) Perudyptes devriesi and a phylogenetic evaluation of the penguin fossil record. Bull Am Mus Nat Hist 337:1–77
Kuchel LJ, Franklin CE (2000) Morphology of the cloaca in the estuarine crocodile, *Crocodylus porosus*, and its plastic response to salinity. J Morphol 245:168–176. Copyright © 2000 Wiley-Liss, Inc. Reprinted with permission from John Wiley and Sons
Kuratani S, Ota KG (2008) Hagfish (Cyclostomata, Vertebrata): searching for the ancestral developmental plan of vertebrates. BioEssays 30(2):167–172
Kuratani S, Kuraku S, Nagashima H (2011) Evolutionary developmental perspective for the origin of turtles: the folding theory for the shell based on the developmental nature of the carapacial ridge. Evol Dev 13(1):1–14. Copyright © 2011 Wiley Periodicals, Inc. Reprinted with permission from John Wiley and Sons
Kutschera U (2005) Predator-driven macroevolution in flying fishes inferred from behavioral studies: historical controversies and a hypothesis. Ann Hist Philos Biol 10:59–78
Lauder GV (1983) Functional design and evolution of the pharyngeal jaw apparatus in euteleostean fishes. Zool J Linnean Soc 77:1–38
Lauder GV (2000) Function of the caudal fin during locomotion in fishes: kinematics, flow visualization, and evolutionary patterns. Am Zool 40:101–122
Lauder GV, Drucker E (2002) Forces, fishes, and fluids: hydrodynamic mechanisms of aquatic locomotion. News Physiol Sci 17:235–240
Lauder GV, Liem KF (1983) The evolution and interrelationships of the actinopterygian fishes. Bull Mus Comp Zool 150:95–197
Lauder GV, Nauen J, Drucker EG (2002) Experimental hydrodynamics and evolution: function of median fins in ray–finned fishes. Integr Comp Biol 42:1009–1017
Lauder GV, Drucker EG, Nauen J et al (2003) Experimental hydrodynamics and evolution: caudal fin locomotion in fishes. In: Bels V, Gasc J–P, Casinos A (eds) Vertebrate biomechanics and evolution. Bios Scientific Publishers, Oxford
Laurin M (2002) Tetrapod phylogeny, amphibian origins, and the definition of the name Tetrapoda. Syst Biol 51:364–369
Laurin M (2010) How vertebrates left the water. University of California Press, Berkeley
Laurin M, Soler–Gijón R (2010) Osmotic tolerance and habitat of early stegocephalians: indirect evidence from parsimony, taphonomy, palaeobiogeography, physiology and morphology. Geol Soc Lond Spec Publ 339:151–179
Lequette B, Verheyden C, Jowentin P (1989) Olfaction in subantarctic seabirds: its phylogenetic and ecological significance. Condor 91:732–735
Lillywhite HB et al (2009) Republished with permission of The Company of Biologists Ltd, from Lillywhite HB, Menon JG, Menon GK et al (2009) Water exchange and permeability properties of the skin in three species of amphibious sea snakes (Laticauda spp.). J Exp Biol 212:1921–1929, Copyright (2009); permission conveyed through Copyright Clearance Center, Inc
Lim J, Fudge DS, Levy N et al (2006) Hagfish slime ecomechanics: testing the gill–clogging hypothesis. J Exp Biol 209:702–710
Lin YS, Wei CT, Olevsky EA et al (2011) Mechanical properties and the laminate structure of *Arapaima gigas* scales. J Mech Behav Biomed Mater 4(7):1145–1152
Lingham–Soliar T (1999) Rare soft–tissue preservation showing fibrous structures in an ichthyosaur from the Lower Lias (Jurassic) of England. Proc R Soc B 266:2367–2373

Lingham–Soliar T (2001) The ichthyosaur integument: skin fibers, a means for a strong, flexible and smooth skin. Lethaia 34:287–302

Linnaeus C (1758) Systema Naturae per regna tria naturae, secundum classes, ordines, genera, species, cum characteribus, differentiis, synonymis, locis. Editio decima, reformata. Laurentius Salvius: Holmiae. ii, 824 pp. Available online at http://www.archive.org/details/systemanaturae01linnuoft

Lockley R (1984) Seabirds of the world. Facts on File, Inc, New York

Loeb LB (1936) The "flight" of flying fish. Science 83:260–261

Lofgren L (1984) Ocean birds. Croom Helm Ltd., Gothenburg

Lohmann KJ, Putman NF, Lohmann CM (2008) Geomagnetic imprinting: a unifying hypothesis of long–distance natal homing in salmon and sea turtles. Proc Natl Acad Sci U S A 105(49):19096–190101. Copyright (2008) National Academy of Sciences, U.S.A

Long JA (1995) The rise of fishes: 500 million years of evolution. Johns Hopkins University Press, Baltimore

Long JA, Gordon M (2004) The greatest step in vertebrate history: a paleobiological review of the fish–tetrapod transition. Physiol Biochem Zool 77(5):700–719

Long JA, Trinajstic K, Young GC et al (2008) Live birth in the Devonian period. Nature 453:650–652

Long JA, Trinajstic K, Johanson Z (2009) Devonian arthrodire embryos and the origin of internal fertilization in vertebrates. Nature 457:1124–1127

Long JA, Hall BK, McNamara KJ et al (2010) The phylogenetic origin of jaws in vertebrates: developmental plasticity and heterochrony. Kirtlandia 57:46–52

Lönnberg E (1904) On the homologies of the different pieces of the compound rhamphotheca. Arkiv für Zoologie 1:473–512

Lopuchowycz VB, Massare JA (2002) Bone microstructure of a Cretaceous ichthyosaur. Paludicola 3:139–147

Lund R (2000) The new actinopterygian order Guildayichthyiformes from the lower Carboniferous of Montana (USA). Geodiversitas 22:171–206

Maher B (2009) Reprinted by permission from Macmillan Publishers Ltd: Nature, Maher B (2009) Evolution: biology's next top model? Nature 458:695–698, Copyright (2009)

Maisey JG (1979) Finspine morphogenesis in squalid and heterodontid sharks. Zool J Linnean Soc 66:161–183

Maisey JG (1996) Discovering fossil fishes. Henry Holt & Co, New York

Mallatt J (1984) Early vertebrate evolution: pharyngeal structure and the origin of gnathostomes. J Zool 204:169–183

Marples BJ (1932) The structure and development of the nasal glands of birds. Proc Zool Soc London 102:829–844

Marshall AT (1989) Intracellular and luminal ion concentrations in sea turtle salt glands determined by x–ray microanalysis. J Comp Physiol B 159:609–616

Martin RA (2006) The origin of modern sharks. Reef Quest. Retrieved 9 Sept 2006

Martin RA (2012) http://www.elasmo-research.org/education/evolution/ancient.htm. Accessed 18 Dec 2012

Martin S (2008) Global diversity of crocodiles (Crocodilia, Reptilia) in freshwater. Hydrobiologia 595:587–591

Massare JA (1994) Swimming capabilities of Mesozoic marine reptiles: a review. In: Maddock L, Bone Q, Rayner JMV (eds) Mechanics and physiology of animal swimming. Cambridge University Press, London

Mazzotti FJ, Dunson WA (1984) Adaptations of *Crocodylus acutus* and Alligator for life in saline water. Comp Biochem Physiol A Physiol 79:641–646

McFarland LZ (1959) Captive marine birds possessing a functional lateral nasal gland (Salt gland). Nature 184:2030–2031

McGowan C (1999) A practical guide to vertebrate mechanics. Cambridge University Press, Cambridge

McGowan C, Motani R (2003) Ichthyopterygia: Handbuch der Paläoherpetologie Part 8. Verlag Dr. Friedrich Pfeil, München

McGowen MR, Spaulding M, Gatesy J (2009) Divergence date estimation and a comprehensive molecular tree of extant cetaceans. Mol Phylogenet Evol 53:891–906

Menon GK, Menon J (2000) Avian epidermal lipids: functional considerations and relationships to feathering. Am Zool 40:540–552, reprinted by permission of Oxford University Press

Meunier FJ (1980) Les relations isopedine – tissu osseux dans le post–temporal et les ecail Jes de la ligne laterale de *Latimeria chalumnae* (Smith). Zool Scr 9:307–317

Meyers RA, Meyers RP (2005) Reprinted from Meyers RA, Myers RP (2005) Mandibular Bowing and Mineralization in Brown Pelicans. The Condor 107(2):445–449 by permission from The Cooper Ornithological Society. Copyright © 2005, The Cooper Ornithological Society. Published by the Cooper Ornithological Society

Meylan AB (1988) Spongivory in hawksbill turtles: a diet of glass. Science 239:393–395

Mi LY, Fritton SP, Basu M, Cowin SC (2005) Analysis of avian bone response to mechanical loading – Part One: distribution of bone fluid shear stress induced by bending and axial loading. Biomech Model Mechanobiol 4:118–131

Mills CA (1936a) Source of propulsive power used by flying fish. Science 83:80

Mills CA (1936b) Propulsive power used by flying fish. Science 83:262

Mok H, Chang H (1986) Articulation of the pelvic spine in acanthopterygian fishes, with notes on its phylogenetic significance. Jpn J Ichthyol 33:145–149

Monroe MH "Australia: The Land Where Time Began" (published on-line http://austhrutime.com/porolepiformes.htm. Accessed 15 May 2014). 2014 Copyright © Austhrutime.com. Reprinted with permission

Montague JJ (1983) A new size record for the saltwater crocodile (*Crocodylus porosus*). Herpetol Rev 14:36–37

Mora C, Tittensor DP, Adl S, Simpson AGB et al (2011) How many species are there on earth and in the ocean? PLoS Biol 9:e1001127

Motani R (2001) Estimating body mass from silhouettes: testing the assumption of elliptical body cross–sections. Paleobiology 27:735–750

Motani R (2002a) Swimming speed estimation of extinct marine reptiles: energetic approach revisited. Paleobiology 28:251–262

Motani R (2002b) Scaling effects in caudal fin kinematics: implication for ichthyosaurian speed. Nature 415:309–312

Motani R (2005) Ichthyosauria: evolution and physical constraints of fish–shaped reptiles. Annu Rev Earth Planet Sci 33:395–420

Motani R (2009) The evolution of marine reptiles. Evol Educ Outreach 2:224–235

Motani R, Rothschild BM, Wahl W Jr (1999) Large eyeballs in diving ichthyosaurs. Nature 402:747

Moyer BR, Rock AN, Clayton DH (2003) Experimental test of the importance of preen oil in rock doves (Columba Livia). The Auk 120(2):490–496 published by the American Ornithologists' Union. Reprinted with permission

Naro–Maciel E, Le M, Fitz Simmons NN, Amato G (2008) Evolutionary relationships of marine turtles: a molecular phylogeny based on nuclear and mitochondrial genes. Mol Phylogenet Evol 49:659–662

Neill WT (1958) The occurrence of amphibians and reptiles in saltwater areas and a bibliography. Bull Mar Sci Gulf Caribb 8:1–9

Nelson JB (2005) Pelicans, cormorants, and their relatives. Oxford University Press, Oxford

Nelson JS (2006) Fishes of the world, 4th edn. Wiley, New York

Neumann D (2006) Type catalogue of the ichthyological collection of the Zoologische Staatssammlung München. Part I: Historic type material from the "Old collection", destroyed in the night 24/25 April 1944. Spixiana 29(3):259–285

Nevitt GA (2008) Sensory ecology on the high seas: the odor world of the procellariiform seabirds. J Exp Biol 211:1706–1713

Nevitt GA, Bonadonna F (2005) Sensitivity to dimethyl sulphide suggests a mechanism for olfactory navigation by seabirds. Biol Lett 1:303–305

Nicolson SW, Lutz PL (1989) Salt gland function in the green sea turtle Chelonia mydas. J Exp Biol 144:171–184

Niedzwiedzki G, Szrek P, Narkiewicz K et al (2010) Tetrapod trackways from the early Middle Devonian period of Poland. Nature 463:43–48
Norberg UM (1985) Flying, gliding, and soaring. In: Hildebrand M, Bramble DM, Liem KF, Wake DB (eds) Functional vertebrate morphology. Harvard University Press, Cambridge, MA
Norberg UM (1995) Wing design and migratory flight. Isr J Zool 41:297–305
Nunn G, Stanley S (1998) Body size effects and rates of cytochrome B evolution in tube–nosed seabirds. Mol Biol Evol 15(10):1360–1371
Oeffner J, Lauder GV (2012) The hydrodynamic function of shark skin and two biomimetic applications. J Exp Biol 215:785–795
Olson SL (1985) The fossil record of birds. In: Farner DS, King JR, Parkes KC (eds) Avian biology. Academic, Orlando
"On the Wings of the Albatross" by Carl Safina (2007) http://ngm.nationalgeographic.com/2007/12/albatross/safina-text. Accessed at 11 Apr 2014. © 2007 National Geographic Society. Reprinted with permission
Onley D, Scofield P (2007) Albatrosses, petrels and shearwaters of the world. Princeton field guides. University Press, Princeton
Ortiz C, Boyce MC (2008) Materials science – bioinspired structural materials. Science 319:1053–1054
Ostrom JH (1976) Archaeopteryx and the origin of birds. Biol J Linn Soc 8:91–182
Owen R (1846) Lectures on the comparative anatomy and physiology of the vertebrate animals. Delivered at the Royal College of Surgeons 1844 and 1846. (no publisher given). London
Park H, Choi H (2010) Republished with permission of The Company of Biologists Ltd, from Park H. Choi H (2010) Investigation of aerodynamic capabilities of flying fish in gliding flight. J Exp Biol 213:3269–3279. Copyright (2009); permission conveyed through Copyright Clearance Center, Inc
Patterson C (1965) The phylogeny of the chimeroids. Philos Trans R Soc Lond B 249:101–219
Patterson C, Johnson GD (1995) The intermuscular bones and ligaments of teleostean fishes. Smithson Contrib Zool 559:1–83
Pavlov V (2006) Dolphin skin as a natural anisotropic compliant wall. Bioinspir Biomim 1:31–40
Pavlov V, Riedeberger D, Rist U, Seibert U (2012) Analysis of the relation between skin morphology and local flow conditions for a fast–swimming Dolphin. In: Tropea C, Bleckman H (eds) Nature–inspired fluid mechanics. Springer, Berlin/Heidelberg
Peaker M, Linzell JL (1975) Salt glands in birds and reptiles. Cambridge University Press, New York
Pennisi E (2011) Manta machines. Science 332:28–29
Pennycuick CT (1967) The strength of the Pigeon's wing bones in relation to their function. J Exp Biol 46:219–233
Pennyquick CJ (1987) Flight of seabirds. In: Croxall JP (ed) Seabirds. Feeding ecology and role in marine ecosystems. Cambridge University Press, Cambridge
Piepenbrink H (1989) Examples of chemical changes during fossilisation. Appl Geochem 4:273–280
Pinkerton JV, Dalkin AC, Crowe SE et al (2010) Treatment of postmenopausal osteoporosis in a patient with celiac disease. Nat Rev Endocrinol 6:167–171
Pollerspöck J (2012) www.shark-references.com, World Wide Web electronic publication, Version 2012 date
Poore CB, Wilson GDF (1993) Marine species richness. Nature 361:597–598
Prince PA, Huin N, Weimerskirch H (1994) Diving depths of albatrosses. Antarct Sci 6(3):353–354
Raikow RJ, Icanovsky L, Bledsoe AH (1988) Forelimb joint mobility and the evolution of wing–propelled diving in birds. Auk 105:446–451
Rasmussen AR (1997) Systematics of sea snakes: a critical review. Symp Zool Soc Lond 70:15–30
Rasmussen AR (2001) Sea snakes. In: Carpenter KE, Niem VH (eds) FAO species identification guide for fishery purposes. The living marine resources of the Western Central Pacific, vol 6. Bony fishes Part 4 (Labridae to Latimeriidae), estuarine crocodiles, sea turtles, sea snakes and marine mammals. FAO, Rome, pp 3987–4008
Rasmussen A, Arnason U (1999a) Phylogenetic studies of complete mitochondrial DNA molecules place cartilaginous fishes within the tree of bony fishes. J Mol Evol 48:118–123

Rasmussen A, Arnason U (1999b) Molecular studies suggest that cartilaginous fishes have a terminal position in the piscine tree. Proc Natl Acad Sci U S A 96:2177–2182
Rasmussen AR, Elmberg J, Gravlund P, Ineich I (2011a) Sea snakes (Serpentes: subfamilies *Hydrophiinae* and *Laticaudinae*) in Vietnam: a comprehensive checklist and an updated identification key. Zootaxa 2894:1–20
Rasmussen AR, Murphy JC, Ompi M, Gibbons JW, Uetz P (2011b) Marine reptiles. PLoS One 6(11):e27373. Copyright © 2011 Rasmussen et al. CC BY 2.5
Rayfield EJ (2007) Finite element analysis and understanding the biomechanics and evolution of living and fossil organisms. Annu Rev Earth Planet Sci 35:541–576
Reid REH (1997) Dinosaurian physiology: the case for "intermediate dinosaurs". In: Farlow JO, Brett–Surman MK (eds) The complete dinosaur. Indiana University Press, Bloomington
Reidenberg JS (2007) Anatomical adaptations of aquatic mammals. Anat Rec 290:507–513. Copyright © 2007, Wiley-Liss, INC. Reprinted with permission from John Wiley and Sons
Reina RD, Jones TT, Spotila JR (2002) Salt and water regulation by the leatherback sea turtle *Dermochelys coriacea*. J Exp Biol 205:1853–1860
Reinhart RH (1953) Diagnosis of the new mammalian order, Desmostylia. J Geol 61:187
Reinhart RH (1982) The extinct mammalian order Desmostylia. Natl Geogr Soc Res Rep 14:549–555
Renaud CB (2011) Lampreys of the world: an annotated and illustrated catalogue of lamprey species known to date. FAO species catalogue for fishery purposes no. 5. FAO, Rome
Renaud CB, Economidis PS (2010) *Eudontomyzon graecus*, a new nonparasitic lamprey from Greece (Petromyzontiformes: Petromyzontidae). Zootaxa 2477:37–48
Retallack GJ (2011) Woodland hypothesis for Devonian tetrapod evolution. J Geol 119:235–258. Copyright © 2011, The University of Chicago Press
Rhee H et al (2009) Reprinted from Rhee H, Horstemeyer MF, Hwang Y et al (2009) A study on the structure and mechanical behavior of the Terrapene carolina carapace: a pathway to design bio–inspired synthetic composites. Mater Sci Eng C 29:2333–2339. Copyright (2009) with permission from Elsevier
Rhodin AGJ (1985) Chondro–osseous development and growth of marine turtles. Copeia 1985:752–771
Rice DW (1998) Marine mammals of the world: systematics and distribution. Soc Mar Mamm Spec Publ 4:213
Richmond ND (1964) The mechanical functions of the testudinate plastron. Am Midl Nat 72:50–56
Rieppel O (1995) The genus Placodus: systematics, morphology, paleobiogeography, and paleobiology. Fieldiana Geol New Ser 31:1–44
Riess J (1986) Locomotion, biophysics of swimming and phylogeny of the ichthyosaurs. Palaeontogr Abt A 192:93–155
Ritter EK (2002) Analysis of sharksucker, *Echeneis naucrates*, induced behavior patterns in the blacktip shark, *Carcharchinus limbatus*. Environ Biol Fish 65:111–115
Roberts TR (1982) Unculi (horny projections arising from single cells), an adaptive feature of the epidermis of ostariophysan fishes. Zool Scr 11:55–76
Robertson GM (1935) The ostracoderm order osteostraci. Science 82:282–283
Robson P, Wright GM, Youson JH et al (2000) The structure and organization of Lamprin genes: multiple–copy genes with alternative splicing and convergent evolution with insect structural proteins. Mol Biol Evol 17(11):1739–1752
Romer AS, Parsons TS (1986) The vertebrate body, 6th edn. Saunders College Publishing, Philadelphia
Ronald K, Gots BL, Lupson JD et al (1991) An annotated bibliography on seals, sea lions, and walrus – supplement 2. International Council for the Exploration of the Sea, Copenhagen
Rosenberger LJ (2001) Pectoral fin locomotion in batoid fishes: undulation versus oscillation. J Exp Biol 204:379–394
Ruben JA (1991) Reptilian physiology and the flight capacity of Archaeopteryx. Evolution 45:1–17
Ruben J (1995) The evolution of endothermy in mammals and birds: from physiology to fossils. Annu Rev Physiol 57:69–95
Ruben JA, Jones TD (2000) Selective factors associated with the origin of fur and feathers. Am Zool 40(4):585–596. Copyright © 2000, Oxford University Press. Reprinted by permission of Oxford University Press

Ruben JA, Jones TD, Geist NR (1998) Respiratory physiology of the dinosaurs. Bioessays 20:852–859. Copyright © 1998 John Wiley & Sons, Inc. Reprinted with permission from John Wiley and Sons
Ruta M, Jeffery JE, Coates MI (2003) A supertree of early tetrapods. Proc R Soc B 270(1532):2507–2516
Ruud JT (1954) Vertebrates without erythrocytes and blood pigment. Nature 173:848–850
Sale A, Luschi P (2009) Navigational challenges in the oceanic migrations of leatherback sea turtles. Proc Biol Sci 276(1674):3737–3745
Salibian A, Montalti D (2009) Physiological and biochemical aspects of the avian uropygial gland. Braz J Biol 69(2):437–446
Sander PM (2000) Ichthyosauria: their diversity, distribution, and phylogeny. Paläontol Ztschr 74:1–35
Sander PM, Chen X, Cheng L et al (2011) Short–snouted toothless ichthyosaur from China suggests late triassic diversification of suction feeding ichthyosaurs. PLoS One 6(5):e19480. Copyright © 2011 Sander et al. CC BY 2.5
Sansom RS (2009) Phylogeny, classification and character polarity of the Osteostraci (Vertebrata). J Syst Palaeontol 7:95–117
Savile DBO (1957) Adaptive evolution in the avian wing. Evolution 11:212–224
Schaefer JT, Summers AP (2005) Batoid wing skeletal structure: novel morphologies, mechanical implications and phylogenetic patterns. J Morphol 264:298–313
Schleich H, Kastle W (1988) Reptile eggshells: SEM atlas. Gustav Fischer Verlag, Stuttgart
Schmidt–Nielsen K, Fange R (1958) Reprinted by permission from Macmillan Publishers Ltd: Nature (Schmidt–Nielsen K, Fange R (1958) Salt glands in marine reptiles. Nature 182:783–785) copyright (1958)
Schmidt–Nielsen K, Sladen WJL (1958) Nasal salt secretion in the Humboldt penguin. Nature 181:1217–1218
Schmidt–Neilsen KP, Hainsworth FR, Murrish DE (1970) Countercurrent heat exchange in the respiratory passages: effect on water and heat balance. Respir Physiol 9(2):9263–9276
Scholander PF, Walters V, Hock R et al (1950) Body insulation of some arctic and tropical mammals and birds. Biol Bull 99:225–236
Schreiber EA, Burger J (2001) Seabirds in the marine environment. In: Schreiber EA, Burger J (eds) Biology of marine birds. CRC Press, Boca Raton
Schreiber RW, Woolfenden GE, Olfenden O et al (1975) Prey capture by the brown pelican. Auk 92:649–654
Schultze HP (1999) The fossil record of the intertidal zone. In: Horn MH et al (eds) Intertidal fishes: life in two worlds. Academic, San Diego
Schultze HP (2010) Gnatostomata, Kiefermünder. Spezielle Zoologie. Springer, Berlin/Heidelberg
Shadbolt L (1908) On the flying fish. Aeronaut J 12:111–114
Shoulejkin W (1929) Airdynamics of the flying fish. Int Rev Ges Hydrobiol Hydrogr 22:102–110
Shubin NH, Daeschler EB, Jenkins FA (2006) The pectoral fin of *Tiktaalik rosae* and the origin of the tetrapod limb. Nature 440:764–771
Shufeldt RW (1890) Contributions to the comparative osteology of arctic and sub–arctic water–birds: Part VIII. J Anat Physiol 25:60–77
Simons ELR (2010) Forelimb skeletal morphology and flight mode evolution in pelecaniform birds. Zoology 113:39–46
Simons ELR, O'Connor PM (2012) Bone laminarity in the avian forelimb skeleton and its relationship to flight mode: testing functional interpretations. Anat Rec Adv Integr Anat Evol Biol 295:386–396. Copyright © 2012 Wiley Periodicals, Inc. Reprinted with permission from John Wiley and Sons
Simons ELR, Hieronymus TL, O'Connor PM (2011) Cross–sectional geometry of the forelimb skeleton and flight mode in pelecaniform birds. J Morphol 272:958–971
Sire J–Y, Marin S, Allizard F (1998) A comparison of teeth and dermal denticles (odontodes) in the teleost *Denticeps clupeoides* (Clupeomorpha). J Morphol 237:237–256
Sivak JG (1980) Reprinted from Trends Neurosci 3, Sivak JG (1980) Avian mechanisms for vision in air and water. Trends Neurosci 3:314–317. Copyright (1980) with permission from Elsevier
Smith M (1926) Monograph of the sea snakes (Hydrophiidae). Wheldon & Wesley, Oxford

Smith ND (2010) Phylogenetic analysis of Pelecaniformes (Aves) based on osteological data: implications for waterbird phylogeny and fossil calibration studies. PLoS One 5(10):e13354

Smith L (2013) Shark Evolution and Classification. Published online: http://saltwaterlife.co.uk/ws/sharkiologist/articles/shark-evolution-and-classification/. Access 15 May 2014. Copyright © 2013, Saltwater Life (www.saltwaterlife.co.uk). Reprinted with permission

Soldaat E, Leopold MF, Meesters EH et al (2009) Albatross mandible at archeological site in Amsterdam, the Netherlands, and WP records of *Diomedea albatrosses*. Dutch Birding 31(1):1–16

Song J, Ortiz C, Boyce MC (2011) Threat–protection mechanics of an armored fish. J Mech Behav Biomed Mater 4(5):699–712

Soons J, Herrel A, Genbrugge A et al (2010) Mechanical stress, fracture risk and beak evolution in Darwin's ground finches (Geospiza). Philos Trans R Soc B 365:1093–1098

Soons J, Herrel A, Aerts P, Dirckx J (2012a) Determination and validation of the elastic moduli of small and complex biological samples: bone and keratin in bird beaks. J R Soc Interface 9(71):1381–1388 by permission of the Royal Society

Soons J, Herrel A, Genbrugge A et al (2012b) Multi–layered bird beaks: a finite–element approach towards the role of keratin in stress dissipation. J R Soc Interface 9(73):1787–1796

Stahl B (1999) Chondrichthyes III: Holocephali. In: Schultze H–P (ed) Handbook of paleoichthyology 4. Verlag Dr. Friedrich Pfeil, München

Stelbrink B, von Rintelen T, Cliff G et al (2010) Molecular systematics and global phylogeography of angel sharks (genus Squatina). Mol Phylogenet Evol 54:395–404

Stettenheim P (1972) The integument of birds. In: Famer DS, King JR (eds) Avian biology, vol 2. Academic, New York

Stettenheim PR (2000) The integumentary morphology of modern birds–an overview. Am Zool 40(4):461–477

Stiassny MLJ, Moore JA (1992) A review of the pelvic girdle of atherinomorph fishes. Zool J Linnean Soc 104:209–242

Sudo S, Tsuyuki K, Ito Y et al (2002) A study on the surface shape of fish scales. JSME Int J Ser C 45:1100–1105

Talevi M, Fernández MS (2012) Unexpected skeletal histology of an ichthyosaur from the Middle Jurassic of Patagonia: implications for evolution of bone microstructure among secondary aquatic tetrapods. Naturwissenschaften 99(3):241–244

Tamiya N, Yagi T (2011) Studies on sea snake venom. Proc Jpn Acad Ser B Phys Biol Sci 87(3):41–52

Taplin LE (1984) Homeostasis of plasma electrolytes, sodium and water pools in the Estuarine crocodile, *Crocodylus porosus*, from fresh, saline and hypersaline waters. Oecologia 63:63–70

Taplin LE (1985) Sodium and water budgets of the fasted estuarine crocodile, *Crocodylus porosus*, in sea water. J Comp Physiol B 155:501–513

Taplin LE, Grigg GC (1981) Salt glands in the tongue of the Estuarine Crocodile *Crocodylus porosus*. Science 212:1045–1047

Taplin LE, Loveridge JP (1988) Nile crocodiles, *Crocodylus niloticus*, and estuarine crocodiles, *Crocodylus porosus*, show similar osmoregulatory responses on exposure to seawater. Comp Biochem Physiol A Comp Physiol 89:443–448

Taub AM, Dunson WA (1967) The salt gland in a sea snake (Laticauda). Nature 215:995–996

Taylor MA (1987) A reinterpretation of ichthyosaur swimming and buoyancy. Palaeontology 30:531–535

Thewissen JGM, Williams EM (2002) The early radiations of Cetacea (Mammalia): evolutionary pattern and developmental correlations. Annu Rev Ecol Syst 33:73–90

Thiel R, Eidus I, Neumann R (2009) The Zoological Museum Hamburg (ZMH) fish collection as a global biodiversity archive for elasmobranchs and actinopterygians as well as other fish taxa. J Appl Ichthyol 25(S1):9–32

Thomson KS (1969) The biology of the lobe–finned fishes. Biol Rev 44:91–154

Thomson KS (1975) On the biology of cosmine. Bull Peabody Mus Nat Hist 40:1–59

Thorne PM, Ruta M, Benton MJ (2011) Resetting the evolution of marine reptiles at the Triassic–Jurassic boundary. Proc Natl Acad Sci U S A 108:8339–8344

Tickell WLN (2000) Albatrosses. Pica Press, Sussex
Tobalske BW (2010) Hovering and intermittent flight in birds. Bioinspir Biomim 5(4):045004
Travers RA (1981) The interarcual cartilage: a review of its development, distribution and value as an indicator in euteleostean fishes. J Nat Hist 15:853–871
Tu MC, Lillywhite HB, Menon JG, Menon GK (2002) Postnatal ecdysis establishes the permeability barrier in snake skin: new insights into barrier lipid structures. J Exp Biol 205: 3019–3030
Turner S (1992) Thelodont lifestyles. In: Mark–Kurik E (ed) Fossil fishes as living animals. Akademia (Tallinn, Estonia). Academy of Sciences of Estonia, Tallinn
Turner S, Burrow CJ (2011) A Lower Carboniferous xenacanthiform shark from Australia. J Vertebr Paleontol 31(2):241–257
Tyler SJ, Ormerod SJ (1994) The dippers. T and AD Poyser, London
Uetzt P (2011) 15: the reptile database. Available: http://www.reptile-database.org. Accessed 23 Sept 2011
Uhen MD (2007) Evolution of marine mammals: back to the sea after 300 million years. Anat Rec 290(6):514–522. Copyright © 2007 Wiley-Liss, Inc. Reprinted with permission from John Wiley and Sons
Van der Brugghen W, Janvier P (1993) Denticles in thelodonts. Nature 364:107
Videler JJ (2005) Avian flight. Oxford University Press, New York
Vogel S (1994) Life in moving fluids. Princeton University Press, Princeton
Vogel S (2003) Comparative biomechanics: life's physical world. Princeton, Princeton University Press. © 2003 by Princeton University Press. Reprinted with permission
Voris HK, Voris HH (1983) Feeding strategies in marine snakes: an analysis of evolutionary, morphological, behavioral and ecological relationships. Am Zool 23(2):411–425
Waite JH, Broomell CC (2012) Changing environments and structure–property relationships in marine biomaterials. J Exp Biol 215:873–883
Walsh SA, MacLeod N, O'Neill M (2008) Spot the penguin: can reliable taxonomic identifications be made using isolated foot bones? In: Walsh S (ed) Automated taxon identification in Systematic. Reproduced with permission of TAYLOR & FRANCIS GROUP LLC in the format Republish in a book via Copyright Clearance Center
Wang X et al (2012) Reprinted from Biochem Biophys Res Commun, 421, Wang X, Zhang Y, Wu Q, Zhang H (2012) Evolutionary landscape of amphibians emerging from ancient freshwater fish inferred from complete mitochondrial genomes, 228–231. Copyright (2012) with permission from Elsevier
Ward DJ, Duffin CJ (1989) Mesozoic chimeroids. 1. A new chimeroid from the Early Jurassic of Gloucestershire. Engl Mesozoic Res 2:45–51
Warham J (1977) Wing loadings, wing shapes, and flight capabilities of Procellariiformes. NZ J Zool 4:73–83
Warham J (1990) The Petrels: their ecology and breeding systems. Academic, London
Watanuki Y, Niizuma Y, Geir WG, Sato K, Naito Y (2003) Stroke and glide of wing–propelled divers: deep diving seabirds adjust surge frequency to buoyancy change with depth. Proc R Soc Lond B 270(1514):483–488 by permission of the Royal Society
Watanuki Y, Wanless S, Harris M et al (2006) Swim speeds and stroke patterns in wing–propelled divers: a comparison among alcids and a penguin. J Exp Biol 209:1217–1230
Westhoff G et al (2005) Reprinted from Zoology 108, Westhoff G, Fry BG, Bleckmann H (2005) Sea snakes (Lapemis curtus) are sensitive to low–amplitude water motions. Zoology 108:195–200, Copyright (2005) with permission from Elsevier
Wetherbee DK, Barlett LM (1962) Egg teeth and shell rupture of the American Woodcock. Auk 79:117
Wiebe KL (2010) Reprinted from Wiebe KL (2010) A supplemental function of the avian egg tooth. Condor 112:1–7 by permission from The Cooper Ornithological Society. Copyright © 2010, The Cooper Ornithological Society. Published by the Cooper Ornithological Society
Wikelski M, Thom C (2000) Reprinted by permission from Macmillan Publishers Ltd: Nature (Wikelski M, Thom C (2000) Marine iguanas shrink to survive El Niño. Nature 403(6765):37–388) copyright (2000)

Wiley ML, Collette BB (1970) Breeding tubercles and contact organs in fishes: their occurrence, structure, and significance. Bull Am Mus Nat Hist 143:143–216
Wiley EO, Johnson GD (2010) A teleost classification based on monophyletic groups. In: Nelson JS, Schultze H–P, Wilson MVH (eds) Origin and phylogenetic interrelationships of teleosts. Verlag Dr. Friedrich Pfeil, München
Wilga CD, Lauder GV (2004) Biomechanics of locomotion in sharks, rays, and chimeras. In: Carrier JC, Musick J, Heithaus M (eds) The biology of sharks and their relatives. CRC Press, Boca Raton, pp 139–164
Williams TD (1995) The penguins. Oxford University Press, Oxford
Wilson MVH (2010) Acanthodii. Access Science Encyclopedia at McGraw–Hill. Accessed on 17 Feb 2010
Wilson MVH, Caldwell MW (1993) New Silurian and Devonien fork–tailed 'thelodonts' are jawless vertebrates with stomachs and deep bodies. Nature 361:442–444
Wilson DS, Tracy CR, Tracy CR (2003) Estimating age of turtles from growth rings: a critical evaluation of the technique. Herpatologica 59(2):178–194
Wyneken J (2001) The anatomy of sea turtles. NOAA technical memorandum NMFS–SEFSC–470. NOAA, Miami
Xian–guang H, Aldridge RJ, David J et al (2002) New evidence on the anatomy and phylogeny of the earliest vertebrates. Proc R Soc B 269(1503):1865–1869
"Yellow-Bellied Sea Snake" http://www.waikikiaquarium.org/experience/animal-guide/reptiles/yellow-bellied-sea-snake/. Last accepted 25 May 2014. Copyright (c) 2014, Waikiki Aquarium.)
Young GC (1986) The relationships of placoderm fishes. Zool J Linnean Soc 88:1–57
Yu X, Zhu M, Zhao W (2010) The origin and diversification of osteichthyans and sarcopterygians: rare Chinese fossil findings advance research on key issues of evolution. Paleoichthyol 24:71–75
Zhu M, Ahlberg PE (2004) The origin of the internal nostril of tetrapods. Nature 432:94–97
Zhu M, Schultze HP (1997) The oldest sarcopterygian fish. Lethaia 30:293–304
Zhu M, Yu XB (2002) A primitive fish close to the common ancestor of tetrapods and lungfish. Nature 418:767–770
Zhu M, Yu X, Janvier P (1999) A primitive fossil fish sheds light on the origin of bony fishes. Nature 397:607–610
Zhu M, Yu XB, Ahlberg PE (2001) A primitive sarcopterygian fish with an eyestalk. Nature 410:81–84
Zhu M, Yu XB, Wang W et al (2006) A primitive fish provides key characters bearing on deep osteichthyan phylogeny. Nature 441:77–80
Zhu M, Zhao WJ, Jia LT et al (2009) The oldest articulated osteichthyan reveals mosaic gnathostome characters. Nature 458:469–474
Zylberberg L, Meunier FOJ, Laurin M (2010) A microanatomical and histological study of the postcranial dermal skeleton in the devonian Sarcopterygian Eusthenopteron foordi. Acta Palaeontol Pol 55(3):459–470

Part II
Biomineralization in Marine Vertebrates

Chapter 2
Cartilage of Marine Vertebrates

Abstract Cartilage is primarily composed of a specialised extracellular matrix synthesised by chondrocytes and contains of the numerous molecules (collagens, polysaccharides, low molecular peptides). This biological material represents unique avascular tissue. This chapter considers the current state-of-the-art biomaterial characterisation of both non-mineralized and mineralized (calcified) cartilages, with respect to the future tissues in biomimetical and biomedical applications.

2.1 From Non-mineralized to Mineralized Cartilage

Non-mineralized Cartilage Cartilage can be defined generally as an internal cellular support tissue high in fibrous protein and mucopolysaccharide content (Cole and Hall 2004; Hall 2005). Historically, cartilage has been defined as a vertebrate tissue that forms part of the skeletal system that is grossly different from mineralized bone both in function and histological composition (Cole 2011). Romer (1964) suggested that cartilage arose in the vertebrates as an embryonic adaptation to stresses and deformations produced by rapid growth. Since then, the view seems to have prevailed amongst most biologists that cartilage is a uniquely vertebrate tissue.

Interestingly, cartilage-like tissues have also been described in a variety of invertebrate lineages. For example, cartilages found within Cephalopod mollusks are remarkably similar to vertebrate hyaline cartilage at a histological level (Cole and Hall 2004). The polychaete family Sabellidae is defined by the presence of a cellular cartilage-like skeleton within its feeding tentacles (see for review Person and Philpott 1963, 1969; Kupriyanova and Rouse 2008). However, principally invertebrates appear to lack any tissue that is biochemically or morphologically identical with vertebrate cellular cartilage (Cattell et al. 2011). Moreover, "comprehensive analysis of gene expression in amphioxus, a basal chordate, suggests that no single invertebrate cell type coexpresses all, or even most, of the genes needed to drive cellular cartilage formation" (Cattell et al. 2011). The hemichordate *Saccoglossus bromophenolosus* possess so called acellular cartilage. This cell-free cartilage is secreted by pharyngeal endodermal cells. It was proposed that this kind of endodermal

H. Ehrlich, *Biological Materials of Marine Origin*, Biologically-Inspired Systems 4,
DOI 10.1007/978-94-007-5730-1_2

secretion is "primarily the ancestral mode of making pharyngeal cartilage in deuterostoes," (Hecht et al. 2008).

Cartilage forms the major supporting structure in very primitive marine vertebrates, like jawless lamprey and hagfish. Both are representatives of cyclostomes and diverged from the mammalian lineage about 500 million years ago (Forey and Janvier 1993). Lamprey and hagfish are well known key species for understanding of numerous developmental mechanisms in vertebrates, including evolution of the neural crest, cartilage, mineralized cartilage, and bones. Interestingly, a large amount of diversity in the highly viscoelastic cartilage forms has been observed within each individuum of these animals. According to Fernandesa and Eyre (1999), in lamprey, "the head and gill regions are cartilaginous in texture and histological appearance. For example, annular cartilage supports the mouth, while branchial cartilages support the gills. Trabecular, piston and pericardial cartilages have also been described," (Fernandesa and Eyre 1999). Experiments using transmission electron microscopy showed that the extracellular matrix of all these fibrillarly-based structures is collagen-free (Fernandesa and Eyre 1999).

In spite of the presence of special structural proteins and differences in extracellular matrix organization, mechanical properties of the lamprey's cartilages are largely similar to those of mammalian cartilages. The dense network of randomly arranged, branched, noncollagenous matrix fibrils constitutes the basis of the extracellular matrix (ECM) of lamprey cartilages (Wright and Youson 1983; Wright et al. 1988). Special attention was paid to *lamprin*, the unique insoluble matrix protein. This was found to constitute 44–51 % of the dry mass of the annular cartilage (Wright et al. 1983; Robson et al. 1993). However, the major matrix protein(s) of pericardial cartilage possess rather noncollagenous, elastin-like proteins than lamprin (Wright et al. 2001).

The branchial basket that supports the lamprey pharynx represents the viscerocranial skeleton. Its development was recently described in the sea lamprey, *Petromyzon marinus* by Martin et al. (2009). The authors reported about the skeletal rods within the branchial basket, which are comprised of chondrocyte stacks. Meanwhile, "the subchordal, parachordal, and trabecular cartilages form as aggregate condensations of polygonal cells. The subchordal and parachordal cartilage condensations form anchor points that tether the skeletal rods to the notochord," (Lakiza et al. 2011). This rod-like midline structure is of mesodermal origin and serves as a primitive axial skeleton.

The body axis of vertebrates, established during gastrulation, is characterized by the formation of the notochord (Stemple 2005). Both structural and compositional features of notochord are similar to that of cartilage. Thus, the composition of the sheath of the lamprey notochord, where the major collagen is type II, resembling that of cartilage of higher animals (see for review Eikenberry et al. 1984). "The collagen fibrils in the notochord sheath are of small, uniform diameter but, in contrast to cartilage, are highly oriented and crystalline" (Eikenberry et al. 1984). It cannot exclude that chondrogenesis in the ancestral vertebrate was determined by the SoxE genes and regulation of cartilage effector proteins. Some authors proposed that "type II collagen arose early in vertebrate evolution as an extracellular matrix protein in cartilage formation" (see for review Lakiza et al. 2011).

Two different types of cartilages have been described in hagfishes. The first one is the "soft cartilage". It is positive detected for Col2a1, contains large hypertrophic chondrocytes and are surrounded by a thin extracellular matrix. The second one, termed as "hard cartilage", possesses no Col2a1, and contains smaller chondrocytes that are surrounded by an abundance of extracellular matrix (see for review Zhang and Cohn 2006; Rychel and Swalla 2007). Probably, the last common ancestor of hagfishes, lampreys, and gnathostomes also contains Col2a1 in the cartilages because this protein is structurally similar in hagfish soft cartilage with that lamprey. Thus, "hagfish fibrillar collagens reveal that type II collagen-based cartilage evolved in stem vertebrates" (Zhang and Cohn 2006).

In summary, the diversity of lamprey cartilages is shown by the following properties:

"a) the '*hard cartilage*' in the dorsal portion of the branchial basket skeleton consists of disorganized polygonal chondrocytes and expresses an elastin-like lamprin;
b) In the branchial and hypobranchial bars, discoidal chondrocytes expressing fibrillar collagen and elastin generate so-called '*soft cartilage*', and
c) '*mucocartilage*' is the main skeletal tissue in the ventral pharynx and oral region.
While biochemically similar to definitive cellular cartilage, mucocartilage is histologically distinct, consisting of scattered mesenchymal cells embedded in a mucopolysaccharide matrix" (Cattell et al. 2011).

There are no doubts that the fossil remnants of exceptionally preserved non-biomineralized vertebrates play important role "in constraining the time of origin of vertebrate cartilages (and thus the genome duplication), and in reconstructing cartilage development in the vertebrate common ancestor" (Sansom et al. 2011). The finding of two Col2A1 orthologues in hagfish and lampreys suggest that genome duplication have occurred on the vertebrate stem. For example, amphioxus (Meulemans and Bronner–Fraser 2007) and tunicates have just one ancestral clade A fibrillar collagen gene as recently reported by Sansom and co-workers (Sansom et al. 2011).

The in-shore hagfish, *Eptatretus burger* possess small nodule-like structures similar to vertebra-like cartilages. These formations are proposed to be "homologous to gnathostome vertebrae, implying that this animal underwent a secondary reduction of vertebrae in most of the trunk" (Ota et al. 2011).

In mammals, three classic types of cartilage have been histologically identified based upon the relative contribution and distribution of fibers within the extracellular matrix. Hyaline cartilage is the typical cartilage that is generally referred to when one thinks of cartilage (Cole 2011). Elastic cartilage contains additional elastin fibers, and fibrocartilage contains regions of organized fibrous tissue within the extracellular matrix (Fig. 2.1).

In contrast to other animal groups, the cartilage of Mammalia is "a specialized connective tissue in which the chondrocytes occupy only 5 % of the volume" (Knudson and Knudson 2001). Chondrocytes are located into the matrix in such

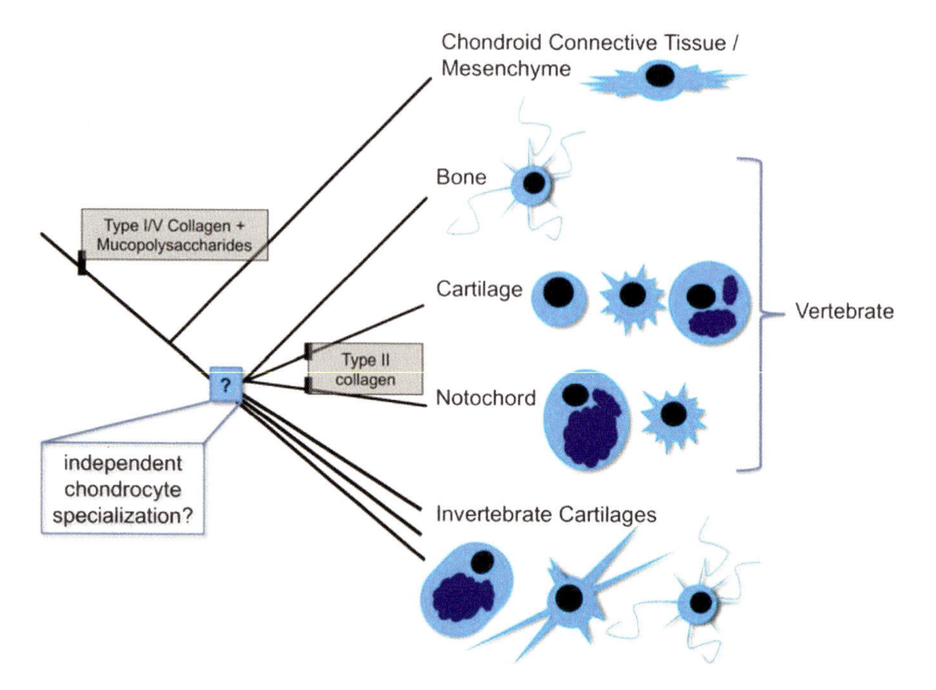

Fig. 2.1 Schematic depicting the proposed and disputed evolutionary relations between various connective tissues according to Cole (2011). Deposition of fibrous collagens and mucopolysaccharides in a structured connective tissue is an ancestral feature. In contrast, type II collagen is known to be specific to the vertebrate radiation, and is present in both vertebrate cartilage and notochord. At present it is unclear whether the chondrocyte has evolved multiple times, or if there was a single origin and subsequent diversification of this cell type (Cole (2011). Reprinted with permission.)

manner that they do not touch other cells. Cell–matrix interactions in the mammalian cartilage are essential for the maintenance of the extracellular matrix. The organic network that contains fibrillar collagen provides the capacity to contain the swelling pressure of the embedded proteoglycans as well as tensile strength to the tissue. Cartilaginous collagen fibrils are represented by collagens type II, IX and XI. Typical representatives of the proteoglycans are biglycan, fibromodulin, lumican and epiphycan, the cell surface syndecans and glypican, the basement membrane proteoglycan, perlecan, and the small leucine-rich proteoglycans decorin. Both decorin and type IX collagen play important role in regulation of collagen fibril formation. One of the large chondroitin sulphate-containing proteoglycans is the *aggrecan* that is expressed during chondrogenesis. The aggrecan is responsible for achievement of osmotic resistance necessary for cartilage to resist compressive loads. "Following its secretion, aggrecan self-assembles into a supramolecular structure with as many as 50 monomers bound to a filament of hyaluronan" (Knudson and Knudson 2001).

Developmentally, cartilage can be followed histologically through three general phases (Hall and Miyake 2000). Prior to cell differentiation, a cellular condensation forms within the mesenchyme. This condensation is known as either the cartilage anlagen (Cameron et al. 2009), protocartilage (Cole and Hall 2009), or cartilage condensation (Hall and Miyake 2000). These cells then begin to secrete the cartilage specific matrix. As development progresses the chondrocytes continue to proliferate and generate extracellular matrix, which then calcifies as the cells enter the hypertrophic phase of differentiation. This calcified cartilage matrix is then replaced by bone – in a process known as endochondral bone formation.

From genetics point of view, in mammals the *Runt* genes are the key players of skeletogenesis because even the stem species of chordates harboured a single Runt gene. It is accepted that the Runt locus duplications occurred during early vertebrate evolution. Probably, "Runt was part of a core gene network for cartilage formation. This network was already active in the gill bars of the common ancestor of cephalochordates and vertebrates, and diversified after Runt duplications occurred during vertebrate evolution" (Hecht et al. 2008).

Mineralized Cartilage Some kind of non-mineralized and non-collagen-based cartilaginous endoskeleton was the earliest skeleton in the vertebrate lineage. In such groups as lancelets, hagfish and lampreys, this skeleton was associated mostly with the pharynx. The collagen-based cartilage arose only after the evolution of collagen type II from earlier simple forms of the collagen's family. "In contrast to animals with completely non-collagenous skeletons, some of the primitive chondrichthyans (such as sharks) were able to form skeletal parts though the process of endochondral ossification" (Obradovic-Wagner and Aspenberg 2011). Principally, the calcified cartilage as a tissue type in the endoskeleton of sharks is suggested to be a primitive vertebrate characteristic (Coates et al. 1998). Matrix calcification in cartilaginous fish is well known (see for review Egerbacher et al. 2006); and is found as cortical mineralisation next to the cartilage surface, deep in the vertebra body, and in the neural arch of vertebrae. It will be described in more detail, with particular emphasis placed on concomitant changes in the organic matrix components which precede or accompany mineralisation but do not result in chondrolysis and ossification. In this way, calcification of cartilage in chondrichthyes provides a model to study direct metaplasia of chondrocytes, morphologically indicated by the remodelling of cartilage matrix and mineral deposition.

According to Ørvig (1951) there are "three principal kinds of calcified cartilage in elasmobranchs:

(a) globular calcification, considered to be an early stage of mineralization both ontogenetically and phylogenetically;
(b) prismatic or granular calcification, the type which forms the tesserae;
(c) areolar calcification, which occurs in the vertebral centra of Euselachii," (Kemp and Westrin 1979).

Endoskeletal tesserae (Applegate 1967) or Kalkplättchen (Roth 1911) of elasmobranchs and holocephalans are formations of calcified tissue which are responsible for rigidity of their cartilaginous skeletons. The neural or hemal arches and centra of vertebrae, as well as the chondrocranium, jaws and visceral arches, in the supporting cartilages of fins and clasper spurs, are possible locations where tesserae in sharks may occur (Applegate 1967). The structure of shark's tesserae with respect to their outer and inner surfaces is not homogenous. It was observed by Kemp and Westrin (1979) using SEM that "calcospherites and hydroxyapatite crystals similar to those commonly seen on the surface of bone are present on the outer surface of the tessera adjacent to the perichondrium. On the inner surface adjoining hyaline cartilage, however, calcospherites of variable size are the predominant surface feature. TEM shows calcification in close association with coarse collagen fibrils on the outer side of a tessera, but such fibrils are absent from the cartilaginous matrix along the underside of tesserae," (Kemp and Westrin 1979).

The tesserate patterns in modern Chondrichthyes are also an example of the primitive characteristic. Calcification phenomenon observed in this fish group "may be derived from an ancestral pattern of a continuous bed of calcified cartilage underlying a layer of perichondral bone as theorized by Ørvig (1951)," (Kemp and Westrin 1979).

The structure and ultrastructure of calcified cartilage in the endoskeletal tesserae have been characterized not only of sharks (Bargmann 1939; Kemp and Westrin 1979; Dean et al. 2008) but also of stingrays (Summers et al. 1998; Dean and Summers 2006; Dean 2007; Dean et al. 2007, 2009a, b, 2010). For example, growth of the jaw cartilage as well as the tessellation of the round stingray *Urobatis halleri* has been recently studied in the Lab of Adam Summers. Both phenomena have been characterized by changes in chondrocyte morphology, their orientation and distribution (Dean et al. 2009a). These researches suggested that "tessellated cartilage growth is made possible by an early organization of isolated surface mineralization centers. These grow appositionally to maintain contact as the underlying uncalcified matrix expands in volume. The tessellated skeleton is therefore an elegant solution to the problem of skeletal growth with continued integrity, but without resorption or remodelling," (Dean et al. 2009a) (Fig. 2.2).

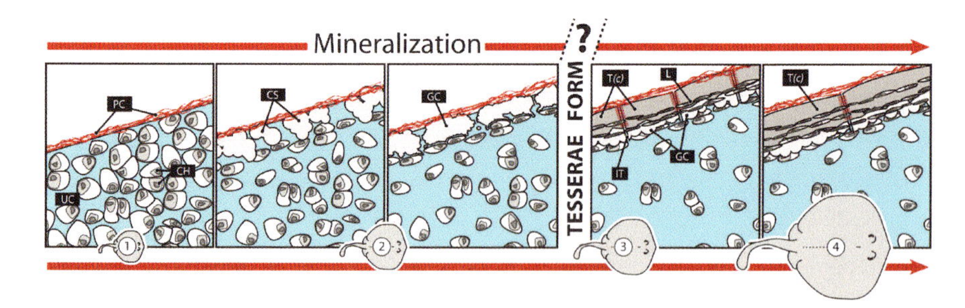

Fig. 2.2 Biomineralization in elasmobranch. Schematic overview on example of the stingray *Urobatis halleri*, and previous works (see Dean et al. 2009a). *Bottom*: Age classes of stingrays for comparison, with age increasing from *left to right*. Abbreviations (*UC*) the uncalcified matrix, is (*CH*) chondrocytes, (*PC*) the perichondrium, (*CS*) spherulitic calcospherites, (*GC*) globular calcifications, *T*(*c*) tesserae shown in cross-section, (*L*) the lacunae, (*IT*) the intertesseral joints (Reprinted from Dean et al. (2009a) with permission John Wiley and Sons. © 2009 The Authors. Journal compilation © 2009 Anatomical Society of Great Britain and Ireland)

In some fish the tesserae represent multi-layered, quite thick structure (see for review Dean and Summers 2006). For example, very complex tesserae have been demonstrated by the myliobatid stingrays (Summers 2000). Evolutionary expediency concerning development of these structures especially in representatives of this cartilaginous fish can be determined by their feeding behaviour. Usually, they are "durophagous", means crushing hard shelled molluscs between cylindrical tooth plates. This manner of feeding necessarily imposes large loads on their own skulls. Correspondingly, some species possess heavily mineralised jaws containing up to six layers of tesserae.

Other interesting structures are mineralised *trabeculae* of cartilaginous fish. These are hollow and structurally similar to that in bone. Their function, however, is to take loads from the tooth plates to the thickened parts of the jaws and skull. The overall diameter of these thinwalled trabeculae is about 800 μm, and the lumen being on the order of 500 μm (Currey 2010). According to Currey (2010), "this is a better arrangement than that in the solid bony trabeculae usual in tetrapods, because it produces a greater stiffness per unit mass than do solid struts. Producing hollow trabeculae does not seem to be part of the adaptive repertoire of bone," (Currey 2010).

Distribution and structural organization of tesserae are also crucial in faster swimming sharks which need increased scleral skeletal support for their eyes. Thus, in the larger, more active shark's species tessellated cartilage provides such kind of support stronger in comparison to regular hyaline cartilage (Pilgrim and Franz–Odendaal 2009).

Although the biomineralization may share some principal similarities with mammalian endochondral calcification concerning the cell organization "(e.g. alignment of flattened chondrocytes at the tissue periphery) and perhaps matrix reorganization (e.g. the expression of alkaline phosphatase and reduced sulfation in zones of mineralization)," (Dean et al. 2009a) it is believed the similarities to end there (Dean et al. 2009a).

Recently, Adam Summers and co-workers (Dean et al. 2010) investigated the ultrastructure of tessellated skeleton without damaging the delicate relationships between constituent tissues or to the tesserae themselves using synchrotron radiation tomography (SRT) as well as cryo-electron microscopy. In this way they observed previously unknown internal structures, namely passages connecting the lacunar spaces – the *intratesseral canaliculi* – within tesserae (Dean et al. 2010). These formations link consecutive lacunar spaces into long lacunar strings. It was shown that these strings radiate outward from the center of tesserae.

The skeletal tissues of elasmobranch fishes definitively illustrate the challenges of studying the principles of hierarchical organization of biological materials (Dean et al. 2009b). From this point of view, "each tessera is a geometric block (hundreds of microns deep and wide in adults), comprised of hydroxyapatite crystals on a collagen scaffold. The skeleton is therefore a fibro-mineral composite (perichondrium-tesseraeintertesseral fibers) wrapping a fiber-reinforced gel (uncalcified cartilage)

Fig. 2.3 Dorsal view of left side of stained embryo of California butterfly ray (*Gymnura marmorata*). *Red* areas indicate calcification, *blue* is cartilage (Adapted from Schaefer and Summers (2005) with permission John Wiley and Sons. Copyright © 2005 Wiley-Liss, Inc.)

with stark transitions between constituent tissues types," (Dean et al. 2010; see for more information Dean et al. 2008, 2009b).

The fibrous connective tissue within the elasmobranch vertebral cartilage is known as "areolar": Thus, it possesses fibers which are arranged in a net with "a weblike infiltration of mineral in a hyaline cartilage matrix that varies in morphology by species," (Porter et al. 2007; see for more information Moss 1977; Ridewood 1921). Some interspecific mineralization patterns observed in Elasmobranchii are variable enough to be of systematic importance (Ridewood 1921). For example, mineral in the vertebrae of the silky shark (*Carcharhinus falciformis*) are covered with a thick crust. However, in the vertebrae of the shortfin mako (*Isurus oxyrinchus*) mineralized plates are localized around the centre (Porter et al. 2006). It is known that in bone significant variation in mineral content is pathological or an interspecific effect. In contrast to bone, mineral in vertebral cartilage of sharks varies within individuals, intraspecifically and interspecifically (Porter et al. 2007) (Fig. 2.3).

2.1.1 Marine Cartilage: Biomechanics and Material Properties

Cartilage is an example of "a viscoelastic material having both fluid and solid characteristics, and therefore displays strain-rate dependent mechanical and material properties" (Porter et al. 2007). Interestingly, the material properties of cartilages of mammalian and of lower vertebrates (e.g. lamprey) origin are very similar. Both are known as useful models for better understanding of fundamental principles that modulate cartilage structure–function relationships *in vivo* and *in vitro* (Courtland et al. 2003). It is well established that even minor changes in mineral content can have drastic effects on material properties in hard tissues. For example, those materials with less mineral are weaker and less stiffer than those with larger mineral content. As reviewed by Porter et al. (2007), "a biological example of this relationship is the rostrum of the Blaineville's beaked whale *Mesoplodon densirostris*, which is composed of 96 % mineral, resulting in an incredibly stiff material (46 GPa). The fin whale *Balaenoptera physalus* tympanic bulla has 14 % less mineral, and is 35 % less stiff," (Porter et al. 2007). Interestingly, in numerous representatives of the beaked whales (*Ziphiidae*), so called *mesorostral* cartilage begins to ossify with the attainment of sexual maturity (Cozzi et al. 2010). This type of cartilage is homologous to the cartilaginous nasal septum of all mammals and determines the development of a very dense and compact bone due to ossification in *Ziphiidae* family (Zotti et al. 2009).

Interest in biomechanical and materials properties of both non-mineralized and mineralized cartilages of marine fish and mammals origin as model organisms is very high. What is the mechanical function of cartilage? According to Mansour (2004), the compliance of this biological material helps to distribute the loads, for example, between opposing bones in a synovial joint specifically. If cartilage were a bone-like strong material, the area of contact would be much smaller, however, the contact stresses at a joint would be much higher.

Cartilaginous fishes perform at functional extremes, below are some examples. Thus, some species of suction-feeding sharks feed on prey using the generation of suction pressure that is determined by the *ceratohyal* cartilage. Usually, this type of cartilage supports the "tongue" and is made by a pair of rod-shaped cartilages. It articulates with the *hyomandibular* cartilage at its proximal end, and with the *basihyal* cartilage at its distal end (Tomita et al. 2011). As observed by Tomita and co-workers, "the ceratohyal cartilages rotate around the hyomandibuloceratohyal articulations ventrally just after opening the mouth during prey capture. The hyoid arch is depressed and the tongue depresses, resulting in an increase in volume of the oral cavity which generates suction pressure." Correspondingly, the hyoid depression is based on activity of special muscles which connect the basihyal cartilage and the pectoral girdle. These muscles, also known as *coracohyoideus coupling* (Wilga et al. 2000) contain the *coracoarcualis* and the *coracohyoideus*. Their function is described in following way: "by shortening the coracohyoideus coupling, the basihyal cartilage is pulled posteriorly and the hyoid arch rotates ventrally, expanding the oral cavity. At this time, negative pressure is generated in the oral cavity, and suction-feeding sharks use this negative pressure to suck in the prey into front of its mouth" (Tomita et al. 2011). It was reported by (Motta et al. 2008) that contraction forces of the coracohyoideus coupling are able to generate the negative pressure in the oral cavity in the nurse shark *Ginglymostoma cirratum* up to 2,100 kPa. It was hypothesized that stronger suction feeders are expected to have stronger contraction forces of the coracohyoideus coupling because they generate larger negative pressures. It was confirmed (Tomita et al. 2011) that "the stiffness of the ceratohyal cartilage is positively correlated with suction feeding".

The nurse shark, the bamboo shark as well as the cookie-cutter shark, are not only suction specialists, they also possess stiffest ceratohyals. Nurse sharks and bamboo sharks (*Chiloscyllium plagiosum*) can generate among the greatest recorded subambient suction pressures when feeding (Motta et al. 2008; Wilga and Sanford 2008). The feeding behaviour of the cookie-cutter shark (*Isistius brasiliensis*) is described as follow: this predator "remove a plug of flesh from the body of a cetaceans or some large fish with strong suction pressure after cutting the flesh using its razor-shaped lower teeth," (Tomita et al. 2011; see for more information Shirai and Nakaya 1992). The zebra shark (*Stegostoma fasciatum*) and the nurse shark (*G. cirratum*) are known suction capture small bony fishes as well as suck out the soft parts of molluscs like moon shells or trumpet shells from their hard mineralized encasements. It was reported about large amounts of moon shell opercula that have been found in the gastric contents of *S. fasciatum* (Tomita et al. 2011).

Giant whale sharks (for example, the *Rhincodon typus*) are the biggest fish in the ocean (Gudger 1941), and, correspondingly, possess skeletons which are many times larger than those of the largest bony fish. It was reported that "shark skeletons endure tens to hundreds of millions of loading cycles in their transoceanic migrations," (Porter et al. 2006; see also Bonfil et al. 2005). It was supposed (Porter et al. 2006) that in some sharks, "the skeleton must resist the high loads that occur during burst swimming," (Porter et al. 2006). These authors investigated both the biochemistry and the material properties of the mineralized cartilage found in the vertebrae of selected elasmobranch species. Similarity with respect to ultimate strength and material stiffness has been observed between mammalian trabecular bone and calcified cartilage of Elasmobranchii studied. It was shown that the "collagen contents are more similar to mammalian bone than to mammalian cartilage, and these vertebrae have mineral fractions equalling that of mammalian bone" (Porter et al. 2006). Moreover, "that vertebral cartilage has bone-like stiffness and strength makes it unlikely that decreased functional demands were a selective force in the abandonment of a bony skeleton by cartilaginous fishes," (Porter et al. 2006).

The functional stiffness of articular cartilage is $|E^*|$, the dynamic elastic modulus, as explained in Fig. 2.4. $|E^*|$ is a function of the rate of deformation and can be determined from cyclic load/displacement data (Loparic et al. 2010). The rate employed should reflect the transient loading-unloading time of normal ambulation (i.e., running or walking). In humans, this is in the range of a few hundred milliseconds. Therefore, the authors performed indentation measurements at a rate of three complete loading/unloading cycles per second, corresponding to a tip unloading time of ~150 ms. Even after hundreds of loading/unloading cycles, they did not observe any progressive change in the load/displacement behavior, persistent residual indentations (which would be indicative of yield and plastic flow), or effects indicative of material fatigue.

Very interesting results were recently reported by Adam Summers (Macesic and Summers 2012). He and co-workers investigated a batoid (skate and ray) appendicular skeletal element, the propterygium, and its response to forces experienced during punting (benthic pelvic fin locomotion). The goals for their study were to

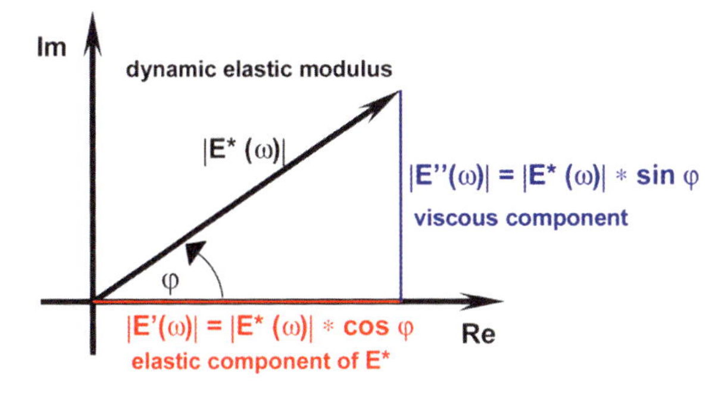

Fig. 2.4 Illustration of stiffness parameter relationships for a viscoelastic material subjected to a cyclic dynamic force or deformation (Reprinted from Loparic et al. 2010, Copyright (2010), with permission from Elsevier). At low frequencies, the magnitudes of force and deformation are out of phase, i.e., they do not reach maximum values simultaneously. This is expressed as the phase angle, φ, between their maximum values. As frequency is increased, φ decreases. In the limit $\varphi = 0$ and $E^* = E'$, i.e., the material behaves as an elastic solid. The out-of-phase behavior is due to the inability of the viscous portions of the material structure to store energy. Thus φ is a measure of energy loss and is also called the loss angle or loss tangent (Reprinted from Loparic et al. 2010, Copyright (2010), with permission from Elsevier)

determine: (1) the mechanical and compositional properties of the propterygium and (2) whether these properties correlate with punting ability.

Using five batoid species of varying punting ability, scientists employed a three-point bending test, finding that propterygium flexural stiffness (33.74–180.16 Nm^2) was similar to values found in bone and could predict punting ability. Variation in flexural stiffness resulted from differences in mineral content (24.4–48.9 % dry mass) and the second moment of area. Propterygia material stiffness (140–2,533 MPa) approached the lower limit of bone despite having less than one-third of its mineral content. This drastically lower mineral content is reflected in the radius-to-thickness ratio of the cross-section (mean ± s.e.m. = 5.5 ± 0.44), which is comparatively much higher than bony vertebrates. This indicates that elasmobranchs may have evolved skeletal elements that increase buoyancy without sacrificing mechanical properties. These results highlight the functional parallels between a cartilaginous and bony skeleton despite dramatic compositional differences, and provide insight into how environmental factors may affect cartilaginous skeletal development (Macesic and Summers 2012).

2.1.2 Marine Cartilage: Tissue Engineering

Cartilage of selachian fish provides a useful model to study direct metaplasia of hyaline cartilage into calcified cartilage (cortical mineralisation) and further on (vertebral body) into bony tissue (Egerbacher et al. 2006). Cartilage as typical avascular tissue shows limited capacity for regeneration and even self-repairs. Correspondingly, this feature represents a significant challenge for biomedical engineers and clinician in the case of the cartilage defects treatment (Ahmed and Hincke 2010).

Furthermore, articular cartilage is localized in the form of cover layer on the heads of joints. Because of this location, it tends to be under high-impact as well as constant mechanical stresses. Due to these both factors, it would be very difficult for this avascular tissue defects to heal through so called self-tissue regeneration. According to the modern strategy, "a thorough understanding of cartilage physiology and evaluation of the critical players in cartilage injury, disease, and repair, are of principal importance in deciding on the design parameters (i.e. biomaterial selection, cell type, signaling molecules) of a bioengineered tissue substitute" (Viala and Andreopoulos 2009).

Nowadays, several modern technologies, such as growth-factor delivery (Lee and Shin 2007; Chung and Burdick 2008), cartilage tissue engineering (Rotter et al. 2007; Greene and Watson 2010; Ehrlich et al. 2010), and stem cell therapy, have been proposed and partially applied for regeneration and repair of articular cartilage defects. For these cases, autologous cartilage transplantation is one of the most promising treatment options (see for review Fan et al. 2012).

Different chemical compounds (structural proteins and polysaccharides), which are originally located within marine fish as well as marine mammals cartilages, have been used in tissue engineering as components of artificially developed hydrogels (Guo et al. 2012) and scaffolds (Brittberg 2010).

Hyaluronic acid (or Hyaluronan) is one of the acidic mucopolysaccharides naturally existing in large quantities in shark skin and whale cartilage. Hyaluronan is a biopolymer of a linear repeating disaccharide unit consisting of β-(1 → 4)-linked D-glucopyranuronic acid and β-(1 → 3)-linked 2-acetamido-2-deoxy-D-glucopyranose. The polysaccharide is present in the synovial fluid of joints, in the extracellular matrices, and scaffolding that comprises cartilage. Hyaluronan is unique amongst other glycosaminoglycans because of "its mechanism of synthesis, its size, and its physico-chemical properties," (Murano et al. 2011). The network-forming, viscoelastic and charge properties of hyaluronan are crucial to most biochemical features of living tissues. Also its location on the cell surface as well as within the pericellular space, is very important. It interacts with other macromolecules such as proteins; and participates in regulating cell behavior during several restorative, pathological and morphogenic processes in the organism. The knowledge of hyaluronan in diseases such as various forms of cancers, arthritis and osteoporosis has led to new impetus in research and development in the preparation of biomaterials for surgical implants and drug conjugates for targeted delivery (Toole 2004).

The next important linear anionic polysaccharide is the chondroitin sulfate (CS) that is also a constituent of proteoglycans. The biopolymer represents a repeating disaccharide unit composed of glucuronic acid (GlcA) and N-acetylated galactosamine (GalNAc). Both are arranged in the [(1 → 4)-β-GlcA-(1 → 3)-β-GalNAc-] sequence. This regular structure is decorated during the biosynthesis by the insertion of sulfate groups at different positions of sugar backbone. As a key component of the connective tissues, CS is well known precautionary drug for joint diseases. Recently, the isolation of CS from cartilage of the lesser spotted dogfish (*Scyliorhinus canicula*), was reported by Gargiulo et al. 2009. This representative the *Scryliohinidae* family of the Carcharhiniformes order (Delabre et al. 1998) is broad distributed around the coast of the Mediterranean Sea, however, has no marketability. Unfortunately, regularly a large number of specimens die because they are erroneously captured during fishing expeditions. "The dead animals are usually thrown back to the sea, wasting the opportunity to isolate chondroitin sulfate from their tissues and use it for commercial preparations" (Gargiulo et al. 2009).

Collagen is a well-known as well as well studied biological material of proteinaceous origin (see also **Chap. 8** in this work). Here, only one example of it use as a scaffold. Recently, Sangsen and co-workers (2012) reported about development of composite material that consist of fish collagen and plant pectin. They used the skins of brown banded bamboo sharks (*Chiloscyllium punctatum*) and citrus fruits for isolation of collagen and pectin, respectively. Crosslinking of these novel scaffolds was performed by chemical crosslinking of shark's collagen using carbodiimide and by pregelation of pectin using calcium sulphate. The materials properties of the hybrid scaffolds obtained were as follow: average pore size of 134.53 ± 52.44 μm for the interconnecting pores; the hardness of 0.591 ± 0.135 N, and a springiness of 0.958 ± 0.022 (Sangsen et al. 2012). The scaffolds were biocompatible as shown by an in vitro cytotoxicity test using C_2C1_2 myoblast cells. The authors propose these fish collagen-pectin scaffolds as composite materials suitable for applications in tissue engineering.

However, is it possible to cultivate cartilage cells from marine vertebrates and use them as model systems, or even directly for tissue engineering? To my best knowledge, there are only few papers related to this topic. For example, as previously reported by Langille and Hall (1988), "cartilage from larval (ammocoetes) and adult (prespawning upstream migrant) lamprey was successfully maintained both when cultured *in vitro* and when grafted *in vivo* on the chorioallantoic membrane of host chick embryos," (Langille and Hall 1988). Moreover, it is possible to cultivate teeth from an adult lamprey under *in vitro* conditions. Histological investigations showed with strong evidence the cellular and structural integrity of both the cultured and grafted cartilages. Also ultrastructural analysis of chondrocytes using TEM confirmed the viability of the cartilage. Additionally, in vitro incorporation of radioactive sulfur into the matrix has shown metabolic activity of this tissue. Thus, "teeth cultured in L15-supplemented media for up to 14 days at either 15 or 20 °C retained their structural and cellular integrity as observed histologically, with no apparent cell outgrowth" (Langille and Hall 1988). There are no doubts that with the successful culture of the selected lamprey's tissues, their biochemistry, physiology and development, are potentially of great importance in better understanding of early vertebrate evolution and opens new ways in tissue engineering of marine cartilages.

Moreover, the phenomenon of *in vitro* mineralization of both lamprey teeth and cartilage in culture is one of the principal questions. The capability of lamprey cartilage to calcify *in vitro* under selected ionic conditions has been also reported (Langille and Hall 1985). Numerous experiments with respect to obtaining of hydroxyapatite have been carried out using adult and larval lamprey cartilage, normally unmineralized. The results of these 12 days- long experiments, which were carried out at 20, 30 or 37 °C are as follow:

- "histochemical analysis revealed a temperature-dependent increase over time in calcium phosphate incorporation into the extracellular matrix (ECM) of adult cartilage;
- Ultrastructural analysis revealed the presence of ECM dense crystalline bodies (40 nm average);
- electron-dense particles (15–20 nm) found in close association with the ECM fibrils in some regions of mineralized cartilage;
- larval cartilage incorporated much less calcium, less uniformly over time than did adult cartilage," (Langille and Hall 1993).

Thus, experimental data confirmed that adult lamprey cartilage mineralize under appropriate conditions *in vitro*. Correspondingly, Langille and Hall (1993) hypothesized that "petromyzonids, or their direct agnathan ancestors, may have possessed mineralized skeletons and that this ability is 'repressed' in extant lampreys owing primarily to the environment they inhabit".

One of the aims of the fish cartilage related tissue engineering is determined by the screening of novel sources of bioactive substances (e.g. chondroitin sulphate) as well as anticancer factors by *in vitro* culture of cartilage cells from cartilaginous fishes. Some experiments were very successful. For example, Shakibaei and De Souza (1997) reported that cultivation of chondrocytes in mass culture yielded a pure chondrocyte population. Also, new fish species like skate (*Raja porasa*) were used to initiate primary cultures of cartilage cells (Fan et al. 2003). Interestingly, this was the first attempt to establish an *in vitro* culture system for cartilage cells of skates. It was reported that the "induced cartilage cells cultured formed a confluent monolayer at day 7" (Fan et al. 2003).

2.1.3 Shark Cartilage: Medical Aspect

The discussion of the use of shark cartilage as a source of anticancer agents started more than 30 years ago (Langer et al. 1976; Lee and Langer 1983). Interest in the cartilage application grew because of "its avascular state coupled with the misconception that sharks do not get cancer" (Patra and Sandell 2012; see for review Lane and Comac 1996). Although crude extracts of shark cartilage are ineffective, some purified components may work as cancer retardants (Ostrander et al. 2004). From biochemical point of view, cartilage-derived inhibitor of neovascularization was accepted to be related to enzymes of the metalloproteinases family (Moses et al. 1990, 1992).

In the current literature, we can find very polar opinions concerning anticancer activity of numerous shark-cartilage-derived products. "The discovery of cancer in sharks and the lack of promising results from the most recent clinical trial with

Neovastat were serious setbacks to its use in cancer treatment" (Patra and Sandell 2012). Thus, it was reported that sharks shown symptoms of both malignant and benign neoplasms. According to Finkelstein (2005), because of absence of really existing antiangiogenic and anti-invasive substances in cartilage, there is no logic for additional justifications for using shark cartilage. The fact that some people believe shark cartilage consumption can cure cancer seems to be an example of pseudoscience. Moreover, such negative outcomes as a diversion of patients from effective cancer treatments as well as increased decline in shark populations have been registered (Ostrander et al. 2004).

However, other authors reported about suppression of carcinogenesis (Sato et al. 2004, 2008), growth of tumor, as well as angiogenesis in animal models (see for review Kitahashi et al. 2012) by oral administration of shark cartilage. Especially, low molecular weight proteins (smaller than 20 kDa) have been suggested to be involved in the beneficial effects by oral administration of shark-cartilage-based products (Bargahi and Rabbani–Chadegani 2008; Kitahashi et al. 2012). These low molecular weight substances exerting antiangiogenesis, immunostimulation, and MMP-9 inhibitory activities. Additional "positive" example is the squalamine (Moore et al. 1993; Li et al. 2002), which is isolated from liver and stomach of the dogfish shark (*Squalus acanthias*). This aminosterol inhibited solid tumor growth and angiogenesis *in vivo* (Sills et al. 1998).

The next example, Neovastat, is a naturally occurring multifunctional antiangiogenic drug (Falardeau et al. 2001). Neovastat (or AE-941) contains a mix of water-soluble components less than approximately 500 kDa derived after homogenisation of the shark cartilage in water followed by sequential extraction to remove inactive and water-insoluble molecules. The drug is prepared by a proprietary manufacturing process developed by Aeterna Laboratories (Quebec, Canada) (Dupont et al. 1997). The effects observed after Neovastat treatments are listed by Patra and Sandell (2012) as follow:

- "it induced a concentration-dependent inhibition of cell proliferation in human umbilical vein endothelial cells (HUVECs) and bovine endothelial cells;
- it inhibited the formation of blood vessels induced by basic fibroblast growth factor (FGF) in the chicken chorioallantoic membrane model;
- it severely inhibited *in vivo* the vascular invasion of bFGF-containing Matrigel implanted in C57BL6 mice fed orally with Neovastat;
- it inhibited lung metastases in the murine Lewis lung carcinoma model," (Patra and Sandell 2012).

In addition, Neovastat shown synergetic effect being combined with, a conventional anticancer agent named *cisplatin*. It exhibited greater anticancer activity than cisplatin alone (Patra and Sandell 2012).

Studies involving the isolation of novel functional biomolecules from shark cartilage are still on-going. For example, as recently reported by O'Connell et al. (2012), Trimethylamine N-oxide (TMAO), extracted from shark cartilage is a kind of a natural osmolyte, that induce protein folding, and counteracts the destabilizing effect of the high concentrations of urea stored by the fish.

2.2 Conclusions

Cartilaginous skeleton in chondrichthyan fishes is an example of the structural element of great significance for better understanding of the vertebrate evolution today. The diversity, biochemistry, structure of the cartilage and the formation of both non-mineralized and mineralized cartilages have been studied extensively in lampreys, hagfish, different chondrichthyes, and in marine mammals. It is proposed that the jawless agnathans contain the core features of cellular cartilage development which are conserved in the most basal extant vertebrates. Thus, according to Cattell et al. (2011), "the branchial basket cartilage of the agnathan lamprey possesses all of the diagnostic histological and biochemical properties of gnathostome cellular cartilage; including stack-of-coins and polygonal morphology, alcian-blue reactivity and fibrillar collagen expression," (Cattell et al. 2011). The cartilaginous skeleton of chondrichthyan fish represents just one of their unique alternatives to body design and adaptation, which evolved in ways that differ from other fish.

Chondrogenesis is the complex biochemical process that initiates during embryogenesis in the vertebrates. It leads to production of cartilage due the condensation of initially loosely arranged mesenchymal cells into compact aggregates. This is followed by assumption of chondrocyte fate through the action of corresponding transcription factors related to the SOX family and bone morphogenetic factor (BMP) intercellular signals (Glimm et al. 2012). Once differentiated, chondrocytes initiate synthesis and secretion of a specialized extracellular matrix. Calcification in higher vertebrates is initiated by the cartilage elements produced in the embryonic stage, within days through chondrocyte death and subsequent colonization of the natural cartilage scaffold with bone producing cells known as osteoblasts.

Nowadays, the mechanical properties of soft and hard cartilage can be successfully determined such high precision methods as compression, indentation, tension, and shear tests. These properties are necessary for any analysis of stress in the tissue; as well as for better understanding of such phenomena as swimming, diving and locomotion of marine vertebrates as one of the basic groups of animals for bioinspiration and biomimetics. The procedures include biological observation, kinematics modeling, mechanism design, prototype implementation, and initial experiments. From experimental and theoretical standpoints, the development of the diverse bioinspired cartilage-based models would be the most intensively studied subjects for biological materials science and biomedicine in the future.

References

Ahmed TA, Hincke MT (2010) Strategies for articular cartilage lesion repair and functional restoration. Tissue Eng B Rev 16(3):305–329

Applegate SP (1967) A survey of shark hard parts. In: Gilbert PW, Mathewson RF, Rall DP (eds) Sharks, skates and rays. The Johns Hopkins Press, Baltimore

Bargahi A, Rabbani–Chadegani A (2008) Angiogenic inhibitor protein fractions derived from shark cartilage. Biosci Rep 28:15–21

Bargmann W (1939) Zur Kenntnis der Knorpelarchitekturen (Untersuchungen am Skelettsystem von Selachiern). Z Zellforsch 29:405–424
Bonfil R, Meyer M, Scholl MC et al (2005) Transoceanic migration, spatial dynamics, and population linkages of white sharks. Science 310:100–103
Brittberg M (2010) Cell carriers as the next generation of cell therapy for cartilage repair a review of the matrix–induced autologous chondrocyte implantation procedure. Am J Sports Med 38:1259–1271
Cameron TL, Belluoccio D, Farlie PG et al (2009) Global comparative transcriptome analysis of cartilage formation *in vivo*. BMC Dev Biol 9:20
Cattell M, Lai S, Cerny R et al (2011) A new mechanistic scenario for the origin and evolution of vertebrate cartilage. PLoS One 6(7):e22474. Copyright: © 2011 Cattell et al. This is an open-access article distributed under the terms of the Creative Commons Attribution License, which permits unrestricted use, distribution, and reproduction in any medium, provided the original author and source are credited
Chung C, Burdick JA (2008) Engineering cartilage tissue. Adv Drug Deliv Rev 60(2):243–262
Coates MI, Sequeira SEK, Sansom IJ (1998) Spines and tissues of ancient sharks. Nature 396:729–730
Cole AG (2011) A review of diversity in the evolution and development of cartilage: the research for the origin of the chondrocyte. Eur Cell Mater 21:122–129
Cole AG, Hall BK (2004) Cartilage is a metazoan tissue; integrating data from non-vertebrate sources. Acta Zool 85:69–80
Cole AG, Hall BK (2009) Cartilage differentiation in cephalopod molluscs. Zoology 112:2–15
Courtland H–W, Wright GM, Root RG et al (2003) Comparative equilibrium mechanical properties of bovine and lamprey cartilaginous tissues. J Exp Biol 206:1397–1408
Cozzi B, Panin M, Butti C et al (2010) Bone density distribution patterns in the rostrum of Delphinids and Beaked Whales: evidence of family–specific evolutive traits. Anat Rec 293:235–242
Currey JD (2010) Mechanical properties and adaptations of some less familiar bony tissues. J Mech Behav Biomed Mater 3(5):357–372. Copyright (2010), with permission from Elsevier
Dean M (2007) Ontogeny, morphology and mechanics of the tessellated skeleton of cartilaginous fishes. J Morphol 268:1066
Dean MN, Summers AP (2006) Mineralized cartilage in the skeleton of chondrichthyan fishes. Zoology 109:164–168
Dean MN, Bizzarro JJ, Summers AP (2007) The evolution of cranial design, diet, and feeding mechanisms in batoid fishes. Integr Comp Biol 47:70–81
Dean MN, Gorb SN, Summers AP (2008) A cryoSEM method for preservation and visualization of calcified shark cartilage (and other stubborn heterogeneous skeletal tissues). Microsc Today 16:48–50
Dean MN, Mull CG, Gorb SN, Summers AP (2009a) Ontogeny of the tessellated skeleton: insight from the skeletal growth of the round stingray *Urobatis halleri*. J Anat 215:227–239. doi:10.1111/j.1469-7580.2009.01116.x. Copyright © 2009 The Authors. Journal compilation © 2009 Anatomical Society of Great Britain and Ireland. Reproduced with permission of Wiley-Liss, Inc
Dean MN, Youssefpour H, Earthman JC et al (2009b) Micro–mechanics and material properties of the tessellated skeleton of cartilaginous fishes. Integr Comp Biol 49:e45
Dean MN, Socha JJ, Hall BK, Summers AP (2010) Canaliculi in the tessellated skeleton of cartilaginous fishes. J Appl Ichthyol 26:263–267. © 2010 Blackwell Verlag, Berlin. Reproduced with permission of Wiley-Liss, Inc
Delabre C, Spruyt N, Delmarre C et al (1998) The complete nucleotide sequence of the mitochondrial DNA of the dogfish, *Scyliorhinus canicula*. Genetics 150:331–344
Dupont É, Brazeau P, Juneau C (1997) Extracts of shark cartilage having an antiangiogenic activity and an effect on tumor progression: process of making thereof. US Patent, 5,618,925
Egerbacher M, Helmreich M, Mayrhofer E et al (2006) Mineralization of hyaline cartilage in the small–spotted dogfish *Scyliorhinus canicula*. L Scripta Med (Brno) 79(4):199–212

Ehrlich H, Steck E, Ilan M et al (2010) Three-dimensional chitin-based scaffolds from Verongida sponges (Demospongiae: Porifera). Part II. Biomimetic potential and application. Int J Biol Macromol 47:141–147
Eikenberry EF et al (1984) Reprinted from: J Mol Biol 176(2), Eikenberry EF, Childs B, Sheren SB, Parry DAD, Craig AS, Brodsky B (1984) Crystalline fibril structure of type II collagen in lamprey notochord sheath 261–277. Copyright (1984), with permission from Elsevier
Falardeau P, Champaigne P, Poyet P et al (2001) Neovastat, a naturally occurring multifunctional antiangiogenic drug, in phase III clinical trials. Semin Oncol 28:620–625
Fan TJ et al (2003) With kind permission from Springer Science + Business Media: Fan TJ, Jin LY, Wang XF (2003) Initiation of cartilage cell culture from skate (*Raja porasa* Günther). Mar Biotechnol (NY) 5(1):64–69. Copyright © 2003, Springer-Verlag New York Inc.
Fan W, Wu C, Miao X et al (2012) Biomaterial scaffolds in cartilage–subchondral bone defects influencing the repair of autologous articular cartilage transplants. J Biomater Appl published online 8 June 2012. doi:10.1177/0885328211431310
Fernandesa RJ, Eyre DR (1999) Reprinted from: Fernandes RJ, Eyre DR (1999) The elastin-like protein matrix of lamprey branchial cartilage. Biochem Biophys Res Commun 261(3):635–640. Copyright (1999) with permission from Elsevier
Finkelstein JB (2005) Sharks do get cancer: few surprises in cartilage research. J Natl Cancer Inst 97:1562–1563
Forey P, Janvier P (1993) Agnathans and the origin of jawed vertebrates. Nature 361:129–134
Gargiulo V et al (2009) Reprinted with permission from: Gargiulo V, Lanzetta R, Parrilli M et al (2009) Structural analysis of chondroitin sulfate from *Scyliorhinus canicula*: a useful source of this polysaccharide. Glycobiology 19:1485–1491, by permission of Oxford University Press
Glimm T, Headon D, Kiskowski MA (2012) Computational and mathematical models of chondrogenesis in vertebrate limbs. Birth Defects Res (Part C) 96:176–192
Greene JJ, Watson D (2010) Septal cartilage tissue engineering: new horizons. Facial Plast Surg 26(5):396–404
Gudger EW (1941) The food and feeding habits of the whale shark, *Rhineodon typus*. J Elisha Mitchell Sci Soc 57:57–72
Guo Y, Yuan T, Xiao Z et al (2012) Hydrogels of collagen/chondroitin sulfate/hyaluronan interpenetrating polymer network for cartilage tissue engineering. J Mater Sci Mater Med 23:2267–2279
Hall BK (2005) Bones and cartilage: developmental skeletal biology. Elsevier/Academic Press, London
Hall BK, Miyake T (2000) All for one and one for all: condensations and the initiation of skeletal development. BioEssays 22:138–147
Hecht J, Stricker S, Wiecha U, Stiege A, Panopoulou G, Podsiadlowski L, Poustka AJ, Dieterich C, Ehrich S, Suvorova J, Mundlos S, Seitz V (2008) Evolution of a core gene network for skeletogenesis in chordates. PloS Genet 4(3):e1000025. doi:10.1371/journal.pgen.1000025. Copyright (c) 2008 Hecht et al. This is an open-access article distributed under the terms of the Creative Commons Attribution License, which permits unrestricted use, distribution, and reproduction in any medium, provided the original author and source are credited
Kemp NE, Westrin SK (1979) Ultrastructure of calcified cartilage in the endoskeletal tesserae of sharks. J Morphol 160:75–101. doi:10.1002/jmor.1051600106. Copyright © 1979 Wiley-Liss, Inc. Reproduced with permission of Wiley-Liss, Inc
Kitahashi T, Ikawa S, Sakamoto A et al (2012) Ingestion of proteoglycan fraction from shark cartilage increases serum inhibitory activity against matrix metalloproteinase–9 and suppresses development of N–nitrosobis(2–oxopropyl)amine– induced pancreatic duct carcinogenesis in hamster. J Agric Food Chem 60:940–945
Knudson CB, Knudson W (2001) Reprinted from: Knudson CB, Knudson W (2001) Cartilage proteoglycans. Semin Cell Dev Biol 12(2):69–78. Copyright (2001), with permission from Elsevier
Kupriyanova EK, Rouse GW (2008) Yet another example of paraphyly in Annelida: molecular evidence that Sabellidae contains Serpulidae. Mol Phylogenet Evol 46:1174–1181

Lakiza O, Miller S, Bunce A, Myung-Jae Lee E, McCauley DW (2011) SoxE gene duplication and development of the lamprey branchial skeleton: insights into development and evolution of the neural crest. Dev Biol 359(1):149–161

Lane IW, Comac L (1996) Sharks still don't get cancer. Avery Publishing Group, New York

Langer R, Brem H, Falterman K et al (1976) Isolation of a cartilage factor that inhibits tumor neovascularization. Science 193:70–72

Langille RM, Hall BK (1985) *In vitro* calcification of lamprey cartilage in hydroxyapatite metastable medium. Anat Rec 112:104A

Langille RM, Hall BK (1988) With kind permission from Springer Science+Business Media: Langille RM, Hall BK (1988) The organ culture and grafting of lamprey cartilage and teeth. In Vitro Cell Dev Biol 24(1):1–8. Copyright © 1988, Tissue Culture Association, Inc.

Langille RM, Hall BK (1993) Calcification of cartilage from the Lamprey *Petromyzon marinus* (L.) *in vitro*. Acta Zool 74:31–41. doi:10.1111/j.1463-6395.1993.tb01218.x. Copyright © 1993, The Royal Swedish Academy of Sciences. Reproduced with permission of John Wiley and Sons

Lee A, Langer R (1983) Shark cartilage contains inhibitors of tumor angiogenesis. Science 221:1185–1187

Lee SH, Shin H (2007) Matrices and scaffolds for delivery of bioactive molecules in bone and cartilage tissue engineering. Adv Drug Deliv Rev 59(4–5):339–359

Li D, Williams JI, Pietras RJ (2002) Squalamine and cisplatin block angiogenesis and growth of human ovarian cancer cells with or without HER–2 gene overexpression. Oncogene 21:2805–2814

Loparic M, Wirz D, Daniels AU, Raiteri R, van Landingham MR, Guex G, Martin I, Aebi U, Stolz M (2010) Micro- and nanomechanical analysis of articular cartilage by indentation-type atomic force microscopy: validation with a gel-microfiber composite. Biophys J 98:2731–2740

Macesic LJ, Summers AP (2012) Flexural stiffness and composition of the batoid propterygium as predictors of punting ability. J Exp Biol 215:2003–2012

Mansour JM (2004) Biomechanics of cartilage. In: Oatis KC (ed) Kinesiology: the mechanics and pathomechanics of human movement. Lippincott Williams & Wilkins, a business of Wolters Kluwer Health, Inc., Baltimore/Philadelphia

Martin WM, Bumm LA, McCauley DW (2009) Development of the viscerocranial skeleton during embryogenesis of the sea lamprey, *Petromyzon marinus*. Dev Dyn 238:3126–3138

Meulemans D, Bronner–Fraser M (2007) Insights from amphioxus into the evolution of vertebrate cartilage. PLoS One 2:e787

Moore KS, Wehrli S, Roder H et al (1993) Squalamine: an aminosterol antibiotic from the shark. Proc Natl Acad Sci U S A 90:1354–1358

Moses MA, Sudhalter J, Langer R (1990) Identification of an inhibitor of neovascularization from cartilage. Science 248(4961):1408–1410

Moses MA, Sudhalter J, Langer R (1992) Isolation and characterization of an inhibitor of neovascularization from scapular chondrocytes. J Cell Biol 119(2):475–482

Moss ML (1977) Skeletal tissues in sharks. Am Zool 17:335–342

Motta PJ, Hueter RE, Tricas TC et al (2008) Functional morphology of the feeding apparatus, feeding constraints, and suction performance in the nurse shark *Ginglymostoma cirratum*. J Morphol 369:1041–1055

Murano E, Perin D, Khan R, Bergamin M (2011) Hyaluronan: from biomimetic to industrial business strategy. Nat Prod Commun 6(4):555–572. Reprinted with permission. Copyright (c) 2011, Natural Product Inc. (NPI)

O'Connell GD, Fong JV, Dunleavy N et al (2012) Trimethylamine N–oxide as a media supplement for cartilage tissue engineering. J Orthop Res 30:1898–1905

Obradovic-Wagner D, Aspenberg P (2011) Where did bone come from? An overview of its evolution. Acta Orthop 82(4):393–398. Copyright © 2011, Informa Healthcare. Reproduced with permission of Informa Healthcare

Ørvig T (1951) Histologic studies of placotlerms and fossil elasmobranchs. I. The endoskeleton, with remarks on the hard tissues of lower vertebrates in general. Ark Zool 2:321–454

Ostrander GK, Cheng KC, Wolf JC, Wolfe MJ (2004) Shark cartilage, cancer and the growing threat of pseudoscience. Cancer Res 64(23):8485–8491

Ota KG et al (2011) Reprinted by permission from Macmillan Publishers Ltd: Ota KG et al (2011) Identification of vertebra-like elements and their possible differentiation from sclerotomes in the hagfish. Nat Commun 2:373. Copyright (2011)

Patra D, Sandell LJ (2012) Antiangiogenic and anticancer molecules in cartilage. Expert Rev Mol Med 14:e10. doi:10.1017/erm.2012.3. Copyright © Cambridge University Press 2012, reproduced with permission

Person P, Philpott DE (1963) Invertebrate cartilage. Ann N Y Acad Sci 109:113–126

Person P, Philpott DE (1969) The nature and significance of invertebrate cartilages. Biol Rev Camb Philos Soc 44:1–16

Pilgrim BL, Franz–Odendaal TA (2009) A comparative study of the ocular skeleton of fossil and modern chondrichthyans. J Anat 214:848–858

Porter ME et al (2006) Republished with permission of The Company of Biologists Ltd., from: Porter ME, Beltrán JL, Koob TJ, Summers AP (2006) Material properties and biochemical composition of mineralized vertebral cartilage in seven elasmobranch species (Chondrichthyes). J Exp Biol 209:2920–2928. doi:10.1242/jeb.02325. Copyright (2006); permission conveyed through Copyright Clearance Center, Inc.

Porter ME et al (2007) Republished with permission of The Company of Biologists Ltd., from: Porter ME, Koob TJ, Summers AP (2007) The contribution of mineral to the material properties of vertebral cartilage from the smooth-hound shark *Mustelus californicus*. J Exp Biol 210:3319–3327. Copyright (2007); permission conveyed through Copyright Clearance Center, Inc.

Ridewood WG (1921) On the calcification of the vertebral centra in sharks and rays. Philos Trans R Soc Lond B Biol Sci 210:311–407

Robson P, Wright GM, Sitarz E et al (1993) Characterization of lamprin, an unusual matrix protein from lamprey cartilage. J Biol Chem 268:1440–1447

Romer AS (1964) Bone in early vertebrates. In: Frost HM (ed) Bone biodynamics. Little, Brown & Co., Boston, pp 13–40

Roth W (1911) Beitrage zur Kenntnis der Strukturverhaltnisse des Selachier–Knorpels. Morphol Jahrb 42:485–555

Rotter N, Bucheler M, Haisch A et al (2007) Cartilage tissue engineering using resorbable scaffolds. J Tissue Eng Regen Med 1(6):411–416

Rychel AL, Swalla BJ (2007) Development and evolution of chordate cartilage. J Exp Zool B Mol Dev Evol 308:325–335

Sangsen Y, Benjakul S, Oungbho K (2012) Fabrication of novel shark collagen–pectin scaffolds for tissue engineering. In: Biomedical Engineering International Conference (BMEiCON), Chiang Mai, Thailand, pp 273–278, 29–31 January 2012

Sansom RS, Gabbott SE, Purnell MA (2011) Decay of vertebrate characters in hagfish and lamprey (Cyclostomata) and the implications for the vertebrate fossil record. Proc Biol Sci 278(1709):1150–1157, by permission of the Royal Society

Sato K, Murata N, Tsutsumi M, Shimizu–Suganuma M et al (2004) Moderation of chemo–induced cancer by water extract of dried shark fin: anti–cancer effect of shark cartilage. In: Sakaguchi M (ed) Developments in food science: more efficient utilization of fish and fisheries products. Elsevier, Oxford

Sato K, Kitahashi T, Itho C et al (2008) Shark cartilage: potential for therapeutic application for cancer. In: Barrow C, Shahidi F (eds) Marine nutraceuticals and functional foods. CRC Press, Boca Raton

Schaefer JT, Summers AP (2005) Batoid wing skeletal structure: novel morphologies, mechanical implications, and phylogenetic patterns. J Morphol 264:298–313

Shakibaei M, De Souza P (1997) Differentiation of mesenchymal limb bud cells to chondrocytes in alginate beads. Cell Biol Int 21(2):75–86

Shirai S, Nakaya K (1992) Functional morphology of feeding apparatus of the cookie–cutter shark *Isistius brasiliensis* (Elasmobranchii, Dalatiinae). Zool Soc 9:811–821

Sills AK Jr, Williams JI, Tyler BM (1998) Squalamine inhibits angiogenesis and solid tumor growth *in vivo* and perturbs embryonic vasculature. Cancer Res 58:2784–2792

Stemple DL (2005) Structure and function of the notochord: an essential organ for chordate development. Development 132:2503–2512

Summers AP (2000) Stiffening the stingray skeleton—an investigation of durophagy in myliobatid stingrays (Chondrichthyes, Batoidea, Myliobatidae). J Morphol 243:113–126

Summers AP, Koob TJ, Brainerd EL (1998) Stingray jaws strut their stuff. Nature 395:450–451

Tomita T, Sato K, Suda K, Kawauchi J, Nakaya K (2011) Feeding of the megamouth shark (Pisces: Lamniformes: Megachasmidae) predicted by its hyoid arch: a biomechanical approach. J Morphol 272:513–524. doi:10.1002/jmor.10905. Copyright © 2011 Wiley-Liss, Inc. Reprinted with permission

Toole BP (2004) Hyaluronan: from extracellular glue to pericellular cue. Nat Rev Cancer 4:528–539

Viala X, Andreopoulos FM (2009) Novel biomaterials for cartilage tissue engineering. Curr Rheumatol Rev 5:51–57. Reprinted by permission of Eureka Science Ltd

Wilga CD, Sanford CP (2008) Suction generation in white–spotted bamboo sharks *Chiloscyllium plagiosum*. J Exp Biol 211:3128–3138

Wilga CD, Wainwright PC, Motta PJ (2000) Evolution of jaw depression mechanics in aquatic vertebrates: insights from chondrichthyes. Biol J Linn Soc 71:165–185

Wright GM, Youson JH (1983) Ultrastructure of cartilage from young adult sea lamprey, *Petromyzon marinus* L: a new type of vertebrate cartilage. Am J Anat 167:59–70

Wright GM, Keeley FW, Youson JH (1983) Lamprin: a new vertebrate protein comprising the major structural protein of adult lamprey cartilage. Experientia 39:495–497

Wright GM, Armstrong LA, Jacques AM et al (1988) Trabecular, nasal, branchial, and pericardial cartilages in the sea lamprey, *Petromyzon marinus*: fine structure and immunohistochemical detection of elastin. Am J Anat 182:1–15

Wright GM, Keeley FW, Robson P (2001) The unusual cartilaginous tissues of jawless craniates, cephalochordates and invertebrates. Cell Tissue Res 304:165–174

Zhang G–J, Cohn MJ (2006) Hagfish and lancelet fibrillar collagens reveal that type II collagen–based cartilage evolved in stem vertebrates. Proc Natl Acad Sci U S A 103:16829–16833

Zotti A, Poggi R, Cozzi B (2009) Exceptional bone density DXA values of the rostrum of a deep–diving marine mammal: a new technical insight in the adaptation of bone to aquatic life. Skelet Radiol ISSN:1432–2161. doi:10.1007/s00256–009–0647–4

Chapter 3
Biocomposites and Mineralized Tissues

3.1 Bone

Abstract The skeletal system of marine vertebrates includes cartilage and bone. Variations in the densities and composition in the bones of marine vertebrates may be related to adaptations for buoyancy and locomotion, or to habitat or phylogeny. Several bone types like cellular and acellular, dermal, perichondral and endochodral bones are discussed in this chapter. Endochondral bone is cancellous endosteal bone formed during growth by a process involving the destructive replacement of cartilage. Special attention is payed to giant endochondral bones of whale origin. Some whale bones are known for their specific properties like high porosity, oil storage, hypermineralization, and high bone density. The strength and flexibility of whale bones makes them an excellent construction material, and they were used for this purpose in ancient Arctic cultures.

Bones represent a family of biological materials with complex, hierarchically organized architecture. These diverse mineralized structures are excellently adapted to the variety of mechanical functions and stresses (Weiner et al. 1999; Beniash 2011). According to modern point of view, "bone is specific to vertebrates, and originated as mineralization around the basal membrane of the throat or skin, giving rise to tooth-like structures and protective shields in animals with a soft cartilage-like endoskeleton" (Obradovic-Wagner and Aspenberg 2011). In his excellent monograph, John Long (1995) described the origin and diversity of bone structures which I will now briefly summarize. Bone can be examined as the calcified tissue that supports the skeleton, external or internal, of vertebrates and shows a broad variety of mechanical adaptations at nano- and microscales (Currey 1984, 2002; Weiner and Wagner 1998; Fratzl et al. 2004). A functionally important mechanical property of bones is stiffness, both in the whole element sense and in the material sense (Horton and Summers 2009). Main components of bone include hydroxylapatite

H. Ehrlich, *Biological Materials of Marine Origin*, Biologically-Inspired Systems 4,
DOI 10.1007/978-94-007-5730-1_3

(HAP) (as inorganic part), nanofibrillar collagen fibres that support the *in vivo* development of mineralised bone, and corresponding vascular tissue that supplies blood to the living cell components of bone. Since publication by Kölliker (1859), the presence of *cellular* and *acellular* types in the bone of early vertebrates is well established. In spite of that the structures of these bone types are similar, the principal difference between them are the spaces in cellular bone for the osteocytes, which occur throughout this hard tissue.

One of the structural features of the earliest forms of acellular bone is their lamination. Corresponding layers were deposited by the dermis. The heterostracans, which are related to early agnathans, possess acellular bone that is called *aspidin* (Tarlo 1963). Acellular bone have been also observed in layered bone from more advanced fish like placoderms, where this unique structure is found in the basal bone layer of the fish plates. The second structural feature founded in the most skeletons of the earliest fish is formation of the bone that is produced by the dermis. Thus, so called *dermal bone* was localized in the head and trunk, as well as the scales and biting surfaces inside the mouth. For example the scales and the dermal bones of dipnoans and crossopterygians were covered with a shiny enameloid layer. The complex layer known as *cosmine* (see also Sect. 4.2 in this work), covered an interconnecting network of flask-shaped cavities in the uppermost bone layer.

Numerous specialised bone types have been developed in various species of early osteichthyans. These bone types differ from each other by the nature and composition of the external layers of the dermal bones, complexity. Their hierarchical structure was determined by corresponding growth and resorption processes which were involved in their development.

Both, *perichondral bone* (a thinly laminated acellular bone) and *endochondral bone*, are characteristic examples of such kind of specialized bones. As reviewed by Obradovic-Wagner and Aspenberg (2011), these bones possess following features and functions:

"(a) Perichondral bone is often found surrounding soft tissue that passes through cartilage, as can be found in placoderms with cartilage braincases where the nerves and arteries passing through the braincase wall can have perichondral bone surrounding them. The endogirdles that support the pelvic and pectoral fins is also surrounded by perichondral bone.

(b) Endochondral bone forms around a cartilage precursor. These bones make up much of the internal skeleton in reptiles and mammals, and forms the arm and leg bones. This type of bone first evolved as a specialised feature in some groups of gnathostomes (osteichthyans and acanthodians)," (http://austhrutime.com/bone.htm; see also Obradovic-Wagner and Aspenberg 2011).

Today, two mechanisms of the bone development are accepted, I mean the intramembranous bone formation and the endochondral bone formation. Bone formation is complex, but the three-dimensional positioning of cells and matrices is straightforward. This process is determined by eventual differentiation of osteoprogenitor cells into either mesenchymal osteoblasts, which synthesize woven bone in random orientation. Also, the surface osteoblasts, which synthesize bone on surfaces in a well oriented lamellar array, play very important role here (see for details Shapiro 2008). I absolutely agree with Elia Beniash that "understanding basic principles of formation, structure and functional properties of different bone types might lead to novel bioinspired strategies for material design, and better treatments for diseases of the mineralized tissues," (Beniash 2011).

Acellular Bone "*Aspidin*" is an example of earliest bone and represents specific acellular, matrix-rich (presumably consisting of collagen fibres) mineralized tissue (for review see Donoghue and Sansom 2002; Donoghue et al. 2006). According to Tarlo (1963), "aspidin was the 'ancestor' of true, cellular bone, which evolved sometime after the origin of vertebrates" (Ruben and Bennett 1987). Acellular bone has a plesiomorphic character in vertebrates found in both primitive craniates and vertebrate lineages (Ørvig 1965, 1989) including teleost fish (Moss 1960, 1961, 1962). In a group of fossil marine jawless vertebrates which lived from 470 to 370 Ma ago and known as pteraspidomorphs, "aspidin is present both in the form of bone of attachment associated with the superficial dentine-enameloid tubercles, and comprises the whole of the underlying middle 'spongy' and basal 'lamellar' layers of the dermoskeleton" (Donoghue 2002; Donoghue et al. 2006).

Mapping acellular bone on the teleost phylogeny suggests an increasing trend toward acellularity, with the superorder Percomorpha containing a little more than 85 % of known acellular bony fishes (Kranenbarg et al. 2005). The multiple origins of acellularity within teleosts (Meunier et al. 2004) indicate a possible selective advantage of this type of bone. However, there is no consensus yet concerning the possible functional role of acellular bone, as the factors that have been investigated such as environment, activity level and gross morphology do not predict the presence of acellularity (Moss and Freilich 1963; Moss 1965). Therefore, the adaptive significance or selective pressures that lead to the repeated evolution of acellular bone in the teleosts remain unclear.

Recently, Horton and Summers (2009) carried out series of comparative tests concerning material properties (elastic modulus) of acellular and cellular bone. They used a three-point bending method to test the hypothesis that the material stiffness of cellular bone is lower than that of acellular bone. These researchers suggested that material properties were a selective pressure in the evolution of acellular bone as specialized skeletal material. The acellular ribs of great sculpin (*Myoxocephalus polyacanthocephalus*) were used in these experiments. It was reported that "contrary to their expectations, acellular bone was not stiffer by virtue of fewer lacunae but instead falls at the very low end of the range of stiffness seen in cellular bone. There remains the possibility that other properties (e.g. fatigue resistance, toughness) are higher in acellular bone" (Horton and Summers 2009).

Thus, according to Donoghue et al. 2006 "it appears that cellular bone evolved from an acellular bone, and acellularity has arisen secondarily in a number of instances, often through distinct developmental pathways". Is it true that especially the collagenous matrix plays the crucial role in bone development? Intriguingly, the alternative pathway has been reported (Meunier and Huysseune 1992), in galeaspids and placoderms, where it was shown that bone can develop through spheritic mineralization in the absence of a collagenous matrix.

Dermal Bone Nowadays, the most diverse collections of dermal elements are found in reptiles like crocodylians. However, in early representatives of many stem gnathostomes (structural-grade ostracoderms) especially the dermal skeleton was once the predominant skeletal system. During the evolution, it has undergone extensive reduction and modification visible in most modern forms today (Sire and Huysseune 2003). As reviewed by Vickaryous and Hall (2008), "for tetrapods, the most obvious remnants include the craniofacial skeleton (dermatocranium or desmocranium), dental tissues, and one or more elements of the pectoral apparatus. The dermal skeleton also includes bones developing within the eye (scleral ossicles), eyelid (palpebral), integument (osteoderms), and across the abdomen (gastralia)".

Additional type of bones, which arise directly through differentiation of mesenchyme, initially compacted in sheets or membranes, are known as *intramembranous bones*. The process of *intramembranous bone formation* on example of craniofacial skeleton is reviewed by Helms and Schneider (2003).Abzhanov et al. (2007) proposed following classification of the intramembranous bones:

(α) "the sesamoid bones, which form in tendons as a result of mechanical stress (such as the patella in the tendon of the quadriceps femoris);
(β) the periosteal bones, which form from connective tissue and add to the thickness of long bones;
(χ) the dermal bones, which form within the dermis of the skin".

Dermal bones tend to be flat and plate-like. Dermal bones of the skull and the pectoral girdle develop within the integument, generally in the lower layer of the dermis (Castanet et al. 2003). These ossifications are frequently penetrated by numerous canals that carried blood vessels and nerves (including the lateral line system of fishes and non-amniote basal tetrapods) to the external bone surface and into the directly overlying integument (see for review Witzmann 2009). As reviewed by Janis et al. (2012), heavy dermal cover of some of early tetrapods possessed various types of dermal mineralized elements like pits and ridges. Interestingly, blood vessels in these animals run through the superficial integument to the epidermis because dermal bone does not block interaction between it and the dermis. This kind of vascularization within dermal sculpture seems to be crucial from physiological point of view. Thus, according to Janis and co-workers (2012), various hypotheses for the presence of sculptured dermal bones in early tetrapods have been proposed:

- "cutaneous respiration;
- protection from desiccation;
- strengthening adaptation;
- mechanical protection of the soft-tissue dermis including vessels, nerves, and thermoregulation".

Furthermore, as recently proposed, the highly vascularized sculptured dermal bones of early tetrapods, "functioned to buffer the respiratory acidosis that would have resulted from an increased duration on land. These animals likely would have lacked adequate means for CO_2 elimination, such as the capacity to achieve the high ventilation rates. This is made possible by costal aspiration, the ability to lose significant amounts of CO_2 via the skin, or the kidney function necessary to increase blood HCO_3^- concentrations to levels required to fully compensate for respiratory acidosis" (see for discussion Janis et al. 2012).

Formation of dermal bone in zebrafish has been observed after amputation in the following manner (Quint et al. 2002):

"After amputation, epithelial cells migrate from the stump to cover the wound region, beneath which a blastema containing undifferentiated proliferative mesenchymal cells forms. Scleroblasts then differentiate within the blastema at the epithelial mesenchymal interface and begin to secrete the matrix that will form the new dermal bone".

Interestingly, specific proteoglycans play important role in development of fish dermal bones. Recent studies report about the isolation of a novel *lectican* gene from zebrafish termed *dermacan* (Kang et al. 2004). Lecticans are representatives of proteoglycans that are located in matrices of many tissues. Their functions are related to modulating of the activities of extracellular signalling molecules, to differentiation of vertebrae, as well as in maintaining the structural integrity of corresponding tissues. The experimental results suggest that in zebrafish dermacan can play a key role in differentiation and morphogenesis of dermal bones in the head skeleton, especially for the opercle, the branchiostegal ray, and dentary (Kang et al. 2004).

Perichondral Bone *Perichondrum* is a thin layer of dense cells growing within outer surface of any cartilage. The braincases of numerous primitive placoderms represent well ossified structures with layers of perichondral bone. It is suggested that perichondrun in the early placoderms, was the only part of the internal skeleton that could ossify. Interestingly, "the unornamented inner side of the placoderm eye capsule is actually two paper-thin layers of perichondral bone, enclosing a space originally filled with cartilage" (Young 2008).

Perichondrium, taken from the cartilaginous part of a rib, can develop into normal hyaline cartilage being placed in a joint (Skoog and Johanason 1976; Engkvist and Ohlsen 1979). Woo et al. (1987) demonstrated that this newly-formed tissue had the hyaline cartilage-like visco-elastic properties. Amiel et al. (1988), one year after perichondral grafting, found *neocartilage* with biochemical and histological features similar to those of normal articular cartilage.

Endochondral Bone Mineralized tissue known as *endochondral bone* is unique because it begins life as cartilage, which serves as template for endochondral bone development. In paper by Zustin et al. (2010), see also Blumer et al. (2008) we can find that "Endochondral bone formation is a challenging process in that the originally completely avascular cartilaginous anlage becomes highly vascularized and is eventually replaced by bone and the marrow cavity". Chondrocytes, osteoblasts and osteoclasts are the main players in formation of the skeleton by endochondral ossification (Karsenty and Wagner 2002). At the same time, "cartilage morphogenesis is a key rate-limiting step in bone development" (Reddi 2000a, b). Furthermore, endochondral bone development is a very complex process that includes a multistep, sequential, developmental cascade with activities of micro RNAs (Nakamura et al. 2011), bone morphogenetic proteins (Retting et al. 2009), endocrine/paracrine factors (e.g., PTHrP, 1,25(OH)(2)D(3), IGF-1, FGFs, and prolactin), as well as requiring the presence of collagens and dietary calcium, silizium and vitamin D (see for review Wongdee et al. 2012). The origin of the calcium phosphate phases and biochemical, structural, and physico-chemical mechanisms in their formation of endochondral bone (Glimcher 2006) as well as the special role of polyphosphates (Omelon et al. 2009) are still intriguing research topics in understanding the principles of endochondral bone biomineralization.

The endochondral ossification can also be accepted as the remodeling of cartilage templates. As briefly described by Scotti et al. (2010):

"this process relies on the specialized morphoregulatory functions of hypertrophic chondrocytes. Hypertrophic chondrocytes derive from the condensation of mesenchymal precursors and produce a type X collagen-rich avascular cartilaginous matrix. At the periphery of this cartilage tissue, the so-called 'borderline' hypertrophic chondrocytes instruct surrounding mesenchymal cells to differentiate into osteoblasts, which results in the formation of a 'bony collar.' In parallel, chondrocytes in the central regions direct mineralization of the hypertrophic cartilage by initiating remodeling via the production of specific matrix metalloproteinases (MMP), and attract blood vessels by releasing vascular-endothelial growth factor (VEGF). The in-growing blood vessels deliver osteoblastic, osteoclastic, and hematopoietic precursors, which mediate resorption of the cartilaginous template and formation of vascularized bone containing the so-called stromal sinusoids, which provide the microenvironment for hematopoiesis".

However, what about the location where endochondral ossification usually occurs? This place is to find in the expanding growth plate, a "dynamic region of the young skeleton located beneath the soft articular cartilage that caps the ends of the growing long bones, and above the mineralized bone itself. These bones grow along their vertical axis through the progressive expansion of the growth plate. Within the active growth plates, the bone elongates as new cartilage forms on its ends" (Omelon et al. 2009). According to Brighton (1994), "older cartilage beneath that new-formed cartilage mineralizes with apatite; it is then resorbed by bone-resorbing cells (osteoclasts) that remove both calcified cartilage and mineralized bone. Finally, osteoblasts build new bone to replace the resorbed calcified cartilage. This is one process that increases the size of the skeleton," (Omelon et al. 2009). Endochondral bones are usually three dimensional.

Variations in the densities and composition of the bone of marine vertebrates may be related to adaptations for buoyancy or locomotion, or to habitat or phylogeny (see for review Tont et al. 1977). The diversity of bones within different clades of marine vertebrates is huge, however, and thus here I wish to concentrate special attention on whale bones as examples of the biggest animal bone known.

3.1.1 Whale Bone: Size, Chemistry and Material Properties

Cetaceans (whales, dolphins and porpoises), particularly the Mysticetes (baleen whales) include some of the largest mammals in the world (Marx 2010). One baleen whale—the blue whale, *Balaenoptera* can reach 30–37 m long, have a girth of 14 m and weigh on average between 90 and 181 metric tons (see for review Cook 1973; Perrin et al. 2002). Correspondingly, the largest bone in the world belongs to this animal. For example, the blue whale's powerful jawbone measures an average of 2.5 m in length and weighs up to 550 kg. Traditionally, bone is seen to exhibit seven levels of structural hierarchy, with striking self-similarity near the bottom levels of hierarchy (see for review Zhang et al. 2011). However, I suggest that in the case of giant whale bones we can speak about an additional the mega-level of structural hierarchy a bone too.

From a materials science perspective, the chemistry, composition, nanostructure, material properties and biomechanics of diverse whale bones are intriguing. However, studies on these bones are a challenging task due to the difficulties of obtaining statistically relevant samples for comparative studies. Below, I would like to discuss several properties, which are very specific for the structure-function relationship in whale bones. Included are such phenomena as hypermineralization, porosity, and the presence of oil.

Ziphiidae, or beaked whales are representatives of odontocetes (toothed whales). Their feeding behavior is mostly teuthophagous. Some species of ziphiids, with dive records at more than 1,800 m, are considered to be deep divers (Tyack et al. 2006). These cetaceans arrive depths at which prey are detected in near-complete darkness by echolocation. They distinguished from other odontocetes by the anatomy of the rostrum that frequently displays extensive changes in the thickness, shape and density of its constituent bones. For example, "the rostral part of the composite calvarium bones of an adult male of the ziphiid species, *Mesoplodon densirostris*, yielded among the highest values for density (2.6 g/cm^3), mineralization (86.7 %), and compactness (99 %) yet reported," (Zioupos et al. 1997).

Probably, the extremely high mineralization rate and physical density of the rostrum (2.612–2.686 g/cm^3) (De Buffénil and Casinos 1995; Zylberberg et al. 1998) in *M. densirostris*, is determined by a drastic reduction of its collagenous fibrillar network. According to Zylberberg et al. (1998), "Mechanical testing of this bone has shown it to be the stiffest (Young's modulus up to 49.6 GPa) and hardest yet examined, as might be expected from its high mineral content," (Zylberberg et al. 1998).

In these hypermineralized secondary osteons, following features has been reported:

- "lamellae were not observed;
- nor mineralized fibrils;
- the mineral rod fragments are composed of stacked assemblies of crystal plates elongated along their c axes and parallel to the lengths of the rods;
- the mineral rods can clearly accommodate a much higher density of crystals than mineralized fibrils commonly found in other bones;

- in the rostrum, these rods appear to be quite closely packed within a relatively sparse framework of thin collagen fibrils;
- the low collagen content results in a low elastic component in the rostral bone, which would account for its great stiffness and hardness, and its low bending strength," (Zylberberg et al. 1998).

For bone crack propagation in another marine mammals (e.g. in manatees, which have all cortical bone, no trabecular and no marrow cavities) see Clifton et al. (2008).

In contrast to turkey leg tendon, where calcium phosphate crystal growth occurring along, but not inside, the fibrils (Traub et al. 1992), in the rostrum of ziphiids, crystal nucleation may occur near the surface of collagen fibrils. Zylberberg et al. (1998) proposed alternative hypothesis where "the rostrum mineral nucleation and growth may have occurred in heavily mineralized fibrils, as in other bones, but with most of the collagen being later removed from the mature mineralized fibrils. This leaves only a loose framework of thin collagen around dense mineral rods," (Zylberberg et al. 1998) (Fig. 3.1).

However, what is the functional importance of this specialized compact rostrum of extant and extinct ziphiids? As proposed by Lambert et al. (2011), "the most convincing hypotheses can be classified into three categories, related to three main functional and ecological features:

- acoustics,
- deep diving,
- and intraspecific fights between males," (Lambert et al. 2011).

In odontocetes, the sound-producing organs in the forehead include air sacs surrounding the phonic lips. These formations may act as acoustic mirrors for echolocation and communication sounds. It is suggested that the production of echolocation sounds occurs in the forehead, however, the transmission is made through the fatty lump of tissue termed as *melon*. There are no doubts that in deep divers like *M. densirostris* and *Ziphius cavirostris* which "have been recorded at depths up to 1,251 and 1,888 m, with dive durations reaching 57 and 85 min, respectively," (Lambert et al. 2011), especially role will play the hydrostatic pressure. Definitively, it strongly reduces the volume of air spaces within animal. It was suggested that "density disparity between forehead tissues and highly compact bone could therefore constitute an alternative acoustic reflector," (Lambert et al. 2011).

In her doctoral thesis (Campbell-Malone 2007) and subsequent collaborative work, Campbell-Malone determined the physical and material properties of the right whale jawbone and used those as the foundation of the first biomechanical models of vessel-whale collisions (Tsukrov et al. 2009) (Fig. 3.2). Unfortunately, one of the most critically endangered whales in the world is the *Eubalaena glacialis*, also known as North Atlantic right whale (Right Whale Consortium 2005; see also Kraus and Rolland (2007)). The deaths statistics resulting from entanglement in fishing gear and vessel-whale collisions, accounted for 27 (67.5 %) of the 40 right whales examined postmortem between 1970 and December 2006. Of those deaths, 21 (52.5 %) were attributed to vessel-whale collisions and at least 9 (22.5 %) of those resulted from blunt contact with the hull of a vessel.

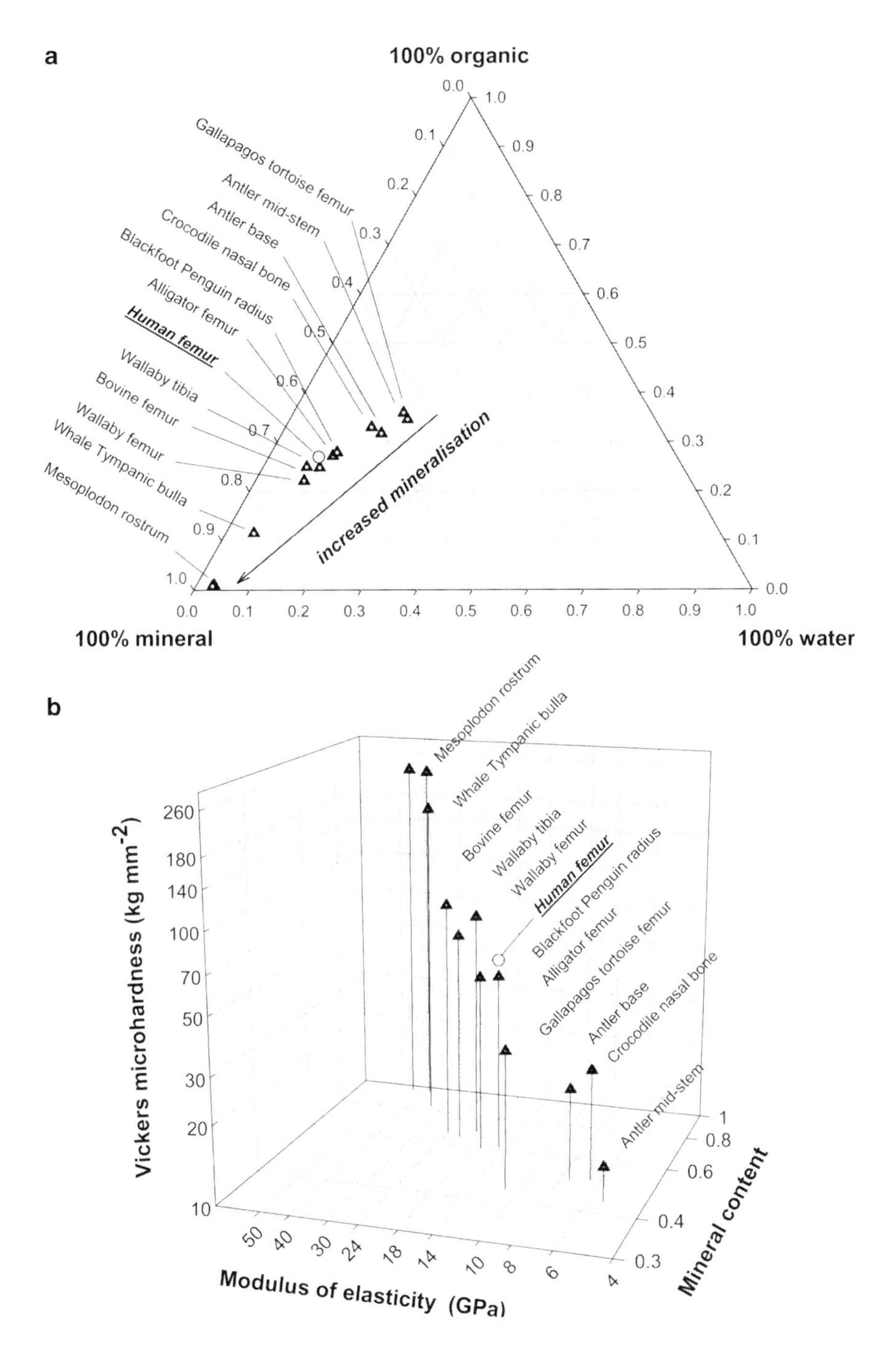

Fig. 3.1 Mineral content, macroscopic Young's modulus and micro-Vickers hardness in selected bone specimens (Adapted from Zioupos 2005). (**a**) The diagram represents the interaction between water, organic matrix and mineral content in dependence of mineral content. (**b**) High mineralisation correlates with increase in the macroscopic modulus of elasticity and the hardness of bone tissue (Zioupos 2005)

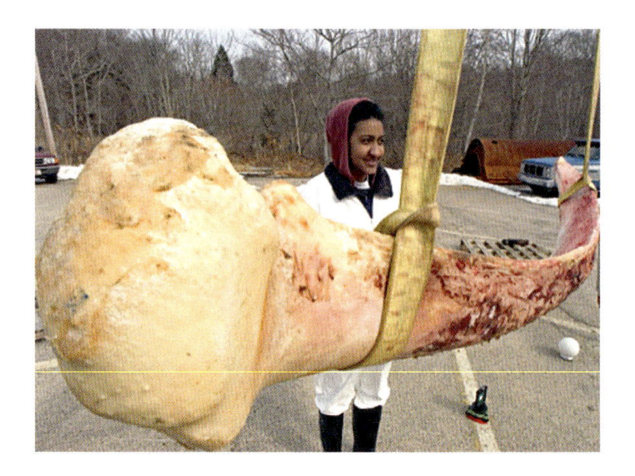

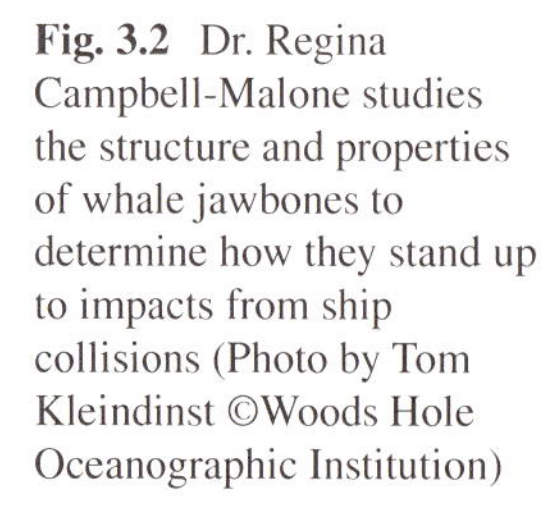

Fig. 3.2 Dr. Regina Campbell-Malone studies the structure and properties of whale jawbones to determine how they stand up to impacts from ship collisions (Photo by Tom Kleindinst ©Woods Hole Oceanographic Institution)

Mechanical testing revealed the anisotropic nature of right whale trabecular bone. The results of mechanical testing of trabecular bone indicated that the rostro-caudal axis is 3–4 times stiffer and 4–5 times stronger than the bucco-lingual and dorso-ventral axes (Campbell-Malone et al. 2008). The Young's modulus for trabecular bone from the right whale mandible was strikingly similar to values from the human mandible (!) (374.3 MPa and 373 MPa respectively). This finding indicates that the structural organization of right whale mandibular bone tissue plays an important role in determining the stiffness and strength of this bone tissue (Campbell-Malone 2007).

Whale bones are surprisingly porous (Fig. 3.3) and light, except for the dense ribs and caudal vertebrae. In paleonthological remains, these "bones can be distinguished from the other osseous materials by a cancellous, osteoporotic-like structure with irregularly distributed rounded porosities with diameters reaching up to 500 μm" (Reiche et al. 2011). The porosity of whale bones is also an issue that influences the bones' mechanical behavior and their subsequent degradation. Only the mandibles and upper limb bones of whales are beam-like, load-bearing structures; and it is only these elements that contain substantial amounts of compact bone. The vertebrae and ribs are composed largely of spongy or trabecular bone with a high porosity. Additional microscale computed tomography of right whale cortical bone in this transition zone revealed it to have the porous structure of trabecular bone (albeit lower porosity) and an average μCT apparent density of 0.501 g/cm^3 (±0.0312). Individual trabeculae were extremely large, on the order of 1 mm in thickness and plate-like columns of trabecular bone were also visible (Campbell-Malone 2007). The greater porosity of whale bones means that they have a high storage capacity of oil while the animal is alive, and that they retain a high proportion of that oil even after the bones have been macerated to remove the soft tissues.

Significant differences in the composition of bones from different parts of the whale skeleton have been reported (for review see Higgs et al. 2011a). For example, even vertebrae show 30–40 % differences in lipid content between different parts of the spine however they possess similar structure. Also different cetaceans species

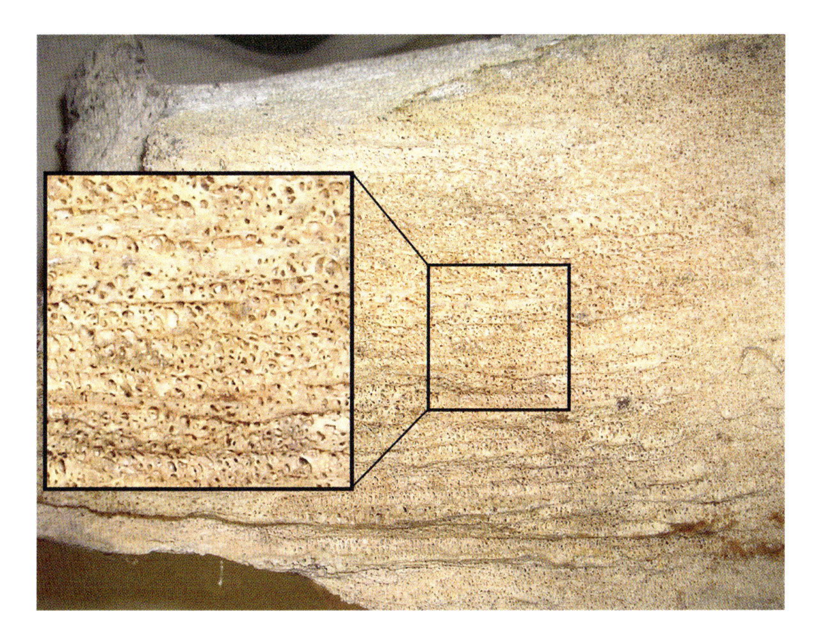

Fig. 3.3 Whale bone possess 1 mm large porous structure (Image courtesy Andre Ehrlich)

distinguish from each other because of the composition of their bones. Here, the lipid contents in the lower jaw of selected whales: 30 % in the rorquals, 20 % in the grey whale and 7.2 % in the odontocete sperm whale. We can note an increase in lipid content towards the rear of the skull from the tip of the upper jaw in the species listed above. Interestingly, about 84 % oil content was measured by Feltmann et al. (1948) for the mandible of a blue whale. It was reported (Zylberberg et al. 1998) that the rostrum bones of toothed whale *M. densirostris* do contain large amounts of lipids. Probably, lipids play their own role in the mineralization process and may contribute to the specific mechanical properties of mineralized tissues of whales.

Whale bones such as ribs with a vertical orientation tend to be more soaked with oil in the lower extremities. A seemingly counterintuitive observation is that the spinous processes of some vertebrae, which stand vertically above the vertebral bodies, and the upper margins of the scapulae are often oily at the uppermost extremities. This presumably reflects the pore size distribution of the spongy bone tissues where oil is drawn into the smallest available pores by capillary forces that can overcome gravity. This wicking of oil into different parts of the bone architecture can be seen in several skeletal elements. Furthermore, oil can be seen to have wicked up into the thick layers of dust lying on the upper surfaces of vertebrae (Fig. 3.4) (Turner-Walker 2012). Thus, it is not surprisingly that in Inuit culture, whale bones were burned directly as well as used for fuel by cutting them to liberate the oil (see Heizer 1963).

The composition of whale oils varies considerably between species but like all oils derived from marine animals it is characterized by a high proportion of

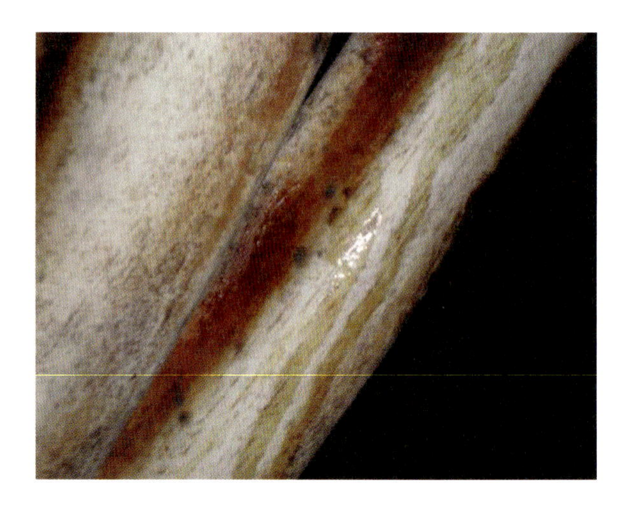

Fig. 3.4 Oil that has run down the surfaces of two minke whale rib (Image courtesy of Gordon Turner-Walker, The University Museum of Bergen – Natural History Collections (Turner-Walker 2012))

triacylglycerols and esters of unsaturated fatty acids—i.e. the fatty acids containing one or more C=C double bonds.

The oil in whale bones plays an important ecological role, especially for deep-sea fauna because the long-lasting skeletons of cetaceans serve as a primary source of lipids for these unique communities. For example, the sipunculan worms *Phascolosoma saprophagicum* use the whale bones directly for food (Gibbs 1987). Some polychaete worms like *Osedax* sp. destroy the bone matrix to obtain nutrition, aided by endosymbiotic heterotrophic bacteria (Rouse et al. 2004; Higgs et al. 2011b). Thus, whale skeletal carcasses may have played significant roles in maintaining deep-sea diversity and facilitating adaptive radiations of highly specialized fauna on the ocean bottom.

3.1.2 Whale Bone Hause

Intriguingly, the whale bone plays very special, but important role in human architecture, culture and engineering from ancient time. Petillon (2008) recently reported some artefacts related to Paleolithic people in Western Europe, which confirm the technical exploitation of whale bone for producing weapons and tools. As summarized by Betts (2007):

"Bowhead bone, in particular, was used as a raw material in the manufacture of essential items, such as architectural superstructures, construction (i.e., digging) implements, transportation components, and hunting equipment. Bone was also used extensively in the production of domestic utensils and figured prominently in the manufacture of ritual paraphernalia," (Betts 2007).

Whale bone houses are found in association with a number of prehistoric and historic cultures throughout much of the coastal regions of eastern Siberia, Alaska, Arctic Canada and Greenland. In the case of the Canadian Arctic, archaeologists

Fig. 3.5 Computer reconstruction of the Thule whalebone house showing the ridgepole design (© 2006 IEEE. Reprinted, with permission, from Levy and Dawson 2006)

have investigated the construction of these houses, as well as how social relations influence their design (Savelle 1997; Patton and Savelle 2006). The bones of baleen whales were used by the Thule culture (AD 1100–1400). For the Thule inhabiting the whaling regions of the central and eastern Canadian Arctic, wood and other building materials were typically very rare, thus entire dwellings tended to be constructed primarily of whale bone. Approximately 10,500 bones, representing almost 1,000 animals, were counted on the shores of six adjacent islands between 1975 and 1985 (McCartney and Savelle 1985) (Fig. 3.5).

The strength and flexibility of whale bones makes them an excellent construction material, however, due to their variable shape and size, they require expert craftsmanship and careful planning. The scarcity of good building materials in the Arctic meant that abandoned sites were likely to be used as a material source for others, with the result that sites rarely exhibit what they once were, even in their remnants.

From the architectural perspective, especially the skeletal fragments with very low meat utility values like mandibles, crania, maxillae/premaxillae and cervical vertebrae, can be accepted as the highest ranked engineering elements have. In contrast, such parts as the hyoid and caudal vertebrae have the highest meat utility values, an, correspondingly, very low architectural utility. As reviewed in papers by Savelle (1997), Savelle and McCartney (2003), and Savelle and Habu (2004), "the results suggest that architectural utility, as opposed to meat utility, was the primary determinant in the production of Thule whale bone assemblages in essentially all contexts, including initial whale processing," (Savelle 1997).

3.1.3 Conclusion

According to the classical definition (Turner 2006), "bone is a multiphase material made up of a tough collagenous matrix intermingled with rigid mineral crystals. The mineral gives bone its stiffness. Without sufficient mineralization, bones will plastically deform under load. Collagen provides toughness to bone making it less brittle so that it better resists fracture," (Turner 2006). Activities of individual genes, evolutionary processes like "fish-to-tetrapod" transition, or physiological loading during life, were factors that determined corresponding changes in whole-bone morphology. The brief analysis of marine vertebrate bones represented above, shows that the complexity of bone's properties arises from the complexity in its hierarchical structure and chemical composition. Unfortunately, acellular and dermal bones are still poorly investigated from a biomaterials point of view. Most attention has been payed to endochondral bone that differs from the other connective tissues because of its greater stiffness and strength, compared to its weight. These properties follow from it being a heterogeneous and anisotropic composite biomaterial (Katz and Bronzino 2000).

The marine vertebrate bones give us numerous tasks related to understanding structure-function relationships, which have been developed over the long course of animal evolution. For example, "de Bufférnil and Casinos (1995) discovered the densest known bone in the rostrum of the ziphiid whale *M. densirostris*, whose skeleton is otherwise greatly lightened," (Taylor 2000). It was suggested (Taylor 2000) that this dense bone provides "sound tracts that are in some way involved with the transmission or reception of echolocation pulses, perhaps at frequencies which males exploit more than females," (Taylor 2000). However, probably, similar to the sexually dimorphic sound-producing crests of hadrosaurian dinosaurs, this phenomenon is related to some special aspect of communication by males within whale's communities. All together, I have no doubts that different types of bones in marine fish, reptilian, birds and mammals hold immense promise for contributing to advances in biological materials science, bionics and biomimetics.

3.2 Teeth

Abstract The ancestor of recent marine vertebrate teeth was a denticle, the tooth-like structure, on the outer body surface of jawless fishes. This chapter covers the diversity of the tooth-like structures (odontodes, keratinized, rostral, pharyngeal, extra-oral and extra-mandibular teeth) and oral "true teeth". Their shape and forms (folded teeth and hypermineralized tooth plates), material properties, as well as a broad variety of biological materials (enameloid, enamel, dentine, petrodentine, isopedine, plicidentine, durodentine, keratine etc.) they are made of in marine vertebrates both within extinct and extant taxa are discussed. Special attention is paid to the structure and material properties of shark and whale teeth, as well as to the unique narwhal and walrus tusks.

Tooth is defined as "*a mineralized hard tissue unit consisting of attachment bony basal pad, dentine or similar dentinous tissue. Sometimes there is a superficial layer of enamel/enameloid formed from a single papilla present only in the oropharyngeal*

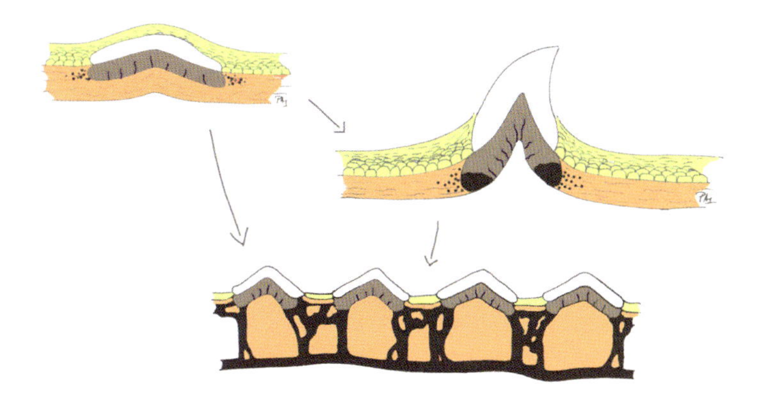

Fig. 3.6 The origin of bone: "precipitation of hydroxyapatite around the basal membrane of the skin gave rise to enamel- and dentine like tissues that formed odontodes, which became the progenitors of teeth and scales. Spread of mineralization deeper in the dermis formed shields consisting of acellular—and later cellular—bone," (Obradovic-Wagner and Aspenberg 2011, copyright © 2011, Informa Healthcare. Reproduced with permission of Informa Healthcare)

cavity, with a distinctive patterned oropharyngeal distribution with associated/connected replacements developing in advance of their requirement," (Fraser et al. 2010). However, so called tooth-like structures *odontodes* have been proposed as the examples of the earliest mineralized structures in the vertebrate lineage. Their origin is still long-debated topic. Probably, odontodes emerged first in the throats of conodonts, the jawless, eel-like creatures with a notochord. From other point of view, they arose rather as dental-like structures in the dermis, arranged closely together to form a protective shield (Fig. 3.6). We cannot exclude that protection from predation and predation was a driving force for their development (see for review Donoghue 2002; Fraser et al. 2010; Obradovic-Wagner and Aspenberg 2011).

According to Fraser et al. (2009), "during evolution, teeth originated deep in the pharynx of ancient and extinct jawless fishes. Later, with the evolution of bony fish, teeth appeared in the mouth, as in most current vertebrates, although some living fishes retain teeth in the posterior pharynx," (Fraser et al. 2009).

The question about the origin of teeth as example of dermal mineralization, or vice versa, is still open. Recent data (reviewed by Fraser et al. 2010) show that teeth arose from epithelium of endoderm, ectoderm, and even a combination of the two, "when properly combined with neural crest derived ectomesenchyme," (Fraser et al. 2010). According to proposed definition:

"*The neural crest is a migratory multipotent embryonic progenitor cell population that emerges from the dorsal neural tube to invade diverse regions of the embryo, giving rise to numerous derivatives in vertebrates including neurons, glia, pigment cells, bone, cartilage and dentine,*" (Fraser et al. 2010).

It is suggested that both denticles (the forerunners of bony skin plates) and early teeth appear to be the product of the same genetic mechanism, regulating epithelial/mesenchyme interactions and able to produce similar structures at different locations. The histological and expression data reported by Debiais-Thibaud et al. (2011) strongly suggest that the same set of *Dlx* genes is responsible to control the development of dermal denticles and teeth from the same developmental platform. Teeth and dermal denticles should to be homologous structures which were developed at several body locations, however through the initiation of a common gene regulatory network. Alternatively, Fraser and Smith (2011) suggested recently "*that the patterning mechanism observed for the oral and pharyngeal dentition is unique to the vertebrate oro-pharynx and independent of the skin system. Therefore, a co-option of a successional iterative pattern occurred in evolution not from the skin, but from mechanisms existing in the oro-pharynx of now extinct agnathans,*" (Fraser and Smith 2011).

In contrast to "true" teeth, denticles are mineralized structures, located on the dermal surface and within the oro-pharyngeal cavity of vertebrates like Elasmobranchii. They consist of dentine and attachment bone. In some cases denticles are covered with superficial layer of enameloid from a single papilla origin (Fraser et al. 2010).

Three types of teeth, based on their origin, have been proposed: jaw, mouth and pharyngeal (Koussoulakou et al. 2009). Diversity of the teeth-like structures (e.g. pharyngeal denticles) and "true teeth", their shape and forms, mechanisms of biomineralization, as well as the broad variety of biological materials (enameloid, enamel, dentine, petrodentine, isopedine, plicidentine, durodentine, keratine etc.) they are made of in marine vertebrates both within extinct and still existing taxa, is amazing and represents a goldmine of interest and bioinspiration for material scientists.

3.2.1 Tooth-Like Structures

Odontodes The term odontode, first introduced by Ørvig (1967) and comprehensively reviewed (Ørvig 1977), includes denticles and all hard tissue units corresponding closely to teeth in their development and structure (Fig. 3.7). According to modern

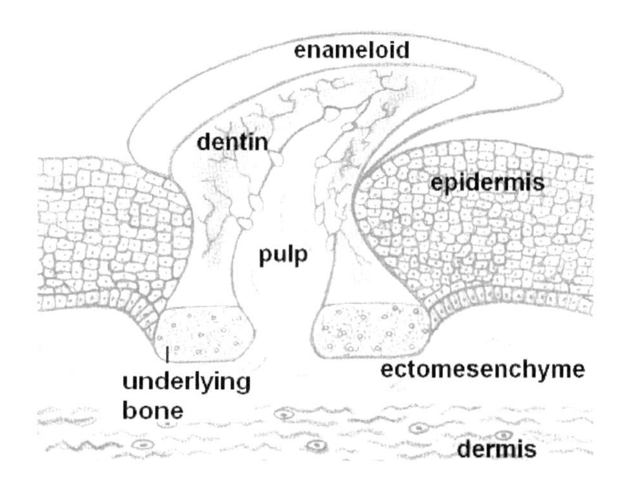

Fig. 3.7 Hypermineralized tissue like enamel or enameloid covered the odontodes, which morphologically looked like the placoid scales of recent sharks (Adapted from Koussoulakou et al. (2009) with permission)

definitions, odontodes are "*all structures that comprise a mineralized hard tissue unit consisting of attachment bony basal pad, dentine or similar dentinous tissue. Occasionally there is a superficial layer of enamel/enameloid formed from a single papilla; odontodes include both teeth and denticles,*" (Fraser et al. 2010). Principal question concerning the appearance and development of odontodes during early vertebrate evolution is logically rooted in the classic problem of anatomical homology (Ørvig 1977). There are numerous locations around the body of lower vertebrates, where odontodes can be observed. For example, covering the dermal surface as in extant rays and sharks, as lining the oropharyngeal cavity associated with gill arches, or dentition in oral and pharyngeal locations (Sire and Huysseune 2003). The perceived homology of all odontodes is suggested using palaeontological data, comparing developmental and structural similarities (see for review Fraser et al. 2010).

Location, morphology and function of odontodes even within one organism can be diverse. For example, the neotropical Andean astroblepid catfish possess two types of odontodes (Fig. 3.8). Thus, Schaefer and Buitrago-Suárez (2002) reported that:

"*Both odontode types in astroblepids conform in structure to dermal teeth of gnathostomes in having dentine surrounding a central pulp cavity covered by a superficial layer of enameloid, but differ from one another in terms of attachment and association with other epidermis features. Type I odontodes are larger (40–50 μm base diameter), generally conical and sharply pointed, occur on the fin rays, and are associated with dermal bone. Type I odontodes attach to an elevated pediment of dermal bone in the fin lepidotrich, and to dermal bone generally in loricarioids fish, via a ring of connective tissue. Type II odontodes are smaller (15–20 μm base diameter) and blunt, occur in the skin of the head, maxillary barbels, nasal flap, and lip margins, and are not associated with dermal bone,*" (Schaefer and Buitrago-Suárez 2002).

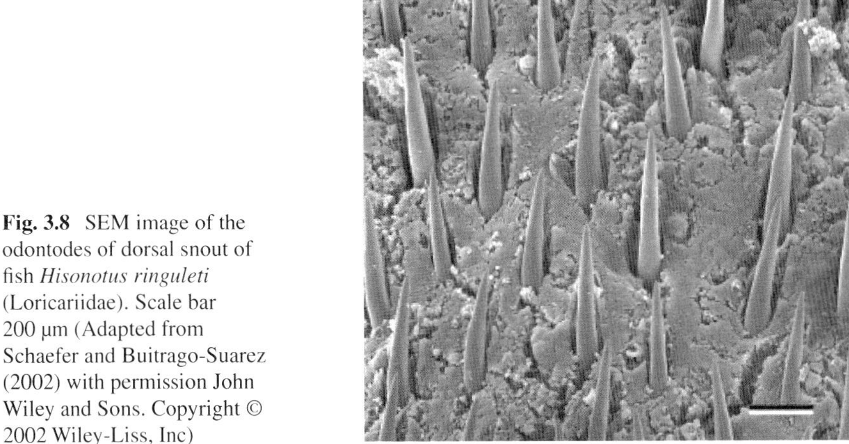

Fig. 3.8 SEM image of the odontodes of dorsal snout of fish *Hisonotus ringuleti* (Loricariidae). Scale bar 200 μm (Adapted from Schaefer and Buitrago-Suarez (2002) with permission John Wiley and Sons. Copyright © 2002 Wiley-Liss, Inc)

Investigations using histological techniques and SEM show that Type II odontodes are associated with special epithelial structures to form a putative mechanosensory organ. It was observed that the odontode base is localized deep in the fish dermis. It was suggested that the "*presence of odontodes on the fin spines and in the skin of the head, lips, and barbels may enhance skin surface adhesion in torrential currents,*" (Schaefer and Buitrago-Suárez 2002).

3.2.2 Keratinized Teeth

Lampreys are jawless, armorless fish without paired appendages. However, their mouth is rigid, conical, and lined with keratinized teeth-like structures. The sucker with 11 or 12 rows of these teeth, arranged in concentric circles and enclosed by an oral hood, is the most distinguishing external features of the adult sea lamprey (*Petromyzon marinus).* Teeth are cyclically replaced by the formation of new teeth from the germinal epithelium, so that in a random sampling, 1 or 2 generations of teeth may be observed (Alibardi and Segalla 2011). So called buccal funnel or suctorial disk is the characteristic teeth-hood arrangement of the animal. The lamprey fastens itself to various substrates and objects using this sucker. The buccal funnel is surrounded by tiny finger-like papillae. Just posterior (interiorly) to the mouth is the buccal cavity. At the back of the buccal cavity is a rasping tongue, used for rasping the flesh of a host. The tongue can be extended forward into the buccal funnel to scrape flesh. A valve called the velum separates the digestive tract from the respiratory pathway. This prevents food from entering the gills, but keeps water flowing over the gills even while feeding (Iwasaki 2002). The tips of the teeth are markedly sharp, and curves inward in the mouth.

Here, I take the liberty to insert the description of the fine structure of the horny teeth of the lamprey, *Entosphenus japonicas*, made by Uehara et al. (1983) as following:

"*Most of the horny teeth consisted of two horny and two nonhorny layers. The primary horny layer was coated with keratin, and the cells were closely packed and intensely interdigitated, joined together by many modified desmosomes. The plasma membrane of the horny cell, unlike the membranes of other vertebrates, was not thickened. The intercellular spaces were filled with electron-dense material. Microridges were seen on the free surface. Structures resembling microridges were found on the underside of the primary horny layer. The secondary horny layer displayed various stages of keratinization. The keratinization started at the apex and developed toward the base. In the early stage of keratinization, the superficial cells became cylindrical and were arranged in a row, forming a dome-shaped line. Their nuclei were situated in the basal part of the cells. The appearance of the nonhorny layers varied according to the degree of keratinization of the horny layers beneath them. The nonhorny cells were joined together by many desmosomes and possessed many tonofilament bundles,*" (Uehara et al. 1983).

Recently, Alibardi and Segalla (2011) reported that the ultrastructural features of the horny teeth of freshwater lamprey *Lenthenteron zanandreai* are reminiscent of similar aspects present in the lamprey *E. japonicus* (Uehara et al. 1983). Electron-dense bundles of keratin but no keratohyaline-like granules accumulated in the cytoplasm of transitional cells that were incorporated in the dense stratum corneum of the *L. zanandreai* tooth. Mature corneocytes were delimited by a cell corneous envelope and formed corneous microridges on the tooth surface. Although the increase in the electron density of the corneous layer suggested the presence of sulfur, the low to absent reaction for sulfhydryl groups indicated that cysteine was largely oxidized to form disulphide bonds in the corneous material of the teeth. A 2-dimensional electrophoretic analysis of the corneous material from the horny teeth showed the presence of acidic proteins, most likely keratins of 45–66 kDa (Alibardi and Segalla 2011). Based on the size, it is likely that acidic and basic non-keratin proteins of 16–20 kDa were also present in the oral mucosa, generally in higher amounts than keratins. This suggests that the low-molecular-weight basic proteins are likely associated with acidic keratins to produce the dense corneous material of the tooth, a process that also occurs in hard skin derivatives of other vertebrates like amphibians (Alibardi 2010a, b).

Thus, the representatives of living agnatha possess, epithelially derived keratin-containing teeth, rather than the epidermal and dermally derived highly mineralized "true teeth" of most other toothed vertebrates. Interestingly, it was possible to culture the teeth from adult lampreys *in vitro* as reported by Langille and Hall (1988). These authors reported that "*teeth cultured in L15-supplemented media for up to 14 days at either 15 or 20 °C retained their structural and cellular integrity as observed histologically, with no apparent cell outgrowth,*" (Langille and Hall 1988). It seems to be necessary to repeat these experiments using modern tissue engineering technologies. In particular the influence of calcium, phosphate and carbonate sources on development of these non-mineralized hard tissue structures should be studied.

3.2.3 Rostral Teeth

Another and very speculative example of "*non true teeth*" is rostral teeth of some sawfishes, which are actually a type of ray. While modern sawfish are related to the Pristidae family, derived from a Greek word meaning "a sawyer or saw"; an ancient sawfish family (Sclerorhynchidae) lived in primitive seas and have long since gone extinct. Members of both families have the specific long, flattened, blade-like toothed rostrum, "a flattened head and trunk, and a shark-like appearance and manner of swimming," (Wiley et al. 2008). The rostral teeth of the Pristidae are implanted in socklets along the margin of a rostrum. This phenomenon—gomphosis (a type of immovable articulation, as of a tooth inserted into its bony socket)—is one of the rarer methods of attachment in pisces. These teeth are reputed to be held in the socklets by cementum, and they grow from persistent pulps (Bradford 1957). It was suggested (Miller 1974) that the rostral teeth do not arise in an alternate way as do most tooth structures, and may indeed not be true teeth.

Between 18 and 23 equally spaced pair of teeth expanding at the base up from the rostrum is the characteristic feature of modern *Pristis microdon* (Seitz 2011). It is suggested that the rostral teeth at the juvenile stage of fish development are not fully erupted and even covered in a sheath of tissue so as not to injure the female. Sawfishes use their specialized rostrum in order to obtain benthic invertebrates and small fishes in sandy/muddy bottoms. The feeding behaviour is well described as follow:

"*They move it in a slashing gesture from side-to-side when attaching/stunning schools of fish, as well as for extracting animals such as molluscs and small crustacean from the benthic sediments*" (Allen 1982).

In pristids, the teeth along the saw are not replaced if they are lost. However, evidence suggests that at least one species of extant saw fishes may have rotated out their rostral teeth. This is *Schizorhiza* ("split root") a fossil genus that include *Schizorhiza stromeri* as a single species. Its fossil remnants are dated as 71 and 65.5 MYR (Kirkland and Aguillón Martínez 2002). This animal is unique among all "saw-snouted" Elasmobranchii because of replacement mechanism for their teeth based on a continuous serrated cover of tooth enamel localized on the rostrum's edge. These teeth with the size about 1–2 cm tall and 4–8 mm wide had a small triangular or rhomboidal shape at the tip, with sharp cutting edges, and a long forked peduncle. The new teeth developed inside this fork. The animal also possess the oral teeth, which were, however, "very small (about 1.5–2.5 mm high and 1–2 mm wide), with a large and recurved central point and keels at the side that formed tiny secondary points" (Kirkland and Aguillón Martínez 2002).

There are different opinions with regards to origin and types of rostral teeth dentine. The first form of dentine was probably orthodentine (Miller 1974). However, Shellis and Berkovitz (1980) reported that that "the osteodentine has a unique arrangement of matrix collagen fibres in recent rostral teeth. The dentine is composed of thick bundles of closely packed, mineralized collagen fibres orientated parallel with the long axis of the tooth," (Shellis and Berkovitz (1980)). Unlike the rostral teeth of modern pristides, the rostral teeth of the cretaceous *Sclerorhynchidae* species have a cup with smooth enameloid (see also Sect. 4.1 in this work).

To my best knowledge, we have no information about dentition mechanisms in rostral teeth of saw fishes. However, the finding of sawfish rostral teeth with pathological deformities represented in Fig. 3.9. demonstrate the existence of very complex biomineralization phenomenon.

3.2.4 Pharyngeal Denticles and Teeth

Pharyngeal denticles were reported for extant placoderms (Johanson and Smith 2005). In contrast to the group of dentate Chondrichthyes, Acanthodii and Osteichthyes, the fossil placoderms lacks a marginal dentition. It was suggested that "during evolution, pharyngeal denticles and teeth are independently derived

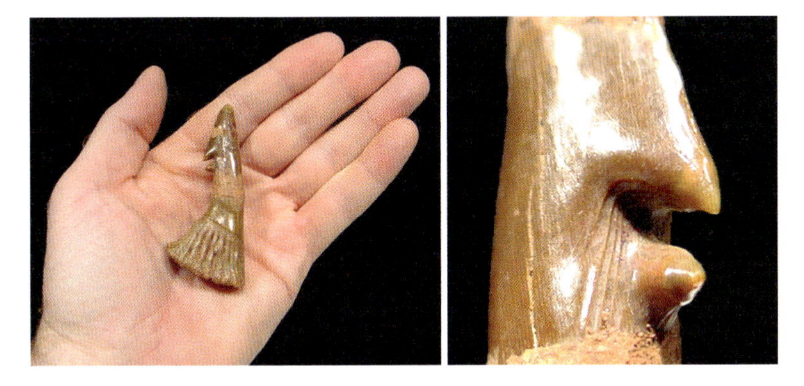

Fig. 3.9 This is a double-barbed rostral tooth from the sawshark *Onchopristis numidus* (Upper Cretaceous, Tegana Formation – Kem Kem, Morocco) (Adapted from PaleoDirect.com (http://www.paleodirect.com/) with permission)

relative to external skin tubercles, and that neither pharyngeal denticles, nor teeth, derive from the latter during the evolution of jawed fishes," (Johanson and Smith 2003; see also Johanson and Smith 2005).

Teeth are found in two main locations in vertebrates: the oral cavity and the pharynx. Most teleosts (bony fishes) have four pairs of gill arches that are just posterior to the bones of the mouth and oral cavity of the individual. Depending on the species, these arches can possess gill rakers that aid in the filtration of fine particles from the water for use as a food source; or for protection of the soft gill tissues. These rakers usually are attached to the epibranchials, packed structures that comprise a portion of the gill arches. In several families of fish, an interesting modification is seen in another bone present in the gill arch, the ceratobranchial element of the fifth arch. This arch is where the pharyngeal teeth are located in the pharynx, from which they get their name (Miller 1999). Pharyngeal teeth, for example, those found in teleost fish, develop from interactions between endoderm and mesenchyme. Evidence from the fossil record indicates that pharyngeal teeth were present in jawless vertebrates and therefore preceded oral teeth (Tucker and Sharpe 2004). Pharyngeal and oral teeth "share the same typical tooth structure: external mineralised layers (made of enamel or enameloid, and dentine) surrounding a central dental mesenchyme," (Debiais-Thibaud et al. 2007).

Cypriniformes are known as a highly diverse clade with more than 3,000 species. All representatives of this group possess "pharyngeal dentition attached to the fifth ceratobranchial, and do not develop oral teeth," (Pasco-Viel et al. 2010) (Fig. 3.10). The way pharyngeal teeth come together during the mastication process varies in different families. In some cyprinids, the teeth interlock laterally and crush food as it passes toward the alimentary canal. Other fishes such as cichlids have ventral pharyngeal plates that are used to crush food against a top plate inside the pharynx (Miller 1999).

Pharyngeal teeth of some fish species possess high biomimetic potential for material scientists. Those from Parrotfish *Scarus rivulatus* are a good example.

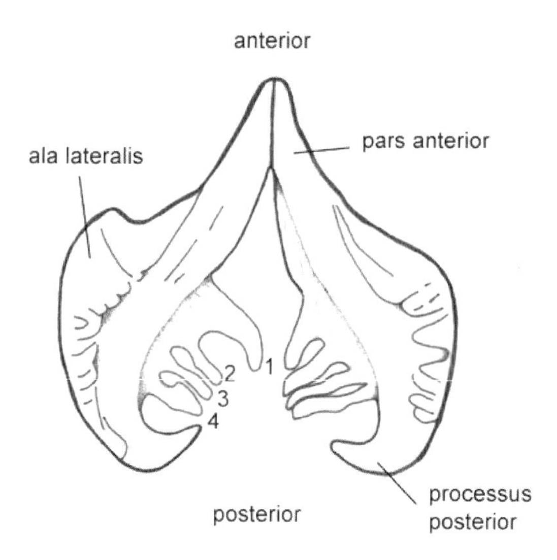

Fig. 3.10 Pharyngeal jaw with teeth (*Carassius carassius*) (Adapted from Rückert–Ülkümen and Yigitbas 2007, Copyright © 2007, TÜBİTAK. Reprinted with permission)

Parrotfish are common fish found in tropical and sub-tropical waters worldwide, and are generally associated with the coral reefs upon which they feed. Parrotfish have a specific diet of epilithic algae which grows on and within the dead coral substratum (Carr et al. 2006). To help feed on the algae, the parrotfish have five tooth bearing bones. Two anterior tooth bearing bones are used to graze on the algae or bite into the coral substratum. The resulting material passes to the back of the throat to the pharyngeal mill made up of two upper tooth bearing bones and one lower tooth bearing bone. The pharyngeal mill fractures the coral fragments and liberates nutrients from the algae. Some species of parrotfish can process up to 5 tonnes of coral per year per individual (Carr et al. 2006).

The mechanical properties of pharyngeal teeth of this fish investigated were hardness, fracture and deformation mechanisms, and wear characteristics. Hardness results show that the enameloid, the mineralised tissue that makes up the outer surface of the teeth, is of a similar hardness to human enamel. It has a microhardness of 315HV, well above the hardness of coral. The fracture and deformation of teeth showed that the enameloid crystallite bundles deformed by acting as units and slipping past other bundles. This enabled the deformation to remain localised so that large deformations could be absorbed and the damage minimised without reducing the structural integrity of the tooth. The principal methods used to minimise fracture of enameloid were microcracking, crack branching and crack deflection. The microstructure is ideally arranged to counter tensile forces and to resist abrasion in apical enameloid and basal enameloid (Carr et al. 2006).

Several mechanisms, working together, provide parrotfish with good mechanical properties including toughness, hardness and wear resistance. These mechanisms rely heavily on the microstructure which is ideally oriented to balance the competing requirements imposed on the teeth during mastication.

3.2.5 Extra-oral and Extra-mandibular Teeth

Fish organs that are expressed out of place are rare, but possible, if we would like to analyse, as example, the occurrence of dermal denticles outside the oral cavity in some species of living teleost fish. According to Sire (2001): "*Not only have teeth been expressed in extra-oral locations, but these new phenotypes have also been maintained during subsequent evolution, suggesting that these denticles most likely represent a selective advantage for their owners*," (Sire 2001).

Therefore, the phenomenon of existence of extraoral denticles can be accepted as important event with respect to their evolutionary, developmental, and functional implications (Sire and Allizard 2001).

Here, I would like to cite the description represented by Jean-Yves Sire:

"*In Atherion elymus, the denticles are particularly numerous on the whole front of the snout, the chin, and the undersides of the lower region of the head, where they are aligned forming a denticulate keel [Fig. 3.11]. Like teeth located in the oral*

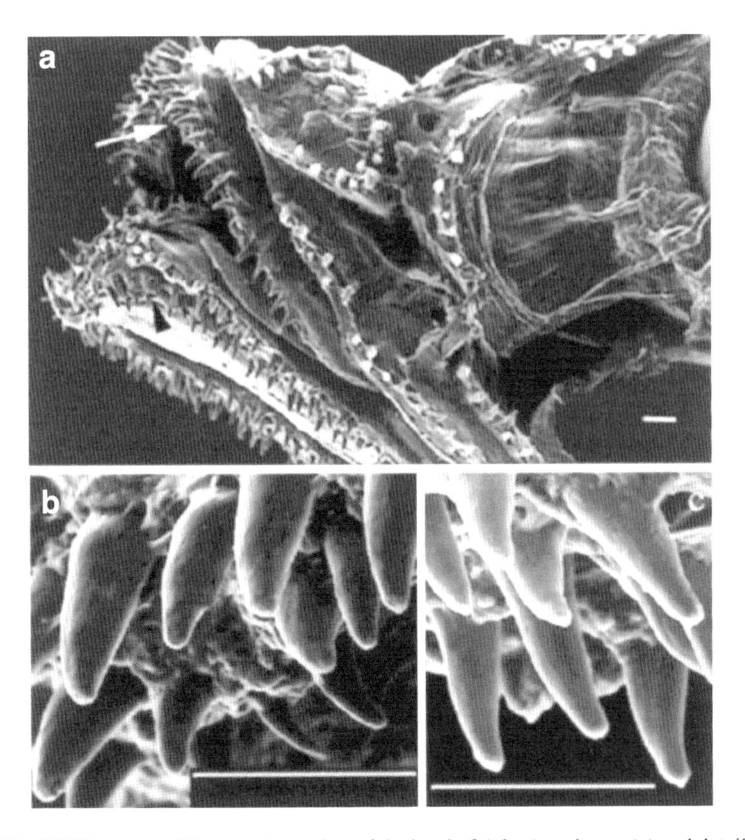

Fig. 3.11 SEM imagery of the anterior region of the head of *Atherion elymus* (**a**) and detail of the teeth, white arrow (**b**) and of the denticles on the chin, *black arrowhead* (**c**). Scale bars 100 μm (Adapted from Sire (2001) with permission John Wiley and Sons. Copyright © 2001 BLACKWELL SCIENCE, INC)

cavity and odontodes, these denticles consist of a pulp cavity surrounded by dentine, itself covered by enameloid. The denticles are held onto the bony surface by means of ligaments and attachment bone. These features make them identical to teeth and odontodes. Moreover, the denticles project beyond the skin, as do teeth, and are certainly replaced as indicated by the observation of several buds in serial sections," (Sire 2001).

From the structural and morphological point of view, the denticles in *A. elymus* are similar to those of teeth inside the oral cavity. According to Sire (2001) these denticles are not evolutionarily related to odontodes of early vertebrates. In spite of homology between odontodes and teeth, the origin of the denticles is not to be found in odontodes, but in teeth. "*The denticles are simply teeth that form outside the mouth, probably derived from a sub-population of odontogenically pre-specified neural crest cells. These 'accidental' extra-oral teeth have arisen independently in these lineages and were selectively advantageous in a hydrodynamic context,*" (Sire 2001). Thus, extra-oral teeth are denticle-like structures that are not organized into functional systems and cover only selected parts of the dermal skeleton.

Except for the extra-oral teeth described above, fish can possess tooth-sets outside their jaws. Such kind of extra-mandibular dentition has been very recently described by Finarelli and Coates (2012). They studied the earliest known holocephalans, *Chondrenchelys problematica*. As described by these authors, the species "possess *a set of individual teeth in conjunction with toothplates external to the mandibular arch and distinct from the mandibular toothplate dentition. The outer surfaces of these extra-mandibular teeth have a pin-cushion appearance, reflecting vertically oriented, closely spaced tubules emerging onto the outermost dentine surface. Chondrenchelys, like other holocephalans, shows no trace of a pharyngeal dentition; yet there is evidence for a dentition external to the mandibular arch, possibly supported on labial cartilages. This extant animal now provides evidence for the odontogenic potential of extra-mandibular neural crest cell populations,*" (Finarelli and Coates 2012).

3.2.6 Vertebrate Oral Teeth

Many osteichthyan fish possess teeth, which are instead regularly patterned and replaced. In these species, a dental lamina morphologically typical for tetrapods, is absent. However, according to the classical view, the development of a sub-epithelial, permanent dental lamina is the crucial step for the initiation of a pattern order for teeth at the mouth margin. Thus, "*classically the oral dentition of teeth regulated into a successional iterative order was thought to have evolved from the superficial skin denticles migrating into the mouth at the stage when jaws evolved,*" (Fraser and Smith 2011).

It was suggested generally that, an ectodermal invagination is the base element for development of the oral cavity of vertebrates. Correspondingly, "*oral teeth are proposed to arise exclusively from ectoderm, contributing to tooth enamel epithelium, and from neural crest derived mesenchyme, contributing to dentin and pulp,*" (Soukup et al. 2008). However, recently, very interesting results have been obtained in experiments with a combination of fate-mapping approaches using transgenic axolotls. Soukup and co-workers (2008) showed with strong evidence the existence of oral teeth derived from both the endoderm and ectoderm. Also the teeth with a mixed ecto/endodermal origin have been identified. It was proposed the dominant role for the neural crest mesenchyme over epithelia in axolotl's tooth initiation. Also, "*from an evolutionary point of view, that an essential factor in teeth evolution was the odontogenic capacity of neural crest cells, regardless of possible 'outside-in' or 'inside-out' influx of the epithelium*" (Soukup et al. 2008).

Multi-layered true teeth, which include also layers of dentine and enameloid, were developed by osteichthyan fish. (Please, note that detailed description of diversity and properties of enameloid, enamel, and dentin are represented in a separate Sect. 4.1 below!).

Because of the broad diversity of marine vertebrate oral teeth, I take the liberty to represent and discuss only teeth with mostly unusual structure (e.g. folded teeth) and material properties (hypermineralized teeth) in this chapter.

3.2.6.1 Folded Teeth

Among crossopterygians and lower tetrapods as well as in the ichthyosaurs, the mosasaurs, varanid-like reptiles, actinopterygian *Lepisosteus* and its closest relatives, and possibly the primitive diapsid *Champsosaurus* (Warren and Davey 1992; Warren and Turner 2006) there are groups that have teeth of peculiar internal structure, involving an infolding of the special type of dentine –orthodentine- of the pulp cavity wall. Such orthodentine is called "folded dentine" (plicidentine), and teeth possessing it are "folded teeth" (Bystrow 1938, 1939; Schultze 1969, 1970) (Fig. 3.12). The increase of the number of folds of dentine correlated with the age of the individual (Bystrow 1938). As reviewed by Maxwell, "*the appearance of dentine folding can be created when radial canals divide the dentine into lobes (e.g., Eurypodus), or dentine spicules of uneven length project into the pulp cavity (e.g., Mylobatis) in the absence of external folding, making the morphology of the pulp cavity alone misleading for predicting the presence of plicidentine,*" (Maxwell et al. 2011).

Are there some adaptive advantages of folded teeth in comparison to classical true teeth? Following properties are under discussion:

- "increased strength of the tooth base without a large increase in the amount of mineralized tissue;
- increased flexibility of the tooth base;
- increased surface area for attachment tissues," (Maxwell et al. 2011; see also Besmer 1947; Peyer 1968; Scanlon and Lee 2002).

May be plicidentine is the characteristic feature of the teeth observed in large kinetic-feeding predators (Scanlon and Lee 2002; Modesto and Reisz 2008).

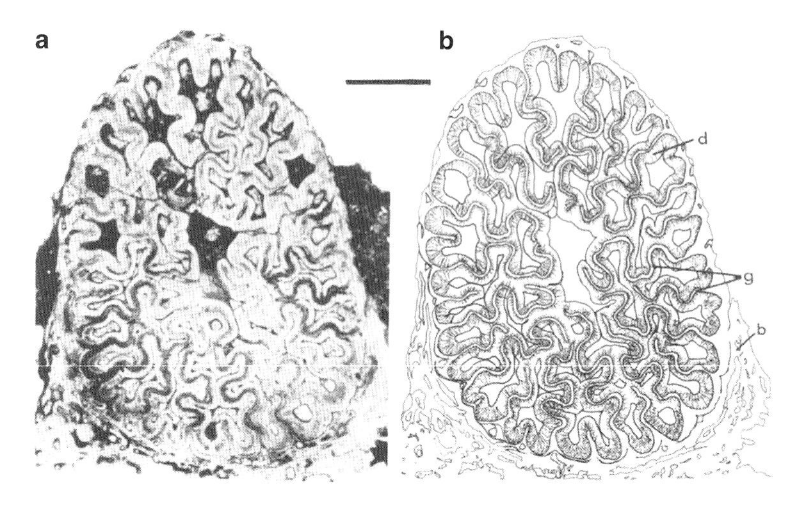

Fig. 3.12 Folded tooth of *Metoposaurus* sp. (**a**) Photograph of tooth cross section. (**b**) Drawing of tooth in (**a**) (Abbreviations: *b* bone, *g* globular zone, *d* dentine) (Adapted from Warren and Davey (1992), copyright © Association of Australian palaeontologist, reprinted by permission of Taylor & Francis Ltd, (www.tandfonline.com) on behalf of The Association of Australian Palaeontologist)

It seems that detailed analysis of biomechanical and material properties of these unusual teeth will give us more understanding into mechanism of this very special kind of biomineralization, as well as in the construction of similar artificial structures.

3.2.6.2 Hypermineralized Tooth Plates

Teeth, which are covered with a complex, hypermineralized tissue, have been described in the cochliodonts, chondrenchelyiform fishes and other Paleozoic chondrichthyans. This was termed as *orthotrabeculine* (Zangerl et al. 1993). According to Didier et al. (1994) "*in many of these archaic forms, the crowns of the teeth were extensively hypermineralized, and the bases of the teeth within a family showed a tendency to fuse,*" (Didier et al. 1994).

Also chimaeroid fishes (Chondrichthyes, Holocephali) possess hypermineralized tissues (Owen 1945) in the form of tooth plates, which grow continuously from the lingual, or posterior, margin throughout the life of the animal. Following pairs of tooth plates have been described in the chimaeroids (Didier et al. 1994):

- "a pair of large mandibular tooth plates in the lower jaw;
- two pairs of tooth plates in the roof of the mouth;
- the posterior palatine and anterior vomerine tooth plates," (Didier et al. 1994).

These structures look like "nipping" blades, some with hypermineralized rods reinforcing the cutting edges (Huber et al. 2008).

Trabecular dentine (Peyer 1968) is an example of hypermineralized tissue that was found in the tooth plates of the Australian ghostshark *Callorhinchus milii*. This specific biological material is overlain by a thin veneer of vitrodentine (Bargmann 1933) that is quickly worn away on exposed surfaces. Also tritoral hypermineralized tissue, termed as "tubular dentine" (Moy Thomas 1939), and "pleromin" (Ørvig 1976), have been described in the large mandibular and palatine tooth plates of some species. These occupy a central position and become exposed on the occlusal surfaces by wear.

Some hypermineralized tissue are located not on the surface of the vomerine tooth plates in chimeroids but deep within them in the form of a small region visible only in sectioned material. Additionally, specific structures containing the hypermineralized tritoral tissue appears as slender rods or "perlstrings" in the interior of the plates of extant chimaeroids like *Chimaera* (Didier et al. 1994).

Another example of hypermineralized tooth plates has been described in lungfish (Dipnoi) (see for review Denison 1974; Kemp 2003). Anne Kemp gives us following description:

"*The dentition develops from initially isolated cusps, arranged in a radiating pattern, into tooth plates based on radiating ridges, sheathed in enamel and containing several distinct forms of dentine and bone. Single cusps, made up of enamel and mantle dentine, grow towards the underlying bone and fuse with bone trabeculae to form a primary tooth plate, with four ridges in the upper jaw and three in the lower. The ridges grow by the addition of new cusps, and the tooth plate grows by the addition of new ridges, up to seven in all, but these processes are only partly responsible for growth of the tooth plate as a whole,*" (Kemp 2002).

Here, we can speak about formation of hierarchical biocomposite structure because with each "*new cusp, a layer of enamel and mantle dentine is added around the external surface of the tooth plate, and layers of additional dentines grow within the pulp cavity to increase the thickness of the tooth plate. These processes are balanced by wear of hard tissue from the occlusal surface of the tooth plate,*" (Kemp 2003).

The *petrodentine*, one of the principal constituents of the tooth plates of lungfish, have been investigated using contact microradiography as well as by light and electron microscopy as reported by Ishiyama and Teraki (1990). This biological material is a highly mineralized tissue noticeably analogous to that of enameloid, and is deposited intermittently in a proximal direction by the sole participation of mesenchymal petroblasts. The petroblasts appear very likely to have prominent biphasic functions of secreting the petrodentine matrix and of eliminating the matrix by resorption. The petrodentine thus may be capable of attaining hypermineralization with the participation mesenchymal cells alone. Mineral crystals constituting the petrodentine are large, hexagonal or similar, but have irregularly shaped columns. Initially these crystals appeared extremely thin, with curled ones frequently noted. The petrodentine of lungfish tooth plates is considered (Ishiyama and Teraki 1990) to be structurally distinct from the pleromin (cosmine) of holocephalian tooth plates because of the difference between their constituent crystals.

3.2.6.3 Shark Teeth

Studying the functional morphology, structure, chemistry and materials properties of unique and diverse shark teeth not only helps to understand the biological role that teeth play in feeding of extinct, extant and living species, but also stimulate progress in biological materials science, biomimetics and development of fish inspired robotics.

In most fish, the teeth are homodont: all the teeth are roughly of similar sizes and shapes. So called "dental lamina" is a structure producing teeth throughout the life of the Chondrichthyans and represent an example of one of the earliest innovations in sharks evolution. Tooth form is highly variable both within and between individuals of sharks. Wide variety of shapes of the extant shark teeth can be represented by teeth "*with triangular serrated cusps, oblique serrated and non-serrated cusps, notched serrated cusps, non-serrated recurved cusps, multicusped teeth, and flattened tooth pavements,*" (Whitenack 2008). For example, the holocephalans and brady-odont (cochliodont) sharks, which feed by crushing hard-shelled molluscs and crustaceans, possess flat and pavement-like teeth, reinforced by tubes of pleuromic dentine. There are numerous papers on tooth terminology and variation in sharks with regard to tooth location and design (see for review Gudger 1937; Applegate 1965; Zangerl 1981; Compagno 1984a, b, 1988; Cappetta 1986; Powlik 1995; Hubbell 1996; Ramsay and Wilga 2007). Thus, to briefly summarize Whitenack and Motta (2010), following types of teeth in sharks:

- "*clutching-type, small teeth, often with lateral cusplets;*
- *tearing-type teeth, which are considered to function best in puncture, have narrow, tall cusps and are usually not serrated.*
- *cutting-type, teeth whose crowns are lingo-labially flattened and widen towards the base;*
- *Molariform teeth, such as those found in the heterodontids and many batoids, fall into 'crushing-type' or 'grinding-type'*," (Whitenack and Motta 2010).

Shark tooth morphology determines their biological role (Whitenack and Motta 2010). It is established that "*teeth with different shape and/or size play different roles during prey catching and handling in elasmobranchs,*" (Lucifora et al. 2001; see also Applegate 1965). Here, some examples. The bonnethead shark *Sphyrna tiburo* (Carcharhiniformes) and the horn sharks *Heterodontus* spp. (Heterodontiformes) are species with heterodonty. Both species possess posterior teeth which are molariform and crush the hard food, however anterior teeth are sharp and efficient in grasping food (Tricas et al. 1997). The term Dental Insertion Angles (DIA), is widely used in elasmobranchs studies and represents an angle between the long axis of a jaw and the axis of a tooth. For example, a relationship between the DIA and the cutting capability of teeth has been reported by Frazzetta (1988) during biomechanical investigations of serrated and smooth-edged shark teeth. As summarized by Whitenack and Motta (2010):

"*Applying artificial blades and teeth from various shark species to compliant materials, the author observed that smooth, slender teeth, such as those of the mako shark Isurus oxyrinchus, appeared better suited for puncturing and piercing. Serrated teeth, such as those of the white shark Carcharodon carcharias, appeared to be more useful for slicing and cutting, although they are more susceptible to binding of the teeth in prey than the smooth, nonserrated teeth,*" (Whitenack and Motta 2010).

In earlier experiments (Abler 1992), it was shown that smooth teeth blades concentrate a large force on the cutting edge, creating high pressure that "crushes" the material beneath it, producing the cut.

Of course, the giant teeth of *Carcharodon megalodon* "as the largest shark to have ever lived," (Pimiento et al. 2010), attract our attention. It was calculated based on its tooth crown height (of about 168 mm) that the total length of this animal reached of more than 16 m (Gottfried et al. 1996) and the weight was between 48,000 and 103,000 kg as estimated by Wroe et al. (2008). Purdy (1996) suggested that this huge fish was an active predator of large whales. Furthermore, "this giant shark first immobilized its leviathan prey before feeding," (Wroe et al. 2008; see for more information Purdy 1996). It seems not to be surprisingly, because the estimated maximums bite force in *C. megalodon* (at 108,514–182,201 N) is truly extraordinary (Wroe et al. 2008). The characteristic features of *C. megalodon* teeth (Fig. 3.13) are listed by Pimiento and co-authors as follow:

Fig. 3.13 Example of the *C. megalodon* tooth (13 cm straight, 18 cm in diagonal) from the Miocene. (Source: a news release issued by the University of New South Wales)

Fig. 3.14 Reconstruction by Vito Bertucci (Late). This jaw likely represents the pinnacle of *C. megalodon* in size

"*large size, triangular shape, fine serrations on the cutting edges, a convex lingual face, a slightly convex to flat labial face, and a large v-shaped neck,*" (Pimiento et al. 2010).

These giant sharp blade-like constructs (Fig. 3.14) together with the bit force of the shark's jaw must be unique for marine vertebrate world. With the aim to test maximum bite force and to examine relationships among their three-dimensional geometry, material properties and function, Wroe et al. (2008) have digitally reconstructed the jaws of a white shark *C. carcharias*. The authors suggest that bite force in this predator may exceed ca. 1.8 tonnes, the highest known for any living species. Probably, these forces may have been an order of magnitude greater still in the *C. megalodon*.

The mechanism of cutting during unidirectional draw in sharks is described as follow:

"*As the tooth is moved across the prey during draw, the tip of the cusp can engage the tissue, creating compression directly under the cusp tip and further adding tension via more bulging,*" (Whitenack and Motta 2010; see also Frazzetta 1988). Numerous factors like the angle of the tooth apex with respect to the prey item, the shape and the position of the teeth within the jaw, the interaction between teeth of the upper and lower jaws are very important. There are no doubts that sharks teeth do not work alone and represent an example of structural and functional complex. Lisa Whitenack concentrates our attention on the following still unanswered questions:

"*How teeth of the upper and lower jaws shear past each other and how teeth of the same jaw affect puncture and cutting during draw?*" (Whitenack and Motta 2010).

These questions are challenging because numerous combinations of blade shapes can affect cutting efficiency (Anderson and LaBarbera 2008). Additionally, some aspects on the structural level must be taken into account. As discussed by Frazzetta (1988).

"*The flexible collagenous attachment of shark teeth may also facilitate cutting during draw as the teeth may pivot anteroposteriorly around obstructions preventing them from 'hanging up' on tough material. Tooth base overlap within the same jaw may transmit forces to linked teeth, as overlapping bases are lashed together with collagenous Sharpey's fibres,*" (Whitenack and Motta 2010).

In a recently published study, Whitenack and Motta (2010) investigated the puncture and draw performance of tearing-type, cutting-type, and cutting–clutching type of extant teeth from ten shark species in detail. Here, some selected results from this study:

- "*Differences in puncturing performance occurred among different prey items;*
- *the majority of teeth were able to puncture different prey items;*
- *differences in puncture performance occurred among tooth types;*
- *broader triangular teeth were less effective at puncturing than narrow-cusped teeth;*
- *no differences between the maximum draw forces and maximum puncture forces,*" (Whitenack and Motta 2010).

During feeding sharks teeth undergo stress, strain, and potentially failure as results of occasional extreme loads. Material properties of these teeth are also determined by their chemistry and the micro- and nanostructure of the corresponding components. From anatomical view, shark teeth contain two zones: the crown and the root or base. Shark teeth are typical examples of nanostructured biocomposites, "*with two distinct structural components: a central core of dentine covered by enameloid, an enamel-like substance formed from both odontoblasts and ameloblasts,*" (Whitenack 2008).

Because of the lack of detailed knowledge, there are no doubts that the biomechanics as well as structural mechanics of the tooth itself and its biological materials (enamel, dentine, cementum) must be studied using modern techniques and approaches. For example, as determined by Whitenack et al. (2010) "*the hardness of both osteodentine and orthodentine of sharks, are 125–181 % higher than the petrodentine of both lungfishes and 10–128 % higher than the dentine of mammals,*" (Whitenack et al. 2010). Note that lungfish, which lack enamel, have petrodentine, a specific hypermineralized dentine (Currey and Abeysekera 2003).

Performance testing of extant and extinct shark teeth, nanoindentation of shark teeth, finite element analysis of tooth morphology, and phylogenetically informed analyses of shark tooth morphology and ecology were employed to elucidate the relationship between performance, ecology, and evolution in recent pioneering studies by Whitenack and co-workers (Whitenack 2008; Whitenack et al. 2010, 2011). Finite element analysis (FEA) (Fig. 3.15) has been carried out to visualize stress distributions of extant and fossil shark teeth during holding, cutting and puncture. Following observations have been reported:

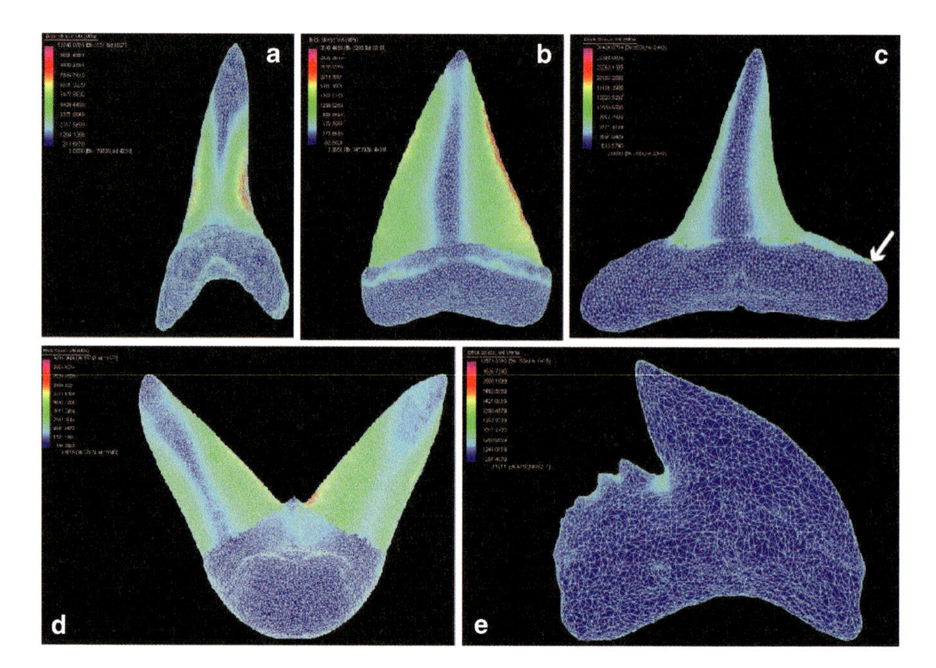

Fig. 3.15 Representative finite element models (FEMs) loaded in puncture. All views are of the labial side of the tooth of different shark species. *Arrow* indicates detine-enamel junction. (**a**) *Carcharhinus limbatus*, (**b**) *Carcharhinus leucas*, (**c**) *Sphyrna mokarran*, (**d, e**) *Hexanchus griseus* (Courtesy of Lisa B. Whitenack)

"*Teeth loaded in puncture have localized stress concentrations at the cusp apex that diminish rapidly away from the apex. When loaded in draw and holding, the majority of the teeth show stress concentrations consistent with mechanically sound cantilever beams. Notches result in stress concentration during draw and may serve as a weak point; however they are functionally important for cutting prey during lateral head shaking behaviour. As shark teeth are replaced regularly, it is proposed that the frequency of tooth replacement in sharks is driven by tooth wear, not tooth failure,*" (Whitenack et al. 2011).

3.2.6.4 Whale Teeth

In contrast to broad variety of placental mammals which possess teeth that are capped with enamel, some groups of marine mammals are toothless. For example, representatives of mysticetes like baleen whales have no teeth. Some odontocetes like pygmy and dwarf sperm whales have enamelless teeth. Fetal narwhals have six pairs of tooth buds, but four pairs disappear before maturity. It was established that narwhals are toothless, however these marine mammals "show vestigial traits indicating that their ancestors had teeth" (Nweeia et al. 2012). Thus, "all toothless and enamelless mammals are descended from ancestral forms that possessed teeth

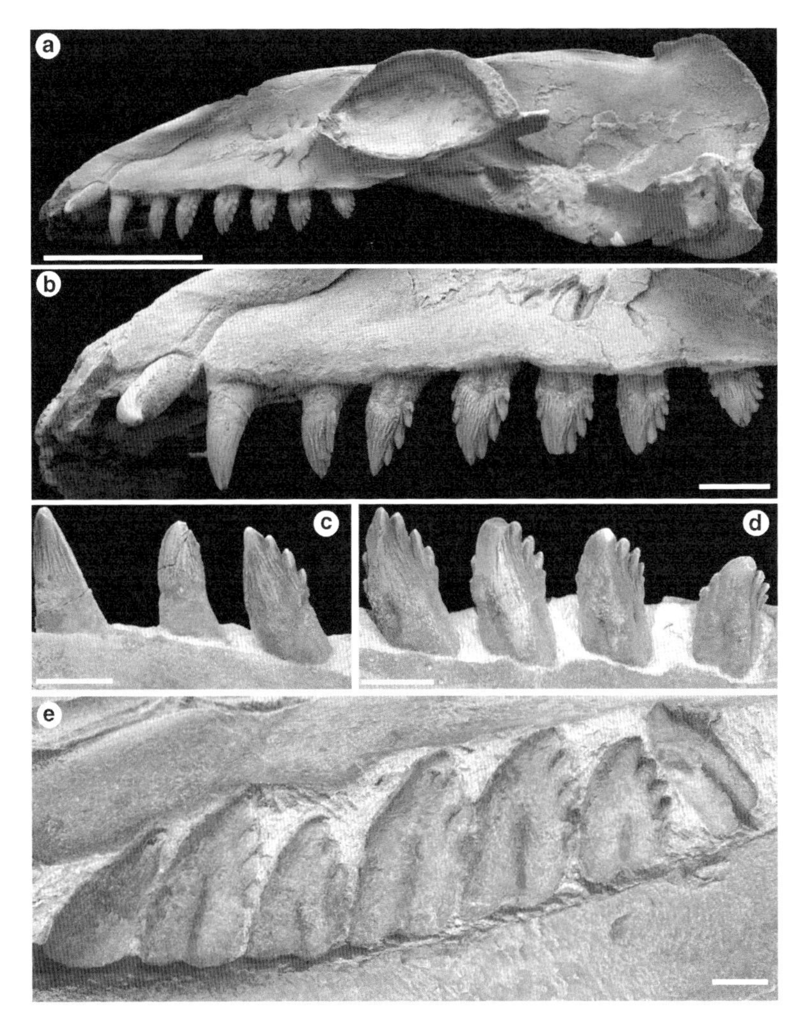

Fig. 3.16 Imagery of recently discovered ancient mystecete *Janjucetus hunderi* gen. et sp. nov. (NMV P216929). (**a**) Left lateral view and (**b**) left upper dentition in buccal view of the holotype skull. Scale bars: (**a**) 100 mm, (**b–e**) 20 mm (Reprinted from Fitzgerald (2006), by permission of the Royal Society)

with enamel," (Meredith et al. 2009; see also Uhen 2010). Recently, Fitzgerald (2006) reported about unique toothed mysticete *Janjucetus hunderi* from the Upper Oligocene Jan Juc Marl, Victoria, Australia that lacked derived adaptations for bulk filter-feeding. Probably, this species is more archaic than any previously described. Because of characteristic morphological features, it was suggested that this whale was a macrophagous predator (Fig. 3.16). It feeding diet included marine reptiles and the extant leopard seal (*Hydrurga leptonyx*). The discovery of this ancient mysticete is of great scientific importance especially for the evolution of filter-feeding. Thus, Fitzgerald suggested that:

"Mysticetes evidently radiated into a variety of disparate forms and feeding ecologies before the evolution of baleen or filterfeeding. The phylogenetic context of the new whale indicates that basal mysticetes were microphagous predators that did not employ filter-feeding or echolocation, and that the evolution of characters associated with bulk filter-feeding was gradual," (Fitzgerald 2006).

All odontoceti teeth are meant for holding prey, not ripping, tearing, or chewing. Odontocetes take individual prey. Their feeding diet contains of cephalopod mollusks like squids and fishes (see for review Rice 1989). World oceans are the habituate for orcas, or killer whales (Delphinidae family), which are known as top predator in all aquatic niches where they occur. These animals have extremely large, curved teeth and consume diverse forms of prey, ranging from small schooling fish to phocid, bearded seals, and cetaceans like beluga, narwhal, and even large baleen whales (Ferguson et al. 2012).

Sperm whale (*Physeter macrocephalus*) ivory is derived from one of the forty-eight curved conical teeth that are located in the lower jaw of the animal (see for review Boschma 1938). These cone-shaped teeth possess a nerve root that goes to the tip as well as a hollow end. The distinguishing characteristic is a clear line of transition between the dentine and cementum layers. Thus, Sperm whale ivory possesses characteristic structure (Fig. 3.17) and is readily identified when a cross-section of a tooth is examined. It has an oval outline and displays a bull's-eye pattern that is formed from a central star-shaped remnant of the pulp cavity. This is surrounded by a central mass of dark concentrically deposited lamellae of dentine, which is in turn covered by a thick peripheral layer of lighter coloured cementum (Espinoza and Mann 2000).

Sperm whale teeth can rich height of approximately 20 cm (see for review Espinoza and Mann 2000). Killer whale (*Orcinus orca*) possesses teeth which are smaller. Conical shape and a small amount of enamel at the teeth tips are common features for these species. Cementum is the biological material that covers the rest of the tooth. Cross sections of sperm whale and killer whale teeth show their rounded forms.

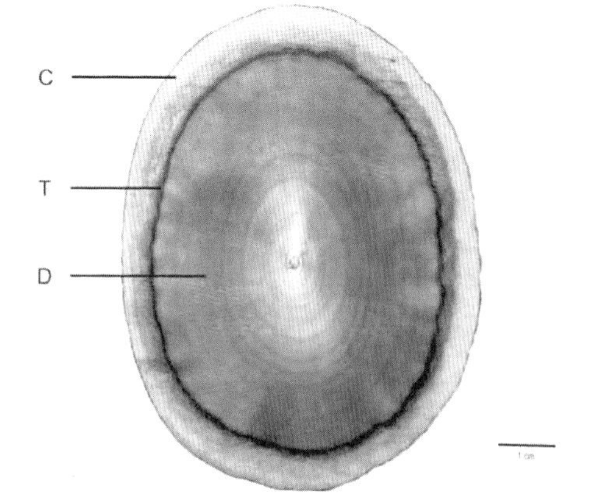

Fig. 3.17 Cementum (*C*), transition ring (*T*), and dentine (*D*) are well visible on the enlarged and enhanced photograph of a cross-section of a sperm whale tooth (Adapted from Espinoza and Mann (2000), reprinted with permission)

According to Espinoza and Mann (2000).

"*killer whale teeth show two slight peripheral indentations. The dentine is deposited in a progressive laminar fashion. As a result of this laminar deposition, killer and sperm whale teeth will show prominent concentric dentine rings in cross-section. Killer whale teeth may also display a faint rosette pattern in the dentine cross-section. The dentine is separated from the cementum by a clearly defined transition ring,*" (Espinoza and Mann 2000).

Rhythmically accreted growth structures ("growth layer groups" GLG) (Perrin and Myrick 1981), analogous to tree rings, are visible in longitudinal cross sections of the toothed whale teeth. As physeterid teeth grow continuously over ontogeny, these structures can be used to ascertain the age of the animal at death (see for review Scheffer and Myrick 1980; Evans and Robertson 2001; Amano et al. 2011). The method of age estimation of toothed cetaceans, based on counts of growth layer groups (GLGs,) of tissue in the teeth, was introduced 35 years ago (Nishiwaki and Yagi 1953). Since then GLG counting has become an important procedure when studying the age-related biology of odontocete populations. The dark and light GLG became visible as opaque layers by transmitted light, however, as translucent by reflected light. Because of the anisotropic features of the translucent layer, it was concluded (Scheffer and Myrick 1980) that it possesses crystalline structure. It was also observed that, the thickness of the newest layer tends to decrease in correlation with increasing age. Interestingly that the layer thickness in females is slightly lower than in males. Possibly, the GLG possess paleoclimatic and nutritional information too. The authors reported that:

"*The dark band layer seems to be accumulated as the result of a good nutritious condition. Partly from these measurements of seasonal progression in thickness of dentin and partly from study of marked whales, it was concluded that only one dark layer a year is deposited,*" (Scheffer and Myrick 1980).

Sperm whales are related to widely distributed cetaceans, which habituate diverse aquatic niches including tropical waters and areas near the polar ice (Rice 1989). Meat, bone and spermaceti oil were the reasons for hunting on these animals in tropical and subtropical waters especially in the 18th century. Partially pelagic sperm whaling had declined by the 1900s. However it started again after World War II. The destruction of the sperm whales population continued until the international moratorium on whaling went into effect in 1986 (Evans 1987). Thus, more than 500,000 specimens were killed.

Finally, I would like to insert here the critical view made by Pichler and co-authors (2001) as follow:

"*Sperm whaling was a dangerous occupation, as depicted in paintings, woodcuts, and engravings of the era. For both commercial whalers and the indigenous people who used stranded animals as a resource, sperm whale teeth were considered valuable trophies or status symbols. Among sailors, the intricate engraving of sperm whale teeth and other ivory developed into a distinctive art form known as scrimshaw. Today, large quantities of scrimshaw and carved indigenous artefacts are found in museums and private collections throughout the world. These collections represent an almost unprecedented historical population sample for a marine species.*

Access to museum collections of sperm whale artefacts may allow assessment of the impact of whaling by enabling examination of genetic diversity in historic (whaling-era) sperm whale populations," (Pichler et al. 2001).

3.2.6.5 Narwhal Tusk

Tusks and teeth differ in their functions, but are similar because of their common origin. Teeth are necessary for food mastication. Tusks have evolved from teeth and are examples of extremely large teeth projecting beyond the lips of the animal. These specialized structures show an evolutionary advantage in some species. Tuskbearing marine mammals include the narwhal and walrus. "*Large teeth and tusks are often used interchangeably to describe ivory,*" (Chen et al. 2008).

Monodon monoceros is the scientific name of narwhal species (Greek for "one tooth, one horn"—a clue to what its "horn" is) that is related to the Monodontidae family (Fig. 3.18). Other representative of this family is the beluga whale (*Delphinapterus leucas*) (see for review Reeves and Tracey 1980; Hay and Mansfied 1989).

According to Kristin Laidre, the "Narwhal FAQ" from the University of Washington (http://staff.washington.edu/klaidre/narwhalfaq.html):

"*these marine mammals typically dive to at least 800 m between 18 and 25 times per day every day for 6 months. Many of these dives go even deeper than 800 m: over half reach at least 1,500 m. Dives to these depths last around 25 min, including the time spent at the bottom and the transit down and back from the surface. In addition*

Fig. 3.18 The Narwhal (*Monodon monoceros*) painted by in Johann Christian Daniel von Schreiber (1774)

to making remarkably deep dives, narwhals also spend a large amount of their time below 800 m (>3 h per day). This is an incredible amount of time at a depth where the pressure can exceed 2200 PSI (150 atmospheres) and life exists in complete darkness. Finally, they can live up to 90 years in the wild, are the only whales that overwinter in the Arctic pack ice. There are about 80,000 of them worldwide," (Laidre, "Narwhal FAQ" published online; for more information Laidre and Heide-Jørgensen 2005, 2011; Laidre et al. 2003).

The very characteristic morphological feature of the male narwhal is its single 2–3 m long tusk, a canine tooth (Nweeia et al. 2012). This unique structure projects from the left side of the upper jaw and forms a left-handed helix (Kingsley and Ramsay 1988; Brear et al. 1990). The left-hand helical nature of tooth development has been hypothesized to be a functional adaptation to maintain the overall concentric center of mass during growth (Nweeia et al. 2009). This hypothesis certainly makes a great deal of sense when one considers the hydrodynamic loads that would develop if the tooth were curved or skewed to one side. The animal with a body length of 4–5 m, and weigh up to 10 kg, can possess tusk up to 3 m long (see for review Best 1981; Milius 2006). A female narwhal are characterized by a straighter and shorter tusk. Also a female with dual tusks have been observed, however it may be an exception. In males and females, the left tooth erupts at the end of the first year (Silverman and Dunbar 1980). The development of two tusks has been also reported (Clark 1871).

As recently described by Coyne (2012):

"Narwhals *have also vestigial teeth that lodge (horizontally) in the maxillary bone. Although they rarely erupt in some males, they don't erupt into the mouth; they lodge between the palatal tissue and underlying maxillary bone. These teeth are very small, from 1 to 30 mm long. They lie behind and to the left of the sockets for the two tusks (only one of which usually erupts),*" (Coyne 2012).

Numerous hypothesis have been proposed (Gervais 1873; Silverman and Dunbar 1980; Best 1981; Gerson and Hickie 1985; Nweeia et al. 2009, 2012) for better understanding of the purpose and function of the tusk. These suggestions include use as a secondary sexual characteristic in males an instrument for breaking ice, a spear for hunting, a ritualistic appendage in establishing male hierarchy, a weapon of aggression between males, a breathing organ, a swimming rudder, a thermal regulator, a tool for digging, and an acoustic organ or sound probe. Examination of tusk anatomy, histology, and biomechanics combined with traditional knowledge of Inuit elders and hunters has revealed features that support a new sensory hypothesis for tusk function (Nweeia et al. 2005).

The microanatomy of the tusk provided insight into potential function. SEM micrographs of the pulpal surface revealed tubule features that are similar in size and shape to the dentin tubules found in masticating teeth (see for details Nweeia et al. 2009). The tubule diameters are similar to those observed in human teeth, but the spacing of these tubules across the pulpal wall is three to five times wider than that seen in human dentin (see for comparison Zaslansky 2008). The polished cross sections show that these tubules run continuously throughout the entire thickness of dentin, just as they do in human dentin. A surprising finding, however, was the presence of tubule orifices on the outer surface of the cementum. This indicates that the dentin tubules communicate entirely through the wall of the tusk with the ocean environment (Nweeia et al. 2009). Thus, narwhal teeth have similar physiology to human teeth, having both pulpal neurons and dentin tubules (Nweeia et al. 2010). The most distinguishing difference is that the tubules in the narwhal tusk are not protected by an overlying layer of enamel. This raises the distinct possibility that the tusk could provide a variety of sensory capabilities. Any stimulus that would result in movement of fluid within these tubules could possibly elicit a response (Fig. 3.19).

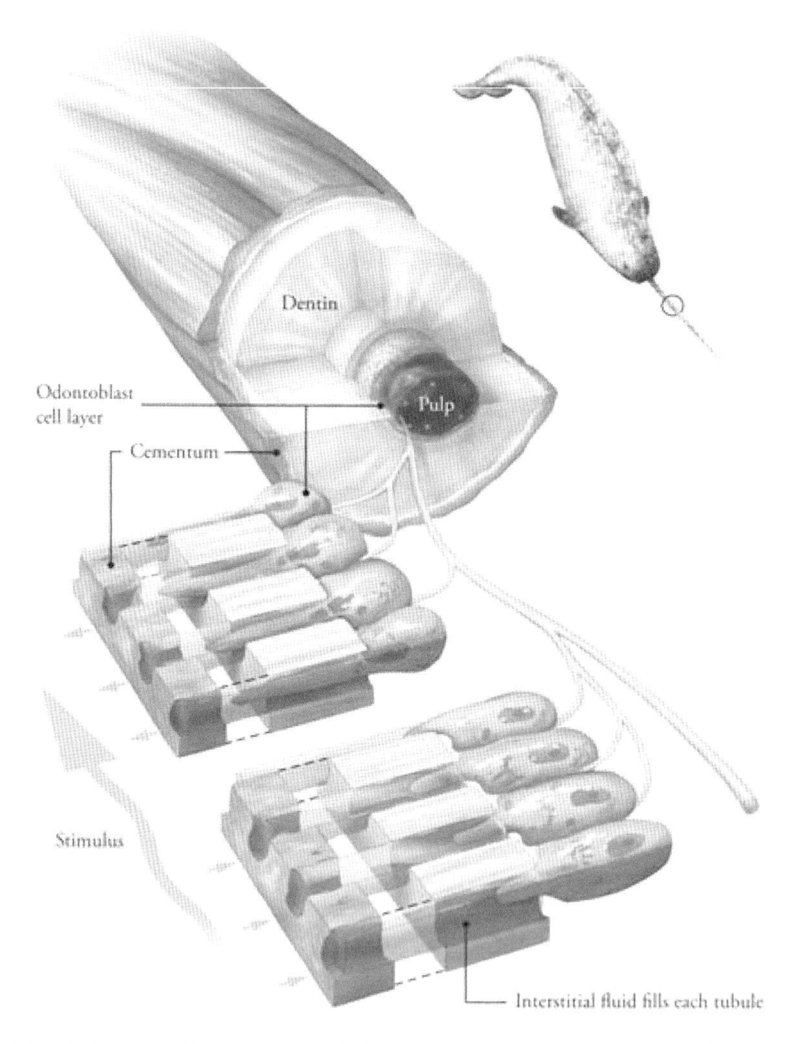

Fig. 3.19 Tusk of male *M. monoceros* as a hydrodynamic sensor (Adapted from Nweeia et al. 2009). The anatomic features of then narwhal tusk provide the potential for sensing stimuli that would result in movement of interstitial fluid within the dentin tubules. This fluid movement stimulates neurons located at cellular odontoblastic layer found at the base of each tubule (Reprinted from Nweeia et al. (2009), with permission of Smithsonian Institution)

These include ion gradients, such as water salinity, pressure gradients caused by dive depth or atmospheric pressure changes, air temperature and movement, or possibly other chemical stimuli specific to food sources or environment. Due to these results, the narwhal dentin is proposed to be a unique example of the naturally designed tissue with sensory abilities to detect, pressure, osmotic gradients and temperature. The role of sexual selection cannot be excluded because erupted tusks are usually associated with males (Nweeia et al. 2005, 2009).

Of the total radius of the narwhal tusk, the dentine occupies about 45 %, the cementum about 15 % and the pulp cavity about 40 % (Brear et al. 1990). Of course, the chemical content, size and morphology of narwhal tusk prompted numerous comparative studies on mechanical and material properties of this interesting biological material (Brear et al. 1990, 1993; Currey et al. 1994; Zioupos and Currey 1996; Nweeia et al. 2009). For example, Currey (2006) suggested three quite different mechanisms for toughening of the narwhal dentine:

- "*One makes use of the difference in properties of fully mineralized dentine and the hypomineralized, interglobular dentine.*
- *The second mechanism makes it easier for a crack to form along the fibrillar direction.*
- *A third, more classical mechanism makes use of the fact that dentine is arranged in layers, coaxial to the long axis of the tusk,*" (Currey 2006).

Microhardness and Young's Modulus measured by nanoindentation both appear to be correlated with the mineral-to-collagen ratios (MCR) in narwhal tooth tissues away from the tusk base. It was showed that these properties and the degree of correlation are location specific (Eidelman et al. 2005). They show tendency to decline in moving from the tip to the base. The modulus demonstrated a strong decline moving from pulp to the outer surface in the two cross sections away from the tip. The flexural strength and work of fracture both increased for dentin when comparing the tusk base to the midsection. The flexural strengths of 95 MPa at mid tusk and 165 MPa at the base compare to approximately 100 MPa for human dentin. These are adaptations for a tooth that must withstand high flexural stresses and deformation rather than the compressive loads of chewing (Nweeia et al. 2009).

As reviewed by Freeman et al. (1998), narwhals have been hunted in Greenland and eastern Canada for centuries, and may have brought the Greenlandic Inuit in close contact with the Norse in Greenland beginning in the tenth century. Narwhal ivory was bartered among Inuit long before European contact. Narwhal tusks were highly valued by European traders in the Middle Ages, who sold the tusks in Europe mislabeled as unicorn horn, sometimes for their weight in gold. The royal throne of Denmark, made in the fifteenth century, is made almost entirely of narwhal ivory. Numerous handmade articles made from narwhal tusks are still present in museums and collections worldwide (Fig. 3.20). However, Inuit in Greenland and Canada used the tusks to create durable and functional tools, especially harpoon fore shafts.

Fig. 3.20 This 96 cm- long walking stick is made from a highly unusual twisted narwhal tusk, probably in late eighteenth—early nineteenth century, in St. Petersburg, Russia. Belonged to the Princes Yusupov (Image is used from http://www.hermitagemuseum.org), Courtesy of The State Hermitage Museum, St. Petersburg, Russia)

3.2.6.6 Walrus Tusk

The two subspecies, the Atlantic walrus (*Odobenus. r. rosmarus*) and the Pacific walrus (*O. r. divergens*), are representatives of the Odobenidae family, which occur in geographically isolated populations. Male Pacific walruses weigh about 800–1,700 kg and are about 2.7–3.6 m long. Female Pacific walruses weigh about 400–1,250 kg and are about 2.3–3.1 m long. Specimens of *Odobenus. r. rosmarus* are slightly smaller: females weigh about 794 kg and reach lengths of 2.4 m, however, males weigh about 908 kg and reach lengths of 2.9 m (see for review von Baer 1837; Murie 1871; Schmidt 1885; Chapskii 1936; Fay 1955, 1982; Outridge et al. 2003; Wilson and Reeder 2005).

Most of these marine mammals possess 18 teeth. Their upper canine teeth are modified into long ivory tusks (Fig. 3.21). The skull does not only have to withstand blows when the tusks hit the substrate, but it also has to withstand the tractive force produced from the tusks when they act as levers. Fay (1955) reported that on average they grow in length of about 36 cm, but they may be up to 100 cm. The feeding diet of walruses includes mostly mollusks. Correspondingly their cheek teeth are specialized for crushing shells. These animals use their tusks for prying shellfish from the sea bottom as well as they can dive to a depth of 70 m to find the prey. However, two other important mechanical functions of Walrus tusks are to aid in hauling out onto land or ice (*Odobenus rosmarus* means 'tooth walking sea horse') (Kastelein and Gerrits 1990), and to work as a chisel to keep holes in the ice open (Fay 1982). These two functions of the tusks could explain the fact that females

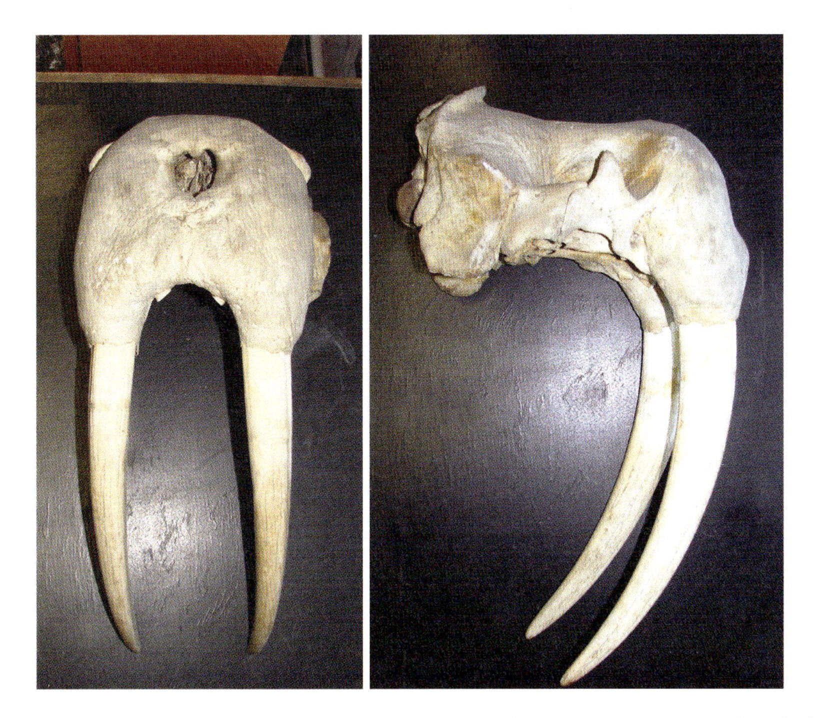

Fig. 3.21 Walrus cranium and tusk in St. Petersburg Zoological Museum (Courtesy of Dr. Andrey Bublichenko)

have tusks as well as males. Belcher (1885) described a Walrus hauling itself out of the water onto an ice-floe: "*It then dug its tusks with a terrific force into the ice that I feared for its brain. Leech-like, the animal hauled itself forward by the enormous muscular power of the neck, repeating the operation until it was secure. The force with which the tusks were struck into the ice appeared not only sufficient to break them, but the concussion was so heavy that I was surprised that any brain could bear it,*" (Belcher 1885). It was also suggested that the tusks can play a social role because animals use them in dominance and mating. Without doubts, these strong formations are dangerous when walruses cross them, however used only secondarily as weapons. Tusking is confirmed due to observations of the scars on the necks and shoulders of adult males (Kastelein and Mosterd 1989; Kastelein et al. 1991).

Unfortunately, there is no detailed information regarding the ultra- and micro-structure as well as material properties of walrus tusk.

Brody et al. (2001) reported that there are some possibilities to identify walrus tusk from bone and teeth of other animals using FT Raman spectroscopy. When viewed on its cross-section, the tusk has a wavy oval outline. The centre of each tusk is filled with an identifying core of "rice bubble"- like osteo-dentine. This core is surrounded by lighter coloured, silky, radially arrayed fibrous primary dentine, and an outer layer of lighter coloured cementum (Fig. 3.22).

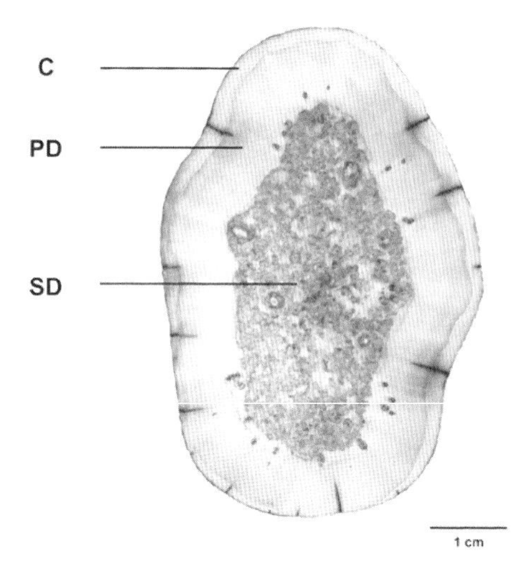

Fig. 3.22 Cementum (*C*), primary dentine (*PD*), and secondary dentine (*SD*) are well visible on the enlarged and enhanced image of a cross-section of walrus tusk (Adapted from Espinoza and Mann (2000), Reprinted with permission)

The root and the tusk proper are the structural components of the walrus tusks. In their classical work, Espinoza and Mann (2000) described the structural organization of this hard tissue as follow:

"*The tip of a walrus tusk has an enamel coating which is worn away during the animal's youth. Fine longitudinal cracks, which appear as radial cracks in cross-section, originate in the cementum and penetrate the dentine. These cracks can be seen throughout the length of the tusk. Whole cross-sections of walrus tusks are generally oval with widely spaced indentations. The dentine is composed of two types: primary dentine and secondary dentine (osteodentine). Primary dentine has a classical ivory appearance. Secondary dentine looks marble or oatmeal-like. This type of secondary dentine is diagnostic for walrus tusk ivory,*" (Espinoza and Mann 2000).

Additionally, a very thick cementum with characteristic cementum rings can be observed on the cross-sections of a walrus tooth. It was suggested that these concentric rings are examples of the phenomenon known as *hypercementosis* (see for review Locke 2008).

Presently, native Inuit are the only peoples allow to harvest and sell walrus ivory. However, previously commercial hunting of walruses for ivory, hides, and blubber has greatly reduced the walrus population. Especially the Eskimo hunted these mammals for food and clothing. In Canada and Russia, this kind of hunting is accepted as a traditional part of their economy.

3.2.7 Conclusion

Teeth in different marine vertebrates have been engineered from unique biological materials by Mother Nature and adapted for broad variety of purposes dependant on the ecology and survival under the corresponding environmental conditions during the long time of metazoan evolution. The answer to the question "where do the true teeth begin?" is still open because of the existence of numerous mineralized teeth-like structures, which are biocomposites that also have hierarchically organized architecture. Moreover, these structures contain Ca-phosphate phases, collagen and different acidic proteins similar to those described in true oral teeth, however in other proportions and molecular orientations at the nanolevel. Thus, even in fish teeth we can differentiate between enameloid and enamel, between dentine and petrodentine, plicidentine, osteodentine, orthodentine etc. because of specificity of their chemistry, structural and material properties. Some of these biological materials are not known for human teeth, whose surfaces at the macrolevel are arranged only in a way that achieves cutting, crushing or grinding of a food supply. Differences in biomechanics between human and marine vertebrate's teeth are, probably, determined by different mechanisms of biomineralization. For example, the crystal shape in mammalian enamel is controlled by growth kinetics, whereas in hypermineralized lungfish dentine the crystal shape is controlled by a protein scaffolding.

The fossilized teeth of extinct marine vertebrates represent a unique source of ancient biological materials and reveal the principles of their structural organization.

3.3 Otoconia and Otoliths

Abstract Otoconia are crystalline biocomposites of calcium carbonate and phosphate phases with localization within the inner ear and vestibular system of marine vertebrates. Thousands of micro-otoconia can exist separately, or can be organized within the inertial mass. The transition from the otoconial mass to the rigid polycrystalline otolith probably occurred in fish by progressive fusion of otoconia from a loose aggregate to a semi-rigid mass. The organic fraction of otoconia and otoliths contains numerous specific proteins, some of which possess collagen domains. Each mineral polymorph of otoconia contains proteins unique to that polymorph. Molecular structure and composition of fish otoliths and their importance in ecological science are discussed.

Otoconia are composite-based microstructures of organic molecules and inorganic crystals formed in the peripheral portion of the vestibular system of marine vertebrates. "*Thousands of small otoconia provide inertial mass and generate shearing forces to allow the cells to sense gravity and the organisms to maintain normal balance,*" (Kawasaki et al. 2009; see also Hughes et al. 2006). Movement of the otoconial layer through action of gravitational or inertial forces activate the underlying mechanosensory hair cells to generate action potentials that are transmitted to the brain (Dror et al. 2010). Thus the otoconia must be located within an ear and vestibular sensory apparatus.

The lancelet is a typical representative of the subphylum Cephalochordata, most closely related to vertebrates. The animal is an interesting model organism because it has neither an ear nor organs that resemble the lateral line organs (Fritzsch 1996; Lacalli 2004). In contrast to lancelet, all craniate vertebrates possess so called angular and linear acceleration sensors. These two discrete sensory systems are the essential components of the ear. Such primitive vertebrates as lampreys and hagfish possess "*the gravistatic receptor is a single macula covered with otoconia*" (Beisel et al. 2005; see also Lewis et al. 1985). It is suggested that that vertebrate ear evolution is primarily the diversification of a statocyst (Beisel et al. 2005), the structure that is known in invertebrates. Definitively, the vertebrate's ear that consists of three different organs (utricle, saccule, lagena), represents more complex sensory system than statocyst (Lewis et al. 1985; Retzius 1884).

Being on the organ level of organization, I wish briefly to describe the role of some of them with respect to otoconia. For example, in Elasmobranchii, the lagena is part of the vestibular system that is responsible for orientation with respect to the gravitation force, as well as for hearing. Anatomically, the lagenar macula appears first in the posterior part of the sacculus of this fish group (Khorevin 2008). Also, the saccular macula found within the sacculus of the fish inner ear is, probably, to be responsible for sound detection in elasmobranchs (Corwin 1981). According to Popper et al. (2003), "*this macula is hypothesized to be stimulated in response to sound waves, as a result of a smaller displacement of the dense overlying otoconial mass relative to the displacement of the fish's body,*" (Mills et al. 2011; see for more information Popper et al. 2003). Under "otoconial mass" we mean some kind of gelatinous matrix that bind together the assemblages of otoconia, calcite, aragonite, vaterite, or calcium carbonate monohydrate crystals. The elasmobranch otoconial mass is analogous to the otolith (see below) of other fish.

In amphibians, the lagena is responsible for vestibular and auditory functions. This structure is localized in the posterior part of the sacculus. The lagena in birds is crucial for their navigation. Lagenar otoconia in birds, including sea birds species, possess magnetic properties and, correspondingly, are supposed to be capable of orienting within the magnetic field of the Earth. Mammals (with exception of Monotremata like platypus and echidna), have no lagena as an "*independent endorgan*" (Khorevin 2008).

Otolithic membrane of utricles, saccules, and lagena represent the next level of organization of the otoconia-containing organs and formations. So called *otoconial comples* is the specialized extremely dense matrix that is localized only within the utricle and saccule. It "*overlies the sensory epithelium and provides inertial mass to generate shearing forces essential for the mechanoreceptors to sense gravity and linear acceleration,*" (Lundberg et al. 2006). Otolithic membrane in various otolithic organs in many animals was found to differ by shape, size, structure, and composition of otoconia (see for review Lychakov 2004). Otolithic membrane of utricle of amphibians and reptiles appears as a thin plate of non-uniform structure. Otolithic apparatus in saccule is a large cobble-stone-like conglomerate of otoconia. Otolithic membrane of lagena looks like a bent plate and is poorly differentiated in amphibians, but well differentiated in reptiles. Thus, transition of vertebrates to the earth surface was accompanied by a fundamental reorganization of otolithic membrane structure. As reported by Lychakov and co-workers (2000):

"*the mass of the otolithic membrane and the length of the animal are power related in elasmobranch fishes. The otoconia of rays tend to be lemon shaped or spherical, however, the dogfish has large, cuboidal (parallelepiped-shaped) endogenous otoconia. The size of the endogenous otoconia does not depend on the size of the animals. No specialized zones were found in rays either on the surface or inside the otolithic membranes containing otoconia of one type or size. The data indicate that the mass of the otolithic apparatus in rays increases on account of the formation of new otoconia,*" (Lychakov et al. 2000).

On the cellular level, otoconia formation occurs outside the cells and therefore depends on secretion of the required assembly components into the endolymphatic spaces. The delicate balance between the organic and inorganic components of otoconia, including their spatial and temporal distribution, determines the growth rate, shape, and composition of the mineral (Dror et al. 2010). The organic fraction of otoconia contains several matrix proteins (see below) that are critical for the nucleation and mineralization of otoconia.

Thus, otoconia can exist as separate microparticles and in the form of otoconial mass. The otoconial mass observed in the endolymphatic sac of rays and sharks represents some kind of loose aggregates. However, teleosts have the rigid polycrystalline otolith. As suggested by Gauldie (1996), this transition from one phase to another is probably occurred by "progressive fusion of otoconia from a loose aggregate to a semi-rigid mass," (Gauldie 1996). Electron microscopy investigations of teleosts otoliths, show that sometimes examples of primitive fused otoconia type of otolith still occur in the otherwise polycrystalline otoliths. The morphological features of the fused otoconia are, probably, dependent of the polymorphs of calcium carbonate which are involved, as well as the particular crystal habit of these formations. It was suggested that "*Ostwald ripening, Keith-Padden spherulitic growth and carbonate cementation are significant in the chemistry of fusion of otoconia in the evolution of the aragonite teleost otolith,*" (Gauldie 1996).

Both otoconia and otoliths can also occur simultaneously in fish organisms. For example, both otoliths and otoconia consisting of crenelated spherules of calcium carbonate have been identified in four species of unrelated teleosts from New Zealand waters: barracuda (*Thyristes atun*), leatherjacket (Parika scaber), the tarakihi (*Cheilodactylus macropterus*), and red cod (*Pseudophycis bacchus*) (Gauldie et al. 1986). Note, that the "*critical function of otoconia and otoliths is to impart inertial movements in response to gravity or linear acceleration, which stimulates the underlying sensory hair cells by deflecting their stereocilia bundles,*" (Deans et al. 2010; see also Hudspeth 2008).

There are numerous publications on "earstones", or otoliths which are defined as "paired calcified structures used for balance and/or hearing in all teleost fishes," (Campana 1999; see also Campana 2004; Tuset et al. 2003). The "chemistry of water" seems to be the crucial point for the basic pathway of the bulk of inorganic elements into the crystalline structure like otolith. Gills or intestine are involved in this transport between the water and the blood plasma of the fish. The next step occurs between the endolymph and the crystallizing otolith. As it was described:

"Water passing over most elements in freshwater fish and the continual drinking of marine fish is the main source of waterborne elements for assimilation via the intestine. Plasma concentrations of Ca are approximately 1/3 of that of marine waters, but flux rates between water and blood are even lower, since only excreted ions are replaced. In saltwater fish, trace elements are probably assimilated from the intestine in direct proportion to their relative concentration in the water, albeit with low efficiency," (Campana 1999; see also Olsson et al. 1998).

The detailed mechanism concerning formation of otoconia and how it is subsequently embedded in the otoconial membrane during inner ear development is still unknown. In contrast to mammals where this process is initiated during embryogenesis and is completed during early postnatal maturation (Lim 1973; Erway et al. 1986), development of the otolith in Teleostei, also being initiated early in otic development, proceeds throughout the life of the animal (Whitfield et al. 2002). It is thought that the origin of calcium in the endolymph is regulated by the activity of plasma membrane associated enzyme called Ca^{2+}-ATPase isoform 2 (PMCA2) that is responsible for the extrusion of Ca^{2+} from the hair cells (Kozel et al. 1998).

Both, molecular structure and composition of fish otoliths is excellently represented in the classical paper by Degens and co-workers in 1969. The summary of this work include some points, which are still correct to this date, thus, I take a liberty to cite them as follow:

"1. Otoliths are mineralogically composed of aragonite. The aragonite fibrils are arranged with their long axis roughly perpendicular to the outer margin of the otoliths. Bands of organic matter intersect the aragonite fibrils transverse to c; the spacing of the bands narrows towards the center of the otoliths.
2. The interrelationship between organic and inorganic matter indicates that the aragonite is formed by epitaxial growth on a protein matrix. Metal ions become coordinated to the oxygen functions displayed on the organic tissue, resulting in the formation of metal ion coordination polyhedra, and bicarbonate becomes linked via hydrogen bridges to amino acids. Subsequent exchange of bicarbonate oxygen for metal ion polyhedra oxygen will stabilize the structure and introduce the nucleation of mineral seeds. Inasmuch as $Ca^{++}O_9$, polyhedra are involved, the mineral form will be aragonite.
3. The mineralized tissue is a fibrous protein with a molecular weight exceeding 150,000. The amino acid composition is biochemically unique and not affected by phylogenetic and environmental events. The term otolin is proposed for this new kind of protein.
4. The variation of total organic matter and the stable isotope distribution in the aragonite can be used as phylogenetic and environmental criteria to distinguish, for example, between freshwater and marine species, to determine migratory tendencies, or to measure the mean temperature at which the fish lived.
5. The compositional variation of otoliths in combination with their ultrastructure suggests that otoliths may function as piezoelectric bodies for the recording of depth and sound," (Degens et al. 1969; see also Morris and Kittleman 1967).

Sound is composed of two major components, the propagating sound pressure wave and particle motion. All fish detect particle motion (the directional component of sound) using their inner ear otoliths (otoconia in case of elasmobranchs) which act as accelerometers (see for review Casper 2006). There have been two proposed pathways of sound to the inner ear. The inner ear otoconia are the main players in the otolithic pathway. Because the density of the shark's body is approximately equal to the surrounding water, sound essentially travels through the shark's body until it comes in contact with structures of a different density in the ear. In teleost fish these structures are solid calcium carbonate otoliths. In elasmobranchs they are represented by otoconia which contain calcium carbonate, with silica nanoparticles of exogenous origin, but localized within gelatinous matrix. As sound moves through the fish body, it comes in contact with these structures which are overlying the sensory hair cells of the inner ear. "Since they are denser than the surrounding tissues, they will lag relative to the rest of the body in the sound field. This lag causes a shearing of the hair cells, thus stimulating the ear," (Casper 2006).

Of course, today, the data regarding to the chemistry and composition of otoliths is amazing, especially with respect to specific proteins, mechanisms of crystal growth, and metal incorporation into otoliths (see for review Melancon et al. 2008). Additionally, novel scientific directions like climate change obtain interesting information from the otolith research. Recently, Checkley et al. (2009) suggested that otoliths in eggs and larvae of White Sea bass (*Atractoscion nobilis*) reared in seawater with elevated CO_2 would grow more slowly than they do in the same water, but with normal concentration of CO_2. Contrary to expectations, the otoliths of fish grown in seawater with high CO_2, and, correspondingly, lower pH and saturation state of $CaCO_3$, were significantly larger than those of experimental animals grown under simulations of present-day conditions. As reported by these authors, "estimated otolith masses were 10 to 14 % and 24 to 26 % greater, respectively, for fish under 993 and 2,558 μatm of CO_2. The dry mass of fish did not vary with CO_2, and thus fish of the same size had larger otoliths when grown under elevated CO_2," (Checkley et al. 2009).

Results reported by Munday et al. (2011) "support the hypothesis that pH regulation in the otolith endolymph can lead to increased precipitation of $CaCO_3$ in otoliths of larval fish exposed to elevated CO_2," (Munday et al. 2011). However, these data also suggested some differences between fish species with respect to sensitivity. This is a very interesting effect for better understanding of the influence of elevated CO_2 on marine biomineralization in the ocean. Increase of the carbon dioxide content in the aquatic environments may regulate the biomineralization-demineralization circle of calcium-based skeletons of diverse marine taxa.

3.3.1 Chemistry and Biochemistry of Otoconia and Otoliths

One of the first chemical analyses of teleost otoliths (statoliths) was carried out by Wicke (1863) who found that the statoliths of cod consisted of 91.1 % inorganic and 8.9 % organic material. Chemical composition of the Trout (*Oncorhynchus mykiss*)

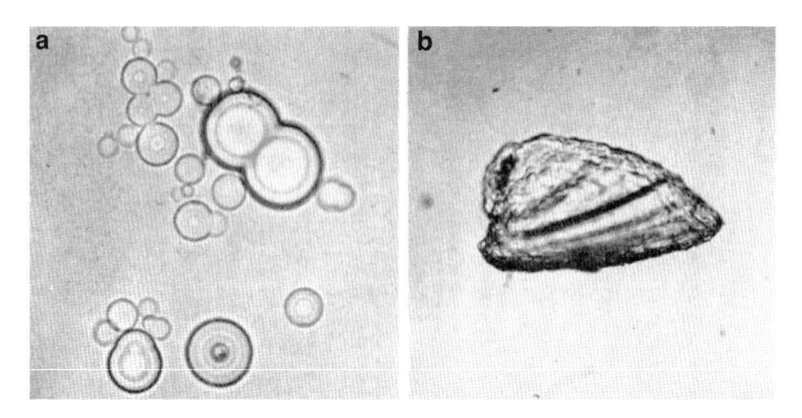

Fig. 3.23 In *L. fluviatilis*, the otoliths consisted mainly of statoconia in the form of transparent spheres, which sizes usually ranged from 2 to 25 μm in diameter. Quite frequently two or three statoconia are fused together, as illustrated in Fig. (**a**). In addition, a few larger bodies (100–250 μm) were found in each labyrinth. The largest of these had a characteristic, somewhat plano-convex shape and a regular layered structure, as seen in Fig. (**b**) (Figure 1 from Carlström (1963). Reprinted with permission from the Marine Biological Laboratory, Woods Hole, MA)

otoliths was reported as follow: proteins (48 %), collagens (23 %), and proteoglycans (29 %). However, in the endolymph these compounds were found in other concentrations (proteins 85 %, collagens 12 %, and proteoglycans 3 %) (Borelli et al. 1994).

The inorganic salt was found by Wicke (1863) to be calcium carbonate admixed with small amounts of magnesium carbonate, calcium phosphate and calcium sulphate. However, otoconia and otoliths are made not only from calcium carbonate and aragonite as usually thought. For example, the lamprey possesses gravity-sensing calcium-phosphate-based biominerals, instead of calcium carbonate. As described by Carlström (1963), in *Lampetra fluviatilis*, the otoliths consisted mainly of statoconia in the form of transparent spheres showing a concentric layering. Their sizes usually ranged from 2 to 25 μm in diameter but were occasionally larger. Quite frequently two or three statoconia were fused together, as illustrated in Fig. 3.23a. In addition, a few larger bodies (100–250 μm) were found in each labyrinth. The largest of these had a characteristic, somewhat plano-convex shape and a regular layered structure, as seen in Fig. 3.23b. These microstatoliths were apparently not formed by fusion of a large number of statoconia as suggested by Studnicka (1912). X-ray diffraction patterns of statoconia, as well as of the microstatoliths, gave nothing but a broad halo, showing that the mineral salt composing these bodies was in a non-crystalline state (see for details Carlström 1963). After heating to 700 °C, a treatment which causes re crystallization, perfectly sharp patterns of apatite, a basic calcium phosphate, were obtained. In the labyrinth of Myxine only statoconia were found. They had the same shape and about the same size as in Lampetra; an apatite diffraction pattern was obtained in this case without any preheating. The diffraction lines were, however, very broadened, showing that the degree of crystallinity was fairly low (Carlström and Engström 1955).

Calcite is also known to constitute the endogenous statoconia of the spined dogfish, *Squalus acanthias* (Vilstrup 1951). Moreover, it was reported that the Zebrafish otolith "switches from an aragonite polymorph to calcite polymorph with the knockout of a single 613-amino acid protein 'Startmaker', named for the bizarre and irregular type of otolith formed when the amino acid sequence of the protein is changed," (Berry 2004; see also Sollner et al. 2003). Fish with these altered otoliths showed disorientation in flowing water.

Surprisingly, also calcium oxalates can be involved in otolith development of some mammals. Recently, Dror et al. (2010) reported for the first time that "calcium oxalate stones are formed in the mouse inner ear of a genetic model for hearing loss and vestibular dysfunction in humans. A missense mutation within the Slc26a4 gene abolishes the transport activity of its encoded protein, pendrin. As a consequence, dramatic changes in mineral composition, size, and shape occur within the utricle and saccule in a differential manner. Although abnormal giant carbonate minerals reside in the utricle at all ages, in the saccule, a gradual change in mineral composition leads to a formation of calcium oxalate in adult mice," (Dror et al. 2010).

Intriguingly, some fish possess magnetic microparticles in their otoliths as well. For example, the saltwater ray has the inner ear gravity receptors that contain mix phases of black coloured magnetic particles with white crystalline otoconia. These are suggested to be some kind of the "*multidomains of magnetite-ilmenite,*" (O'Leary et al. 1981). Also, the ventral region of sacculus of the Fiddler Ray, or guitar fish (order Rhinobatiformes) contains magnetite particles localized within the otolithic mass. As reported by Vilches-Troya et al. (1984): "the exogenous magnetite particles differed from the endogenous otoconia both in their capacity of orienting to magnetic fields, and their difference in mass due to the higher atomic weight of iron. In addition to the normal gravistatic function of the sacculus, two additional receptor functions are hypothesized based upon the differences between the endogenous and exogenous otoconia," (Vilches-Troya et al. 1984).

These authors suggested that geomagnetic orientation of these animals is determined by a geomagnetic field that could induce magnetite displacements detectable by the hair cells. "Alternatively, the greater atomic weights of magnetite, relative to that of otoconia, could result in gravitational and linear acceleration, which differed in different regions of the macula," (Vilches-Troya et al. 1984).

However, more surprisingly is the fact that some elasmobranchs, presumably those having a wide ductus endolymphaticus, may use sea-sand as statoconia, a unique feature within vertebrates (see for review Mills et al. 2011). Thus, Retzius (1881) observed irregular grains in the labyrinth of *Acanthias*, and Stewart (1903–1906) seems to have been the first to identify these particles as sand. Exogenous statoconia are also known to occur in different sharks and ray species. While some authors state that the statoconia of these species consist of nothing but sand, others have found both sand and endogenous statoconia and only endogenous statoconia in the same species. This discrepancy may be due partly to varying proportions of the two constituents in different species (Carlström 1963). "The otolithic apparatus of the dogfish can also increase in mass by addition of small grains of sand, which enter the labyrinth and are incorporated into the otolithic membrane," (Lychakov et al. 2000).

Recently, Mills et al. (2011) reported about the nature and origin of exogenous otoconia from inner ear of *Heterodontus portusjacksoni,* the fish species known also as Port Jackson shark. Surprisingly, these authors identified within the otoconial mass of the specimens silicon dioxide particles of exogenous origin, which were bound within a carbon matrix. These siliceous structures are suggested to play the similar role to the otoconia found in other species of elasmobranchs.

The outer surface of otoconia and otoliths consists mostly of precipitated $CaCO_3$, in contrast to the inner core matrix that include glycoproteins (termed Otoconins) and proteoglycans (Thalmann et al. 2001; Lundberg et al. 2006; Deans et al. 2010). In case of calcium carbonates, their crystals exist in three major polymorphs:

- "calcite (found in mammals and birds);
- aragonite (found in amphibians and fish);
- vaterite (found in primitive jawfish such as garfish)" (Deans et al. 2010; see also Ross and Pote 1984).

It is generally believed that in otoconia and otoliths the major matrix proteins, which bind calcium and make up the organic core, are responsible for diversity of calcium carbonate polymorphs (Pote and Ross 1991; Schipiani 2003). Principally, it is suggested that "the organic matrix of otoconia/otoliths serves as a framework for the inorganic $CaCO_3$ crystallites to deposit and grow," (Xu et al. 2010; see also Zhao et al. 2007).

Here, I would like to represent a short list of corresponding main matrix proteins (for review see also Xu et al. 2010):

- Otoconin 90 (Oc90/95), a highly glycosylated glycoprotein to be found in mammals and birds (Wang et al. 1998; Verpy et al. 1999);
- Otoconin 22, in amphibians (Pote et al. 1993; Yaoi et al. 2001);
- Otoconin 54, in primitive jawfish (Pote and Ross 1991);
- (Omp) the otolith matrix protein – in teleost fish (Murayama et al. 2000, 2002, 2004, 2005);
- otolin-1, in fish (Murayama et al. 2002; Davis et al. 1995; Deans et al. 2010);
- Starmaker, in fish (Sollner et al. 2003);
- matrix macromolecule-64, in fish (Tohse et al. 2008);
- Otoc1 (zOc90), in fish (Petko et al. 2008);
- Sparc, precerebellin-like protein (Cblnl) and neuroserpin, in fish (Kang et al. 2008).

The role of these proteins is very important. For example, morpholino knockdown of Oc90 orthologs in fish lead to an aberrant otolith phenotype (Petko et al. 2008). Similarly, targeted deletion of Oc90 in mice results in balance deficits due to absent or abnormal (few and large) otoconia (Zhao et al. 2008). Also, knockdown of Starmaker, Sparc and Otoc1, OMP, and otolin-1, lead to various abnormalities in zebrafish otolith growth (Kang et al. 2008; Murayama et al. 2004, 2005; Sollner et al. 2003). The otoconia organic matrix disappeared if the predominant crystal protein otoconin-90 (Oc90) is absent (Zhao et al. 2007).

An interesting question concerns the role of collagen domains within some of these proteins (Davis et al. 2002). Thus, the otolin-1 extracted from the otoliths of the chum salmon (*Oncorhynchus keta*), possess sequences of two tryptic peptides, which are showed high homology with fragments of a saccular collagen. Furthermore, "cloning of a cDNA coding for otolin-1 revealed that the deduced amino-acid sequence contained a collagenous domain in the central part of the protein" (Murayama et al. 2002).

We can speak about existence of the site-specific calcification of otoconia especially in the macula where unique ionic microenvironment of the endolymph near its epithelium occur. However, most of otoconia-related proteins have been reported in other structures of the inner ear. It is suggested "that proteins are critical in sequestering calcium for crystallization in the calcium-poor endolymph," (Lundberg et al. 2006).

3.3.2 Practical Applications of the Fish Otoliths

Ever since 1899, when Reibisch (1899) demonstrated that fish otoliths could be used for accurate age determinations, these bodies have been subjected to extensive studies. In his Science paper, Pannella (1971) reported about "the early-stage annual rings in otoliths from some cold-temperate fish, which consist of thin growth bands, the number of which corresponds to that of the days in a year. This indicates that growth takes place by daily increments. Other recurrent patterns show a fortnightly and monthly periodicity. Spawning rings are microscopically distinguishable from winter rings," (Panella 1971).

Today, it is established that the composition of these rings or growth marks (Fig. 3.24), is determined by numerous endogenous and exogenous factors. Fablet and co-workers (Fablet et al. 2007a, b) proposed to use these characteristics as the basis for exploiting them "as biological archives to define environmental proxies (e.g., for instance to reconstruct temperature series) or to reconstruct individual life traits (e.g., individual age and growth information or migration paths)," (Fablet et al. 2007a, b).

The big potential of fish otoliths for reconstructing the environmental history of individual animal and for distinguishing among them, stimulate amazing interest of numerous scientific groups to study the elemental composition of fish otoliths today (see for review Elsdon et al. 2008). Elemental analytics of fish otolith have been used to differentiate among fish stocks, to infer migration, to reconstruct temperature history, validate age interpretations through radiochemical dating, and detect anadromy, the phenomenon known as the migration of fish, from salt water to fresh water, as well as to detect chemical marks applied through mass marking (Campana 1999, 2001; Patterson 1999; Campana and Thorrold 2001; Campana et al. 1995, 2006; Fablet et al. 2007a, b).

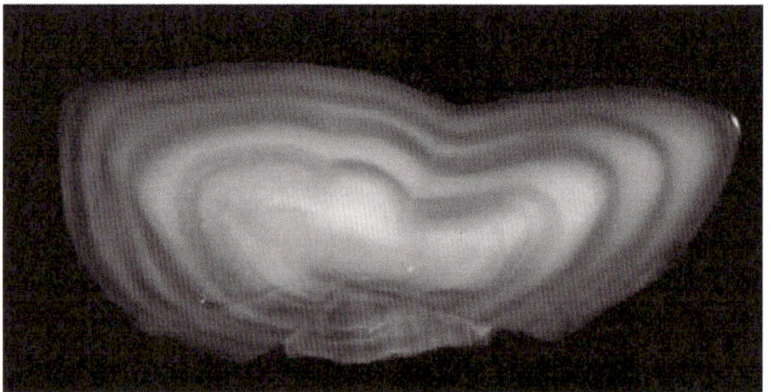

Otolith growth

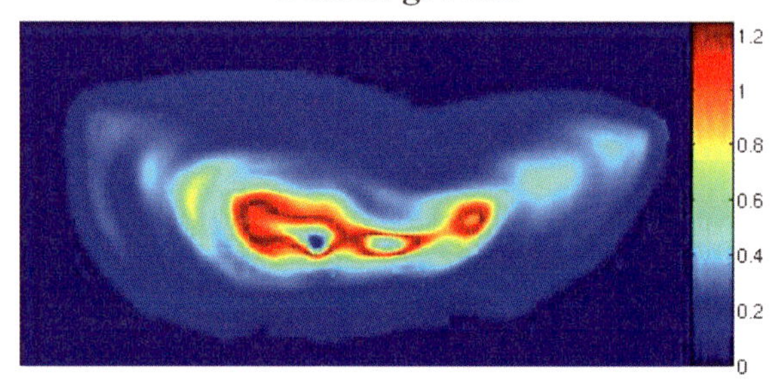

Fig. 3.24 The section through the 5-year pollock (*Pollachius pollachius*) otolith: image and quantitative analysis of the 2D otolith growth (Reprinted from Fablet et al. 2009, Copyright 2009, with permission from Elsevier)

3.3.3 Conclusion

The main principal question of my personal interest regarding otoconia and otolith research is determined by the existence of these biocomposites within Arctic and Antarctic fish specimens. It is well known that fish that live in the polar oceans survive at low temperatures by virtue of antifreeze plasma proteins (AFPs) in the blood that bind to ice crystals and prevent these from growing, up to the level of −1.5 °C (Marshall et al. 2004). Although our own investigations, showed with strong evidence the presence of crystalline hydroxyapatite within bones, teeth and scales of numerous ice fish species, it is established that both reduced bone density and decreased mineralization of their skeleton are examples of the characteristic features. Moreover, these fishes are used as appropriative model organisms to study a reduction in bone mineral density termed as osteopenia in human. However, the following open question still exists: if the skeleton of this fish is really poorly-mineralized, what about their otoliths?

Previously, it was reported that otoliths of some Antarctic ice fish species grew very slowly (Townsend 1980; Radtke and Targett 1984) and contain aragonite and vaterite (Avallone et al. 2003). Also, the "high concentration of calcium-binding proteins in matrices suggests that, in these Antarctic fish, all otoliths are involved in calcium metabolism. This specialization might have occurred as an adaptation to the exceptionally stressing environmental conditions," (Motta et al. 2009). In spite of the fact that AFPs can be used as a model platform to understand biomineralization processes, there is lack of knowledge about their possible role in calcification phenomena at temperatures near the freezing point. The possible mechanisms and kinetics of this *psychrophilic calcification* in particular attracts my attention.

3.4 Egg Shells of Marine Vertebrates

Abstract Of the many marine vertebrates, only reptilia and birds lay eggs. Egg shell is an example of a unique calcium carbonate-based composite with diverse thicknesses and physical properties. Correspondingly, there are different categories of egg shells with respect to their rigidity (soft, flexible, rigid egg shells) and several basic types (testudoid, crocodiloid, dinodauroid, ornitoid, geckoid). The types of hard egg shells are partially based on geometrical crystal growth considerations. In contrast to well-studied mineral components of marine reptilian egg shells, the organic matrix proteins of these taxa are poorly investigated; despite the fact that proteins play a crucial role in the biomineralization of eggshells.

The egg of marine vertebrates, like reptiles, birds and formerly dinosaurs, is "first and foremost an embryonic chamber, through which gas exchange must occur and from which certain ions such as calcium have to be available for embryo development," (Solomon and Gain 1996; see also Deeming and Thompson 1991). Additional functions such as protection against bacterial invasion and mechanical stress are determined by the multi-layered egg shell construct (Bain 1990). The only biomineralized layered structure of the egg is the egg shell, whose ultrastructure varies considerably (see for review Erben 1970; Schleich and Kastle 1988; Mikhailov 1991; Winkler 2006). Thus, the calcareous shell unit is an example of a "complex bioceramic" (Hincke et al. 2012), is made of calcite, aragonite (Roberts and Sharp 1985) or vaterite (Lakshminarayanan et al. 2005), and is the most basic structural unit of an amniote egg. The physical properties like structure, breaking strength and rigidity of both organic and inorganic components are genetically controlled features of a shell (see for review Young 1950; Bain 1990; Hincke et al. 2012). However, early amniote eggs apparently lacked a calcified eggshell. From a phylogenetic analysis of the eggshells of extant amniotes, Stewart (1997) concludes that deposition of calcium carbonate crystals over the fibrous eggshell membrane is a derived character of sauropsids (extant "reptiles" and birds), and not a primitive amniote feature. A brittle calcified eggshell has evolved independently in archosaurs (dinosaurs, crocodilians, birds), turtles (chelonians), and some gekkonid lizards, leading to a variety of eggshell structures and differences in crystalline form (Oftedal 2002). Of course, to find fossilized eggshell with exceptionally well preserved organic matrix is an event in contrast to the calcium carbonate structures which are usually well preserved in fossilized material. The conclusion that early amniote eggs were not rigidly calcified is consistent with the absence of fossil eggs from the Carboniferous, Permian, and early Triassic, a period of about 100 Ma (Oftedal 2002).

Some authors proposed some kind of classification of eggs that is based on the physical properties of the eggshell. Thus, there are three categories known like soft, flexible, or rigid (see for review Packard et al. 1981; Hirsch 1983; Bain 1990; Wangkulangkul et al. 2000). The proportion of inorganic to organic matter determined the eggshell rigidity. Rigid eggshell has more dominated calcium carbonate phase in comparison to organic matter incorporated into this structure. Conversely, eggshell that contains more organic matter than hard crystalline material are related to the soft and flexible categories.

Soft eggshell: Poorly mineralized eggshells with well-developed organic matrix. Representatives of snakes, lizards, and tuataras lay such kind of eggs. "These eggs collapse and shrivel after the animal hatches, and are therefore unlikely to be identified or even preserved in the fossil record," (Wilson et al. 2014).

Flexible eggshell: These shells are more calcified in comparison to soft shells. Reptiles, including some snakes, lizards, and turtles, lay eggs with flexible shells.

Rigid eggshell: Well mineralized, robust eggshells have been reported for eggs of dinosaurs, all crocodilians, some turtles and geckos, as well as birds (Jackson et al. 2004). One of the characteristic example—thick, multiple-shelled turtle eggs (Ewert et al. 1984).

The diversity between these egg shell categories is intriguing even within one group of marine vertebrates. Packard et al. (1982) divided the eggshells of turtles on the basis of structural features into three categories:

1. Flexible-shelled eggs laid by sea turtles of the families *Cheloniidae* and *Dermochelyidae*;
2. Flexible- shelled eggs laid by most emydids and chelydrids;
3. Rigid shelled eggs laid by testudinids, trionychids, and kinosternids (Packard et al. 1984).

According to Hincke et al. (2012), most generally, eggshell types can be categorized as membrane-like (snakes and lizards), pliable (most turtles) and rigid (some turtles, some gecko, all crocodiles, all birds and dinosaurs). There are also such definitions as pliable-shelled (see for examples Deeming and Whitfield 2010) and parchment-shelled eggs like those of pterosaurus species (Unwin and Deeming 2008; Lü et al. 2011). Most lizards and snakes have parchment-shelled eggs, with scattered and variable sites of calcium deposition (Oftedal 2002). These eggs increase in mass, volume, and surface area because of their ability to absorb water during incubation time. Fossil egg shells have been reported nearly worldwide, especially from Upper Cretaceous and Tertiary deposits, and they have been assigned, according to their microstructure and biomineralization, to turtles, crocodiles, dinosaurs and birds (reviewed by Hirsch and Packards 1987). Five basic types of hard eggshells of fossilized and living amniotic vertebrates may be distinguished (Mikhailov 1991, 1997a, b) (Fig. 3.25):

1. *Testudoid type* (Chelonia). The shell unit consists of a single ultrastructural zone with regular spherulithic growth of aragonite crystals (radial aragonite structure) and organic core in the base.

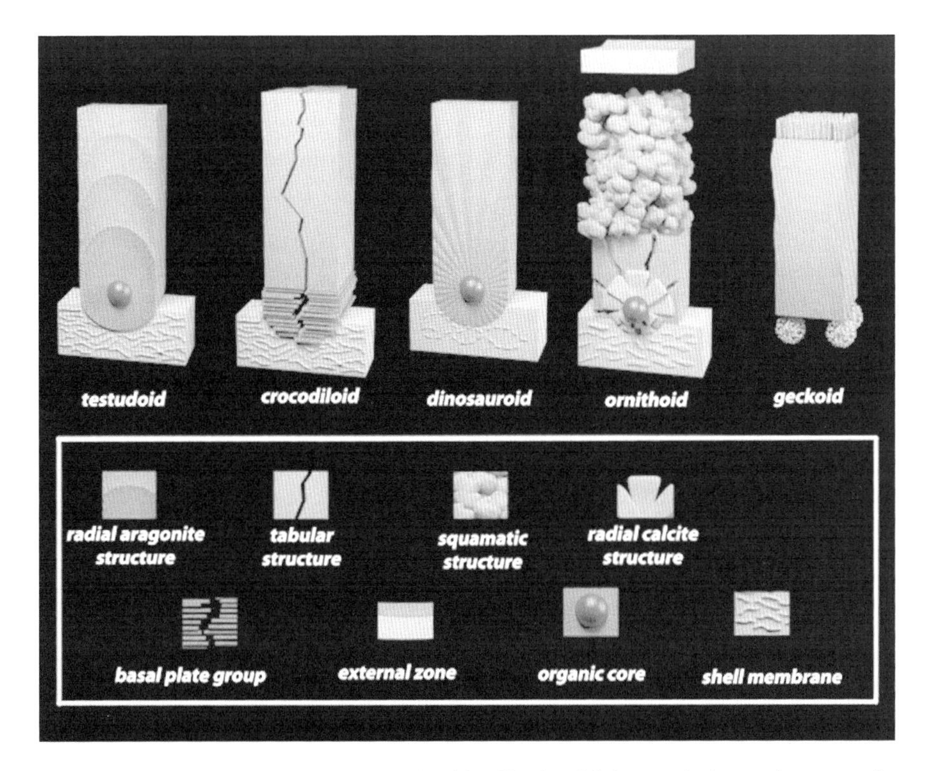

Fig. 3.25 Five basic types of hard eggshells of fossilized and living amniotic vertebrates may be distinguished according to Mikhailov (1991) (Image designed by Alexei Rusakov)

2. *Crocodiloid type* (Crocodilia). Much of the shell unit consists of a single ultrastructural zone with irregular radial growth of tabular crystallite aggregates (tabular structure), and with a basal plate group (rosette of plates) in the base; the organic core is absent.
3. *Dinosauroid type* (Dinosauria: Sauropoda and Ornithischia). The shell unit consists of one ultrastructural zone with more or less regular spherulithic (or prismatic) growth of tabular crystalline aggregates (tabular structure) and organic core in the base; there is possibly a basal plate group in the form of an eisospherite.
4. *Ornithoid type* (Aves; some Dinosauria). The shell unit consists of at least three ultrastructural zones: (i) a zone with spherulithic growth of minute platy crystallites and their petalloid aggregates (radial calcite structure); (ii) a zone with spherulithic growth of tabular crystalline aggregates (tabular structure); and (iii) a polycrystalline zone with squamatic elements (squamatic structure, true spongy layer); the external zone (iv) is present in most avian eggshells. The organic core and the basal plate group in the form of an eisospherite occur in the base of shell unit.
5. *Geckoid type* (Gekkota). The shell unit consists of two or three zones whose ultrastructure is still not understood, the organic core is absent.

The types of hard egg shells are partially based on geometrical crystal growth considerations, which have been proposed for the deposition of the crocodilian, testudinian and avian eggshells (Silyn-Roberts and Sharp 1986). As described by these authors:

"In each shell column, crystal deposition is initiated at a single location, from which growth fans out at all angles to the shell normal. In both calcitic and aragonitic shells, growth is in the [001] direction, resulting in an increase in the degree of (001) preferred orientation with distance from nucleation. Where there is unhindered crystal growth, the shells show crystalline fracture morphology, and the degree of texture that develops is a simple function of the column radius. This type of growth makes up the whole of the testudinian shell, the inner 30–40 % of the thick ratite shells, and the cone layer of the other avian shells," (Silyn-Roberts and Sharp 1986).

Most physiological, biochemical, nutritional, structural and morphological studies have been carried out on avian eggs, and mostly on the egg of the domestic chicken. This is of course because of its ready availability and commercial importance as a nutritious food for human consumption. In contrast, there is much less known regarding eggs and eggshell of other birds or nonavian animals (snakes, lizards, turtles, crocodiles and dinosaurs) (Hincke et al. 2012).

3.4.1 Eggshells of Marine Reptilia

The egg is the most vulnerable stage in a reptile's life. Immobile and often lacking parental care, it is susceptible to predation and exposed to the prevailing environmental conditions. Suitable substrate temperature and moisture are vital for successful development of the embryo and subsequent hatchling performance and behaviour (Packard and Packard 1988). Flexible-shelled eggs of turtles exchange water with the environment of the nest cavity. These water exchanges dramatically affect changes in mass of the eggs over the course of incubation. Eggs incubated on relatively wet substrates gain water and mass during incubation, while those incubated on relatively dry substrates lose water and mass over the course of development (Miller et al. 1987). Therefore, the reptilian advancement of the hard eggshell was a clear advantage. For the first time, the reproductive process was not tied to sources of water. The development of the amniote eggshell made the impermeability to fluids possible. At the same time, pores allowed the embryo to inhale oxygen and exhale carbon dioxide. Dinosaurs used this basic reptilian design, while the modern birds took this same format with added complexity (Pollinger 1997).

The structural peculiarities of the turtles egg shells are well studied, and for detailed analysis I recommend the following papers published by Solomon and Baird 1976, 1977; Packard and Packard 1979, 1980; Packard 1980; Solomon and Watt 1985; Acuña–Mesén 1984, 1989; Woodall 1984; Chan and Solomon 1989; Carthy 1992; Sahoo et al. 1996a, b; Mahanty and Sahoo 1999; Kitimasak et al. 2003; Al-Bahry et al. 2009. The finding of the double layered (calcareous and

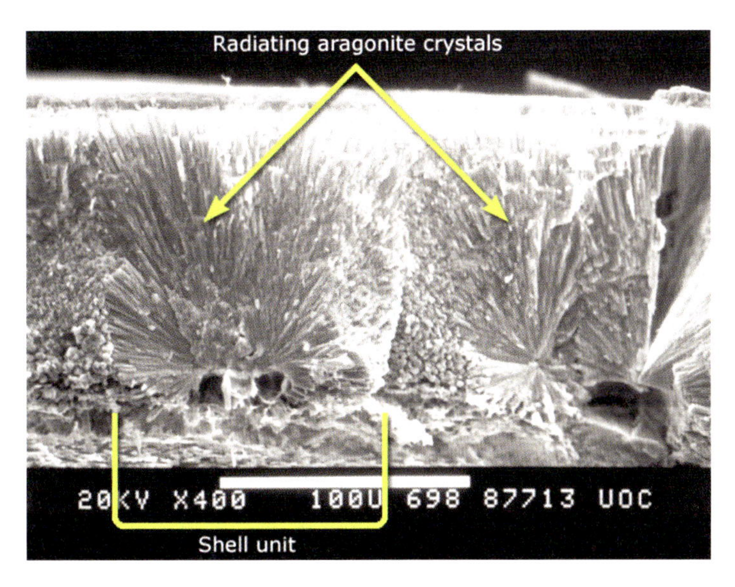

Fig. 3.26 SEM image of the Trionyx spiniferus turtle eggshell. The radial organization of aragonite crystals and slightly discernible separation of eggshell membrane and crystalline layer are well visible (Wilson et al. 2014; Copyright © 2014, by the Regents of the University of California. Reprinted with permission)

fibrous layers) ultrastructure of turtle eggshell, confirms previous investigations on other reptilian eggshells (Packard and Hirsh 1986) (see as example Fig. 3.26). Thus, flexible eggshells from numerous turtle's species possess an inner membrane layer and an outer mineral layer of approximately equal thickness. Feder et al. (1982) measured the permeability of both layers to H_2O and O_2 at various levels of eggshell hydration. These authors reported following results:

"Both the mineral layer of the eggshell and the shell membrane offer significant resistance to diffusion of water vapor and oxygen in eggshells of the snapping turtle, *Chelydra serpentina*. Conductance to water vapor increases in both the membrane and mineral layer with increasing hydration of the eggshell, but conductance to oxygen decreases under similar conditions. Removal of the mineral layer increases conductance to oxygen in moist and dry eggshells, but decreases conductance at intermediate levels of dehydration. Removal of the mineral layer consistently increases conductance to water vapor. The eggshell membrane accounts for 24–76 % of overall resistance to diffusion of water vapour," (Feder et al. 1982).

Of course, the porosity of turtle egg shells also plays a crucial role in this case. As reported by Packard and co-workers (1979), "flexible-shelled eggs from snapping turtles (*Chelydra serpentina*) have conductances to water vapor that are 55 times higher than predicted for avian eggs of similar size, whereas rigid-shelled eggs of softshell turtles (*Trionyx spiniferus*) have conductances that are only five times higher than expected for comparable eggs of birds," (Packard et al. 1979).

Here, we must note that reptilian eggs have much higher effective pore areas than those of birds. The functional significance of this structural difference is well explained in the following way:

"The relatively high porosities of these reptilian eggs presumably facilitate the transport of oxygen and carbon dioxide into eggshells in later stages of incubation when air trapped inside nest chambers may become hypoxic and hypercapnic, yet does not seem to lead to excessive transpiration of water vapor owing to the high humidities in nests where incubation occurs," (Packard et al. 1979; see also Doody 2011).

No difference concerning biomineralization of the eggshells from wild and captive animals have been reported (Solomon and Baird 1976; Baird and Solomon 1979). However, species related specialization occur. For example, aragonite is the only calcium carbonate phase that was found within Olive ridley's (*Lepidochelys olivacea*) eggshell. However, other species possess eggshells that consist of calcite, vaterite and aragonite. Intriguingly, the eggshells of the leatherback (*Dermochelys coriacea*), has all three phases together (Sahoo et al. 1996a, b). The analysis of the ultrastructure and elements of three marine turtle eggshells (*Chelonia mydas*, *D. coriacea*, and *Eretmochelys imbricata*) showed that the eggshell had three layers: the outer cuticle layer or the crystalline layer, the middle layer, and the inner fibrous layer (Nuamsukon et al. 2009). The outer layer was thick and had porosity appearing like the clusters of branching needle-like crystals, the middle layer was compact thick, while the inner layer was compact thin. The eggshell thicknesses of *C. mydas*, *D. coriacea* and *E. imbricata* were 108.66 ± 1.74 μm, 114.86 ± 0.37 μm, and 98.73 ± 3.56 μm, respectively (Nuamsukon et al. 2009). In the loggerhead marine turtle (*Caretta caretta*) three eggshell layers were also recognized (Al-Bahry et al. 2011). Here, the detailed description:

"The outer calcareous layer consists of loose nodular units of different shapes and sizes with loose attachment between the units, resulting in numerous spaces and openings. Each unit consists of $CaCO_3$ crystals in aragonite (99 %) and calcite (1 %). The middle layer has several strata with numerous openings connecting the calcareous and the inner shell membrane. Crystallites of the middle layer are a mix of amorphous material with aragonite (62 %) and calcite (38 %). The inner shell membrane has numerous reticular fibers mixed predominantly with halite (NaCl) and small amounts of sylvite," (Al-Bahry et al. 2011).

It is suggested (Hirsch 2001; Jackson and Varricchio 2003) that prolonged egg retention, often resulting from physiological or environmental stress, can be the reason for development of heavy mineralized multilayered eggshells of some extant amniotes. However, as reported in at least nine extant species (see for review Jackson et al. 2004) this category of eggs is relatively common in some hard-shelled turtle eggs. Very special feature have been reported for eggshells of the turtles *Rhinoclemmys areolata*. These possess complex pores with basal openings partially occluded with crystallites. According to (Ewert et al. 1984):

"Each of the abnormally thick eggshells has the usual membrane and mineral layer encased within a second membrane and mineral layer and, in one case, within yet a third membrane and mineral layer. Information on oviducal function and the structure of these and most previously reported multiple-shelled turtle eggs allows the hypothesis that such eggs arise from oviducal retention of normally shelled eggs of one clutch that become reshelled during shelling of eggs ovulated to form the next clutch. This is likely only in species that lay more than one clutch per season," (Ewert et al. 1984).

Similar phenomenon occurs, probably, in dinosaurs that laid multiple-shelled eggs during several clutches per season (Ewert et al. 1984).

As the eggshell is made of significant amounts of calcium, what are the sources for the mineral required for its growth? Early in the growth phase during the second half of incubation, turtle embryos initially obtain calcium from the egg yolk. The yolk is quickly depleted of calcium, which must then be mobilized from the eggshell during the last trimester (Packard 1994; Sahoo et al. 1998). Calcium is the major inorganic constituent of sea turtle eggshell (20–21 %) (Solomon and Baird 1976; Sahoo et al. 1998), and the majority required for embryogenesis is derived from this source (60 %—Sahoo et al. 1998; 62 %—Bustard et al. 1969; 75 %—Simkiss 1962). Consequently, calcium depletion could be expected to influence not only embryonic development (Solomon and Baird 1980) but also eggshell structure (Sahoo et al. 1996a, b). There are no significant difference reported concerning calcium content of sea turtle eggshells in loggerhead (*Caretta caretta*), flatback (*Natator depressus*), hawksbill (*Eretmochelys imbricata*), and green (*Chelonia mydas*) are not significantly different (Phillott et al. 2006).

Intriguingly, the phenomenon of calcium depletion from eggshells after fungal invasion of sea turtle eggs is also described (see for review Phillott et al. 2006). Solomon and Baird (1980) suggested hyphal penetration of the eggshell and eggshell membranes may impair gaseous exchange, invade embryonic tissue, and/or impede normal embryonic development by depleting the amount of calcium in the eggshell. Phillott and Parmenter (2001) concluded that if hyphal impediment of gas exchange occurs, the severity of its influence would depend upon the size and location of the fungal growth, the sea turtle species, and the egg size. Fungi are capable of penetrating the eggshell and invading embryonic tissue (Phillott 2002; Phillott et al. 2004). Solomon and Baird (1980) observed fungal hyphae between the soft shell membrane and crystalline shell layer in green sea turtle eggs. They concluded that the high calcium content of these hyphae, in conjunction with their proximity to the calcified eggshell, suggested fungi may be extracting calcium from the eggshell, thereby causing a deficiency in the embryo and impairing normal development.

Calcification for the formation of the eggshell occurs during development in reptilian embryos. In contrast to mammals and birds which show higher level of parental care, such extant reptiles as crocodiles, turtles, squamates, and the tuatara has developed the diverse range of strategies to prolong the egg state (see for review Rafferty and Reina 2012). The biological expediency to arrest development in reptiles enables their embryos to withstand a changing incubation environment in a variety of ecological situations. This ability helps animals to synchronize hatching with seasonal periods, too. As recently formulated by Rafferty and Reina (2012):

"Developmental arrest is a critically important reproductive strategy in the large range of egg-laying animals that provide no parental care after oviposition, because it provides their eggs a mechanism to respond to changing environmental conditions during embryonic development," (Rafferty and Reina 2012).

Several researchers including Ewert (1985) and Miller (1982, 1985) studied the phenomenon of developmental arrest in chelonians in details.

In turtles, the oviduct seems to be the space where limited oxygen exchange might occur as a result of shell mineralization and "the shell pores filling with oviducal fluid," (Rafferty and Reina 2012; see also Andrews and Mathies 2000). Heulin et al. (2002) reported that pre-ovipositional calcification of turtle eggs becomes complete "when the embryos reach gastrulae and enter pre-ovipositional arrest approximately 7 days after ovulation," (Rafferty and Reina 2012). Mineralization of the eggshell may regulate the respiratory gas exchange needed for further development of the embryo (Packard et al. 1977; Guillette 1982). After the availability of oxygen to the embryo is enhanced, the reduction in eggshell thickness can be also associated with extended egg retention (Ewert et al. 1984; Heulin et al. 2002). Additionally, the "eggshell provides a source of calcium for developing turtle embryos, and reducing the degree of eggshell calcification in order to achieve greater O_2 exchange may decrease embryo fitness," (Rafferty and Reina 2012; see also Andrews and Mathies 2000; Bilinski et al. 2001). After calcification of turtle eggs, both the embryonic vitelline membrane and inner shell membranes must adhere to one another. In this way the embryonic growth and gas diffusion can proceed. The relationship between membrane fusion and is discussed by Andrews and Mathies (2000). It was postulated that "species that lay thinner-shelled pliable eggs typically do not use developmental arrest after oviposition and as a result they have shorter incubation periods than those laying thicker, more rigid, brittle-shelled eggs that do," (Rafferty and Reina 2012; see also Ewert 1985).

In contrast to well-studied mineral components of marine reptilian egg shells, the organic matrix proteins of this group of animals are poorly investigated. However, especially proteins play an important role in the eggshells biomineralization. Lakshminarayanan et al. (2005) reported about the isolation of a *pelovaterin*. This glycine-rich peptide with 42 amino acid residues and three disulfide bonds was extracted from eggshells of a soft-shelled turtle *Pelodiscus sinensis*. It was shown that pelovaterin induced the formation of a metastable vaterite phase *in vitro*. The microarchitecture of vaterite crystals is dependent on concentrations of pelovaterin added as follow: "the floret-shaped morphology formed at a lower concentration (approximately 1 μM) was transformed into spherical particles at higher concentrations (>500 μM)," (Lakshminarayanan et al. 2008). Additionally it was shown the entropy-driven dependence of the self-aggregation of this peptide in solution in the form of micellar nanospheres. The development of these micelles is principally possible because molecules of pelovaterin possess a large hydrophobic core and a short hydrophilic N-terminal segment. Although pelovaterin show functional similarity to human ß-defensins, structurally they are absolutely different. Probably, the micellar state of pelovaterin determine the induction and stabilization of the metastable mineral phase by altering the interfacial energy. Intriguingly, "pelovaterin not only stabilizes a thin film of metastable vaterite, but exhibits strong antimicrobial activity against two pathogenic gram-negative bacteria, *Pseudomonas aeruginosa* and *Proteus vulgaris*" (Lakshminarayanan et al. 2008).

Regrettably, there are only few reports (Jenkins 1975; Ferguson 2010) on egg shells of crocodiles, and no detailed information regarding salt water species *C. porosum* except a single paper by Griggs and Beard (1985). Roberts and Sharp (1985) investigated the orientation of calcium carbonate phases in their egg shells by using X-ray diffractometry. In all Crocodylia species the calcitic microarchitecture that develops is one in which the basal plane of the unit cell in each of the two crystal systems tends to lie parallel to the shell surface. Preferred orientation reaches a maximum at the exterior surface, and its development through the shell is normally uninterrupted. While many reptiles have parchment-shelled eggs, crocodilian eggs have a calcareous shell, like birds, which is pierced by many pores. Unlike avian eggs, however, the shell membrane in crocodile eggs is quite thick (Grigg and Beard 1985).

The structure of the crocodylian egg shell seems to be very complex. Here, the detailed description made by Ferguson (1982):

"The alligator (*Alligator mississippiensis*) egg consists of the following layers from outside in: (1) an outer densely calcified layer (100–200 μm thick) consisting of small vertically stacked calcite crystals orientated with their crystallographic c axes at right angles to the shell surface; (2) a honeycomb layer (300–400 μm thick) consisting of horizontally stacked calcite crystals with their crystallographic c axes parallel to the shell surface; (3) an organic layer (approximately 10 μm thick) containing a higher percentage of organic matrix to calcite crystals, and through which the shell cleaves and falls away from the egg shortly before hatching; (4) a mammillary layer (20–30 μm thick) which is more pronounced in the central opaque region of the shell and which attaches the latter to the eggshell membrane; and (5) an eggshell membrane (150–250 μm thick) consisting of an interwoven mesh of fibres, separated from the albumen by an amorphous limiting membrane, in which there are numerous pores," (Ferguson 1982).

However, in spite of complex, multilayered mineral-containing structure of the eggshell, are there some naturally occurring processes, which can lead to demineralization of these constructs? Indeed, two factors like organic acids produced by nest micro-organisms and presence of hydrated carbon dioxide can influence the integrity of the eggshell. Ferguson (1982) proposed terms like "erosion craters and cratered pore orifices," (Ferguson 1982) and described this phenomenon as follow:

"These erosion craters expose at their broad bases the underlying honeycomb layer which contains large numbers of vesicular holes, interconnecting with other cavities throughout the shell. Thus, development of erosion craters renders the entire egg more porous as incubation proceeds. In addition, these craters, together with the mobilization of calcite crystals out of the organic layer for use in embryonic mineralization, progressively weaken the shell causing it to crack and cleave off the eggshell membrane and mammillary layer, thus facilitating hatching. No air space is present in the alligator egg, neither are there any chalazae. The embryo has the usual amniote arrangement of extra-embryonic membranes which are firmly fused along two longitudinally opposite ribs, anchoring the embryo in a constant position within the shell," (Ferguson 1982).

Thus, according to Ferguson (1982) the "structure of the alligator egg is beautifully adapted to both its function and the nesting biology of the animal".

3.4.2 *Egg Shells of Sea Birds*

Nathusius von Koenigsborn (1821–1899) was the first to employ fossil material in his study of the eggshell structure (Tyler 1964). Serious investigations of fossil eggs and eggshell remains started in 1923, after the sensational finds of the American Museum of Natural History Expedition in Mongolia (Hirsch 1994). The domesticated chicken (*Gallus gallus)* has been proposed and used as the classical model for understanding of the mechanisms of avian eggshell formation and biomineralization. The avian egg is considered to represent the most advanced amniotic egg in oviparous vertebrates. The shell is a complex bioceramic that regulates the exchange of metabolic gases and water, and its properties are exquisitely fine-tuned to the environment of a given species. In their excellent work, Hincke et al. (2012) reviewed recently eggshell ultrastructure and microstructure from very modern point of view that summarized data from recent proteomic, genomic, and transcriptomic analyses. Because there are numerous excellent books and reviews on avian egg shell morphology, structure, formation and biomineralization including the role of matrix proteins (Romanoff and Romanoff 1949; Tyler 1964; Erben and Newesely 1972; Board 1982; Packard and Packard 1984; Hamilton 1986; Arias et al. 1993; Mikhailov 1997a, b; Fernandez et al. 2001; Nys et al. 1999, 2004; Dauphin et al. 2006; Rose and Hincke 2009; Hincke et al. 2012), I find necessary to discuss here only some principal points.

When complete, the avian eggshell has a well-defined structure that is described as follows from the inside (egg white side) to the outside (external surface): (i) the mammillae (or mammillary body/cone layer), (ii) the palisades (or palisade layer) comprising the thickest layer of the shell, and (iii) the transitional vertical crystal layer. Finally, a thin non-calcified cuticle layer coats the eggshell. The transitional, inner zone of the cuticle contains spherical aggregates of hydroxyapatite (see for details Hincke et al. 2012).

There are no doubts that different bird species possess egg shells with different thicknesses. From this point of view, the egg shell of such sea birds as penguins is of special interest. Penguins use little nesting material on rocky substrates to incubate their eggs for a long period of time. In spite of these hard conditions, for example, studies on Magellanic Penguins (*Spheniscus magellanicus*) showed that only 2.6 % of 10,023 eggs were broken or cracked. Egg shells for this species are very thick, they "averaged 0.81 mm without the egg membranes and are at least 56 % thicker than expected for bird eggs of similar mass," (Boersma et al. 2004). The source of additional calcium for penguins seems to be the mollusc shells. It was suggested that the "thick eggshells of penguins, along with selective ingestion of mollusc shells, appears to be an adaptive response that reduces egg breakage," (Boersma et al. 2004). Principally, the eggs of penguins, guillemots, cormorants, and murres show the highest proportions of shell mass among the seabirds' eggshells (Schönwetter 1960). The ultrastructure of the egg shells of some sea birds also possess specific pattern. For example, transverse and tangential ground sections through the egg shells of albatrosses (*Diomedea exulans*) show in the column layer a characteristic "globular pattern" (Schmidt 1967).

3.4.3 Conclusion

The eggshell of both marine reptiles and birds is hierarchically structured complex bioceramic composite made of a mineral part (>95 % $CaCO_3$) and an organic matrix of very complex composition (1–3.5 %; Hincke et al. 2012). The microstructure and composition of eggshell varies across its thickness. Especially marine reptiles possess broad variety of egg shells with different physical properties like stiffness and porosity. Unfortunately, very little attention is still paid to investigations of the material properties and biomechanics of these unique structures. Eggshells of sea birds have also been studied mostly from ecological point of view. I am very hopeful that this situation will be rectified in the near future.

3.5 Biomagnetite in Marine Vertebrates

Abstract Magnetite is the most important magnetic mineral on Earth. It occurs in continental and oceanic crust as a primary or secondary mineral in igneous, sedimentary and low- and high-grade metamorphic rocks. Biogenic magnetite, an example of nanomagnetism, occurs in microorganisms, invertebrates and vertebrates species. Among the marine vertebrates that use magnetite are fish, turtles, sea birds and cetaceans. Since Lowenstam (1962) reported biogenic magnetite in the radular teeth of chitons, the idea that biologically synthesized magnetite particles may form the core of the animal magnetic sense became the working hypothesis of a number of behavioural, neurological and physiological studies. For magnetite crystals to function as magnetoreceptors in animals, the magnetite presumably needs to contact the nervous system.

Magnetite is a cubic mineral with inverse spinel structure that has the structural formula Fe^{3+} $(Fe^{2+}Fe^{3+})O_4$, and possess ferromagnetic properties. The unit cell is represented by eight tetrahedral sites filled with Fe^{3+} cations and sixteen octahedral sites, half of which are filed with Fe^{2+} cations and the other half with Fe^{3+} cations (Davila 2005). According to Muheim (2004):

"A magnetic field is a form of stored energy, but in ferrimagnets arises from microscopic currents associated with electrons in the atoms of permanently magnetic material. 97–99 % of the Earth's magnetic field is due to the main field from electric currents in the outer core. Part of the remaining 1–2 % of the crustal field is from magnetized rock in the crust, and part from the external fields of ionized particles in the upper atmosphere and solar wind," (Muheim 2004).

The phenomenon observed by numerous animal taxa to navigate over long distances during migration and homing using the magnetic field of the Earth, have been formulated by Viguier (1882) as hypothesis. This is based on suggestions concerning the existence of magnetic sense in living organisms. This hypothesis seems to be true, as today biogenic magnetite is found in numerous organisms including bacteria, invertebrates (insect) (Gould et al. 1978; Maher 1998) and vertebrates (fish, birds, mammals) (Phillips 1996; Kirschvink et al. 2001; Holland et al. 2008); and as fossil remnants in sedimentary deposits and soils. Bacterial magnetosomes (Mann et al. 1984; Gorby et al. 1988; Schüler 1999; Matsunaga and Sakaguchi 2000; McCartney et al. 2001; Komeili 2007; Faivre and Schüler 2008) and their fossils, (magnetofossils), can significantly contribute to the bulk magnetic properties and the remnant magnetization recorded in marine sediments (Petersen et al. 1986; Fassbinder et al. 1990). Heinz Adolf Lowenstam was the first to identify magnetite in the radula teeth of chitons (marine molluscs), showing that life had also devised mechanisms to synthesize magnetite by using biochemical processes (Lowenstam 1962).

When studying biogenic magnetite (Kirschvink and Gould 1981), it is paramount to determine the size of the particles produced by, or localized within the cell, and/or organism (Fig. 3.27). Usually, the main subdivisions are as follow:

- "multi-domain (MD);
- pseudo-single-domain (PSD),
- magnetically stable single-domain (SD),
- and superparamagnetic (SP),

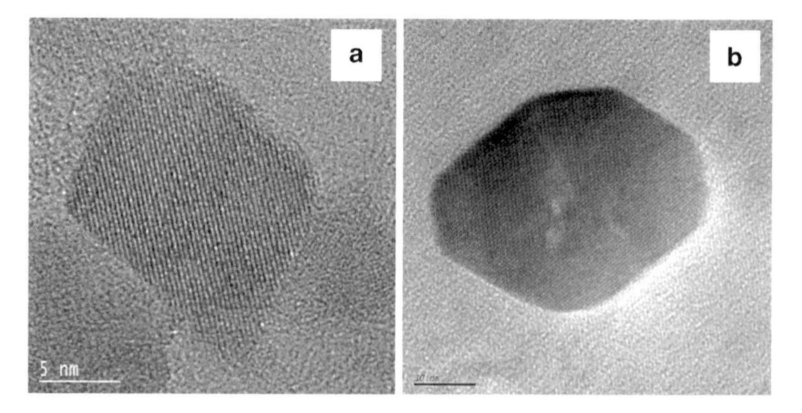

Fig. 3.27 Nanomorphology of cubo-octahedral structure of inorganic (*left*) and biomagnetite (*right*) (Adapted from Faivre (2004), reprinted with permission of Damien Faivre and Nicolas Menguy)

which are determined by the number of magnetic domains contained within the particle (the domain state). The domain state determines the magnetic properties which will, in turn, be selected according to physiological requirements and the role the magnetite particles play in the organism, (Davila 2005).

Magnetic particle with uniform direction of spontaneous magnetization are the basic elements of corresponding magnetic domains. However, situations where different domains within a single particle may have different directions are also known. Correspondingly, an MD particle can show zero remanence if the magnetization directions cancel each other. On the other hand, a magnetically stable SD is always magnetized to saturation and has a remanent magnetization at room temperature. It must be noted that because to their small volume, SP particles loose "their remanence in time spans of seconds to nano-seconds, and can be considered, for practical reasons, to acquire only an induced magnetization in the presence of a magnetic field," (Davila 2005).

Biogenic magnetite can be generated by a wide spectrum of mechanisms, which differ in the degree of control the organism has over the mineralization process (Schüler 2006). Thus, Lowenstam clearly defined biologically controlled (BCM) and biologically induced (BIM) mineralization. He describes BIM as the process with the least biological control. In the case of magnetite, BIM is often related to dissimilatory iron-reducing bacteria. During respiration, these absorb ferric ions (Fe^{3+}) in the form of amorphous ferric oxy-hydroxide and export ferrous ions (Fe^{2+}) into the environment, where they interact with excess ferric oxy-hydroxide resulting in extracellular magnetite precipitates that usually resemble those formed inorganically (Frankel and Blakemore 1991). Thus BIM of magnetite occurs through chemical changes in the environment due to biological activity. The mechanisms governing the formation of such particles are seemingly altered by evolution and natural selection to a low extent; hence BIM of magnetite has likely remained practically invariant throughout geological time (Davila 2005).

BCM of magnetite, on the other hand, is to be found both in bacterial and eukaryote cells with "a remarkable level of control of the particle size, shape, composition and structure" (Davila 2005; see for review Schüler 2006). Contrary to what occurs in BIM processes, the degree of control that organisms exhibit in BCM processes is only achievable as a result of long time of evolution of the biological functions in which these particles are involved.

Single biogenic magnetite nanoparticles (Wei et al. 2011) seem to be the basic structural unit in prokaryote. In 1975, Richard Blakemore discovered a new type of bacterium with an unusual preference to move along the local geomagnetic field lines (Blakemore 1975). These magnetotactic bacteria (MTB) have since then been thoroughly studied (DeLong et al. 1993) and the nature of their magnetotactic behavior is now fairly well understood, and linked to the presence of intracellular chains of magnetically stable single-domain (SD) magnetic crystals. Cells of magnetotactic bacteria include of one or more chains of magnetically stable SD crystals of magnetite (iron-oxide type) or greigite (iron-sulfide type), the so-called magnetosomes, which often appear enveloped by a phospholipid membrane (see for review Davila 2005; Schüler 2006) (Fig. 3.28).

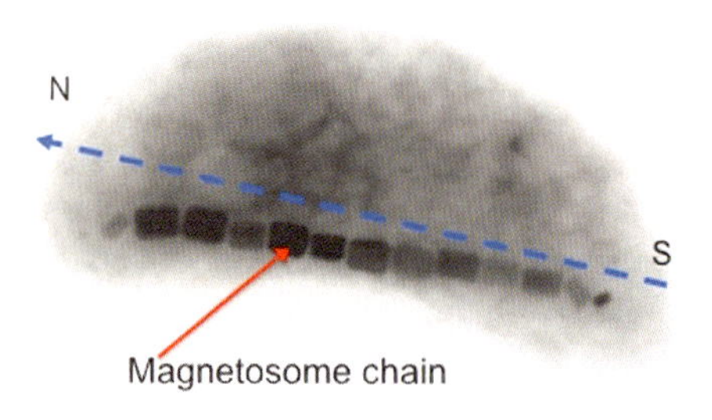

Fig. 3.28 Magnetosome within bacteria (Courtesy of Prof. Dirk Schüler)

Far from being a rarity, MTB are ubiquitous in aquatic environments, suggesting that magnetotaxis may be a common trait among prokaryote organisms inhabiting localized chemical gradients. As there are microaerophilic organisms that inhabit the oxic-anoxic transition zone (OATZ) of marine and fresh-water sediments or water columns, the evolutionary advantage of magnetotaxis is, perhaps, not difficult to understand. According to Davila (2005), when a MTB is released outside of the OATZ, due for example to turbulence in the water column, it finds itself suddenly in a high, potentially toxic oxygen-rich environment. The bacterium needs to move towards a more appropriate environment as quickly as possible to ensure survival. It is in this situation when the intracellular chains of magnetic particles play their evolutionary role: the geomagnetic field lines, inclined with respect to the Earth's surface (except in the equator, where they run parallel to the surface), will exert a torque on the intracellular chain until it is oriented parallel to them. This magnetic torque is in turn transferred to the cellular body; then, the bacterium needs only to swim along the field lines until it reaches the OATZ of the sediment or the water column. It is necessary to note that the orientation itself is passive and only due to the magnetic torque exerted on the chain of magnetic particles. The movement of the bacterium is due to the use of its flagellum, and it is not magnetically driven, hence the term passive magnetotactic orientation, or magnetotaxis (Davila 2005).

Magnetotactic bacteria are, however, not a subject of this book, therefore I strongly recommend the readers to obtain more detailed information about this unique aspect of biomineralization including such topics as iron transport (Baeuerlein and Schüler 1995; Schüler and Baeuerlein 1998), genetics (Schüler 2008; Jogler and Schüler 2009; Lohße et al. 2011). Additional studies include molecular mechanisms that regulate the biomineralization of magnetite crystals with a typical morphology, shape and size (Scheffel et al. 2006; Kolinko et al. 2012), their deposition within intracytoplasmic membrane vesicles (Faivre et al. 2007), and magnetosome particles aggregate into regular chains (Bazylinski and Schüler 2009). Also recommended is visiting the Magnetolab website of Dirk Schüler (http://magnetolab.bio.lmu.de/de/prof_dirk_schueler/index.html). He and co-workers also work on the production and functionalization of magnetosome particles for their use in various biotechnological applications (Bäuerlein et al. 2001; Lang et al. 2007).

In contrast to magnetosomes of prokaryotic origin, SD crystals of magnetite with about 50 nm in diameter have been isolated from eukaryotic forms including animals. These nanoparticles "permanently magnetized bar magnets that twist into alignment with the earth's magnetic field if allowed to rotate freely. Single-domain magnetite crystals might transduce geomagnetic field information to the nervous system in several different ways. One possibility is that such crystals exert pressure or torque on secondary receptors (such as stretch receptors, hair cells or mechanoreceptors) as the particles attempt to align with the geomagnetic field. Alternatively, the movement of intracellular magnetite crystals might open ion channels directly if, for example, cytoskeletal filaments connect the crystals to the channels," (Lohmann and Johnsen 2000; see also Davila 2005).

According hypothesis (Walker et al. 1997), in animals, the functional activity of magnetite nanocrystals as magnetoreceptors supposes immediate contact with the nervous system. However, the strong evidence of such kind of contacts on anatomical level is still absent.

Some organisms possess so called superparamagnetic nanostructures, which differ from typical single-domain crystals in size (they are smaller) and have different magnetic properties (Hanzlik et al. 2000). As described by Lohmann and Johnsen (2000):

"One characteristic is that the magnetic axis of a stationary superparamagnetic crystal can move about to track the direction of an ambient, earth-strength field. By contrast, the magnetic axis of a single-domain crystal is fixed and stable under the same conditions, and the crystal itself must rotate physically to track the field. Superparamagnetic crystals generate fields strong enough to attract or repel adjacent crystals," (Lohmann and Johnsen 2000).

The magnetic sense of animals has generally been related to orientation and navigation purposes. The two most common orientation mechanisms observed in animals, based on magnetic field parameters, have been described: The inclination compass for directional information, and the navigational map for positional information (Davila 2005). With few exceptions, the biological mechanism underlying magnetic field perception in animals is still unknown. However, it has been suggested that inclusions of biogenic magnetite particles connected to nerve structures can provide a suitable transducer mechanism of the geomagnetic field (see for review Kirschvink et al. 2001).

Previously, Kramer (1961) suggested that animals transported to an unfamiliar site need both a map sense to determine the direction of the displacement from home, and a compass to find that direction. The *Map-and-Compass model* involves a two-step process for orientation (Davila 2005):

1. the animal establishes its position relative to the loft with the help of the map;
2. a compass system is used to locate the direction that will lead it home.

Over the past 60 years, it has been shown that several groups of animals have developed biological mechanism to extract map and compass information using geomagnetic field (Wiltschko and Wiltschko 1995; Fischer et al. 2001; Freake et al. 2006).

The geomagnetic field is relatively stable over biological time scales and is axial, with the magnetic field lines roughly directed north-south and symmetric in both hemispheres. This provides a reliable, static reference system for orientation and navigation. Alternatively, magnetic anomalies within the Earth's crust can also be recognized and used as reference features. Taking advantage of these properties of the geomagnetic field, some groups of animals have developed a biological magnetic compass, similar to the magnetic compass used by humans to locate the north magnetic pole. The magnetic compass has been described as an axial compass (also known as Inclination compass) for migratory birds and homing pigeons (Wiltschko and Wiltschko 1972; Walcott and Green 1974), and is based on the axial course of the geomagnetic field lines on the Earth's surface. The magnetic compass is used as a reference system and as a mechanism to maintain steady courses during homing and migrations. Therefore animals able to discriminate the minute but steady changes of the inclination angle and the intensity of the geomagnetic field can potentially establish their latitudinal position. To date, several models for position determination based on magnetic field parameters have been proposed (Davila 2005). Lohmann et al. (1999) proposed that sea turtles use a combination of intensity and inclination, as independent coordinates for map information. Contours of equal magnetic intensity and inclination form a grid that can potentially be used as a bi-coordinate position-finding system over areas of the Atlantic Ocean, where sea turtles spend most of their life cycle. This model cannot, however, be generalized since isolines of magnetic inclination and intensity intersect each other at high angles only over local regions of the Earth's surface (Davila 2005). In regions where the isolines are near-parallel to each other, or where the magnetic landscape is dominated by crustal magnetic anomalies, the bi-coordinate model is not viable for position determination (Walker et al. 2002).

Other model of position determination supposes systematic measurements of the intensity and in the direction of intensity gradient of the Earth's main field. The idea that intensity may be a component of the navigational map system of animals is based "on the observation that homing pigeons are disoriented when released at magnetic anomalies," (Davila 2005; see also Walcott 1978). On the other hand, homing pigeons show area-wide distributions of counterclockwise orientation errors, which are symmetrical about the line of intensity slope through the loft (Gould 1982) supporting the involvement of this field parameter in position determination (Walker 1998; Walker et al. 2002).

Yet a third model takes into account the regular patterns of magnetic anomalies originating from the hard substrates in the ocean crust, and produced during sea-floor spreading (Kirschvink et al. 1986). Although these anomalies are not present in the continental crust, they could potentially be used by marine animals to guide long-distance migrations.

Despite the fact that these three models take into consideration different magnetic field parameters, they are not mutually exclusive. So far there exists no evidence that all animals capable of magnetic field perception use the same sources of information based on the geomagnetic field for navigation and orientation. Furthermore, different groups of animals may have developed different magneto-receptor systems, each of them designed to obtain information from different field parameters according to their necessities and the local conditions of the geomagnetic field (Davila 2005).

3.5.1 Magnetite in Marine Fish

The first experimental results, which confirm the existence of the magnetic sense in the elasmobranch fishes and in salmon, were reported thirty years ago by Kalmijn (1982) and Quinn (1980), respectively. The *magnetotopotaxis* dependent movement by the sharks and their use of the Earth's magnetic field for navigation was described by Klimley (1993). As suggested by Walker et al. (2007), "navigating animals might use the topography of the residual magnetic field to construct a familiar area map," (Walker et al. 2007). After that, the magnetic material was found in teleosts (Hanson and Westerberg 1986) as well as in such marine fish species as *Thunnus albacares*, *T. alalunga* (Walker et al. 1984), *Scomber scombrus, Clupea harengus, Sarda sarda* and, in species migrating from brackish to freshwater and back: *Oncorhynchus tschawytscha* (Kirschvink et al. 1985b), *O. nerka* (Walker et al. 1988), *O. keta* (Yano et al. 1997), *Anguilla anguilla*, and in species inhabiting inland water bodies: *Cyprinus carpio* and *Perca fluviatilis* (see for review Formicki et al. 2002). Geomagnetic imprinting by fish as they swim away from natal areas has been proposed to generate magnetic preferences used in subsequent return migrations (Lohmann et al. 2008). Thus, magnetically stable SD magnetite particles have been extracted from the tissue of several fish species (Mann et al. 1988; Walker et al. 1984). These particles closely resemble the ones typically found in magnetotactic bacteria. In that respect, Walker et al. (1997) and Diebel et al. (2000) imaged a candidate magnetoreceptor structure in the basal lamina of the olfactory epithelium in the rainbow trout, *Oncorhynchus mykiss*. The candidate magnetoreceptor was interpreted as a chain of SD particles of the same size and shape as bacterial magnetosomes. However, a detailed characterization of the fine-structure of the chain and its connection to the surrounding tissue was not possible. Furthermore, the authors could only identify one cell containing the candidate magnetoreceptor structure. This lack of information does not allow for modeling of the functioning of the putative magnetoreceptor mechanism, hence the involvement of these particles in magnetic field perception in the rainbow trout remains as a working hypothesis.

Walker and co-workers (1988) studied magnetoreception in different life stages like fry, yearlings and smolts of the sockeye salmon, *O. nerka*. The authors reported following results:

- "significant quantities of SD have been localized within magnetite in connective tissue from the ethmoid region of the skull of adult (4-year-old) animals;
- the ontogenetic study revealed an orderly increase in the amount of magnetic material in the same region of the skull, but not in other tissues, of sockeye salmon fry, yearlings and smolts;
- SD magnetite particles suitable for use in magnetoreception are produced throughout life in the ethmoid region of the skull in *O. nerka*;
- there are enough particles present in the skulls of the fry to mediate their responses to magnetic field direction;
- by the smolt stage, the amount of magnetite present in the front of the skull is sufficient to provide the fish with a magnetoreceptor capable of detecting small changes in the intensity of the geomagnetic field," (Walker et al. 1988).

Interesting results have been reported about existence of specific area within trout nose (the region of location of the olfactory lamellae) that contains cells with magnetite particles. Moreover, it was shown that these magnetite-containing cells are innervated by the ros V nerve. It cannot be excluded that these cells "function as magnetoreceptors and relay information to the brain through the trigeminal nerve. Because reversals of field direction did not elicit responses from units in the ros V nerve, the putative magnetite receptors have been hypothesized to detect field intensity, a parameter that is potentially useful in a map sense," (Lohmann and Johnsen 2000; see also Bohannon 2007).

In the chapter above, I reported about the finding of magnetite microparticles within the vestibular apparatus of guitar fish (O'Leary et al. 1981; Vilches-Troya et al. 1984). Thus, the location place for magnetite in fish seems to be different for different species. For example, magnetite has been located in Sockeye Salmon (*O. nerca*) "near the basal lamina of the olfactory epithelium, the area innervated by the trigeminal nerve," (Muheim 2001; see for more information Mann et al. 1988; Walker et al. 1988, 1997). Here, some additional locations for SD magnetite nanoparticles within other fish species as follow (Diebel et al. 2000):

- dermethmoid cartilage of the skull, Chinook salmon (*O. tshawytscha*), Yellowfin Tuna (*Thunnus albacares*) (Walker et al. 1984; Kirschvink et al. 1985b);
- the olfactory lamellae in Rainbow Trout (*Oncorhynchus mykiss*) (Walker et al. 1997).

Interestingly, the fish species *Scombridae* and *Salmonidae* are the only examples, which confirm direct evidence of SD magnetite in vertebrates (Ueda et al. 1986; Cadiou and McNaughton 2010).

3.5.2 Magnetite in Marine Reptiles

Long-distance migrations are impressive and characteristic behavioral features of most sea turtle species. As reported by Lohmann and Lohmann (2006), "the journey that loggerhead turtles (*Caretta caretta*) take from their nesting beaches in Japan to their feeding areas near Baja California and back is the longest migration known for a marine animal," (Lohmann and Lohmann 2006). Thus, sea turtles are known to undergo migrations between nesting beaches and feeding grounds separated in some cases by thousands of kilometers (Carr 1967; Lohmann et al. 1999). How such navigation is accomplished has not been determined. By virtue of magnetoreception, however, turtles could potentially detect several parameters of local magnetic fields that could be used as components of a map sense (Gould 1985). Some geomagnetic parameters, such as field line inclination (dip angle), horizontal field intensity, and vertical field intensity vary with latitude (Skiles 1985). Any of these could be used as one component of a map for determining location with respect to a goal (e.g. a nesting beach). Turtles might also have the ability to detect and remember parameters of localized magnetic anomalies unique to specific sites such as feeding, mating, or nesting grounds. It was reported that complete darkness is not a barrier for hatchlings of leatherback sea turtle (*D. coriacea*) because of their excellent orientation using geomagnetic field. These observations suggest that "light-dependence is not a universal feature of vertebrate magnetic compasses," (Lohmann and Lohmann 1993). Young turtles that reside in coastal feeding grounds make a more sophisticate use of Earth's magnetic field. Instead of simply responding to boundaries delineated by magnetic fields, older turtles appear to learn the magnetic topography of their feeding grounds (Perry 1982). Magnetic cues aide them in navigating back to particular sites. These magnetic maps were shown to exist by subjecting juvenile green turtles (*Chelonia mydas*) to magnetic fields situated in locations to the north or south of their territory (Perry 1982). Turtles tethered in water close to their familiar feeding areas that were exposed to the northern fields swam south, and those exposed to a field in the south swam north (Perry 1982). This seems to indicate that the turtles were gaining positional information from the Earth's magnetic field and using this as a map to aid in their navigation (Perry 1982). It is not known if adults use this magnetic map for homing in on their natal beaches. It is reasonable to theorize that turtles imprint on the magnetic fingerprint for their birth territory, and use this information to guide them back to these regions when it comes time to reproduce (Lohmann and Lohmann 2006). If used, these magnetic maps may also be combined with chemical and visual landmarks to provide more accurate navigation (Lohmann and Lohmann 2006).

There are numerous works regarding magnetite and related topics in turtles (Kirschvink 1980; Perry et al. 1985; Mathis and Moore 1988; Lohmann 1991; Lohman and Lohman 1994; Courtillot et al. 1997; Lohmann et al. 2001, 2004). It seems clear that turtles use a ferromagnetic receptor, however, how sea turtles can sense the Earth's magnetic field in detail remains a mystery.

3.5.3 Magnetite in Sea Birds

There are two hypothesis concerning bird's "magnetic feeling". According to the first suggestion, specialized iron-rich cells are responsible for the phenomenon. The second theory "involves the spin of electrons, which is known to be affected by magnetic fields. Because this process requires light, molecules in birds' eyes may be receptive to magnetism," (Hamzelou 2012; see for more information Phillips 1996; Deutschlander and Muheim 2010). Thus, the magnetic compass of several bird species has been characterized as a so-called inclination compass, since it is based "on the axial course of the field lines and their inclination in space," (Davila 2005). According to the inclination compass model, the polarity of the vector is irrelevant: reversing the horizontal component has the same effect as reversing the vertical component. Reversal of both components, implying an inversion in the polarity of the field while maintaining the direction of the field lines, does not alter the behavior of the birds (Wiltschko and Wiltschko 1972). Behavioral and electrophysiological experiments in migrating and homing birds suggest that the inclination compass of birds may be related to the optic system (Wiltschko and Wiltschko 1981).

Physiological aspects of magnetite-based receptors as well as their distribution in birds are described in detail by Beason and Semm (1987). For example, "in the bobolink *[Dolichonyx oryzivorus]*, a migratory bird. In this case, magnetic material thought to be magnetite has been detected in an area of the upper beak. As in the trout, the region that contains the putative magnetite appears to be innervated by the ophthalmic branch of the trigeminal nerve. Specific neurons in the trigeminal ganglion, to which the ophthalmic nerve projects, respond to changes in vertical field intensity as small as about 0.5 % of the Earth's field," (Lohmann and Johnsen 2000; see also Semm and Beason 1990).

Recently, Cadiou and McNaughton (2010) described differences between birds and fish concerning the putative magnetoreceptor cells as follow:

(i) "in birds, magnetite is located in nerve terminals, whereas in the fish it appears to be located in the cell soma;
(ii) birds appear to use SPM magnetite, while fish use single-domain magnetite;
(iii) the magnetoreceptors are in the beak of birds but in the olfactory mucosa in fish," (Cadiou and McNaughton 2010).

Very recently, Treiber et al. (2012) showed, however, that the iron-rich cells with location in the rostro-medial upper beak of the pigeon *Columbia livia* are not magnetosensitive neurons. They are rather microphagous! Here, the sensation as reported: "Ultrastructure analysis of these cells, which are not unique to the beak, showed that their subcellular architecture includes ferritin-like granules, siderosomes, haemosiderin and filopodia, characteristics of iron-rich macrophages," (Treiber et al. 2012). Thus, this novel publication necessitates a traditional view for the true magnetite-dependent magnetoreceptors in birds. Recording from the brainstem within conscious pigeons, it was shown the presence of specialized neurons in the pigeon's brain that "encode the inclination angle and intensity of the geomagnetic field," (Vignieri 2012). These authors suggested that birds "can develop an internal model of geopositional latitude to facilitate spatial orientation and navigation based on magnetoreception," (Vignieri 2012; see also Wu and Dickman 2012).

The debate about optical pumping boils down to whether or not there is magnetite in the retina of Vertebrates. Walker reported that in Fish (Kirschvink et al. 1985a) and note that none of the 'critical' experiments using Rf magnetic fields on birds was done using double-blind techniques (Kirschvink et al. 2010). Numerous papers have been published on magnetite and magnetoreception in birds (Kirschvink 1982; Alerstam and Hogstedt 1983; Wiltschko and Wiltschko 1988; Edwards et al. 1992; Munro et al. 1997; Lohmann and Johnsen 2000; Wiltschko et al. 2002, 2009; Ritz et al. 2000, 2004; Davila et al. 2003; Mouritsen and Ritz 2005; Cadiou and McNaughton 2010; Falkenberg et al. 2010), however, with dominance to pigeons (Viguier 1882; Keeton 1971; Walcott and Green 1974; Bookman 1977; Leask 1977; Hanzlik et al. 2000; Winklhofer et al. 2001; Fleissner et al. 2003; Mora et al. 2004; Biro et al. 2007; Tian et al. 2007; Dennis et al. 2007; Treiber et al. 2012; Wu and Dickman 2012) and migratory songbirds (see for review Wiltschko and Wiltschko 1972; Beason et al. 1995; Wiltschko et al. 2002; Deutschlander and Muheim 2010). Unfortunately, to my best knowledge, there are no papers on magnetite in sea birds.

It was suggested (Wyeth 2010) that marine birds tracking odor plumes to find prey, burrows, or islands in conditions with variable wind direction could potentially switch to a magnetic compass as the next best sensory option, when visual cues for orientation are limited. Importantly, detecting this navigational strategy depends on testing the animals in a situation with unreliable primary cue(s) that will invoke the switch to a magnetic compass sense. Previous tests have not supported a role for magnetoreception in the wandering albatross' navigation (Bonadonna et al. 2005) without focusing observations during periods of wind heading variation (if they occur). Thus, any differences in navigational behaviors affected by the magnetic manipulations may have been overwhelmed by data collected in periods of steady wind, when a magnetic compass sense may not be used in navigation. Consideration of this hypothesis for any particular animal should begin with measurements of locomotory patterns, the location of navigational targets, and the sensory cues animals use to detect those targets. If situations arise where the animals appear to be abandoning the primary cues and switching to another cue prior to reaching a stationary navigational goal, then magnetic and control manipulations in those situations can test whether a magnetic compass sense plays a role in short distance navigation (Wyeth 2010).

3.5.4 *Magnetite in Cetaceans*

The idea to find and to extract the probably very large example of biomagnetite from whales, is definitively intriguing. However, to date there are no reports on any findings. Although the interest in studies on magnetoreception in mammals is high (Nemec et al. 2001), there are only few publications on magnetic senses in these marine mammals. Thus, it was reported about identification of some magnetite-like materials within brains of a Cuvier's beaked whale (*Ziphius cavirostris*) (Cranford et al. 2008), bottlenose dolphin (*Tursiops truncates*), Dall's purpoise (*Phocoenoides dalli*), and the humpback whale (*Megaptera novaeangliae*) (Bauer et al. 1986). In her work that was extended by Kirschvink and co-workers (Kirschvink et al. 1985a, b; Kirschvink 1990) on magnetic sense in cetaceans, Klinowska (1985) suggested as follow: "healthy whales that strand themselves alive must have made a serious navigational mistake and that analyzing the circumstances surrounding such strandings might identify the sensory modality responsible for the error," (Walker et al. 1992). Klinowska found correlation between stranding positions of the animals and areas where magnetic minima intersect the coast of Great Britain. Similar experiments have been carried out by Kirschvink and his colleagues on the east coast of the United States. These researchers developed methods that demonstrated statistically reliable correlations of whale's stranding sites with locations where magnetic minima intersected the coast (Weisburd 1984). Result: "total intensity variations of as little as 50 nano Tesla (nT; 0.1 % of the total field) were sufficient to influence stranding location" (Walker et al. 1992; see also Kirschvink 1990; Kirschvink et al. 1985a, b, 1986). These results confirm the hypothesis that mysticete species "possess magnetic sense that they use to guide migration" (Walker et al. 1992).

3.5.5 *Conclusion*

There are no doubts, that numerous behavioral experiments that have been carried out during last decades (for review see Eder et al. 2012) give us evidence for the existence of a magnetic sense in the animal's world, including representatives of marine fauna. According to Joseph Kirschvink, "all matrix-mediated biomineral systems in higher animals could have evolved from the magnetite system during the Cambrian explosion" (Kirschvink et al. 2001; see also Kirschvink and Hagadorn 2000). Today, only the light-dependent and magnetite-based magnetoreception theories are accepted as those which try to explain the phenomenon to detect Earth-strength magnetic fields by living organisms. Further progress requires development of procedures that will distinguish between magnetic contaminants and biologic precipitates in tissue. Very recently, Michael Winklhofer and co-workers represented an effective approach how to isolate and to characterize potential magnetite-based magnetoreceptor cells from the trout olfactory epithelium (Eder et al. 2012b). They used confocal reflectance imaging, that allow to visualize a μm-sized intracellular structures of iron-rich crystals as bright reflective spots localized close to the cell membrane. These authors summarized that "the magnetic inclusions are found to be firmly coupled to the cell membrane, enabling a direct transduction of mechanical stress produced by magnetic torque acting on the cellular dipole in situ. These specialized cells clearly meet the physical requirements for a magnetoreceptor capable of rapidly detecting small changes in the external magnetic field," (Eder et al. 2012).

3.6 Biohalite

Abstract Cranial exocrine formations that secrete hyperosmotic electrolyte solutions are known as salt glands. They supplement renal ion excretion in some nonmammalian vertebrates. NaCl is the main component of the secreted electrolytes in most animals including sear birds and sea reptiles. The secretory epithelium of salt glands contains salt secreting principal cells. In this chapter, I take the liberty to introduce the new term "*biohalite*" for the NaCl-based biomineral originating from the salt glands.

Halite, the natural form of salt (NaCl), is a very common and well-known mineral (Finger and King 1978). It is found in solid masses, and as a dissolved solution in the oceans and salt lakes. However, marine vertebrates including fish, reptiles and birds possess unique organs, the salt glands, which are responsible for production of NaCl. Moreover, they are "able to produce a highly concentrated salt solution, making it possible to tolerate drinking sea water," (Schmidt-Nielsen and Fange 1958a). Schmidt-Nielsen and Fänge (1958b) discovered this phenomenon. The physiology of salt glands in numerous representatives of tetrapod and elasmobranch fish (Fänge and Fugelli 1962) with respect to better understanding of the principal mechanisms by which these glands facilitate the net secretion of sodium (or potassium) chloride, has been studied in great detail during last 60 years (Peaker and Linzell 1975; Gerstberger and Gray 1993; Shuttleworth and Hildebrandt 1999; Hildebrandt 2001; Dantzler and Bradshaw 2009; Holmgren and Olsson 2011; Babonis and Brischoux (2012). Interesting results has been reported in comparative studies between marine and freshwater species (Babonis and Evans 2011). Studies include work looking at the role of water-regulatory proteins in modulating the secretory output of the salt glands, and researching the diversity in the composition of the secreted fluids. In addition, such topics as function of salt glands under various environmental conditions, bacterial infections of salt glands, their phenotypic plasticity as well as the regulatory mechanisms of secretion by various neurological and endocrine agents, have been studied. Additional work has also been done to investigate the combined "osmoregulatory function of salt glands and other organs," (Babonis and Brischoux 2012; see also Babonis et al. 2011). Interestingly, although the anatomy, structure and function of these glands has been quite well studied, there have been no reports on NaCl as form of a biomineral that is formed because of activity of the corresponding cells. Here, I would like to introduce the term "*biohalite*" for the NaCl-based biomineral that originates from the salt glands.

3.6.1 Diversity and Origin of Salt Glands in Marine Vertebrates

Fish Observations of marine shark's species in rivers suggest their specific ability for acclimatization in aquatic niches with different levels of salinity. Also experiments under laboratory conditions has demonstrated that marine elasmobranchs "do have the capacity to acclimate to changes in salinity through independent regulation of Na^+, Cl^-, and urea levels," (Hazon et al. 2003). Thus, some stenohaline marine elasmobranchs can be accepted as partially euryhaline. However, "once optimally adapted to fresh water, recolonization of sea water by elasmobranchs is problematic due to the loss of urea synthetic capacity and renal structures for urea retention," (Ballantyne and Robinson 2010).

Sharks like *Squalus acanthius* can effectively maintain osmotic homeostasis on the level of urea concentrations ranging between 300- and 400-mM, and corresponding marine osmolalities of 900–1,000 mosmol/kg H_2O (Zeidel et al. 2005). Also the rectal gland that is isotonic with the plasma in this fish species can maintain salt balance without losing urea by secreting a NaCl-rich (500 mM) and urea-poor (18 mM) fluid. An investigation into structural features of apical and basolateral membranes from shark rectal glands showed that their epithelial cells are permeable to water and not to urea (Zeidel et al. 2005). For example, it was reported that "the basolateral membrane urea permeability is fivefold lower than would be anticipated for its water permeability," (Zeidel et al. 2005). These results show the important role of basolateral membranes as selective barrier within the rectal gland of sharks (see also Silva et al. 1990).

Dinosauria and Reptiles All dinosaurs were probably uricotelic animals which had to use the extrarenal way of excreting excess of monovalent ions. It is suggested (Osmolska 1979) that they were able to use the nasal salt gland for this purpose. Its presence may have been especially important for unloading the excess of potassium ions ingested by large herbivores with their vegetarian food, or for getting rid of sodium ions by herbivores living in a saline environment.

A salt-secreting gland was first demonstrated in marine reptiles by Schmidt-Nielsen and Fänge (1958b). These scientists showed that a gland situated in the orbit of the eye was capable of elaborating a secretion with twice the sodium concentration of sea water. Thus, the salt gland is an organ for excreting excess salts in such recent marine reptilians as crocodiles (Dunson 1970; Taplin and Grigg 1981; Taplin et al. 1982; Cramp et al. 2007, 2008, 2010a, b), iguanas (Dunson 1969; Hazard 2004), sea snakes (Taub and Dunson 1967) as well sea turtles (Ellis and Abel 1964; Abel and Ellis 1966; Kooistra and Evans 1976; Hudson and Lutz 1986; Lutz 1997; Reina and Cooper 2000; Reina et al. 2002).

Saltwater crocodiles (*Crocodylus porosus*) possess very specialized lingual salt glands. Peculiarities of osmotic challenge based on function of these glands are still unknown. Recently, Cramp et al. (2010a, b) studied the regulation and distribution of the Na^+/K^+-ATPase (NKA) pump. These authors paid special attention to the *a*-(catalytic) subunit in the salt glands of *C. porosus* specimens, which were permanently acclimated during six months to freshwater (FW) or 70 % seawater (SW). Here, the results reported by these authors:

- "the NKA was immunolocalised to the lateral and basal membrane of secretory cells;
- the NKA alpha-subunit was 2-fold more abundant in SW-acclimated C. porosus salt glands.
- NKA gene expression was elevated in the salt glands of SW- vs FW-acclimated crocodiles.
- no increase in the specific activity of NKA in SW-acclimated animals;
- the proportion of tissue oxygen consumption rate attributable to NKA activity was not different between SW- and FW-acclimated animals;
- the salt glands of SW-acclimated animals were larger than those of FW-acclimated animals;
- there were significantly more mitochondria per unit volume in secretory tissue from SW-acclimated animals;
- crocodiles possess the capacity to moderate NKA activity following prolonged exposure to SW" (Cramp et al. 2010a, b).

It can be suggested (Cramp et al. 2008) that the salt glands of saltwater crocodiles are phenotypically plastic, both from morphological and physiological view and play crucial adaptive role in acclimatization of these unique reptiles to different levels of salinity.

Interesting phenomenon has been reported by Hazard (2001) concerning the ability of nasal salt glands of some lizards to secrete potassium as well as sodium, in the form of chloride or bicarbonate. A single lizard, the Galapagos lizard (*Amblyrhyiichus*

Fig. 3.29 Sneezing Galapagos lizard (*Amblyrhyiichus cristatus*) occasionally blows a spray of salt droplets (Image courtesy http://www.pbase.com/image/108002180) (*above*) and http://www.flickr.com/photos/abizeleth/3629306987/ (*below*)

cristatus) is marine reptilian. It lives on the beaches of the Galapagos Islands where it eats seaweeds and occasionally blows a spray of droplets out through its nostrils, reminiscent of a puff of smoke from a "fiery dragon" (Fig. 3.29). This endemic reptile can secrete as much as 95 % of its sodium (Dunson 1969).

Shoemaker et al. (1972) showed that the salt glands of the desert iguana (*Dipsosaurus dorsalis*) responded only to stimulation by alkali metal ions (sodium, potassium, and rubidium). Hazard (2001) reported about experiments with these reptiles where the factors responsible for initiating secretion by the gland and the rates of cation and anion secretion were investigated. During several days, desert iguanas were given combinations of ions, and secreted salt was collected daily and analyzed for Na, K, Cl, and bicarbonate. Cation secretion, with the maximum cation secretion rate of 4.4 ± 0.38 µmol/g/day, ranged from 24 to 100 % potassium. Surprisingly, even high NaCl loads did not abolish secretion of K ions in the form of chloride. It was hypothesized that the nuance of the response of the salt gland of *Dipsosaurus* may be related to the ecological importance of both potassium and chloride of the dietary origin for herbivorous desert lizards (Hazard 2001).

Except for the brief period of egg-laying, five species of turtle are strictly marine: namely, the Ridley's turtle (*Lepidochelys olivacea*), green turtle (*Chelonia mydas*), the hawksbill turtle (*L. imbricata*), the loggerhead turtle (*Caretta caretta*), and the leatherback turtle (*Dermochelys coriacea*) (Holmes and McBean 1964). The salt gland appears to be the predominant route of Na and K excretion in the marine turtle *C. mydas mydas*. The kidney of this reptile is suggested not to be capable effectively maintain a positive water balance under electrolyte loads presented by food and sea water itself. As reported by Nicolson and Lutz (1989):

"the salt gland fluid of the marine turtle *C. midas* was protein-free, and was mainly composed of Na^+ and Cl^-, in similar relative concentrations to those in sea water. It had substantial amounts of K^+, Mg^{2+} and HCO^{3-}, and negligible amounts of glucose," (Nicolson and Lutz 1989).

Seabirds According to hypothesis (Fernandez and Gasparini 2000), avian salt glands evolved from the nasal glands of reptiles in the late Paleozoic. These glands were localized "immediately under the skin in supraorbital depressions of the frontal bone in the skull of *Charadriiform* birds, but in other groups they may be located above the palate or within the orbit of the eye," (Hughes 2003; see also Fernandez and Gasparini 2000). Probably, such ancient birds as *Ichthyornis* and *Hesperornis* lived in a marine habitat because their skulls possess similar depressions (Marples 1932).

Fish is the main diet of numerous piscivorous marine birds and contain water of a lower salinity than seawater (see for review Goldenstein 2002). In contrast to this diet, marine invertebrates such as crustaceans and mollusks are in osmotic equilibrium with seawater. Correspondingly this kind of feed requires an effective method of salt excretion in sea birds using specialized glands (Schmidt-Nielsen 1960).

Jobert (1869) was the first to describe the avian salt gland as being formed of two distinct segments with separate drainage ducts. The avian salt gland is a countercurrent system that concentrates the secreted salt solution. The capillaries are arranged so that the flow of blood is in the direction opposite to the flow of secretory fluid. This flow maintains a minimum concentration gradient between blood and the tubular lumen along the entire length of the tubule. The excretion of electrolytes by the salt gland is assumed to be dependent on both cholinergic innervation and the presence of adrenocortical hormones (Kühnel 1972). It was reported: "the great number of autonomic nerves in close contact with the secretory cells indicates that the salt gland of birds is controlled by the nervous system," (Kühnel 1972; see also Schmidt-Nielsen 1960) (Fig. 3.30).

The activity of the salt gland is an all-or-none phenomenon (Schmidt-Nielsen et al. 1957). If there is an osmotic load, the gland secretes; in the absence of an osmotic load, the gland is at complete rest. In this intermittency, the gland differs from the kidney which produces urine continuously. The concentration of salt in the nasal secretion is always very high, and remains fairly constant for each species. For example, as reviewed by Schmidt-Nielsen (Schmidt-Nielsen 1960), in the cormorant, which eats fish that is relatively low in salt content, the concentration in the secretion from the gland is about 500–550 mEq/L. The petrel is a bird with pronounced oceanic habits. It spends most of its life at sea and comes to land only to breed. It lives on planktonic organisms, mostly crustaceans, which it picks off the surface of the ocean as it flies by. The planktonic organisms impose a considerable salt load. It is no surprise, therefore, that in the petrel the salt concentration of the nasal fluid is higher than in any other bird that has been examined, up to 1,100 or 1,200 mEq/L.

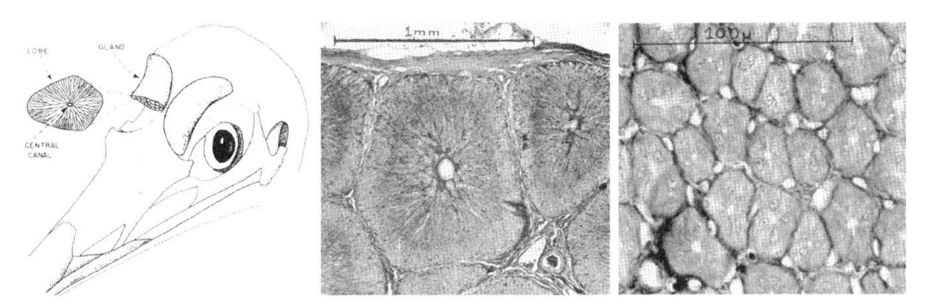

Fig. 3.30 The salt glands of the gull of longitudinal lobes of about 1 mm in diameter (*left*), each lobe has a central canal with the branching secretory tubules arranged radially around it (Drawing by M. Cerame-Vivas). In cross section the lobe of the salt-secreting gland (*middle*) shows the central canal and the radial arrangement of the secretory tubules. In higher magnification the peripheral ends of the secretory tubules appear as closed tubes without any similarity to the glomerular apparatus of the kidney. The secreting tubules (*right*) are interspersed with capillaries (Adapted from Schmidt-Nielsen 1960)

No other gland in higher vertebrates, except the mammalian kidney, can produce fluids which are concentrated to this degree. The concentration limit for electrolytes in the kidney of man is about 400 mEq/L, in the rat 600 mEq/L, in the kangaroo rat, 1,500 mEq/L, and in the champion concentrator, the North African rodent *Psammomys*, 1,900 mEq/L (Schmidt-Nielsen 1960). Thus, the salt gland compares favorably to the kidney in its concentrating ability.

Osmotically free water is the product generated due salt gland secretion. The phenomenon is briefly reviewed by Hughes (2003) and Hughes et al. (2007) as follows:

"Acclimation to saline induces interstitial water and Na move into cells. When the bird drinks seawater, Na enters the plasma from the gut, and plasma osmolality (Osm(pl)) increases. This induces water to move out cells expanding the extracellular fluid volume (ECFV). Both increases in Osm(pl) and ECFV stimulate salt gland secretion. The augmented intracellular fluid content should allow more rapid expansion of ECFV in response to elevated Osm(pl), and facilitate activation of salt gland secretion. To fully utilize the potential of the salt glands, intestinally absorbed NaCl must be reabsorbed by the kidneys. Thus, Na uptake at gut and renal levels may constrain extrarenal NaCl secretion," (Hughes 2003).

3.6.2 Salt Glands: From Anatomy to Cellular Level

The common anatomical feature for all tetrapods is that their salt glands are localized within cephalic area. However the anatomical position of these formations varies among animal groups. Thus, according to Babonis and Brischoux (2012) three following cephalic areas are "currently recognized:

(1) nasal glands in extinct archosaurs, extant birds, and lizards;
(2) orbital glands in turtles;
(3) oral glands in extant crocodiles and snakes," (Babonis and Brischoux 2012).

It was observed, that salt glands of sea birds are 10–100 times larger in size than in non-marine species. The salt concentration of corresponding fluids "varies between species and increases as the glands hypertrophy in response to elevated salt intake" (Suepaul et al. 2010; see also Schreiber and Burger 2001). The anatomical description of sea bird salt glands have been described as follow:

"They are compound tubular glands with a well-developed lobular structure, and a variable countercurrent arrangement of capillary blood flow relative to the flow of secretion within tubules. This allows controlled excretion of salt ions from the bloodstream into collecting ducts, which open into the nasal cavity," (Suepaul et al. 2010; see also by Gerstberger and Gray 1993).

Fine structure, innervation and functional control of marine vertebrate's salt glands are well described (see for review Gerstberger and Gray 1993). The first detailed morphologic description of the avian salt gland at the electron microscopic level was by Doyle (1960) using specimens of the Great Black-Backed gull (*Larus marinus*) and the petrel (*Oceanodroma leucorrhoa*).

Concerning the embryological origin of the salt glands, Marples (1932) stated that the gland is eventually formed by branching of the two main ducts arising from the rudiment of the nasal cavity, and then growing backward to the final position. Histological studies performed with the Adelie penguin (*Pygoscelis adeliae*) allowed the first rudiments of supraorbital salt glands to be traced back to "solid crescent structures on either side of the nasal cartilage," (Stonehouse 1975) growing posteriorly to develop dorsal to the eye (Herbert 1975). Thus, the glandular matrix develops from the ducts as branched tubules radiating from a central canal.

The structure of the salt gland is essentially the same in all birds (Babonis et al. 2009). It is roughly triangular in cross section, and is divided into lobules by septa of dense fibrous connective tissue. The salt glands are principally hierarchically structured. Thus, "each of these lobules consists of many branched tubules which join a central collecting canal. The secretory tubules are lined by wedge shaped columnar cells," (Kühnel 1972; see also Babonis et al. 2009). On the ultrastructural level, characteristic infoldings of the plasma membrane are well visible in electron microscope (see, for example, Kühnel 1972). The basal striations, which can be observed using light microscopy, are associated with the mitochondria (Fig. 3.31). Closely packed microtubules are located parallel to each mitochondria. As described by Kühnel (1972) in detail:

"Usually there exist two or three capillaries, running from the central region of the lobule towards the periphery. Unmyelinated nerve fibers are observed regularly outside the basement membrane of the tubule cells and the central duct cells in the perivascular space, as well as independently from the blood vessels. Structurally, these interstitial nerves consist of several axons (0.4-1.8 μm thick) surrounded by a Schwann cell sheath," (Kühnel 1972).

Salt glands as example of cephalic glands principally differ from similar structures because of high specialization of their secretory epithelium that is represented exclusively by salt secreting principal cells. Some species of marine snakes are the best examples for this (Dunson et al. 1971; Dunson and Dunson 1974; Babonis et al. 2009).

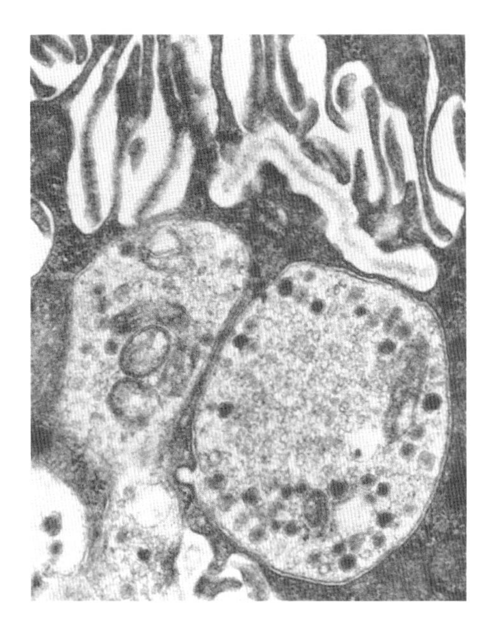

Fig. 3.31 The base of a secretory tubule cell possess two nerve terminals situated in special furrows. The extensive infoldings of the basal plasma membrane are localized in the upper part of the image (magnification ×50,000; Adapted from Kühnel 1982, with kind permission of Springer Science+Business Media)

3.6.3 Conclusion

A thorough overview on salt glands with attention on their nonhomologous origin has been done by Nicolson and Lutz (1989). These supraorbitally located glands represent one of the most effective organs in the vertebrate kingdom involved in the epithelial transport of ions (sodium and chloride) against a marked concentration gradient (Gerstberger and Gray 1993). Thus, the organs used as salt glands in different organisms are nonhomologous and have apparently evolved independently (Peaker and Linzell 1975). Ion secretion occurs in the rectal glands of sharks (Burger and Hess 1960), the sublingual glands of snakes (Dunson 1968; Dunson et al. 1971), the nasal glands of birds and lizards, and the postorbital glands of turtles (Peaker and Linzell 1975). Despite their nonhomologous origins, the salt glands of these different organisms are all characterized microscopically by principal cells with large quantities of mitochondria and convoluted lateral, and sometimes basal, membranes (Ellis and Abel 1964; Peaker and Linzell 1975). The similarity in the structure of the salt glands of different species suggests that there may be similarities in salt gland operation as well. There is evidence of hypertrophy in response to long-term increases in salt loading in both birds and turtles (Peaker and Linzell 1975; Cowan 1969, 1971). Unfortunately, there is lack of knowledge about the visualization of biohalite production by secreting cells at the nanolevel. However, I hope that the recently established method to obtain primary culture of avian salt gland secretory cells (Lowy et al. 1989) could be used for these kinds of experiments in the future. Furthermore, the biomineralogical community needs data on both chemical composition and nanostructural organization of biohalite that can be obtained from different taxa of marine vertebrates.

3.7 Pathological Biomineralization in Marine Vertebrates

Abstract The formation of crystalline structures in pathological mineralization follows the same principles as normal biomineralization. The mechanism of formation of various stones in organisms of marine vertebrates remains unclear. Stones consist of several kinds of minerals like struvite, calcium carbonate and phosphate; either alone or in combination. Different types of calculi (dental, nasal, renal, vaginal, cystic) as well as stones (uroliths, nephroliths, enteroliths, faecoliths) observed in fish, reptilians and marine mammals are described and discussed in this chapter.

According to the classical definition proposed by Catherine Skinner "Biominerals may be deposited within the organism, and within its immediate surroundings or environment, as a result of the metabolism of the living creature," (Ehrlich et al. 2008; see for more information Skinner 2000). Biominerals of the vertebrates origin are divided into two main types (Skinner 2000), which are:

(i) essential, a normal part of the expected physiology of animal systems, such as the mineral matter found in bones and teeth

and

(ii) unexpected, and undesired, or pathologic mineral deposits including stones (renal, pancreatic, kidney), calculi (pancreatic and urinary),cystoliths, bladder stones; gallstones rhinoliths (calculus present in the nasal cavity); tonsilloliths (oropharyngeal concretions); vaginoliths—vaginal calculi; cardiolytes, cutaneous calculi, enteroliths; sialoliths—salivary submandibular and parotid gland stones; ptyaliths—calculus in a salivary glands; dental calculi (see for review Ehrlich et al. 2008). "The formation of crystals in pathological mineralization follows the same principles as normal calcifications," (Ehrlich et al. 2008; see for more information Magalhaes et al. 2006). The mechanism of formation of the calcium carbonate stones remains unclear. In some species, it is thought that formation occurs due to precipitation surrounding an organic nidus, such as proteins, cellular debris, crystals, or foreign bodies (Confer and Panciera 1995).

Struvite seems to be the next mineral often seen in different pathological cases. Struvite ($NH_4MgPO_4 \cdot 6H_2O$) is a magnesium ammonium phosphate hexahydrate. Struvite crystals are the orthorhombic and appear as brownish-white, or white to yellowish pyramidal crystals. Also platy mica-like forms of struvite have been described. With Mohs hardness of 1.5–2 and a low specific gravity of 1.7, it is related to the soft minerals. Struvite is well soluble in acids, however sparingly soluble in neutral and alkaline conditions (see for review Ferraris et al. 1986).

In some cases, known for humans and animals, ammonia-producing microorganisms are responsible for formation of struvite urinary stones and crystals. Also high magnesium/plant-based diets lead to this kind of pathological biomineralization (Osborne et al. 1985).

Marine vertebrates possess numerous pathologic mineral structures. Some of these examples of abnormal biomineralization (Smith 1982) are represented and discussed below.

Calculi The term calculus is derived from the Latin for "pebble" and is defined as an abnormal concretion occurring within the body. Formation of calculi is known as lithiasis (Greek: *lithos* = stone).

Dental Calculi In contrast to well reported cases of pathological conditions on teeth in primates and mostly in humans, similar situations are particularly very rare in marine mammals, and in wild animals in general. Only few species of cetaceans provided scanty records of their dental anomalies. Recently, Loch and co-workers (2011) carried out a systematic study of dental pathology in dolphins. Abnormalities such as mineralized calculus deposits were diagnosed in the delphinids *Delphinus capensis*, *Orcinus orca*, *Steno bredanensis*, *Sotalia guianensis*, *Lagenodelphis hosei*, *Stenella coeruleoalba*, *Tursiops truncatus*, and *Pseudorca crassidens*. Unfortunately, the aetiology of calculus accumulation in these marine animals remains unknown.

Nasal Calculi Nasal Calculi were observed by Curry et al. (1994) in the lumens of posterior nasal sacs and the anterior nasofrontal sacs (for anatomical details see Mead 1975) of eight dolphin specimens: one *Australophocaena dioptrica*, one *Phocoena spinipinnis*, three *P. phocoena*, and three *P. dalli*. Although several of these calculi were completely surrounded by a thin tissue layer, the main fraction of calculi was embedded within the trabeculate areas and within tiny diverticula within epithelial layer of the sacs. No inflammation has been observed in these areas of the nasal tissue. On cross sections of the calculi isolated from *P. dalli* some surface crystals and a "layer of alternating medium tan to lighter tan bands of concentric laminations with a light tan unoriented central core" (Curry et al. 1994) are visible. The chemical composition of these calculi was as follow: 20 % apatite and 80 % calcium carbonate (Curry et al. 1994).

Up today, there are no data which could explain the mechanisms of abnormal mineralization within nasal sacs of phocoenid species. However, following suggestions have been reported as summarized by Curry et al. (1994):

- "abnormal mineralization may be the result of normal tissue healing, following cell necrosis or ectopic mineralization due to excessive supersaturation;
- compression of nasal diverticula may cause water condensation within the upper respiratory tract of dolphins. Foreign particulate matter from inspired air or seawater may remain in the biological fluids found in the nasofrontal and posterior nasal sacs, and supersaturation and subsequent mineralization involving chemical alteration of the foreign substance may occur;
- the calculi are dystrophic calcifications of desiccated mucosal secretions and debris combined with components of seawater," (Curry et al. 1994; see for more information Smith 1982; Simkiss and Wilbur 1989; Coulombe et al. 1965).

Renal Calculi Simpson and Gardner (1972) and Howard (1983) reported about findings of renal calculi of unidentified composition in bottlenose dolphins (*Tursiops truncatus*) and an unspecified beaked whale, respectively. Cowan et al. (1986) observed calcium phosphate- and calcium oxalate-based obstructive calculi within the ureter of a Pacific white-sided dolphin (*Lagenorhynchus obliquidens*). However, widely known cases of renal calculi in delphinids are uric acid-based renoliths, which have been isolated from captive specimens of bottlenose dolphins (Miller 1994; Reidarson and McBain 1994; Townsend and Ringway 1995), and. in Pacific white-sided dolphin (*Lagenorhynchus obliquidens*) (Boehm et al. 1997).

Vaginal Calculi Vaginal calculi have been reported from *Delphinus delphis* by Harrison (1969), who described them as hard, flattened formations containing both organic and inorganic material; and suggested that they were remnants of vaginal plugs of coagulated seminal fluid. Sawyer and Walker (1977) reported nine instances of vaginal calculi from sexually mature dolphins. They demonstrated that these calculi are composed of four types of calcium phosphate compounds also found in mammalian bone. In other species struvite, not calcium phosphate, was the main component of these kinds of calculi. For example, the finding of struvite as a major component in two vaginal calculi of Peruvian dusky dolphin (*Lagenorhynchus obscurus*) suggested an infectious aetiology (Van Bressem et al. 2000). Struvite calculus was reported also in the vagina of a bottlenose dolphin (*T. truncatus*) by McFee and Carl (2004). Benirschke et al. (1984) also described vaginal calculi from a common dolphin. The morphological diversity of vaginal calculi of the *D. delphis* was thoroughly represented (Fig. 3.32). It was suggested that "bone remnants from nonexpelled aborted fetuses may play a pivotal role in vaginal calculi formation," (Burdett and Osborne 2010; see also Woodhouse and Rennie 1991).

A large number of vaginal calculi were recently observed for the first time in a juvenile harbor porpoise (*Phocoena phocoena*) stranded on Whidbey Island, Washington. Histologic examination of the urinary tract revealed mucosal hyperplasia most likely attributable to the calculi (Norman et al. 2011). The calculi were numerous (>30) (see Fig. 3.33) and composed completely of struvite. On culture they yielded *Enterococcus* spp., a bacterium not usually associated with struvite urolith formation in domestic animals.

Cystic Calculi Usually, such factors as excess amounts, or inappropriate types of dietary protein as well as water deprivation lead to development of cystic calculi (Nutter et al. 2000). The occurrence of these calculi in a variety of wild and captive turtle species have been reported. During a postmortem examination of a western spiny softshell turtle (*Apalone spiniferus hartwegi*), a calculus ($24 \times 21 \times 12$ mm) with a dry weight of approximately 5 g was discovered in the urinary bladder. Chemically, the calculus was a mixture of apatite, struvite, and unidentifiable crystals. This is the first report of a cystic calculus in a wild chelonian (McKown 1998).

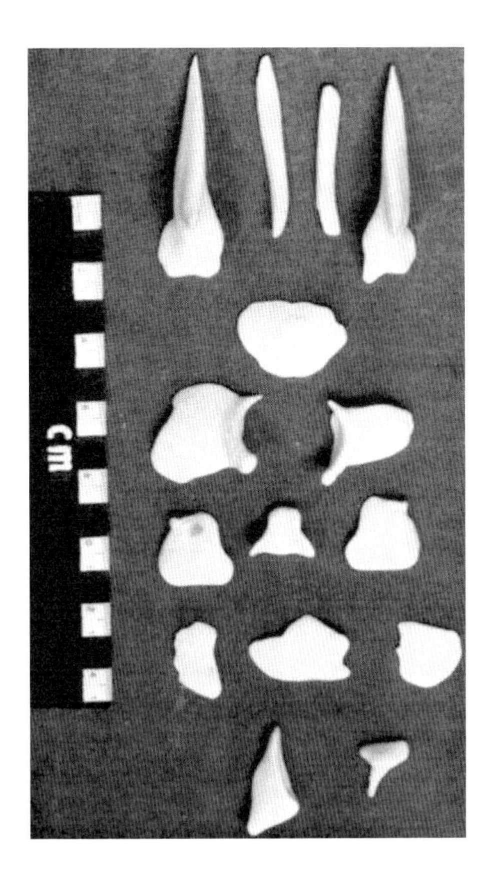

Fig. 3.32 Vaginal calculi recovered from beach cast *Delphinus delphis* (Adapted from Woodhouse, Rennie (1991). Reprinted with permission. © 1991, Wildlife Disease Association)

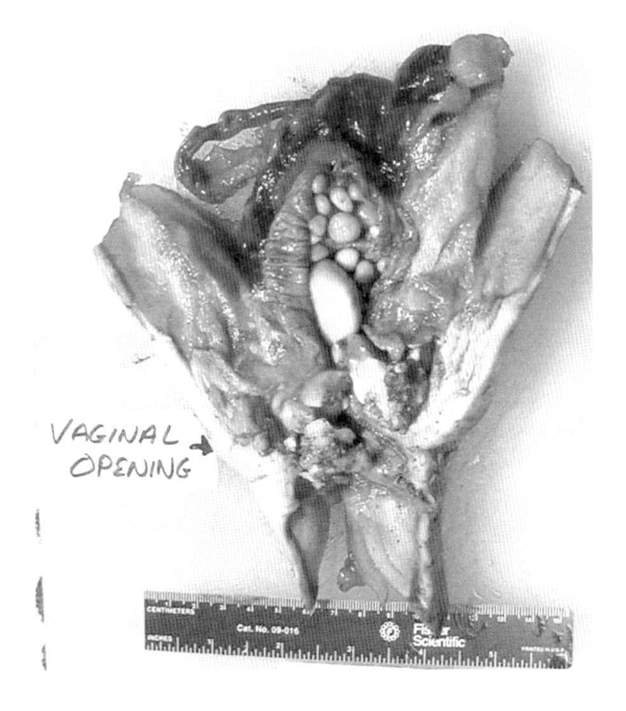

Fig. 3.33 Multiple struvite-based calculi in the urogenital slit and vagina of a harbor porpoise (*Phocoena phocoena*) (Reprinted with permission from Norman et al. (2011). Copyright 2011 by American Association of Zoo Veterinarians)

Uroliths (Bladder Stones) Uroliths (bladder stones)"are abnormal mineral salt concretions found in the urinary tract," (Dennison et al. 2007; see also Berenyi 1972). Several factors including such as such as suboptimum temperatures, acute or chronic inflammatory conditions, dehydration, or purine-rich (animal protein) diets seems to be responsible for their development. Of course, the phenomenon is species dependent. Kidney stones in reptiles may be 100 % monosodium urate, or they may be calcium oxalate (from feeding a diet high in plant foods containing oxalate), or calcium phosphate (Kaplan 1997). The biochemistry of their formation is briefly described by Kaplan (1997) that monosodium urate is the form in which uric acid usually presents in the blood. Both substances are largely insoluble in water. The phenomenon of hyperuricemia develops when the serum concentration of either free uric acid or urate salts is elevated. One of the characteristic marks of this disease is so called tophi, the crystals which may circulate through the blood or even the synovial fluid. Finally, tophi can be observed in tissues, organs and joints throughout the body. Crystals of monosodium urate affect gout, however the pseudogout, another acute inflammatory state, is caused by crystals other than sodium urates (Kaplan 1997).

Of course, urine supersaturation with respect to uric acid can determine the formation and development of uric acid uroliths. However, how this supersaturation can occur? One of possible ways is related to urease-producing microorganisms, which are responsible for corresponding infection. In this case, the mechanism can be described as follow: "These bacteria split urea, leading to an increase in urinary ammonia concentration and pH, which favors precipitation of struvite crystals," (Dennison et al. 2007). Relatively uncommon in marine mammals, uroliths are most frequently reported in cats and dogs where they are generally in the form of struvite and calcium oxalate (Dennison et al. 2007). However, two calculi were found in the urogenital sinus of a 70 kg female sand tiger shark (*Odontaspis taurus*). The calculi were white in color, rough on the surface, and spherical in shape. Crystallographic examination revealed that they were composed of struvite (80 %) and calcium phosphate (15 % carbonate apatite). Approximately 5 % of the stone matrix consisted of blood and protein; a distinct bacterial nidus was not present microscopically (Walsh and Murru 1987).

Reports concerning findings of uroliths in marine mammals include urate nephrolithiasis in a harbour seal (*Phoca vitulina*) (Stroud 1979; Greig et al. 2005) and in a river otter (*Lontra canadensis*) (Grove et al. 2003); struvite penile urethrolithiasis in a pygmy sperm whale (*Kogia breviceps*) (Harms et al. 2004); urinary calculus in pygmy killer whale *Feresa attenuate* (Zerbini et al. 1997); two cases of calcium carbonate nephrolithiasis in West Indian manatee (*Trichechus manatus*) (Moliner 2005; Keller et al. 2008); and urate calculi in bottlenose dolphins (*T. truncatus*) (Miller 1994; Venn-Watson et al. 2010a; Schmitt and Sur 2012).

Uroliths of the uric acid origin have been previously reported in several species of odontocetes. Usually, they have been located primarily in the ureter or kidneys of these animals (Miller 1994; Reidarson and McBain 1994; Townsend and Ringway 1995; Boehm et al. 1997). Probably, the biggest struvite-based stone was observed in the penile urethra of a stranded pygmy sperm whale (*Kogia breviceps*) as reported by Harms and co-workers (2004). The multinodular structured construct was 16 cm long and 3.7 cm diameter and weighed 208 g. Intriguingly, the ureasepositive bacterium *Klebsiella oxytoca* that is occasionally associated with struvite urolith formation in domestic animals, was isolated from this pathological biomineral (Harms et al. 2004).

There are some postulates reported concerning uroliths formation. Here, I take the liberty to list them as follow:

"-struvite calculi can form in any portion of the urinary tract: kidney, ureter, urinary bladder, or urethra:

- Struvite urolith formation requires urine magnesium ammonium phosphate supersaturation, which is favored by urinary tract infection with urease-positive bacteria, alkaline urine, genetic predisposition, and high-protein diet;
- Urease produced by bacteria hydrolyzes urea to form two molecules of ammonia and a molecule of carbon dioxide;
- The ammonia molecules react with water to form ammonium and hydroxyl ions, alkalinizing the urine;
- Struvite precipitation is promoted both by the increased ammonium ion concentration and increased pH," (Harms et al. 2004; see also Osborne et al. 1986).

Thus, urate nephrolithiasis can be caused by uric acid, ammonium acid urate, or sodium acid urate calculi. Both uric acid and ammonium acid urate calculi have been reported in dolphins (Venn-Watson et al. 2010a, b).

Nephroliths Information about nephroliths in marine mammals is scare. For example, Dennison et al. (2007) reported about two structures. The first contains of ammonium urate and varying in size from 1 to 15 mm in diameter in the California sea lion (*Zalophus californianus*). The second one, observed in the northern elephant seal (*Mirounga angustirostris*), contains uric acid and ammonium urate and was between 1–10 mm in diameter. These authors hypothesized that "unknown metabolic derangements due to morphologic or physiologic differences may have played a role in both cases," (Dennison et al. 2007). All nephroliths observed in bottlenose dolphins *T. truncatus* were characterized as 100 % ammonium acid urate (Venn-Watson et al. 2010a).

Enteroliths These biominerals of pathological origin are usually located in the intestinal lumen of humans and mammals like dogs, horses, and cats (see for review Yuki et al. 2006). Also some marine mammals possess enterolithic concretions. From structural view, these formations include a central nidus with concentric layers of mineral phases deposited around it. According to Burdett and Osborne (2010):

> "the progressive accumulation of the mineral layers results in partial or total obstruction of the intestinal lumen. Multiple factors lead to these obstructions. Included are diet, genetics, alkaline intestinal environments, and ingestion of foreign objects that cannot be digested, such as hairs, nails and other metal objects, as well fragments of stingray spine," (Burdett and Osborne 2010).

Examination of representative portions of the dolphin enterolith specimen showed that both calcium phosphate carbonate (85 % of the outer shell of the stone) and struvite (15 %) were the main components there. However the mineral phase's content within different layers, which were observed within the enterolith, was not homogenous. For example "the largest layer of mineral that completely surrounded the nidus was composed of 95 % calcium phosphate carbonate and 5 % struvite. Additionally, a band within the internal portion of the enterolith consisted of 45 % calcium phosphate carbonate and 55 % brushite," (Burdett and Osborne 2010). Interestingly, the chemical composition of this enterolith was similar with that of vaginal calculi isolated from other delphinid species (Sawyer and Walker (1977); Woodhouse and Rennie 1991).

Faecoliths Marine leatherback turtles not only swallow wood, feathers, sand and seaweed, but are also known to swallow plastic bags and man-made garbage such as the twine and polystyrene (e.g. Brongersma 1969; Den Hanog and Van Nierop 1984). In some cases, these materials play a role of the nucleation side for biominerals formation. The large male leatherback beached at Harlech in North Wales, UK (Eckert and Luginbuhl 1988) had a hard, clay-like ball at the junction between the small and large intestines, while the rectum of a turtle beached at Midway Atoll in the north-western Hawaiian Islands contained a hard, smooth, ovoid ball (Davenport et al. 1993). The occurrence of hard masses in the intestine of chelonians is not uncommon either, but these are usually formed from masses of chitinous pans of insects or accumulated indigestible cellulose fibres. Such masses can cause constipation, though intestinal parasitic nematodes may help to break the masses down. The faecolith from chelonian turtle reported by Davenport et al. (1993) consisted, however, of biomineralized faecal material. The mineral was found to be struvite.

Interestingly, much material of anthropogenic origin (plastics in sheet and linear form, plus other packaging materials and monofilament nylon) was incorporated into the faecolith structure. It is hypothesized that the formation of struvite stems from the interaction of the leatherback's osmotic physiology with the metabolism of faecal bacteria. The hind fluid is likely to contain relatively high concentrations of magnesium, calcium and sulphate ions, but little sodium or chloride. However, the ammonium and phosphate ions of struvite are presumably derived from the faecal bacteria (Davenport et al. 1993). While the formation of the faecolith may be pathological, it could alternatively be an adaptive response to package garbage (whether natural or man-made).

3.7.1 Conclusion

The pathological biomineralization is related to poorly studied but very intriguing phenomenon. Of course, we can accept that this kind of biomineralization is an example of "uncontrolled pathological crystallization resulting in painful or even life threatening conditions such as calculi formation, development of gout or arteriosclerosis, tissue calcification associated with cancer, etc.," (Königsberger and Königsberger 2006). In contrast to well investigated pathological biomineralization in humans (see for review Jahnen-Dechent 2004; Giachelli 2005; Wesson and Ward 2007; Golub 2011), the state of the art on pathological biomineralization in marine vertebrates is on the embryonic stage. Today, we can only suggest that several cellular components are involved in their pathological mineralization, and that there is some evidence suggesting pathological mineralization is a regulated process.

3.8 Silica-Based Minerals in Marine Vertebrates

Abstract Silicon, in the form of silicic acid, is a fundamental nutrient for marine invertebrates like diatoms, silicoflagellates, radiolaria, and sponges, all of which polymerize it to build skeletons of biogenic silica. Occurrence of silica-based minerals in mammals and human is usually determined by pathological processes like silica urolithiasis. The only example of the presence of silica in the form of chalcedony that is known is in marine elasmobranch fish. Chalcedony is a microfibrous (microcrystaline with a fibrous structure) variety of quartz, and was identified within electric organ of the marine skate *Psammobatis extenta.*

Silicon dioxide (silica) of biological origin is known as biosilica. It is widely distributed biomineral in plants and animals and possesses broad variety of morphological forms on nano-, micro- and macroscales. Our interest on silica is based on the following reasons (Ehrlich et al. 2010):

(a) "it is definitely the first and the oldest natural bio-skeleton;
(b) it has unique mechanical properties;
(c) it has extremely high specific surface area, and, therefore, adsorption properties for dissolved components in external milieu," (Ehrlich et al. 2010).

Numerous examples of biosilica-based structures observed in both lower and higher plants, bacteria, yeast, fungi, protists, sponges, molluscs, ascidians, crustaceans, brachiopods, terrestrial mammals and human are recently discussed (see for review Ehrlich 2010, 2011).

Reports on silica-containing minerals in marine vertebrates are not common. Here, I take the liberty to analyse several publications concerning findings of silica in representatives of Elasmobranchii. I mean the *Rajidae* family, known usually as skates (Parago 2001). One of the species, the *Psammobatis extenta*, is "endemic to the continental shelf of the western South Atlantic, ranging from Cabo Frio, Rio de Janeiro, Brazil (22°56′S) to Patagonia, Argentina (~45°S)," (Rocha et al. 2010). Surprisingly, the electric organs of this fish, which produce

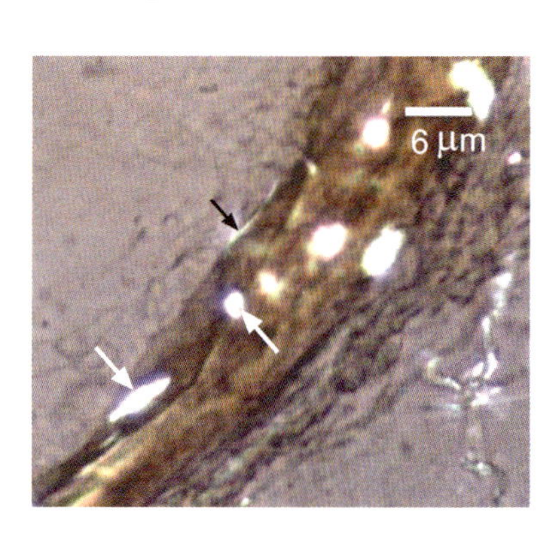

Fig. 3.34 Images of cryostat sections from electric organs of *P. extenta* at high magnification with crossed-polarizers show silica particles within innervated region (*black arrows*). Silica particles within electrocyte are also visible (*white arrows*) (Reprinted from Prado Figueroa et al. (2008), Copyright (2008), with permission from Elsevier)

weak electrical discharges to the surrounding environment, are the place for silica localization. These organs possess specialized cells known as electrocytes (Prado Figueroa 2005; Prado Figueroa et al. 2008; Prado Figueroa 2011). They were described as follow: "Electrocytes are cup-shaped cells, highly polarized, with an anterior, concave, innervated face and a posterior, convex, noninnervated face that shows a very large system of caveolae," (Prado Figueroa et al. 2008). Based on the evidence of aluminium (Barrera et al. 2001) and silicon accumulation in electrocytes from *P. extenta* by SEM and energy dispersive X-ray microanalysis (EDS), Prado Figueroa and co-workers recently reported the presence of silica in the form of chalcedony. This crystalline mineral was detected in the cytoplasm and synaptic regions of electrocytes using a mineralogical microscope. The localization of silica in the subcellular fractions of electric organs of *P. extenta* during oxidative stress as well as in selected cryostatic sections (Fig. 3.34) has been shown for the first time (Prado Figueroa et al. 2005; Prado Figueroa and Cesaretti 2006). Although the origin of chalcedony is frequently debated in scientific community (Heaney 1993), its discovery within electric organ of skate is definitively intriguing.

It was suggested previously (Efremov 1940) that such factors as pH and Eh can play an important role by SiO_2 accumulation. According to hypothesis of Prado Figueroa et al. (2008), probably, this very special kind of silicification implies the death of the cell and the nerves, revealing that some specific conditions of pH and Eh are necessary for this process to occur. As reported: "In different situations involving electric organ cryostatic sections without treatment (from living fishes), silica was identified along the curves of the nerves. This occurrence raises the possibility of a fluid phase moving through the nerves which resulted in precipitation when suitable conditions were reached" (Prado Figueroa et al. 2008) (Fig. 3.35).

Fig. 3.35 Chalcedony crystal from electric organ of *P. extenta* (Image courtesy of Maria Prado Figueroa)

3.8.1 Conclusion

To my best knowledge, there are no reports on examples of pathological biomineralization of marine vertebrates with respect to silica. Probably, this situation is determined by the non-siliceous diet of these animals in contrast to herbivorous animals such as sheep and cattle where siliceous urinary calculi have been observed. The consumption of forage with a high content of plant silica seems to be the reason for development of silica calculi with urethral obstruction in about 4 % of range steers wintered in the northwestern plains of North America (Ehrlich et al. 2010). Although one report on a silica eating marine animal exists, there is no information about finding of some siliceous bodies within the organism. Sponges, and especially demosponges, associated with coral reef are the components of the feeding diet of the Hawksbill turtles (*Eretmochelys imbricata*). Demosponges possess horny, proteinaceous skeletons with huge amounts of siliceous spicules. There are no doubts that these sponges are some kind of protein rich diet for turtles. As reported by Meylan (1988) in her *Science* publication: "dissociated siliceous spicules were found embedded in the intestinal epithelia of hawksbills. No morphological adaptations were identified that might facilitate the handling of spicules. In other spongivores, spicule-compacting organs, modifications of masticatory structures, and copious mucus production are thought to serve this purpose," (Meylan 1988).

Therefore, the occurrence of silica as chalcedony within electric organs of *P. extent* seems to be thus far the only example for marine vertebrates. Definitively, this phenomenon must be investigated in detail in the near future.

References

Abel JH, Ellis RA (1966) Histochemical and electron microscopic observations on the salt secreting glands of marine turtles. Am J Anat 118:337–357

Abler WL (1992) The serrated teeth of tyrannosaurid dinosaurs, and biting structures in other animals. Paleobiology 18:161–183

Abzhanov et al (2007) Reproduced with permission Development 134: Abzhanov A, Rodda SJ, McMahon AP and Tabin CJ (2007) Regulation of skeletogenic differentiation in cranial dermal bone. Development 134:3133–3144

Acuña–Mesén RA (1984) La ultraestructura superficial de la cascara del huevo de la tortuga marina *Lepidochelys olivacea* Eschscholtz. Brenesia 22:299–308

Acuña–Mesén RA (1989) Anatomia microscopica de la cascara del huevo de la tortuga Carey *Eretmochelys imbricata*. Brenesia 31:33–41

Al–Bahry SN, Mahmoud IY, Al–Amri IS et al (2009) Ultrastructural features and elemental distribution in eggshell during pre and post hatching periods in the green turtle, *Chelonia mydas* at Ras Al–Hadd, Oman. Tissue Cell 41:214–221

Al–Bahry SN, Mahmoud IY, Melghit K et al (2011) Analysis of elemental composition of the eggshell before and after incubation in the loggerhead turtle (*Caretta caretta*) in Oman. Microsc Microanal 17(3):452–60. Copyright © 2011, Microscopy Society of America. Reprinted with permission

Alerstam T, Hogstedt G (1983) The role of the geomagnetic field in the development of birds' compass sense. Nature 306:463–465

Alibardi L (2010a) Cornification in the claw of the amphibian *Xenopus laevis* and comparison with claws in amniotes. Ital J Zool 179:399–409

Alibardi L (2010b) Cornification of the beak of *Rana dalmatina* tadpoles suggests the presence of basic keratin associated proteins. Zool Stud 49:51–63

Alibardi L, Segalla A (2011) The process of cornification in the horny teeth of the lamprey involves proteins in the keratin range and other keratin–associated proteins. Zool Stud 50(4):416–425

Allen GR (1982) A field guide to inland fishes of Western Australia. University of Western Australia Press, Perth

Amano M, Yamada TK, Brownell RL Jr et al (2011) Age determination and reproductive traits of killer whales entrapped in ice off Aidomari, Hokkaido. Jpn J Mammol 92:275–282

Amiel D, Coutts RD, Harwood FL et al (1988) The chondrogenesis of rib perichondrial grafts for repair of full thickness articular cartilage defects in a rabbit model: a one year postoperative assessment. Connect Tissue Res 18:27–39

Anderson PSL, LaBarbera M (2008) Functional consequences of tooth design: effects of blade shape on energetics of cutting. J Exp Biol 211:3619–3626

Andrews RM, Mathies T (2000) Natural history of reptilian development: constraints on the evolution of viviparity. Bioscience 50:227–238

Applegate SP (1965) Tooth terminology and variation in sharks with special reference to the sand shark, *Carcharias taurus rafinesque*. Contib Sci Los Angel Cty Mus 86:3–18

Arias JL, Fink DJ, Xiao S et al (1993) Biomineralization and eggshells: cell–mediated acellular compartments of mineralized extracellular matrix. Int Rev Cytol 145:217–250

Avallone B, Balassone G, Balsamo G et al (2003) The otoliths of the antarctic teleost *Trematomus bernacchii*: scanning electron microscopy and X–ray diffraction studies. J Submicrosc Cytol Pathol 35(1):69–76

Babonis LS, Brischoux F (2012) Perspectives on the convergent evolution of tetrapod salt glands. Integr Comp Biol 52(2):245–256. doi:10.1093/icb/ics073. By permission of Oxford University Press

Babonis LS, Evans DH (2011) Morphological and biochemical evidence for the evolution of salt glands in snakes. Comp Biochem Physiol A Mol Integr Physiol 160:400–411

Babonis LS, Hyndman KA, Lillywhite HB et al (2009) Immunolocalization of Na^+/K^+–ATPase and $Na^+/K^+/2Cl^-$ cotransporter in the tubular epithelia of sea snake salt glands. Comp Biochem Physiol A Mol Integr Physiol 154:535–540

Babonis LS, Miller SN, Evans DH (2011) Renal responses to salinity change in snakes with and without salt glands. J Exp Biol 214:2140–2156
Baeuerlein E, Schüler D (1995) Biomineralisation: iron transport and magnetite crystal formation in *Magnetospirillum gryphiswaldense*. J Inorg Biochem 59(2):107
Bain MM (1990) Eggshell strength: a mechanical/ultrastructural evaluation. Dissertation, University of Glasgow, Scotland
Baird T, Solomon SE (1979) Calcite and aragonite in the eggshell of *Chelonia mydas* L. J Exp Mar Biol Ecol 36:295–303
Ballantyne JS, Robinson JW (2010) With kind permission from Springer Science+Business Media: Ballantyne JS, Robinson JW (2010) Freshwater elasmobranchs: a review of their physiology and biochemistry. J Comp Physiol B 180(4):475–493. Copyright © 2010, Springer-Verlag
Bargmann W (1933) Die Zahnplatten von Chimaera monstrosa. Zeit Zell Mikr Anat 19:537–561
Barrera F, Schmitd G, Prado Figueroa M (2001) Electrocytes presence of aluminum in weakly electric fish (Rajidae) from Bahía Blanca. Reunión Anual de la Sociedad Argentina de Neuroquímica. Cell Mol Neurobiol 21:126
Bauer GB, Fuller M, Perry A et al (1986) Magnetoreception and biomineralization of magnetite in cetaceans. In: Kirshvink JL, Jones DS, McFadden BJ (eds) Magnetite biomineralization and magnetoreception in living organisms. Plenum Press, New York
Bäuerlein E, Schüler D, Reszka R et al (2001) Specific magnetosomes, method for the production and use thereof. US patent 6 251 365 B1
Bazylinski DA, Schüler D (2009) Biomineralization and assembly of the bacterial magnetosome chain. Microbe 4:124–130
Beason RC, Semm P (1987) Magnetic responses of the trigeminal nerve system of the bobolink (*Dolichonyx oryzivorus*). Neurosci Lett 80:229–234
Beason RC, Dussourd N, Deutschlander ME (1995) Behavioural evidence for the use of magnetic material in magnetoreception by a migratory bird. J Exp Biol 198:141–145
Beisel KW, Wang-Lundberg Y, Maklad A, Fritzsch B (2005) Development and evolution of the vestibular sensory apparatus of the mammalian ear. J Vestib Res 15:225–241
Belcher E (1885) The last of the Arctic voyages; being a narrative of the expedition in H. M. S. Assistance, under the command of Captian Sir Edward Belcher, C. B., in search of Sir John Franklin, during the years 1852-53-54. Lovell Reeve, London
Beniash E (2011) Biominerals—hierarchical nanocomposites: the example of bone. WIREs Nanomed Nanobiotechnol 3: 47–69. Copyright © 2010 John Wiley & Sons, Inc. Reprinted with permission from John Wiley and Sons
Benirschke KJ, Henderson JR, Sweeny JC (1984) A vaginal mass, containing fetal bones, in a common dolphin, Delphlnus delphis. In: Perrin WF, Brownel RL Jr, DeMaster DP (eds) Reproduction in whales, dolphins and porpoises. Reports of the International Whaling Commission, Special Issue 6, Cambridge, UK
Berenyi M (1972) Models for the formation of uric acid and urate stones. Int Urol Nephrol 4:199–204
Berry C (2004) Hearing the sermons in stones. QJM 97(2):109–110, by permission of Oxford University Press
Besmer A (1947) Die Triasfauna der Tessiner Kalkalpen XVI. Beiträge zur Kenntnis des Ichthyosauriergebisses. Schweiz Palaeontol Abh 65:1–21
Best RC (1981) The tusk of the narwhal (*Monodon monoceros* L.): interpretation of its function (Mammaia: Cetacea). Can J Zool 59:2386–2393
Betts MW (2007) The Mackenzie Inuit whale bone industry: raw material, tool manufacture, scheduling, and trade. ARCTIC 60(2):129–144. doi:10.14430/arctic238. Reprinted with permission
Bilinski JJ, Reina RD, Spotila JR et al (2001) The effects of nest environment on calcium mobilization by leatherback turtle embryos (*Dermochelys coriacea*) during development. Comp Biochem Physiol Part A Mol Integr Physiol 130:152–162

Biro D, Freeman R, Meade J (2007) Pigeons combine compass and landmark guidance in familiar route navigation. Proc Natl Acad Sci USA 104:7471–7476
Blakemore RP (1975) Magnetotactic bacteria. Science 19:377–379
Blumer MJF, Longato S, Fritsch H (2008) Structure, formation and role of cartilage canals in the developing bone. Ann Anat 190:305–315
Board RG (1982) Properties of avian eggshells and their adaptive value. Biol Rev 57:1–28
Boehm JR, Greenwell MG, Coe F (1997) Dietary management in the treatment of uric acid urolithiasis in a Pacific white–sided dolphin (*Lagenorhynchus obliquidens*). Proc Int Assoc Aqua Anim Med 28:134–135
Boersma PD, Rebstock GA, Stokes DL (2004) Why penguin eggshells are so thick. The Auk 121(1):148–155. Published by the American Ornithologists' Union
Bohannon J (2007) Michael Walker: seeking nature's inner compass. Science 5852(318):904–907
Bonadonna F, Bajzak C, Benhamou S et al (2005) Orientation in the wandering albatross: interfering with magnetic perception does not affect orientation performance. Proc R Soc Lond B Biol Sci 272:489–495
Bookman MA (1977) Sensitivity of the homing pigeon to an earth–strength magnetic field. Nature 267:340–342
Borelli G, Mayer–Gostan N, De Pontual H et al (1994) Biochemical relationships between endolymph and otolith matrix in the trout (*Oncorhynchus mykiss*) and turbot (*Psetta maxima*). Hear Res 79(1–2):99–104
Boschma H (1938) On the teeth and some other particulars of the sperm whale (*Physeter macrocephalus* L.). Temminckia 3:151–278
Bradford EW (1957) The structure of rostral teeth and the rostrum of Pristis Microdon. J Dent Res 36:663–668
Brear K, Currey JD, Pond CM et al (1990) The mechanical properties of the dentine and cement of the tusk of the narwhal *Monodon monoceros* compared with those of other mineralized tissues. Arc Oral Biol 35:615–621
Brear K, Currey JD, Kingsley MCS et al (1993) The mechanical design of the tusk of the narwhal (*Monodon monoceros*: Cetacea). J Zool 230:411–423
Brighton CT (1994) Bone formation and repair. Brighton CT, Friedlaender GE, Lane JM, (eds). American Academy of Orthopaedic Surgeons, Rosemount
Brody RH, Edwards HGM, Pollard AM (2001) Chemometric methods applied to the differentiation of Fourier–transform Raman spectra of ivories. Anal Chim Acta 427:223–232
Brongersma LD (1969) Miscellaneous notes on turtles. Proc Kronic Ned Akad Weten Ser C 72:90–102
Burdett LG, Osborne CA (2010) Enterolith with a stingray spine nidus in an Atlantic Bottlenose dolphin (*Tursiops truncatus*). J Wildl Dis 46(1):311–315. Copyright © 2010, Wildlife Disease Association. Published By American Association of Zoo Veterinarians. Reprinted with permission
Burger JW, Hess WN (1960) Function of the rectal gland in the spiny dogfish. Science 131:670–671
Bustard HR, Simkiss K, Jenkins NK et al (1969) Some analyses of artificially incubated eggs and hatchlings of green and loggerhead sea turtles. J Zool Lond 158:311–315
Bystrow AP (1938) Zahnstruktur der Labyrinthodonten. Acta Zool Stockh 19:387–425
Bystrow AP (1939) Zahnstruktur der Crossopterygier. Acta Zool Stockh 20:283–338
Cadiou H, McNaughton PA (2010) Avian magnetite-based magnetoreception: a physiologist's perspective. J R Soc Interf 7(Suppl 2):S193–S205. By permission of the Royal Society
Campana SE (1999) Chemistry and composition of fish otoliths: pathways, mechanisms and applications. MEPS 188:263–297. Copyright © 1999 Inter-Research. Reprinted with permission
Campana SE (2001) Accuracy, precision and quality control in age determination, including a review of the use and abuse of age validation methods. J Fish Biol 59:197–242
Campana SE (2004) Photographic atlas of fish otoliths of the Northwest Atlantic Ocean. NRC Research Press, Ottawa
Campana SE, Thorrold SR (2001) Otoliths, increments, and elements: keys to a comprehensive understanding of fish populations? Can J Aquat Sci 58:30–38

Campana SE, Gagne JA, Mclaren JW (1995) Elemental fingerprinting of fish otoliths using Id–Icpms. Mar Ecol Progress Ser 122:115–120
Campana SE, Jones C, McFarlane GA et al (2006) Bomb dating and age validation using the spines of spiny dogfish (*Squalus acanthias*). Environ Biol Fish 77:327–336
Campbell-Malone R (2007) Biomechanics of North Atlantic right whale bone: mandibular fracture as a fatal endpoint for blunt vessel-whale collision modeling. Doctoral thesis in biological oceanography, Massachusetts Institute of Technology/Woods Hole Oceanographic Institution, Cambridge, MA, USA, p 257
Campbell-Malone R, Barco SG, Pierre-Yves Daoust PY et al (2008) Gross and histologic evidence of sharp and blunt trauma in North Atlantic right whales (*Eubalaena glacialis*) killed by vessels. J Zoo Wildlife Med 39(1):37–55
Cappetta H (1986) Types dentaires adaptatifs chez les sélaciens actuels et post–paléozoïques. Palaeovertebrata 16(2):57–76
Carlström D (1963) A crystallographic study of vertebrate otoliths. Biol Bull 125:441–463
Carlström D, Engström H (1955) The ultrastructure of statoconia. Acta Otolaryngol 45:14–18
Carr A (1967) So excellent a fishe. Natural History Press, New York
Carr A, Kemp AR, Tibbetts IR et al (2006) Microstructure of pharyngeal tooth enameloid in the parrotfish *Scarus rivulatus* (Pisces:Scaridae). J Microsc 221:8–16
Carthy RR (1992) Scanning electron microscopy (SEM) of loggerhead (*Caretta caretta*) eggshell structure. In: Proceedings of the eleventh annual workshop on sea turtle biology and conservation, Jekyll Island, Georgia, 26 February–2 March 1991. Compiled by M. Salmon and J. Wyneken. NOAA Tech. Memo. NMFS–SEFC–302, pp 143–144
Casper BM (2006) The hearing abilities of elasmobranch fishes. Dissertation (Ph.D.), University of South Florida, FL, USA. Copyright © 2006, Casper BM. Reprinted with permission
Castanet J, Francillon-Vieillot H, Ricqlès ADE, Zylberberg L (2003) The skeletal histology of the Amphibia. In: Heatwole H, Davies M (eds) Amphibian biology, vol 5, Osteology. Surrey Beatty and Sons, Pty. Ltd, Chipping Norton, pp 1598–1683
Chan E–H, Solomon SE (1989) The structure and function of the eggshell of the leatherback turtle (*Dermochelys coriacea*) from Malaysia, with notes on attached fungal forms. Anim Technol 40:91–102
Chapskii KK (1936) The walrus of the Kara Sea. Results of an investigation of the life history, geographical distribution, and stock of walruses in the Kara Sea. Trans Arct Inst 67:1–124
Checkley DM Jr., Dickson AG, Takahashi M, Radich JA, Eisenkolb N, Asch R (2009) Elevated CO_2 enhances Otolith growth in young fish. Science 324(5935):1683. Copyright © 2009, American Association for the Advancement of Science. Reprinted with permission from AAAS
Chen et al (2008) Reprinted from Chen P-Y, Lin AYM, Lin Y-S, Seki Y, Stokes AG, Peyras J, Olevsky EA, Meyers MA, McKittrick J (2008) Structure and mechanical properties of selected biological materials. J Mech Behav Biomed Mater 1(3):208–226. Copyright (2008), with permission from Elsevier
Clark JW (1871) On the skeleton of a Narwhal (*Monodon monoceros*) with two fully developed tusks. Proc Zool Soc Lond VI 2:41–53
Clifton KB, Reep RL, Mecholsky JJ Jr (2008) Quantitative fractography for estimating whole bone properties of manatee rib bones. J Mater Sci 43(6):2026–2034
Compagno LJV (1984a) FAO species catalogue. Sharks of the world. An annotated and illustrated catalogue of shark species known to date, part 1: hexanchiformes to lamniformes. Food and Agriculture Organization of the United Nations, Rome, 249
Compagno LJV (1984b) FAO species catalogue. Sharks of the world. An annotated and illustrated catalogue of shark species known to date, part 2: Carcharhiniformes. Food and Agriculture Organization of the United Nations, Rome
Compagno LJV (1988) Sharks of the order Carcharhiniformes. Princeton University Press, Princeton
Confer A, Panciera R (1995) The urinary system. In: Carlton W, McGavin MD (eds) Thompson's special veterinary pathology, 2nd edn. Mosby–Year Book, St. Louis
Cook J (1973) Blue whale: vanishing leviathan. Dodd Mead & Co., New York

Corwin JT (1981) Peripheral auditory physiology in the lemon shark: evidence of parallel otolithic and non-otolithic sound detection. J Comp Physiol 142:379–390

Coulombe HN, Ridgway SH, Evans WE (1965) Respiratory water exchange in two species of porpoise. Science 149:86–88

Courtillot V, Hulot G, Alexandrescu M et al (1997) Sensitivity and evolution of sea–turtle magnetoreception: observations, modelling and constraints from geomagnetic secular variation. Terra Nova 9:203–207

Cowan FBM (1969) Gross and microscopic anatomy of the orbital glands of *Malaclemys* and other *emydine* turtles. Can J Zool 47:723–729

Cowan FBM (1971) The ultrastructure of the lachrymal 'salt' gland and the harderian gland in the euryhaline Malaclemys and some closely related stenohaline emydines. Can J Zool 49:691–697

Cowan DF, Walker WA, Brownwell RL (1986) Pathology of small cetaceans stranded along southern California beaches. In: Bryden MM, Harrison R (eds) Research on dolphins. Clarendon Press, Oxford

Coyne JA (2012) Mysteries of evolution: the narwhal's "tusk," or rather, tooth. Published on-line http://whyevolutionistrue.wordpress.com/2012/04/22/mysteries-of-evolution-the-narwhals-tusk-or-rather-tooth/. Accessed 15 May 2014. Copyright (c) 2012, Jerry Coyne

Cramp RL, Hudson NJ, Holmberg A et al (2007) The effects of saltwater acclimation on neurotransmitters in the lingual salt glands of the estuarine crocodile, *Crocodylus porosus*. Regul Pept 140:55–64

Cramp RL, Meyer EA, Sparks N et al (2008) Functional and morphological plasticity of crocodile (*Crocodylus porosus*) salt glands. J Exp Biol 211:1482–1489

Cramp RL, De Vries I, Anderson WG (2010a) Hormone–dependent dissociation of blood flow and secretion rate in the lingual salt glands of the estuarine crocodile, *Crocodylus porosus*. J Comp Physiol B Biochem Syst Environ Physiol 180:825–834

Cramp RL et al (2010b) Republished with permission of The Company of Biologists Ltd, from Cramp RL, Hudson NJ, Franklin CE (2010) Activity, abundance, distribution and expression of Na+/K+–ATPase in the salt glands of *Crocodylus porosus* following chronic saltwater acclimation. J Exp Biol 213:1301–1308. Copyright (2010); permission conveyed through Copyright Clearance Center, Inc

Cranford TW, McKenna MF, Soldevilla MS et al (2008) Anatomic geometry of sound transmission and reception in Cuvier's beaked whale (*Ziphius cavirostris*). Anat Rec 291:353–378

Currey JD (1984) Comparative mechanical properties and histology of bone. Integr Comp Biol 24(1):5–12

Currey JD (2002) Bones: structure and mechanics. Princeton University Press, Princeton

Currey JD (2006) Bones: structure and mechanics, chapter 6.3. Enamel. Princeton University Press, Princeton

Currey JD, Abeysekera RM (2003) The microhardness and fracture surface of the petrodentine of Lepidosiren (Dipnoi), and of other mineralized tissues. Arch Oral Biol 48:439–447

Currey JD, Brear K, Zioupos P (1994) Dependence of mechanical properties on fibre angle in narwhal tusk, a highly oriented biological composite. J Biomech 27:885–897

Curry BE et al (1994) The occurrence of calculi in the nasal diverticula of porpoises (Phocoenidae). Mar Mamm Sci 10(1):81–86. Copyright © 2006, John Wiley and Sons. Reprinted with permission

Dantzler WH, Bradshaw SD (2009) Osmotic and ionic regulation in reptiles. In: Evans DH (ed) Osmotic and ionic regulation: cells and animals. CRC Press, Boca Raton

Dauphin Y, Cuif JP, Salomé M et al (2006) Microstructure and chemical composition of giant avian eggshells. Anal Bioanal Chem 386:1761–1771

Davenport J, Balazs GH, Faithfull JV et al (1993) A struvite faecolith in the leatherback turtle *Dermochelys coriacea vandelli*: a means of packaging garbage? Herpetol J 3:81–83

Davila AF (2005) Detection and function of biogenic magnetite. Dissertation, LMU München, Fakultät für Geowissenschaften, München. Copyright © 2005, Fernandez Davila A. Reprinted with permission

Davila AF, Fleissner G, Winklhofer M et al (2003) A new model for a magnetoreceptor in homing pigeons based on interacting clusters of superparamagnetic magnetite. Phys Chem Earth 28:647–652

Davis JG, Oberholtzer JC, Burns FR et al (1995) Molecular cloning and characterization of an inner ear–specific structural protein. Science 267:1031–1034

Davis JG, Oberholtzer JC, Burns FR et al (2002) Molecular cloning and characterization of an inner ear–specific structural protein. Eur J Biochem 269(2):688–696

De Buffrénil V, Casinos A (1995) Observations histologiques sur le rostre de *Mesoplodon densirostris* (Mammalia, Cetacea, Ziphiidae): le tissu osseux le plus dense connu. Ann Sci Nat Zool Paris 16(13):21–32

Deans MR, Peterson JM, Wong GW (2010) Mammalian Otolin: a multimeric glycoprotein specific to the inner ear that interacts with otoconial matrix protein Otoconin-90 and Cerebellin-1. PLoS ONE 5(9):e12765. Copyright © 2010 Deans et al. This is an open-access article distributed under the terms of the Creative Commons Attribution License, which permits unrestricted use, distribution, and reproduction in any medium, provided the original author and source are credited

Debiais-Thibaud M, Borday-Birraux V, Germon I, Bourrat F, Metcalfe CJ, Casane D, Laurenti P (2007) Development of oral and pharyngeal teeth in the medaka (Oryzias latipes): comparison of morphology and expression of eve1 gene. J Exp Zool (Mol Dev Evol) 308B:693–708. © 2007 Wiley-Liss, Inc

Debiais–Thibaud M, Oulion S, Bourrat F et al (2011) The homology of odontodes in gnathostomes: insights from Dlx gene expression in the dogfish, *Scyliorhinus canicula*. BMC Evol Biol 11:307

Deeming DC, Thompson MB (1991) Gas exchange across reptilian eggshells. In: Deeming DC, Ferguson MWJ (eds) Egg incubation: its effects on embyronic development in birds and reptiles. Cambridge University Press, Cambridge

Deeming DC, Whitfield TR (2010) Effect of shell type on the composition of chelonian eggs. Herpetol J 20(7):165–171

Degens ET et al (1969) With kind permission from Springer Science+Business Media: Degens ET, Deuser WG, Haedrich RL (1969) Molecular structure and composition of fish otoliths. Mar Biol 2(2):105–113. Copyright © 1969 Springer

DeLong EF, Frankel RB, Bazylinski DA (1993) Multiple evolutionary origins of magnetotaxis in bacteria. Science 259:803–806

Den Hanog JC, Van Nierop MM (1984) A study on the gut contents of six leathery turtles *Dermochelys coriacea* (Linnaeus) (Reptilia: Testudines: Dermochelyidae from British waters and from the Netherlands. Zool Verh Leiden 20:1–36

Denison RH (1974) The structure and evolution of teeth in lungfishes. Fieldiana Geol 33:31–58

Dennis TE, Rayner MJ, Walker MM (2007) Evidence that pigeons orient to geomagnetic intensity during homing. Proc Biol Sci 274:1153–1158

Dennison S, Gulland F, Haulena M et al (2007) Urate nephrolithiasis in a northern elephant seal (*Mirounga angustirostris*) and a california sea lion (*Zalophus californianus*). J Zoo Wildl Med 38(1):114–120. doi:10.1638/05-121.1. Copyright © 2007, Wildlife Disease Association. Published By American Association of Zoo Veterinarians. Reprinted with permission

Deutschlander ME, Muheim R (2010):Magnetic orientation in migratory songbirds. In: Breed MD, Moore J (eds) Encyclopedia of animal behavior. Academic, Oxford

Didier DA, Stahl BJ, Zangerl R (1994) Development and growth of compound tooth plates in *Callorhinchus milii* (chondrichthyes, holocephali). J Morphol 222: 73–89. © 1994 Wiley-Liss, Inc. Reprinted with permission

Diebel CE, Proksch R, Green CR et al (2000) Magnetite defines a vertebrate magnetoreceptor. Nature 406:299–302

Donoghue PC (2002) Evolution of development of the vertebrate dermal and oral skeletons: unraveling concepts, regulatory theories, and homologies. Paleobiology 28(4):474–507

Donoghue PCJ, Sansom IJ (2002) Origin and early evolution of vertebrate skeletonization. Microsc Res Tech 59:185–218

Donoghue PCJ, Sansom IJ, Downs JP (2006) Early evolution of vertebrate skeletal tissues and cellular interactions, and the canalization of skeletal development. J Exp Zool B Mol Dev Evol 306(3):278–294. doi:10.1002/jez.b.21090. Copyright © 2006 Wiley-Liss, Inc., A Wiley Company. Reprinted with permission
Doody JS (2011) Environmentally cued hatching in reptiles. Integr Comp Biol 51(1):49–61
Doyle WL (1960) The principal cells of the salt gland of marine birds. Exp Cell Res 21:386–393
Dror AA, Politi Y, Shahin H, Lenz DR, Dossena S, Nofziger C, Fuchs H, de Angelis MH Paulmichl M, Weiner S, Avraham KB (2010) Calcium oxalate stone formation in the inner ear as a result of an Slc26a4 mutation. J Biol Chem 285:21724–21735. © 2010 The American Society for Biochemistry and Molecular Biology. Reprinted with permission
Dunson WA (1968) Salt gland secretion in the pelagic sea snake Pelamis. Am J Physiol 215:1512–1515
Dunson W (1969) Electrolyte excretion by the salt gland of the Galápagos marine iguana. Am J Physiol 216:995–1002
Dunson WA (1970) Some aspects of electrolyte and water balance in three estuarine reptiles, the diamond back terrapin, American and 'salt water' crocodiles. Comp Biochem Physiol 32A:161–174
Dunson WA, Dunson MK (1974) Interspecific differences in fluid 497 concentration and secretion rate of sea snake salt glands. Am J Physiol 227:430–438
Dunson WA, Packer RK, Dunson MK (1971) Sea snakes: an unusual gland under the tongue. Science 173:437–441
Eckert KL, Luginbuhl C (1988) Death of a giant. Mar Turtl News 43:1–3
Eder SHK, Cadiou H, Muhamad A, McNaughton PA, Kirschvink JL, Winklhofer M (2012) Magnetic characterization of isolated candidate vertebrate magnetoreceptor cells. Proc Nat Acad Sci 109(30):12022–12027. Copyright (2012) National Academy of Sciences, USA. Reprinted with permission
Edwards H, Schnell G, DuBois R et al (1992) Natural and induced remanent magnetism in birds. Auk 109:43–56
Efremov JA (1940) Taphonomy: new branch of paleontology. Pan-Am Geo 174:81–93
Ehrlich H (2010) Biological materials of marine origin. Springer, Heidelberg
Ehrlich H (2011) Silica biomineralization in sponges. In: Reitner J, Thiel V (eds) Encyclopedia of geobiology. Springer, Dordrecht, pp 796–808
Ehrlich H et al (2008) Reprinted from Ehrlich H, Koutsoukos PG, Demadis KD et al (2008) Principles of demineralization: modern strategies for the isolation of organic frameworks. Part I. Common definitions and history. Micron 39(8):1062–1091. doi:10.1016/j.micron.2008.02.004. Copyright (2008), with permission from Elsevier
Ehrlich H et al (2010) Reprinted with permission from Ehrlich H, Demadis KD, Pokrovsky OS et al (2010) Modern views on desilicification: biosilica and abiotic silica dissolution in natural and artificial environments. Chem Rev 110(8):4656–4689. Copyright (2010) American Chemical Society
Eidelman N, Eichmiller FC, Zhang Y et al (2005) Position–resolved structural and mechanical properties of Narwhal tusk dental tissues, Abstract. The Preliminary Program for IADR/AADR/CADR 83rd General Session, Baltimore, MD, USA, 9–12 March 2005
Ellis RA, Abel JH (1964) Intercellular channels in the salt–secreting glands of marine turtles. Science 144:1340–1342
Elsdon TS, Wells BK, Campana SE et al (2008) Otolith chemistry to describe movements and life–history parameters of fishes: hypotheses, assumptions, limitations and inferences using five methods. Oceanogr Mar Biol Ann Rev 46:207–330
Engkvist O, Ohlsen L (1979) Reconstruction of articular cartilage with free autologous perichondrial grafts. An experimental study in rabbits. Scand J Plast Reconstr Surg 13(2):269–274
Erben HK (1970) Ultrastrukturen und mineralisation rezenter und fossiler eischalen bei vogeln und reptilien. Biomin Forschsber 1:1–65
Erben HK, Newesely H (1972) Kristalline bausteine und mineralbestand von kalkigen eischalen. Biomin Forschsber 6:32–48

Erway LC, Purichia NA, Netzler ER et al (1986) Genes, manganese, and zinc in formation of otoconia: labeling, recovery, and maternal effects. Scan Electron Microsc 4:1681–1694

Espinoza EO, Mann MJ (2000) Identification guide for ivory and ivory substitutes, 3rd edn. Ivory Identification, Inc., Richmond. Reprinted with permission

Evans P (1987) Natural history of whales and dolphins. Facts on File, New York

Evans K, Robertson K (2001) A note on the preparation of sperm whale (*Physeter macrocephalus*) teeth for age determination. J Cet Res Man 3:101–107

Ewert MA (1985) Embryology of turtles. In: Gans C, Billett F, Maderson P (eds) Biology of the reptilian. Wiley, New York

Ewert MA, Firth SJ, Nelson CE (1984) Normal and multiple eggshells in batagurine turtles and their implications for dinosaurs and other reptiles. Can J Zool 62(9):1834–1841. © 2008 Canadian Science Publishing or its licensors. Reproduced with permission

Fablet R, Daverat F, de Pontual H (2007a) Unsupervised bayesian reconstruction of individual life histories chronologies from otolith signatures: case study of Sr:Ca transects of eel 436 (*Anguilla anguilla*) otoliths. Can J Fish Res Aquat Sci 64:152–165

Fablet R et al (2007b) Reprinted from Fablet R, Pujolle S, Chessel A, Benzinou A, Cao F (2007) 2D Image-based reconstruction of shape deformation of biological structures using a level-set representation. Comput Vision Image Understand 111(3):295–306. with permission from Elsevier

Fablet R, Chessel A, Carbini S et al (2009) Reconstructing individual shape histories of fish otoliths: a new imagebased tool for otolith growth analysis and modeling. Fish Res 96:148–159

Faivre D (2004) Propriétés cinétiques, minéralogiques et isotopiques de la formation de nanomagnétites a basse temperature: implication pour la détermination de critères de biogénicité. Dissertation, University of Paris, Paris, France

Faivre D, Schüler D (2008) Magnetotactic bacteria and magnetosomes. Chem Rev 108(11): 4875–4898

Faivre D, Böttger L, Matzanke B et al (2007) Intracellular magnetite biomineralization in bacteria proceeds via a distinct pathway involving membrane–bound ferritin and ferrous iron species. Angew Chem Int Ed 46(44):8647–8652

Falkenberg G, Fleissner G, Schuchardt K et al (2010) Avian magnetoreception: elaborate iron mineral containing dendrites in the upper beak seem to be a common feature of birds. PLoS ONE 5(2):e9231

Fänge R, Fugelli K (1962) Osmoregulation in chimaeroid fishes. Nature 196:689

Fassbinder JWE, Stanjek H, Vali H (1990) Occurrence of magnetic bacteria in soil. Nature 343:181–183

Fay FH (1955) The Pacific walrus (*Odobenus rosmarus divergens*): spatial ecology life history, and population. PhD thesis, University of British Columbia, Vancouver

Fay FH (1982) Ecology and Biology of the Pacific Walrus, *Odobenus rosmarus divergens* Illiger. N Am Fauna 74:1–279. US Department of Interior, Fish and Wildlife Service, Washington, DC

Feder ME et al (1982) Reprinted from Feder ME, Satel SL, Gibbs AG (1982) Resistance of the shell membrane and mineral layer to diffusion of oxygen and water in flexible-shelled eggs of the snapping turtle (*Chelydra serpentina*). Resp Physiol 49(3):279–291. Copyright (1982), with permission from Elsevier

Feltmann CF, Slijper EJ, Vervoort W (1948) Preliminary researches on the fat–content of meat and bone of blue and fin whales. Proc R Neth Acad Arts Sci 51:604–615

Ferguson MWJ (1982) The structure and composition of the eggshell and embryonic membranes of *Alligator mississippiensis*. Trans Zool Soc Lond 36:99–152. Copyright © 1982 The Zoological Society of London. Reprinted with permission

Ferguson MWJ (2010) The structure and composition of the eggshell and embryonic membranes of *Alligator mississippiensis*. Trans Zool Soc Lond 36:99–152

Ferguson SH, Higdon JW, Westdal KH (2012) Prey items and predation behavior of killer whales (*Orcinus orca*) in Nunavut, Canada based on Inuit hunter interviews. Aqu Biosyst 8:3

Fernandez M, Gasparini Z (2000) Salt glands in a *Tithonian metriorhynchgid crocodyliform* and their physiological significance. Lethaia 33:269–276

Fernandez MS, Moya A, Lopez L (2001) Secretion pattern, ultrastructural localization and function of extracellular matrix molecules involved in eggshell formation. Matrix Biol 19:793–803
Ferraris G, Fuess H, Joswig W (1986) Neutron diffraction study of $MgNH_4PO_4•6H_2O$ (struvite) and survey of water molecules donating short hydrogen bonds. Acta Cryst 42:253–258
Finarelli JA, Coates MI (2012) First tooth–set outside the jaws in a vertebrate. Proc R Soc B 279:775–779. Copyright © 2012, The Royal Society. Reprinted with permission from The Royal Society
Finger LW, King HE (1978) A revised method of operation of the single–crystal diamond cell and refinement, of the structure of NaCl at 32 kbar. Am Mineral 63:337–342
Fischer JH, Freake MJ, Borland SC et al (2001) Evidence for the use of a magnetic map by an amphibian. Anim Behav 62(1):1–10
Fitzgerald EMG (2006) A bizarre new toothed mysticete (Cetacea) from Australia and the early evolution of baleen whales. Proc Biol Sci 273(1604):2955–2963. doi:10.1098/rspb.2006.3664. Copyright © 2006 The Royal Society
Fleissner G, Holtkamp–Rötzler E, Hanzlik M et al (2003) Ultrastructural analysis of a magnetoreceptor in the beak of homing pigeons. J Comp Neurol 458:350–360
Formicki K, Tański A, Winnicki A (2002) Effects of magnetic field on the direction of fish movement under natural conditions. General Assembly URCI, Maastricht, pp 1–3
Frankel RB, Blakemore RP (1991) Iron biominerals. Plenum Press, New York
Fraser GJ, Smith M (2011) Evolution of developmental pattern for vertebrate dentitions: an oro-pharyngeal specific mechanism. J Exp Zool Mol Dev Evol 316B:99–112, 2011. © 2010 Wiley-Liss, Inc
Fraser GJ, Hulsey CD, Bloomquist RF, Uyesugi K, Manley NR et al (2009) An ancient gene network is co-opted for teeth on old and new jaws. PLoS Biol 7(2):e1000031. doi:10.1371/journal.pbio.1000031. Copyright © 2009 Fraser et al. This is an open-access article distributed under the terms of the Creative Commons Attribution License, which permits unrestricted use, distribution, and reproduction in any medium, provided the original author and source are credited
Fraser GJ et al (2010) Reproduced from Fraser GJ, Cerny R, Soukup V, Bronner-Fraser M, Streelman JT (2010) The odontode explosion: the origin of tooth-like structures in vertebrates. Bioessays 32(9):808–817. Copyright © 2010 WILEY Periodicals, Inc
Fratzl P, Gupta H, Paschalis E, Roshger P (2004) Structure and mechanical quality of the collagen-mineral nano-composite in bone. J Mater Chem 14(14):2115–2123
Frazzetta TH (1988) The mechanics of cutting and the form of shark teeth (Chondrichthyes, Elasmobranchii). Zoomorphology 108:93–107
Freake MJ, Muheim R, Phillips JB (2006) Magnetic maps in animals – a theory comes of age? Quart Rev Biol 81:327–347
Freeman MMR, Bogoslovskaya L, Caulfield RA et al (1998) Inuit, whaling, and sustainability. Altamira Press, Walnut Creek
Fritzsch B (1996) Similarities and differences in lancelet and craniate nervous systems. Isr J Zool 42:147–160
Gauldie RW (1996) Fusion of Otoconia: a Stage in the development of the Otolith in the evolution of fishes. Acta Zool 77:1–23. Copyright © 1996, The Royal Swedish Academy of Sciences. Reprinted with permission
Gauldie RW, Dunlop D, Tse J (1986) The simultaneous occurrence of otoconia and otoliths in four teleost fish species. N Z J Mar Freshw Res 20:93–99
Gerson HB, Hickie JP (1985) Head scarring on male narwhals (*Monodon monoceros*): evidence for aggressive tusk use. Can J Zool 63(9):2083–2087
Gerstberger R, Gray DA (1993) Fine structure, innervation and functional control of avian salt glands. Int Rev Cytol 144:129–215
Gervais P (1873) Remarques sur la Dentition du Narval. J Zool 2:498–500
Giachelli CM (2005) Inducers and inhibitors of biomineralization: lessons from pathological calcification. Orthod Craniofac Res 8(4):229–231

Gibbs PE (1987) A new species of Phascolosoma (Sipuncula) associated with a decaying whale's skull trawled at 880 m depth in the southwest Pacific. NZ J Zool 14:135–137
Glimcher MJ (2006) Bone: nature of the calcium phosphate crystals and cellular, structural, and physical chemical mechanisms in their formation. Rev Mineral Geochem 64:223–282
Goldenstein DL (2002) Water and salt balance in seabirds. In: Schreiber EA, Burger J (eds) Biology of marine birds. CRC Press, Boca Raton, pp 467–480
Golub EE (2011) Biomineralization and matrix vesicles in biology and pathology. Semin Immunopathol 33(5):409–417
Gorby YA, Beveridge TJ, Blakemore RP (1988) Characterization of the bacterial magnetosome membrane. J Bacteriol 170:834–841
Gottfried MD, Compagno LJV, Bowman SC (1996) Size and skeletal anatomy of the giant megatooth shark *Carcharodon megalodon*. In: Klimley AP, Ainley DG (eds) Great white sharks: the biology of Carcharodon carcharias. Academic, San Diego
Gould JL (1982) The map sense of pigeons. Nature 296:205–211
Gould JL (1985) Are animal maps magnetic? In: Kirschvinkn JL, Jones DS, MacFadden BJ (eds) Magnetite biomineralization and magnetoreception in organisms. Plenum Press, New York
Gould JL, Kirschvink JL, Defieyes KS (1978) Bees have magnetic remanence. Science 201:1026–1028
Greig DJ, Gulland FMD, Kreuder C (2005) A decade of live California sea lion (*Zalophus californianus*) strandings along the central California coast: causes and trends, 1991–2000. Aquat Mamm 31:11–22
Grigg G, Beard L (1985) Water loss and gain by eggs of *Crocodylus porosus*, related to incubation age and fertility. In: Grigg G, Shine R, Ehmann H (eds) Biology of Australasian frogs and reptiles. Surrey Beatty & Sons Pty Limited, Chipping Norton, pp 353–359
Grove RA, Bildfell R, Henny CJ et al (2003) Bilateral uric acid nephro–lithiasis and ureteral hypertrophy in a free–ranging river otter (*Lontra canadensis*). J Wildl Dis 39:914–917
Gudger EW (1937) Abnormal dentition in sharks, Selachii. Bull Am Mus Nat His 73:249–280
Guillette LJ (1982) The evolution of viviparity and placentation in the high elevation, Mexican lizard *Sceloporus aeneus*. Herpetology 38:94–103
Hamilton RMG (1986) The microstructure of the hen's eggshell: a short review. Food Microstruct 5:99–110
Hamzelou J (2012) Mystery of bird navigation system still unsolved. Published online http://www.newscientist.com/article/dn21688-mystery-of-bird-navigation-system-still-unsolved.html#.U63jUhZwIQI. Accessed 15 May 2014. Newscientist Life. Copyright © 2012, Reed Business Information Ltd. Reprinted with permission
Hanson M, Westerberg H (1986) Occurrence of magnetic material in teleosts. Comp Biochem Physiol 86A:169–172
Hanzlik M, Heunemann C, Holtkamp–Rötzler E et al (2000) Superparamagnetic magnetite in the upper–beak tissue of homing pigeons. Biometals 13:325–331
Harms CA et al (2004) Struvite penile urethrolithiasis in a pygmy sperm whale (*Kogia breviceps*). J Wildl Dis 40(3):588–593. Copyright © 2004, Wildlife Disease Association. Published By American Association of Zoo Veterinarians. Reprinted with permission
Harrison RJ (1969) Reproduction and reproductive organs. In: Anderson HT (ed) The biology of marine mammals. Academic, New York
Hay KA, Mansfied AW (1989) Narwhal Monodon monoceros Linnaeus, 1758. In: Ridgway SH, Harrison R (eds) Handbook of marine mammals, vol 4. Academic, San Diego
Hazard L (2001) Ion secretion by salt glands of desert iguanas (*Dipsosaurus dorsalis*). Physiol Biochem Zool 74(1):22–31
Hazard LC (2004) Sodium and potassium secretion by iguana salt glands: acclimation or adaptation? In: Alberts A, Carter RL, Hayes WB, Martins E (eds) Iguanas: biology and conservation. University of California Press, Berkley, pp 84–93
Hazon N et al (2003) Reprinted from Hazon N, Wells A, Pillans RD et al (2003) Urea based osmoregulation and endocrine control in elasmobranch fish with special reference to euryhalinity.

Comp Biochem Physiol Part B Biochem Mol Biol 136(4):685–700. Copyright (2003), with permission from Elsevier

Heaney PJ (1993) A proposed mechanism for the growth of chalcedony. Contrib Mineral Petrol 114:66–74

Heizer RF (1963) Fuel in primitive society. J R Anthropol Inst G B Irel 93:186–194

Helms JA, Schneider RA (2003) Cranial skeletal biology. Nature 423(6937):326–331

Herbert CF (1975) In: Stonehouse B (ed) The biology of penguins. Macmillan & Co., London/Basingstone

Heulin B, Ghielmi S, Vogrin N et al (2002) Variation in eggshell characteristics and in intrauterine egg retention between two oviparous clades of the lizard *Lacerta vivipara*: insight into the oviparity–viviparity continuum in squamates. J Morphol 252:255–262

Higgs ND, Little CTS, Glover AG (2011a) Bones as biofuel: the composition of whale bones with implications for deep–sea biology and palaeoanthropology. Proc R Soc B 278:9–17

Higgs ND, Glover AG, Dahlgren TG et al (2011b) Bone–boring worms: characterizing the morphology, rate, and method of bioerosion by *Osedax mucofloris* (Annelida, Siboglinidae). Biol Bull 22:307–316

Hildebrandt JP (2001) Coping with excess salt: adaptive functions of extrarenal osmoregulatory organs in vertebrates. Zoology 104:209–220

Hincke MT, Nys Y, Gautron J et al (2012) The eggshell: structure, composition and mineralization. Front Biosci 17:1266–1280

Hirsch KF (1983) Contemporary and fossil chelonian eggshells. Copeia 1983:382397

Hirsch KF (1994) The fossil record of vertebrate eggs. In: Donovan SK (ed) The palaeobiology of trace fossils. Wiley, New York

Hirsch KF (2001) Pathological amniote eggshell–fossil and modern. In: Tanke DH, Carpenter K (eds) Mesozoic vertebrate life. Indiana University Press, Bloomington/Indianapolis

Hirsch KF, Packard MJ (1987) Review of fossil eggs and their shell structure. Scan Microsc 1:383–400

Holland RA, Kirschvink JL, Doak TG et al (2008) Bats use magnetite to detect the earth's magnetic field. PLoS ONE 3:e1676

Holmes WN, McBean RL (1964) Some aspects of electrolyte excretion in the green turtle, *Chelonia mydas mydas*. J Exp Biol 41:81–90

Holmgren S, Olsson C (2011) Autonomic control of glands and secretion: a comparative view. Auton Neurosci 165:102–112

Horton JM, Summers AP (2009) Republished with permission of The Company of Biologists Ltd, from Horton JM, Summers AP (2009) The material properties of acellular bone in a teleost fish. J Exp Biol 212:1413–1420. Copyright (2009); permission conveyed through Copyright Clearance Center, Inc

Howard EB (1983) Miscellaneous diseases. In: Howard EB (ed) Pathobiology of marine mammal diseases, vol 2. CRC Press, Boca Raton

http://austhrutime.com/bone.htm. Accessed 15 May 2014. 2014 Copyright © Ahttp://usthrutime.com/. Reprinted with permission

Hubbell GS (1996) Using tooth structure to determine the evolutionary history of the white shark. In: Klimley AP, Ainley DG (eds) The biology of the white shark, *Carcharodon carcharias*. Academic, San Diego

Huber DR, Dean MN, Summers AP (2008) Hard prey, soft jaws and the ontogeny of feeding mechanics in the spotted ratfish *Hydrolagus colliei*. J R Soc Interface 5:941–952

Hudson DM, Lutz PL (1986) Salt gland function in the leatherback sea turtle, *Dermochelys coriacea*. Copeia 1986:247–249

Hudspeth AJ (2008) Making an effort to listen: mechanical amplification in the ear. Neuron 59:530–545

Hughes MR (2003) Reprinted from Hughes MR (2003) Regulation of salt gland, gut and kidney interactions. Comp Biochem Physiol Part A Mol Integr Physiol 136(3):507–524. Copyright (2003), with permission from Elsevier

Hughes I, Thalmann I, Thalmann R (2006) Mixing model systems: using zebrafish and mouse inner ear mutants and other organ systems to unravel the mystery of otoconial development. Brain Res 1091:58–74
Hughes MR, Kitamura N, Bennett DC et al (2007) Effect of melatonin on salt gland and kidney function of gulls, *Larus glaucescens*. Gen Comp Endocrinol 151:300–307
Ishiyama M, Teraki Y (1990) The fine structure and formation of hypermineralized petrodentine in the tooth plate of extant lungfish (*Lepidosiren paradoxa* and *protopterus* sp.). Arch Histol Cytol 53(3):307–321
Iwasaki SI (2002) Evolution of the structure and function of the vertebrate tongue. J Anat 201:1–13
Jackson FD, Varricchio DJ (2003) Abnormal, multilayered eggshell in birds: implications for dinosaur reproductive anatomy. J Vertebr Paleontol 23:699–702
Jackson FD, Garrido A, Schmitt JG et al (2004) Abnormal, multilayered titanosaur (Dinosauria: Sauropoda) eggs from in situ Clutches at the Auca Mahuevo Locality, Neuquen Province, Argentina. J Vertebr Paleontol 24(4):913–922
Jahnen–Dechent W (2004) Lot's wife's problem revisited: how we prevent pathological calcification. In: Baeuerlein E (ed) Biomineralization, 2nd edn. Wiley–VCH, Weinheim
Janis CM, Devlin K, Warren DE, Witzmann F (2012) Dermal bone in early tetrapods: a palaeophysiological hypothesis of adaptation for terrestrial acidosis. Proc Biol Sci 279(1740):3035–3040. doi:10.1098/rspb.2012.0558, by permission of the Royal Society
Jenkins N (1975) Chemical composition of the eggs of the crocodile (*Crocodylus novaeguineae*). Comp Biochem Physiol 51:891–895
Jobert M (1869) Récherches anatomiques sur les glandes nasals des oiseaux. Ann Sci Nat Zool 11:349–368
Jogler C, Schüler D (2009) Genetics, genomics, and cell biology of magnetosome formation in magnetotactic bacteria. Annu Rev Microbiol 63:501–521
Johanson Z, Smith MM (2003) Placoderm fishes, pharyngeal denticles, and the vertebrate dentition. J Morphol 257:289–307. Copyright © 2003 Wiley-Liss, Inc
Johanson Z, Smith MM (2005) Origin and evolution of gnathostome dentitions: a question of teeth and pharyngeal denticles in placoderms. Biol Rev Camb Philos Soc 80(2):303–345
Kalmijn AJ (1982) Electric and magnetic field detection in elasmobranch fishes. Science 218:916–918
Kang JS, Oohashi T, Kawakami Y, Bekku Y, Izpisúa Belmonte JC, Ninomiya Y (2004) Characterization of dermacan, a novel zebrafish lectican gene, expressed in dermal bones. Mech Dev 121(3):301–312
Kang YJ, Stevenson AK, Yau PM et al (2008) Sparc protein is required for normal growth of zebrafish otoliths. J Assoc Res Otolaryngol 9:436–451
Kaplan M (1997) Reptile rehabilitation. In: Lowell Ackerman DVM (ed) The biology husbandry, and health care of reptiles. TFH Publishing, Neptune City
Karsenty G, Wagner EF (2002) Reaching a genetic and molecular understanding of skeletal development. Dev Cell 2:389–406
Kastelein RA, Gerrits NM (1990) The anatomy of the walrus head (*Odobenus rosmarus*) part 1: the skull. Aquat Mamm 16(3):101–119
Kastelein RA, Mosterd P (1989) The excavation technique for molluscs of Pacific Walruses (*Odobemus rosmarus divergens*) under controlled conditions. Aquat Mamm 15(1):3–5
Kastelein RA, Gerrits NM, Dubbeldam JL (1991) The anatomy of the Walrus Head (*Odobenus rosmarus*), part 2: description of the muscles and of their role in geeding and haul–out behaviour. Aquat Mamm 17(3):156–180
Katz JL, Bronzino JD (2000) The biomedical engineering handbook, chapter 18, 2nd edn. CRC Press LLC, Boca Raton
Kawasaki et al (2009) Reproduced with permission of Annual Review of Kawasaki et al (2009) Biomineralization in humans: making the hard choices in life. Annu Rev Genet 43:119–142. by Annual Reviews, http://www.annualreviews.org

Keeton WT (1971) Magnets interfere with pigeon homing. Proc Natl Acad Sci USA 68:102–106
Keller M, Moliner JL, Vasquez G et al (2008) Nephrolithiasis and pyelonephritis in two West Indian Manatees (*Trichechus manatus* spp.). J Wildl Dis 44(3):707–711
Kemp A (2002) Growth and hard tissue remodelling in the dentition of the Australian lungfish, *Neoceratodus forsteri* (Osteichthyes: Dipnoi). J Zool 257:219–235
Kemp A (2003) Reprinted from Kemp (2003) Ultrastructure of developing tooth plates in the Australian lungfish, *Neoceratodus forsteri* (Osteichthyes: Dipnoi). Tissue Cell 35:401–426. Copyright (2003), with permission from Elsevier
Khorevin VI (2008) The lagena (the third otolith endorgan in vertebrates). Neurophysiology 40:142–159
Kingsley MCS, Ramsay MA (1988) The spiral in the tusk of the narwhal. Arctic 41:236–238
Kirkland JI, Aguillón Martínez MC (2002) Schizorhiza: a unique sawfish paradigm from the Difunta Group, Coahuila, Mexico. Rev Mex Cienc Geol 19(1):16–24
Kirschvink JL (1980) Magnetic material in turtles: a preliminary report and request. Marine Turtle Newlett 15:7–9
Kirschvink JL (1982) Birds, bees, and magnetism. Trends Neurosci 5:160–167
Kirschvink JL (1990) Geomagnetic sensitivity in cetaceans:an update with live stranding records in the United States. In: Thomas JA, Kastelein RA (eds) Sensory abilities of cetaceans:laboratory an field Evidence. Plenum Press, New York
Kirschvink JL, Gould JL (1981) Biogenic magnetite as a basis for magnetic field detection in animals. BioSystems 13:181–201
Kirschvink JL, Hagadorn JW (2000) A grand unified theory of biomineralization. In: Bäuerlein E (ed) The biomineralization of nano– and microstructures. Wiley–VCH Verlag GmbH, Berlin
Kirschvink JL, Jones DS, MacFadden BJ (1985a) Magnetite biomineralization and magnetoreception in organisms:a new biomagnetism, volume 5 of Topics in geobiology. Plenum Publ, New York
Kirschvink JL, Walker MM, Chang S–B et al (1985b) Chains of single–domain magnetite particles in chinook salmon, *Oncorhynchus tschawytscha*. J Comp Physiol 157A:375–381
Kirschvink JL, Dizon AE, Westphal JA (1986) Evidence from strandings for geomagnetic sensitivity in cetaceans. J Exp Biol 120:1–24
Kirschvink et al (2001) Reprinted from Kirschvink JL, Walker MM, Diebel CE (2001) Magnetite-based magnetoreception. Curr Opin Neurobiol 11(4):462–7. Copyright (2001), with permission from Elsevier
Kirschvink JL, Winklhofer M, Walker MM (2010) Biophysics of magnetic orientation: strengthening the interface between theory and experimental design. J R Soc Interface 7:S179–S191
Kitimasak W, Thirakhupt K, Moll DL (2003) Eggshell structure of the Siamese narrow–headed turtle *Chitra chitra* Nutphand, 1986 (Tetundise: Trionchidae). Sci Asia 29:95–98
Klimley AP (1993) Highly directional swimming by scalloped hammerhead sharks, *Sphyrna lewini*, and subsurface irradiance, temperature, bathymetry, and geomagnetic field. Mar Biol 117:1–22
Klinowska M (1985) Cetacean stranding sites relate to geomagnetic topography. Aquat Mamm 1:27–32
Kolinko S, Jogler C, Katzmann E et al (2012) Single–cell analysis reveals a novel uncultivated magnetotactic bacterium within the candidate division OP3. Environ Microbiol 14(7): 1709–1721
Kölliker A (1859) On the different types in the microscopic structure of the skeleton of osseous fish. Proc Biol Sci 9:656–688
Komeili A (2007) Molecular mechanisms of magnetosome formation. Annu Rev Biochem 76:351–366
Königsberger E, Königsberger L (2006) Solubility phenomena related to normal and biomineralization processes. In: Königsberger E, Königsberger L (eds) Biomineralization – medical aspects of solubility. Wiley, Chichester. Copyright © 2006, John Wiley and Sons. Reproduced with permission of J. Wiley in the format Republish in a book via Copyright Clearance Center
Kooistra TA, Evans DH (1976) Sodium balance in the green turtle, *Chelonia mydas*, in seawater and freshwater. J Comp Physiol 107:229–240

Koussoulakou DS, Margaritis LH, Koussoulakos SL (2009) A curriculum vitae of teeth: evolution, generation, regeneration. Int J Biol Sci 5(3):226–243

Kozel PJ, Friedman RA, Erway LC et al (1998) Balance and hearing deficits in mice with a null mutation in the gene encoding plasma membrane Ca^{2+} – ATPase isoform 2. J Biol Chem 273:18693–18696

Kramer G (1961) Long–distance orientation. In: Marshall AJ (ed) Biology and comparative physiology of birds. Academic, London

Kranenbarg S, van Cleynenbreugel T, Schipper H, van Leeuwen J (2005) Adaptive bone formation in acellular vertebrae of sea bass (*Dicentrarchus labrax* L.). J Exp Biol 208(18):3493–3502

Kraus DS, Rolland RM (2007) The urban whale: North Atlantic right whales at the crossroads. Kraus DS, Rolland RM (eds). Harvard University Press, Cambridge, MA

Kühnel W (1972) With kind permission from Springer Science+Business Media: Kühnel W (1972) On the innervation of the salt gland. Zeitschrift für Zellforschung und Mikroskopische Anatomie 134(3):435–438. Copyright © 1972, Springer-Verlag

Lacalli TC (2004) Sensory systems in amphioxus: a window on the ancestral chordate condition. Brain Behav Evol 64:148–162

Laidre KL, Heide-Jørgensen MP (2005) Winter feeding intensity of narwhals. Mar Mamm Sci 21(1):45–57

Laidre KL, Heide-Jørgensen MP (2011) Life in the lead: extreme densities of narwhals in the offshore pack ice. Mar Ecol Prog Ser 423:269–278

Laidre KL, Heide-Jørgensen MP, Dietz R, Hobbs RC, Jørgensen OA (2003) Deep-diving by narwhals, *Monodon monoceros*: differences in foraging behavior between wintering areas? Mar Ecol Prog Ser 261:269–281

Lakshminarayanan R, Jin EO, Loh XJ et al (2005) Purification and characterization of a vaterite–inducing peptide, pelovaterin, from the eggshells of *Pelodiscus sinensis* (Chinese soft–shelled turtle). Biomacromolecules 6(3):1429–1437

Lakshminarayanan R et al (2008) Reprinted with permission from Lakshminarayanan R, Vivekanandan S, Samy RP et al (2008) Structure, self-assembly, and dual role of a β-Defensin-like peptide from the Chinese soft-shelled turtle eggshell matrix. J Am Chem Soc 130(14):4660–4668. Copyright 2008 American Chemical Society

Lambert et al (2011) Reproduced from Lambert O, de Buffrénil V, de Muizon C (2011) Rostral densification in beaked whales: diverse processes for a similar pattern. (La densification du rostre des baleines à bec : des processus variés pour un résultat similaire). Comptes Rendus Palevol 10(5–6):453–468. Copyright © 2011 Académie des sciences. Published by Elsevier Masson SAS. All rights reserved

Lang C, Schüler D, Faivre D (2007) Synthesis of magnetite nanoparticles for bio– and nanotechnology: genetic engineering and biomimetics of bacterial magnetosomes. Macromol Biosci 7(2):144–151

Langille RM, Hall BK (1988) With kind permission from Springer Science+Business Media: Langille RM, Hall BK (1988) The organ culture and grafting of lamprey cartilage and teeth. In Vitro Cell Dev Biol 24(1):1–8. Copyright © 1988, Tissue Culture Association, Inc

Leask MJM (1977) A physicochemical mechanism for magnetic field detection by migrating birds and homing pigeons. Nature 267:144–145

Levy R, Dawson P (2006) Reconstructing a thule whalebone house using 3D imaging. J IEEE MultiMed 13:78–83

Lewis ER, Leverenz EL, Bialek WS (1985) The vertebrate inner ear. CRC Press, Boca Raton, p 248

Lim DJ (1973) Formation and fate of the otoconia. Scanning and transmission electron microscopy. Ann Otol Rhinol Laryngol 82:23–35

Loch C, Grando LJ, Kieser JA et al (2011) Dental pathology in dolphins (Cetacea: Delphinidae) from the southern coast of Brazil. Dis Aquat Org 94:225–234

Locke M (2008) Structure of ivory. J Morphol 269(4):423–450

Lohße A, Ullrich S, Katzmann E et al (2011) Functional analysis of the magnetosome island in *Magnetospirillum gryphiswaldense*: The mamAB operon is sufficient for magnetite biomineralization. PLoS ONE 6(10):e25561

Lohman K, Lohman CMF (1994) Acquisition of magnetic directional preference in hatchling loggerhead sea turtles. J Exp Biol 190:1–8

Lohmann KJ (1991) Magnetic orientation by hatchling loggerhead sea turtles (*Caretta caretta*). J Exp Biol 155:31–49

Lohmann KJ, Johnsen S (2000) Reprinted from Lohmann KJ, Johnsen S (2000) The neurobiology of magnetoreception in vertebrate animals. Trends Neurosci 23(4):153–159. Copyright © 2000, with permission from Elsevier

Lohmann KJ, Lohmann CMF (1993) A light-independent magnetic compass in the Leatherback Sea Turtle. Biol Bull 185(1):149–151. Copyright © 1993, The Marine Biological Laboratory. Reprinted with permission

Lohmann CMF, Lohmann KJ (2006) Reprinted from Lohmann CMF, Lohmann KJ (2006) Sea turtles. Cur Biol 16(18):R784–R786. Copyright © 2006, Elsevier Ltd., (Under an Elsevier user license), with permission from Elsevier

Lohmann KJ, Hester JT, Lohmann CMF (1999) Long distance navigation in sea turtles. Ethol Ecol Evol 11:1–23

Lohmann KJ, Cain SD, Dodge SA et al (2001) Regional magnetic fields as navigational markers for sea turtles. Science 294:364–366

Lohmann KJ, Lohmann CMF, Ehrhart LM et al (2004) Geomagnetic map used in sea–turtle navigation. Nature 428:909–910

Lohmann KJ, Putman NF, Lohmann CMF (2008) Geomagnetic imprinting:a unifying hypothesis of long–distance natal homing in salmon and sea turtles. Proc Natl Acad Sci USA 105:19096–19101

Long JA (1995) The rise of fishes – 500 million years of evolution. University of New South Wales Press/Johns Hopkins University Press, Sydney/Baltimore

Lowenstam HA (1962) Magnetite in denticle capping in recent chitons (polyplacophora). Geol Soc Am Bull 73:435–438

Lowy RJ, Dawson DC, Ernst SA (1989) Mechanism of ion transport by avian salt gland primary cell cultures. Am J Physiol 256:R1184–R1191

Lü J, Unwin DM, Deeming DC et al (2011) An egg–adult association, gender, and reproduction in pterosaurs. Science (New York) 331(6015):321–324

Lucifora LO, Menni RC, Escalante AH (2001) Analysis of dental insertion angles in the sand tiger shark, *Carcharias taurus* (Chondrichthyes: Lamniformes). Cybium Int J Ichtyol 25(1):23–31. Copyright © 2001 Société Française d'Ichtyologie

Lundberg YW et al (2006) Reprinted from Lundberg YW, Zhao X, Yamoah EN (2006) Assembly of the otoconia complex to the macular sensory epithelium of the vestibule. Brain Res 1091(1):47–57. Copyright (2006), with permission from Elsevier

Lutz P (1997) Salt, water and pH balance in the sea turtle. In: Lutz P, Musick J (eds) The biology of sea turtles. CRC Press, Boca Raton, pp 343–361

Lychakov DV (2004) Evolution of otolithic membrane. structure of otolithic membrane in amphibians and reptilians. J Evol Biochem Physiol 40:331–342

Lychakov DV et al (2000) Reprinted from Lychakov DV, Boyadzhieva-Mikhailova A, Christov I, Evdokimov II (2000) Otolithic apparatus in Black Sea elasmobranchs. Fish Res 46(1–3): 27–38. Copyright (2000), with permission from Elsevier

Magalhaes MCF, Marques PAAP, Correia RN (2006) Biomineralization – medical aspects of solubility. Wiley, Chichester

Mahanty P, Sahoo G (1999) Ultrastructural and biochemical study of egg shell calcium utilization during embryogenesis in the Olive Ridley (*Lepidochelys olivacea*) sea turtles. In: 19th annual sea turtle symposium, South Padre Island, Texas, USA, pp 112–113

Maher BA (1998) Magnetite biomineralization in termites. Proc R Soc Lond Ser B 265:733–737

Mann S, Frankel RB, Blakemore RP (1984) Structure, morphology and crystal growth of bacterial magnetite. Nature 310:405–407

Mann S, Sparks NH, Walker MM, Kirschvink JL (1988) Ultrastructure, morphology and organization of biogenic magnetite from sockeye salmon, *Oncorhynchus nerka*: implications for magnetoreception. J Exp Biol 140:35–49
Marples J (1932) The structure and development of the nasal glands of birds. Proc Zool Soc London 102(4):829–844
Marshall CB, Fletcher GL, Davies PL (2004) Hyperactive antifreeze protein in a fish. Nature 429:153
Marx FG (2010) The more the merrier? A large cladistic analysis of mysticetes, and comments on the transition from teeth to baleen. J Mammal Evol 18:77–100. doi:10.1007/s10914-010-9148-4
Mathis A, Moore FR (1988) Geomagnetism and the homeward orientation of the box turtle *Terrapene Carolina*. Ethology 78:265–274
Matsunaga T, Sakaguchi T (2000) Molecular mechanism of magnet formation in bacteria. J Biosci Bioeng 90:1–13
Maxwell EE, Caldwell MW, Lamoureux DO (2011) The structure and phylogenetic distribution of amniote plicidentine. J Vertebr Paleontol 31(3):553–561. Reprinted by permission of Taylor & Francis Ltd. http://www.tandf.co.uk/journals
McCartney AP, Savelle JM (1985) Thule Eskimo whaling in the central Canadian Arctic. ArcAnthropol 22(2):37–58
McCartney MR, Lins U, Farina M et al (2001) Magnetic microstructure of bacterial magnetite by electron holography. Eur J Mineral 13:685–689
McFee WE, Carl AO (2004) Struvite calculus in the vagina of a bottlenose dolphin (*Tursiops truncatus*). J Wildl Dis 40:125–128
McKown RD (1998) A cystic calculus from a wild western spiny softshell turtle (*Apalone Trionyx spiniferus hartwegi*). J Zool Wildl Med 29(3):347
Mead JG (1975) Anatomy of the external nasal passages and facial complex in the Delphinidae (Mammalia: Cetacea). Smith Contr Zool 207:1–72
Melancon S, Fryer BJ, Gagnon JE et al (2008) Mineralogical approaches to the study of biomineralization in fish otoliths. Min Magaz 72:627–637
Meredith RW, Gatesy J, Murphy WJ, Ryder OA, Springer MS (2009) Molecular decay of the tooth gene enamelin (ENAM) mirrors the loss of enamel in the fossil record of placental mammals. PLoS Genet 5(9):e1000634. doi:10.1371/journal.pgen.1000634. Copyright © 2009 Meredith et al. This is an open-access article distributed under the terms of the Creative Commons Attribution License, which permits unrestricted use, distribution, and reproduction in any medium, provided the original author and source are credited
Meunier FJ, Huysseune A (1992) The concept of bone tissue in osteichthyes. Neth J Zool 42:445–458
Meunier FJ, Sorba L, Béarez P (2004) Presence of vascularized acellular bone in the elasmoid scales of *Micropogonias altipinnis* (Osteichthyes, Perciformes, Sciaenidae). Cybium 28:25–31
Meylan A (1988) From Meylan A (1988) Spongivory in Hawksbill turtles: a diet of glass. Science 239(4838):393–395. Reprinted with permission from AAAS
Mikhailov KE (1991) Classification of fossil eggshells of amniotic vertebrates. Acta Palaeont Polonica 36:193–238
Mikhailov KE (1997a) Avian eggshells: an Atlas of scanning electron micrographs, British Ornitologists'. Club Occasional Publications, Nr.3, 88 p
Mikhailov KE (1997b) Fossil and recent eggshell in amniotic vertebrates: fine structure, comparative morphology and classification. Spec Papers Palaeontol (56):1–80
Milius S (2006) That's one weird tooth. Sci News 169:186
Miller WA (1974) Observations on the developing rostrum and rostral teeth of sawfish: Pristis perotteti and P. cuspidatus. Copeia 1974(2):311–318
Miller JD (1982) Embryology of marine turtles. Dissertation, University of New England, Armidale, New South Wales, Australia
Miller JD (1985) Embryology of marine turtles. In: Gans C, Billett F, Maderson P (eds) Biology of the reptilian. Wiley, New York

Miller GW (1994) Diagnosis and treatment of uric acid renal stone diseases in *Tursiops truncatus*. In: Abstracts of the international association for aquatic animal medicine proceedings, Vallejo, vol 25, pp 22
Miller JM (1999) Morphometric variation in the pharyngeal teeth of zebrafish (Danio rerio Cyprinidae) in response to varying diets. Master dissertation, Texas Tech University, Lubbock, USA
Miller K, Packard GC, Packard MJ (1987) Hydric conditions during incubation influence locomotor performance of hatchling snapping turtles. J Exp Biol 127:401–412
Mills M, Rasch R, Siebeck UE, Collin SP (2011) Exogenous material in the inner ear of the adult Port Jackson Shark, *Heterodontus Portusjacksoni* (Elasmbranchii). Anat Rec 294:373–378. Copyright ©2005, IOS Press All rights reserved
Modesto SP, Reisz RR (2008) New material of Colobomycter pholeter, a small parareptile from the Lower Permian of Oklahoma. J Vertebr Paleontol 28:677–684
Moliner JL (2005) Renal lithiasis and pyelonephritis in two West Indian manatees (*Trichechus manatus* sp). In: Abstracts of the international association for aquatic animal medicine proceedings, Seward, Alaska, USA, vol 34, pp 52
Mora CV, Davison M, Wild JM et al (2004) Magnetoreception and its trigeminal mediation in the homing pigeon. Nature 432:508–511
Morris WR, Kittleman LR (1967) Piezoelectric property of otoliths. Science 19:368–370
Moss ML (1960) Osteogenesis and repair of acellular teleost bone. Anat Rec 136:246–247
Moss ML (1961) Studies on the acellular bone of teleost fish. I. Morphological and systematic variations. Acta Anat 46:343–362
Moss ML (1962) Studies of acellular bone of teleost fish. 2. Response to fracture under normal and acalcemic conditions. Acta Anat 48:46–60
Moss ML (1965) Studies of acellular bone of teleost fish. 5. Histology and mineral homeostasis of fresh-water species. Acta Anat 60:262–276
Moss ML, Freilich M (1963) Studies of acellular bone of teleost fish. 4. Inorganic content of calcified tissues. Acta Anat 55:1–8
Motta CM, Avallone B, Balassone G, Balsamo G, Fascio U, Simoniello P, Tammaro S, Marmo F (2009) Morphological and biochemical analyses of otoliths of the ice-fish *Chionodraco hamatus* confirm a common origin with red-blooded species. J Anatomy 214:153–162. © 2009 The Authors. Journal compilation © 2009 Anatomical Society of Great Britain and Ireland
Mouritsen H, Ritz T (2005) Magnetoreception and its use in bird navigation. Curr Opin Neurobiol 15:406–414
Moy Thomas JA (1939) The early evolution and relationships of the elasmobranchs. Biol Rev 14:1–26
Muheim R (2001) Animal magnetoreception – models, physiology and behaviour. Introductory paper no 128. Department of Ecology, Animal Ecology, Lund University, Lund. Copyright © 2000, Muheim R
Muheim R (2004) Magnetic orientation in migratory birds. Dissertation, Lund University, Lund. Copyright © 2004, R. Muheim. Reprinted with permission
Munday PL, Hernaman V, Dixson DL, Thorrold SR (2011) Effect of ocean acidification on otolith development in larvae of a tropical marine fish. Biogeosciences 8:1631–1641. Copyright © 2011 Munday et al. This work is distributed under the Creative Commons Attribution 3.0 License
Munro U, Munro JA, Phillips JB et al (1997) Evidence for a magnetite–based navigational map in birds. Naturwissenschaften 84:26–28
Murayama E, Okuno A, Ohira T, Takagi Y, Nagasawa H (2000) Molecular cloning and expression of an otolith matrix protein cDNA from the rainbow trout, *Oncorhynchus mykiss*. Comp Biochem Physiol 126B:511–520
Murayama E, Takagi Y, Ohira T, Davis JG, Greene MI and Nagasawa H (2002) Fish otolith contains a unique structural protein, otolin-1. European Journal of Biochemistry 269:688–696. Copyright © 2002, John Wiley and Sons.

Murayama E, Takagi Y, Nagasawa H (2004) Immunohistochemical localization of two otolith matrix proteins in the otolith and inner ear of the rainbow trout, *Oncorhynchus mykiss*: comparative aspects between the adult inner ear and embryonic otocysts. Histochem Cell Biol 121:155–166

Murayama E, Herbomel P, Kawakami A et al (2005) Otolith matrix proteins OMP–1 and Otolin–1 are necessary for normal otolith growth and their correct anchoring onto the sensory maculae. Mech Dev 122:791–803

Murie J (1871) Researches upon the anatomy of Pinnipedia. Part I. on the Walrus (*Trichechus rosmarus*, Linn.). Trans Zool Soc (Lond) 7:411–464

Nakamura Y, Inloes JB, Katagiri T et al (2011) Chondrocyte–specific microRNA–140 regulates endochondral bone development and targets Dnpep to modulate bone morphogenetic protein signaling. Mol Cell Biol 31(14):3019–3028

Nemec P, Altmann J, Marhold S et al (2001) Neuroanatomy of magnetoreception: the superior colliculus involved in magnetic orientation in a mammal. Science 294:366–368

Nicolson SW, Lutz PL (1989) Reproduced with permission Nicolson SW, Lutz PL (1989) Salt gland function in the green sea turtle *Chelonia mydas*. J Exp Biol 144:171–184. Copyright © 1989, The Company of Biologists Limited

Nishiwaki M, Yagi T (1953) On the age and the growth of teeth in a dolphin, (*Prodelphinus caeruleoalbus*). Sci Rep Whales Res Inst (Tokyo) 8:133–146

Norman SA, Garner MM, Berta S et al (2011) Vaginal calculi in a juvenile habor porpoise (*Phocoena phocoena*). J Zool Wildlife Med 42:335–337

Nuamsukon S, Chuen–Im T, Rattanayuvakorn S et al (2009) Thai marine turtle eggshell: morphology, ultrastructure and composition. J Micr Soc Thai 23(1):52–56

Nutter FB, Lee DD, Stamper MA et al (2000) Hemiovariosalpingectomy in a loggerhead sea turtle (*Caretta caretta*). Vet Rec 146:78–80

Nweeia M, Eichmiller F, Orr J (2010) The narwhal tooth sensory organ system and its evolutionary and ecological significance. International Polar Year, Oslo science conference, 8–12 June, 2010

Nweeia MT, Eidelman N, Eichmiller FC et al (2005) Hydrodynamic sensor capabilities and structural resilience of the male Narwhal tusk. In: Abstract presented at the 16th biennial conference on the biology of marine mammals, San Diego, CA, 13 December 2005

Nweeia MT, Nutarak C, Eichmiller FC et al (2009) Considerations of anatomy, morphology, evolution, and function for narwhal dentition. In: Krupnik I, Lang MA, Miller SE (eds) Smithsonian at the poles. Smithsonian Institution Scholarly Press, Washington

Nweeia MT, Eichmiller FC, Hauschka PV et al (2012) Vestigial tooth anatomy and tusk nomenclature for *Monodon monoceros*. Anat Rec (Hoboken) 295(6):1006–1016

Nys Y, Hincke M, Arias JL et al (1999) Avian eggshell mineralization. Poult Avian Biol Rev 10:143–166

Nys Y, Gautron J, Garcia–Ruiz JM et al (2004) Avian eggshell mineralization: biochemical and functional characterization of matrix proteins. CR Palevol 3:549–562

O'Leary DP, Vilches–Troya J, Dunn RF et al (1981) Magnets in guitarfish vestibular receptors. Cell Mol Life Sci 37:86–88

Obradovic-Wagner D, Aspenberg P (2011) Where did bone come from? An overview of its evolution. Acta Orthopaed 82(4):393–398. Copyright © 2011, Informa Healthcare. Reproduced with permission of Informa Healthcare

Oftedal OT (2002) The origin of lactation as a water source for parchment–shelled eggs. J Mammary Gland Biol Neoplasia 7(3):253–266

Olsson PE, Kling P, Hogstrand C (1998) Mechanisms of heavy metal accumulation and toxicity in fish. In: Langston WJ, Bebianno MJ (eds) Metal metabolism in aquatic environments. Chapman and Hall, London

Omelon S, Georgiou J, Henneman ZJ et al (2009) Control of vertebrate skeletal mineralization by polyphosphates. PLoS ONE 4(5):e5634. Copyright: © 2009 Omelon et al. This is an open-access article distributed under the terms of the Creative Commons Attribution License, which permits unrestricted use, distribution, and reproduction in any medium, provided the original author and source are credited

Ørvig T (1965) Palaeohistological notes. 2: certain comments on the phylogenetic significance of acellular bone in early lower vertebrates. Ark Zool 16:551–556
Ørvig T (1967) Phylogeny of tooth tissues: evolution of some calcified tissues in early vertebrates. In: Miles AEW (ed) Structural and chemical organization of teeth, vol I. Academic, London
Ørvig T (1976) Palaeohistological notes. 3. The interpretation of pleromin (pleromic hard tissue) in the dermal skeleton of Psammosteid heterostracans. Zool Scr 5:35–47
Ørvig T (1977) A survey of odontodes ('dermal teeth') from developmental, structural, functional and phyletic points of view. In: Mahala Andrews S, Miles RS, Walker AD (eds) Problems in vertebrate evolution. Academic, New York, pp 53–75
Ørvig T (1989) Histologic studies of ostracoderms, placoderms and fossil elasmobranchs. 6. Hard tissues of Ordovician vertebrates. Zool Scr 18:427–446
Osborne CA, Polzin DJ, Abdullahi SU et al (1985) Struvite urolithiasis in animals and man: formation, detection, and dissolution. Adv Vet Sci Comp Med 29:1–101
Osborne CA, Klausner JS, Polzin DE et al (1986) Etiopathogenesis of canine struvite urolithiasis. Vet Clin North Am Small Anim Pract 16:67–86
Osmolska H (1979) Nasal salt gland in dinosaurs. Acta Paleont Polonica 25:205–215
Outridge PM, Davis WJ, Stewart REA et al (2003) Investigation of the stock structure of Atlantic Walrus (*Odobenus rosmarus rosmarus*) in Canada and Greenland using dental Pb isotopes derived from local geochemical environments. Arctic 56:82–90
Owen R (1945) Odontography: a treatise on the comparative anatomy of the teeth, vols I, 11. Hippolyte Bailliere, London
Packard MJ (1980) Ultrastructural morphology of the shell and shell membrane of eggs of common snapping turtles (*Chelydra serpentina*). J Morphol 165:187–204
Packard MJ (1994) Patterns of mobilization and deposition of calcium in embryos of oviparous, amniotic vertebrates. Israel J Zool 40:481–492
Packard MJ, Hirsh KF (1986) Scanning electron microscopy of eggshells of contemporary reptiles. Scan Electron Microsc 4:1581–1590
Packard MJ, Packard GC (1979) Structure of the shell and tertiary membranes of eggs of soft–shell turtles (*Trionyx spiniferus*). J Morphol 159:131–144
Packard GC, Packard MJ (1980) Evolution of the cleidoic egg among reptilian antecedents of birds. Am Zool 20:351–362
Packard MJ, Packard GC (1984) Comparative aspects of calcium metabolism in embryonic reptiles and birds. In: Seymour RS (ed) Respiration and metabolism of embryonic vertebrates. Dr. w. Junk, The Hague
Packard GC, Packard MJ (1988) The physiological ecology of reptilian eggs and embryos. In: Gans C, Huey RB (eds) Biology of the reptilia, vol 16. Ecology B, Defense and Life History. Alan R. Liss, New York, pp 523–605
Packard GC, Tracy CR, Roth JANJ (1977) The physiological ecology of reptilian eggs and embryos and the evolution of viviparity within the class reptilia. Biol Rev (Camb) 52:71–105
Packard GC et al (1979) Reprinted from Packard GC, Taigen TL, Packard MJ, Shuman RD (1979) Water-vapor conductance of testudinian and crocodilian eggs (class reptilia). Resp Physiol 38(1):1–10. Copyright (1979), with permission from Elsevier
Packard MJ, Packard GC, Boardman TJ (1981) Patterns and possible significance of water exchange by flexible–shelled eggs of painted turtles (*Chrysemys picta*). Physiol Zool 54:165–178
Packard MJ, Packard GC, Boardman TJ (1982) Structure of eggshells and water relations of reptilian eggs. Herpetologica 38:136–155
Packard MJ, Hirsch KF, Iverson JB (1984) Structure of shells from eggs of kinosternid turtles. J Morphol 181:9–20
Panella G (1971) Fish otoliths: daily growth layers and periodical patterns. Science 173(4002):1124–1127. Copyright © 1971, American Association for the Advancement of Science. Reprinted with permission from AAAS
Parago C (2001) Contribuição à taxonomia do gênero *Psammobatis* Günther, 1870 (*Chondrichthyes, Rajidae*): Caracterização das espécies do subgênero I de McEachran (1983) com base em padrões de coloração e espinulação. 52 p. Dissertação (Mestrado). Universidade Federal do Rio de Janeiro, UFRJ, Rio de Janeiro

Pasco-Viel E, Charles C, Chevret P, Semon M, Tafforeau P et al (2010) Evolutionary trends of the pharyngeal dentition in cypriniformes (Actinopterygii: Ostariophysi). PLoS ONE 5(6):e11293. doi:10.1371/journal.pone.0011293. Copyright © 2010 Pasco-Viel et al. This is an open-access article distributed under the terms of the Creative Commons Attribution License, which permits unrestricted use, distribution, and reproduction in any medium, provided the original author and source are credited

Patterson WP (1999) Oldest isotopically characterized fish otoliths provide insight to Jurassic continental climate of Europe. Geology 27:199–202

Patton AK, Savelle JM (2006) The symbolic dimensions of whale bone use in Thule winter dwellings. Études/Inuit/Studies 30:137–161

Peaker M, Linzell JL (1975) Salt glands in birds and reptiles. Cambridge University Press, New York

Perrin WF, Myrick AC Jr (eds) (1981) Age determi nation of toothed whales and sirenians. Rep Int Whal Commun (Spec. Issue No. 3):1–229

Perrin W, Wursig B, Thewissen JGM (2002) Encyclopedia of marine mammals. Academic, Boston

Perry A (1982) Magnetite in the green turtle. Pac Sci 36:514

Perry A, Bauer GB, Dizon AE (1985) Magnetoreception and biomineralization of magnetite in amphibians and reptiles. In: Kirschvink JL, Jones DS, MacFadden BJ (eds) Magnetite biomineralization and magnetoreception in organisms. Plenum Press, New York

Petersen N, von Dobeneck T, Vali H (1986) Fossil bacterial magnetite in deep–sea sediments from the south atlantic ocean. Nature 320(6064):611–615

Petillon J–M (2008) First evidence of a whale–bone industry in the western European Upper Paleolithic: Magdalenian artifacts from Isturitz (Pyrénées–Atlantiques, France). J Human Evol 54(5):720–726

Petko JA, Millimaki BB, Canfield VA et al (2008) Otoc1: a novel otoconin–90 ortholog required for otolith mineralization in zebrafish. Dev Neurobiol 68:209–222

Peyer B (1968) Comparative odontology. University of Chicago Press, Chicago

Phillips JB (1996) Magnetic navigation. J Theor Biol 180:309–319

Phillott AD (2002) Fungal colonisation of sea turtle nests in eastern Australia. Dissertation, Central Queensland University

Phillott AD, Parmenter CJ (2001) The influence of diminished respiratory surface area on survival of sea turtle embryos. J Exp Zool 289:317–321

Phillott AD, Parmenter CJ, Limpus CJ (2004) The occurrence of mycobiota in eastern Australian sea turtle nests. Mem Queensl Mus 49:701–703

Phillott AD, Parmenter CJ, McKillup SC (2006) Calcium depletion of eggshell after fungal invasion of sea turtle eggs. Chel Conserv Biol 5(1):146–149

Pichler FB, Dalebout ML, Baker CS (2001) Nondestructive DNA extraction from sperm whale teeth and scrimshaw. Mol Ecol Notes 1:106–109. Copyright © 2005, John Wiley and Sons

Pimiento C, Ehret DJ, MacFadden BJ, Hubbell G (2010) Ancient nursery area for the extinct giant shark megalodon from the Miocene of Panama. PLoS ONE 5(5):e10552. doi:10.1371/journal.pone.0010552. Copyright © 2010 Pimiento et al. This is an open-access article distributed under the terms of the Creative Commons Attribution License, which permits unrestricted use, distribution, and reproduction in any medium, provided the original author and source are credited

Pollinger ML (1997) Mineralogy and microstructure of dinosaur eggshells. Dissertation, Texas Tech University

Popper AN, Fay RR, Platt C, Sand O (2003) Sound detection mechanisms and capabilities of teleost fishes. In: Colin SP, Marshall NJ (eds) Sensory processing in aquatic environments. Springer, New York, pp 3–38

Pote KG, Ross MD (1991) Each otoconia polymorph has a protein unique to that polymorph. Comp Biochem Physiol B 98:287–295

Pote KG, Hauer CR III, Michel H et al (1993) Otoconin–22, the major protein of aragonitic frog otoconia, is a homolog of phospholipase A2. Biochemistry 32:5017–5024

Powlik JJ (1995) On the geometry and mechanics of tooth position in the white shark *Carcharodon carcharias*. J Morphol 226:277–288

Prado Figueroa M (2005) Distribución Cuantitativa del Malondialdehido entre las Fracciones Subcelulares Obtenidas por Centrifugación Diferencial de Órganos Eléctricos de peces de la familia Rajidae y Topografía del nAChR. III Jornadas de Bioquímica y Biología Molecular de Lípidos y Lipoproteínas, Bahía Blanca, Argentina, p 101

Prado Figueroa M (2011) The growth of chalcedony (nanocrystalline silica) in electric organs from living marine fish. In: Mastai Y (ed) Advances in crystallization processes. InTech, Rijeka, pp 285–300

Prado Figueroa M, Cesaretti NN (2006) Silicificación en ó rganos eléctricos de peces vivientes del estuario de Bahía Blanca. In: IV Congreso Latinoamericano de Sedimentología y XI Reunión Argentina de Sedimentología, San Carlos de Bariloche, Argentina. Limarino y DF Rosetti, GD Veiga, CO, p 184

Prado Figueroa M, Barrera F, Cesaretti NN (2005) Si^{4+} and chalcedony precipitation during oxidative stress in Rajidae electrocyte: a mineralogical study. In: 41th annual meeting. Argentine society for biochemistry and molecular biology research, Pinamar, Argentina, (Biocell 29), p 231

Prado Figueroa M et al (2008) Reprinted from Prado Figueroa M, Barrera F and Cesaretti NN (2008) Chalcedony (a crystalline variety of silica): biogenic origin in electric organs from living *Psammobatis extenta* (family Rajidae). Micron 39(7):1027–1035. Copyright (2008), with permission from Elsevier

Purdy RW (1996) Paleoecology of fossil white sharks. In: Klimley AP, Ainley DG (eds) Great white sharks: the biology of *Carcharodon carcharias*: vol 67. Academic, San Diego

Quinn TP (1980) Evidence of celestial and magnetic compass orientation in lake migrating sockeye salmon fry. J Comp Physiol 137A:243–248

Quint E, Smith A, Avaron F, Laforest L, Miles J, Gaffield W, Akimenko M-A (2002) Bone patterning is altered in the regenerating zebrafish caudal fin after ectopic expression of sonic hedgehog and bmp2b or exposure to cyclopamine. PNAS 99(13):8713–8718. Copyright (2002) National Academy of Sciences, USA. Reprinted with permission

Radtke RL, Targett TF (1984) Rhythmic structural and chemicalpatterns in otoliths of Antarctic fish *Notothenia larseni*: their application to age determination. Polar Biol 3:203–210

Rafferty AR, Reina RD (2012) Arrested embryonic development: a review of strategies to delay hatching in egg-laying reptiles. Proc Biol Sci B 279(1737):2299–2308. doi:10.1098/rspb.2012.0100, by permission of the Royal Society

Ramsay JB, Wilga CD (2007) Morphology and mechanics of the teeth and jaws of white–spotted bamboo sharks (*Chiloscyllium plagiosum*). J Morphol 268:664–682

Reddi AH (2000a) Initiation and promotion of endochondral bone formation by bone morphogenetic proteins: potential implications for Avian Tibial Dyschondroplasia. Poult Sci 79:978–981

Reddi AH (2000b) Reprinted from Principles of Tissue Engineering, 2nd ed.: Reddi AH (2000) Morphogenesis and tissue engineering. In: Lanza R, Langer R, Vacanti JP (eds) Principles of tissue engineering, 2nd edn. Academic, San Diego. Copyright (2000), with permission from Elsevier

Reeves RR, Tracey S (1980) Monodon monoceros. Mamm Species 127:1–7

Reibisch J (1899) Über die Eizahl bei Pleronectes platessa und die Altersbestimmung dieser Form aus den Otolithen. Wiss Meeresuntcrsuch. Abt Kid N F 4:231–248

Reiche et al (2011) Reproduced from Reiche I, Müller K, Staude A, Goebbels J, Riesemeier H (2011) Synchrotron radiation and laboratory micro X-ray computed tomography—useful tools for the material identification of prehistoric objects made of ivory, bone or antler. J Anal Atomic Spectrom 26:1802–1812. With permission of The Royal Society of Chemistry

Reidarson TH, McBain J (1994) Ratio of urine levels of uric acid to creatinine as an aid in diagnosis of urate stones in bottlenose dolphins. Proc Int Assoc Aqua Anim Med 25:21

Reina RD, Cooper PD (2000) Control of salt gland activity in the hatchling green sea turtle, *Chelonia mydas*. J Comp Physiol B 170:27–35

Reina RD, Jones TT, Spotila JR (2002) Salt and water regulation by the leatherback sea turtle *Dermochelys coriacea*. J Exp Biol 205:1853–1860

Retting KN, Song B, Yoon BS et al (2009) BMP canonical Smad signaling through Smad1 and Smad5 is required for endochondral bone formation. Development 136:1093–1104
Retzius G (1881) Das Gehörorgan der Wirbelthiere, vol I. Samson and Wallin, Stockholm
Retzius G (1884) Das Gehörorgan der Wirbeltiere: II. Das Gehörorgan der Amnioten. Samson und Wallin, Stockholm
Rice D (1989) The sperm whale *Physeter macrocephalus* Linnaeus 1758. In: Ridgway SH, Harrison R (eds) Handbook of marine mammals. Academic, London
Right Whale Consortium (2005) North Atlantic right whale consortium photo–id, sightings, genetics, contaminants and necropsy database. New England Aquarium, Boston
Ritz T, Adem S, Schulten K (2000) A model for vision–based magnetoreception in birds. Biophys J 78:707–718
Ritz T, Thalau P, Phillips JB et al (2004) Resonance effects indicate a radical–pair mechanism for avian magnetic compass. Nature 429:177–180
Roberts HS, Sharp RM (1985) Prefered orientation of calcite and aragonite in the reptilian eggshells. Proc R Soc Lond B Bio 255:445–455
Rocha F, Oddone MC, Gadign OBF (2010) Egg capsules of the little skate, *Psammobatis extent* (Garman, 1913) (Chondrichthyes, Rajidae). Braz J Oceanogr 58(3):251–254
Romanoff AL, Romanoff AJ (1949) The Avian egg. Wiley, NewYork
Rose ML, Hincke MT (2009) Protein constituents of the eggshell: eggshell–specific matrix proteins. Cell Mol Life Sci 66(16):2707–2719
Ross MD, Pote KG (1984) Some properties of otoconia. Philos Trans R Soc Lond B Biol Sci 304:445–452
Rouse GW, Goffredi SK, Vrijenhoek RC (2004) Osedax: bone–eating marine worms with dwarf males. Science 305:668–671
Ruben JA, Bennett AA (1987) The evolution of bone. Evolution 41(6):1187–1197. Evolution: international journal of organic evolution by Society for the Study of Evolution. Reproduced with permission of Society for the Study of Evolution, in the format Republish in a book via Copyright Clearance Center
Rückert–Ülkümen N, Yigitbas E (2007) Pharyngeal teeth, lateral ethmoids, and jaw teeth of fishes and additional fossils from the late Miocene (Late Khersonian/Early Maeotian) of Eastern Paratethys (Yalova, Near Üstanbul, Turkey). Turk J Earth Sci 16:211–224
Sahoo G, Mohapatra BK, Sahoo RK et al (1996a) Ultrastructure and characteristics of eggshells of the Olive Ridley turtle (*Lepidochelys olivacea*) from Gahirmatha, India. Acta Anat 156:261–267
Sahoo G, Mohapatra BK, Sahoo RK et al (1996b) Contrasting ultrastructures in the eggshells of olive ridley turtles (*Lepidochelys olivacea*) from Gahirmatha in Orissa. Curr Sci 70:246–249
Sahoo G, Sahoo RK, Mohanty–Hejmadi P (1998) Calciummetabolism in olive ridley turtle eggs during embryonic development. Comp Biochem Physiol Part A 121:91–97
Savelle (1997) Reprinted from Savelle JM (1997) The role of architectural utility in the formation of zooarchaeological whale bone assemblages. J Archaeol Sci 24(10):860–885. Copyright (1997), with permission from Elsevier
Savelle JM, Habu J (2004) A processual investigation of a Thule whale bone house, Somerset Island, Arctic Canada. Arct Anthropol 41(2):204–221
Savelle JM, McCartney AP (2003) Prehistoric bowhead whaling in the Bering Strait and Chukchi sea regions of Alaska: a zooarchaeological assessment. In: McCartney AP (ed) Indigenous ways to the present: native whaling in the Western Arctic, Canadian Circumpolar Institute: studies in whaling no. 6., pp 167–184
Sawyer JE, Walker WA (1977) Vaginal calculi in the dolphin. J Wildl Dis 13:346–348
Scanlon JD, Lee MSY (2002) Varanoid–like dentition in primitive snakes (Madtsoiidae). J Herpetol 36:100–106
Schaefer SA, Buitrago-Suárez UA (2002) Odontode morphology and skin surface features of Andean astroblepid catfishes (Siluriformes, Astroblepidae). J Morphol 254:139–148. Copyright © 2002 Wiley-Liss, Inc. Reprinted with permission from John Wiley and Sons
Scheffel A, Gruska M, Faivre D et al (2006) An acidic protein aligns magnetosomes along a filamentous structure in magnetotactic bacteria. Nature 440:110–114

Scheffer VB, Myrick AB Jr (1980) A review of studies to 1970 of growth layers in the teeth of marine mammals. Report of the International Whaling Commission (Special Issue 3), Cambridge, UK, pp 51–63. Copyright © 1980, International Whaling Commission. Reprinted with permission

Schipiani E (2003) Otoconin–22 and Calcitonin: a novel modality of regulating calcium storages in lower vertebrates? Endocrinology 144:3285–3286

Schleich H, Kastle W (1988) Reptile eggshells. SEM Atlas, Stuttgart

Schmidt M (1885) Das Walross (*Trichechus rosmarus*). D Zool Gart Frankf 1:1–16

Schmidt WJ (1967) Das "globularmuster" im eischalenkalk von Diomedea. Z Zellforsch 77:518–533

Schmidt–Nielsen K (1960) The salt–secreting glands of marine birds. Circulation 21:955–967

Schmidt-Nielsen K, Fange R (1958a) The function of the salt gland in the brown pelican. The Auk 75(3):282–289. Published by the American Ornithologists' Union

Schmidt-Nielsen K, Fänge R (1958b) Salt glands in marine reptiles. Nature 182: 783

Schmidt–Nielsen K, Jörgensen CB, Osaki H (1957) Secretion of hypertonic solutions in marine birds. Fed Proc 16:113–114

Schmitt TL, Sur RG (2012) Treatment of ureteral calculus obstruction with laser lithotripsy in an Atlantic Bottlenose Dolphin (*Tursiops truncatus*). J Zoo Wildl Med 43:101–109

Schönwetter M (1960) Handbuch der oologie. Akademie Verlag, Berlin

Schreiber EA, Burger J (2001) Biology of marine birds. CRC Press, Boca Raton

Schüler D (1999) Formation of magnetosomes in magnetotactic bacteria. J Mol Microbiol Biotechnol 1:79–86

Schüler D (2006) Magnetoreception and magnetosomes in bacteria. (ed) Microbiology monographs, vol 3. Springer, Heidelberg

Schüler D (2008) Genetics and cell biology of magnetosome formation in magnetotactic bacteria. FEMS Microbiol Rev 32(4):654–672

Schüler D, Baeuerlein E (1998) Dynamics of iron uptake and Fe_3O_4 biomineralization during aerobic and microaerobic growth of *Magnetospirillum gryphiswaldense*. J Bacteriol 180(1):159–162

Schultze H–P (1969) Die faltenzähne der rhipidistiiden crossopterygier, der tetrapoden und der Actinopterygier–Gattung Lepisosteus; nebst einer beschreibung der zahnstruktur von onychodus (*Struniiformer Crossopterygier*). Palaeontograph Ital, New Series 35 65:63–137

Schultze H–P (1970) Folded teeth and the monophyletic origin of tetrapods. Amer Mus 2408:1–10

Scotti C, Tonnarelli B, Papadimitropoulos A et al (2010) Recapitulation of endochondral bone formation using human adult mesenchymal stem cells as a paradigm for developmental engineering. Proc Natl Acad Sci USA 107:7251–7256. Copyright (2010) National Academy of Sciences, USA. Reprinted with permission

Seitz JC (2011) Freshwater sawfish ichthyology at the Florida museum of natural history. http://www.flmnh.ufl.edu/fish/Gallery/Descript/Freshwatersawfish/Freshwatersawfish.htm

Semm P, Beason RC (1990) Responses to small magnetic variations by the trigeminal system in bobolinks. Brain Res Bull 25:735–740

Shapiro F (2008) Bone development and its relation to fracture repair. The role of mesenchymal osteoblasts and surface osteoblasts. Eur Cell Mater 15:53–76

Shellis RP, Berkovitz BKB (1980) Reprinted from Shellis RP and Berkovitz BKB (1980) Dentine structure in the rostral teeth of the sawfish Pristis (Elasmobranchii). Arch Oral Biol 25(5):339–343. Copyright © 1980, with permission from Elsevier

Shoemaker VH, Nagy KA, Bradshaw SD (1972) Studies on the control of electrolyte excretion by the nasal gland of the lizard *Dipsosaurus dorsalis*. Comp Biochem Physiol 42A:749–757

Shuttleworth TJ, Hildebrandt JP (1999) Vertebrate salt glands: short– and long–term regulation of function. J Exp Zool 283:689–701

Silva P, Solomon RJ, Epstein FH (1990) Shark rectal gland. In: Fleischer S, Fleischer B (eds) Methods in enzymology, cellular and subcellular transport: epithelial cells. Academic, New York

Silverman HB, Dunbar MJ (1980) Aggressive tusk use by the narwhal (*Monodon monoceros* L.). Nature 284:57–58
Silyn-Roberts H, Sharp RM (1986) Crystal growth and the role of the organic network in eggshell biomineralization. Proc R Soc Lond B 227(1248):303–324. By permission of the Royal Society. Copyright © 1986, The Royal Society
Simkiss K (1962) The sources of calcium for the ossification of the embryos of the giant leathery turtle. Comp Biochem Physiol 7:71–79
Simkiss K, Wilbur KM (1989) Biomineralization, cell biology and mineral deposition. Academic, San Diego
Simpson JG, Gardner MB (1972) Comparative microscopic anatomy of selected marine mammals. In: Ridgway SH (ed) Mammals of the sea: biology and medicine. Charles C. Thomas Publisher, Springfield
Sire J-Y (2001) Teeth outside the mouth in teleost fishes: how to benefit from a developmental accident. Evol Dev 3:104–108. Copyright © 2001 Wiley-Liss, Inc. Reprinted with permission from John Wiley and Sons
Sire JY, Allizard F (2001) A fourth teleost lineage possessing extra–oral teeth: the genus atherion (teleostei; atheriniformes). Eur J Morphol 39(5):295–305
Sire J–Y, Huysseune A (2003) Formation of dermal skeletal and dental tissues in fish: a comparative and evolutionary approach. Biol Rev Camb Philos Soc 78:219–249
Skiles DD (1985) The geomagnetic field: its nature, history, and biological relevance. In: Kirschvink JL, Jones DS, MacFadden BJ (eds) Magnetite biomineralization and magnetoreception in organisms. Plenum Press, New York
Skinner HCW (2000) Minerals and human health, in EMU Notesin mineralogy. In: Varughan DJ, Wogelius RA (eds) Environmental mineralogy, vol 2. Eotvos University Press, Budapest
Skoog T, Johanason SH (1976) The formation of articular cartilage from free perichondrial grafts. Plast Reconstr Surg 57:1–6
Smith LH (1982) Abnormal mineralization. In: Nancollas GH (ed) Biological mineralization and demineralization. Springer, New York
Sollner C, Burghammer M, Busch–Nentwich E et al (2003) Control of crystal size and lattice formation by starmaker in otolith biomineralization. Science 302:282–286
Solomon SE, Baird T (1976) Studies on the egg shell (oviducal and oviposited) of *Chelonia mydas* L. J Exp Mar Biol Ecol 22:145–160
Solomon SE, Baird T (1977) Studies on the soft shell membranes of the egg shell of *Chelonia mydas* L. J Exp Mar Biol Ecol 27:83–92
Solomon SE, Baird T (1980) The effect of fungal penetration on the eggshell of the green turtle. In: Brederoo P, de Priester W (eds) Proceedings of the seventh European congress on electron microscopy. Seventh European Congress on Electron Microscopy Foundation, Leiden, pp 434–435
Solomon SE, Gain M (1996) The normal eggshell. In: Proceedings of the national breeders roundtable, pp 42–53. Copyright (c) 1996, Poultry Science Association, Inc. Reprinted with permission
Solomon SE, Watt JM (1985) The structure of the egg shell of the latherback turtle (*Dermochelys coriacea*). Anim Technol 36:19–27
Soukup V et al (2008) Reprinted by permission from Macmillan Publishers Ltd: Nature, Soukup V, Epperlein H-H, Horácek I, Cerny R (2008) Dual epithelial origin of vertebrate oral teeth. Nature 455:795–798. Copyright (2008)
Stewart C (1903–1906) On the membranous labyrinths of *Echinorchinus*, *Cestracion* and *Rhina*. J Linn Soc Zool 29:439–442
Stewart JR (1997) Morphology and evolution of the egg of oviparous amniotes. In: Sumida SS, Martin KLM (eds) Amniote origins: completing the transition to land. Academic, San Diego
Stonehouse B (1975) The biology of penguins. MacMillan, London/Basingstone. Copyright (c) 1975, Palgrave Macmillan. Reprinted with permission
Stroud RK (1979) Nephrolithiasis in a harbor seal. J Am Vet Med Assoc 175:924–925

Studnicka FK (1912) Die otoconien, otolithen und cupulae terminalis urn gehörorgan von ainmocoetes und von petromyzon. Anal Anz 42:529–562
Suepaul RB, Alley MR, van Rensburg MJ (2010) Salt gland adenitis associated with bacteria in blue penguins (*Eudyptula minor*) from hauraki gulf (Auckland, New Zealand). J Wildl Dis Jan 46(1):46–54. Copyright © 2010, Wildlife Disease Association. Reprinted with permission
Taplin LE, Grigg GC (1981) Salt glands in the tongue of the estuarine crocodile *Crocodylus porosus*. Science 212:1045–1047
Taplin LE, Grigg GC, Harlow P et al (1982) Lingual salt glands in *Crocodylus acutus* and *C. johnstoni* and their absence from *Alligator mississippiensis* and *Caiman crocodilus*. J Comp Physiol 149:43–47
Tarlo LBH (1963) Aspidin; the precursor of bone. Nature 199:46–48
Taub AM, Dunson WA (1967) The salt gland in a Sea Snake (Laticauda). Nature 215:995–996
Taylor MA (2000) Functional significance of bone ballastin in the evolution of buoyancy control strategies by aquatic tetrapods. Hist Biol: An Int J Paleobiol 14(1–2):15–31. Reprinted by permission of Taylor & Francis Ltd, http://www.tandf.co.uk/journals
Thalmann R, Ignatova E, Kachar B et al (2001) Development and maintenance of otoconia: biochemical considerations. Ann NY Acad Sci 942:162–178
Tian L, Xiao B, Lin W et al (2007) Testing for the presence of magnetite in the upper–beak skin of homing pigeons. Biometals 20:197–203
Tohse H, Takagi Y, Nagasawa H (2008) Identification of a novel matrix protein contained in a protein aggregate associated with collagen in fish otoliths. FEBS J 275:2512–2523
Tont SA, Pearcy WG, Arnold JS (1977) Bone structure of some marine vertebrates. Mar Biol 39:191–196
Townsend DW (1980) Microstructural growth increments in some Antarctic fish otoliths. Cybium 3e Ser 8:17–23
Townsend FI, Ringway S (1995) Kidney stones in Atlantic bottlenose dolphins (*Tursiops truncatus*): composition, diagnosis and therapeutic strategies. Proc Int Assoc Aquat Anim Med 26:2–3
Traub W, Arad T, Weiner S (1992) Growth of mineral crystals in turkey tendon collagen fibers. Connect Tissue Res 28(1–2):99–111
Treiber CD et al (2012) Reprinted by permission from Macmillan Publishers Ltd: Nature. Treiber CD, Salzer MC, Riegler J, Edelman N, Sugar C, Breuss M, Pichler P, Cadiou H, Saunders M, Lythgoe M, Shaw J, Keays DA (2012) Clusters of iron-rich cells in the upper beak of pigeons are macrophages not magnetosensitive neurons. Nature 484(7394):367–370. Copyright (2012)
Tricas TC, McCosker JE, Walker TI (1997) Sharks field guide. In: Taylor LR (ed) Sharks and rays. Harper Collins, London
Tsukrov I, DeCew JC, Baldwin K, Campbell-Malone R, Moore MJ (2009) Mechanics of the right whale mandible: full scale testing and finite element analysis. J Exp Mar Biol Ecol 374:93–103
Tucker A, Sharpe P (2004) The cutting–edge of mammalian development; how the embryo makes teeth. Nat Rev Genet 5:499–508
Turner CH (2006) Bone strength: current concepts. Ann N Y Acad Sci 1068:429–446. Copyright © 2006, John Wiley and Sons. Reproduced with permission
Turner–Walker G (2012) The removal of fatty residues from a collection of historic whale skeletons in Bergen: an aqueous approach to degreasing. www.museum.nantes.fr/…/G.%20Turner-Walker
Tuset VM, Lombarte A, Assis CA (2003) Otolith atlas for the western Mediterranean, north and central eastern Atlantic. Scientia Marina 72S1:7–198
Tyack PL, Johnson M, Aguilar Soto N et al (2006) Extreme diving of beaked whales. J Exp Biol 209:4238–4253
Tyler C (1964) Wilhelm von Nathusius 1821–1899 on the avian egg–shells. The Berkshire Printing Co. Ltd., Reading
Ueda K, Maeda Y, Koyama M et al (1986) Magnetic remanences in salmonid fish. Bull Jpn Soc Sci Fish 52:166–170
Uehara K et al (1983) With kind permission from Springer Science+Business Media: Uehara K, Miyoshi S, Toh H (1983) Fine structure of the horny teeth of the lamprey, *Entosphenus japonicas*. Cell Tissue Res 231(1):1–15. Copyright (c) 1983, Springer

Uhen MD (2010) The origin(s) of whales. Annu Rev Earth Planet Sci 38:189–219
Unwin DM, Deeming DC (2008) Pterosaur eggshell structure and its implications for pterosaur reproductive biology. Zitteliana B 28:199–207
Van Bressem MF, Van Waerebeek K, Siebert U et al (2000) Genital diseases in the peruvian dusky dolphin (*Lagenorhynchus obscurus*). Comp Pathol 122(4):266–277
Venn–Watson S, Smith CR, Johnson S et al (2010a) Clinical relevance of urate nephrolithiasis in bottlenose dolphins *Tursiops truncatus*. Dis Aquat Organ 89(2):167–177
Venn–Watson SK, Townsend FI, Daniels RL et al (2010b) Hypocitraturia in common Bottlenose Dolphins (*Tursiops truncatus*): assessing a potential risk factor for urate nephrolithiasis. Comp Med 60:149–153
Verpy E, Leibovici M, Petit C (1999) Characterization of otoconin-95, the major protein of murine otoconia, provides insights into the formation of these inner ear biominerals. Proc Natl Acad Sci USA 96:529–534
Vickaryous MK, Hall BK (2008) Development of the dermal skeleton in *Alligator mississippiensis* (Archosauria, Crocodylia) with comments on the homology of osteoderms. J Morphol 269: 398–422. Copyright © 2007 Wiley-Liss, Inc., A Wiley Company. Reprinted with permission
Vignieri S (2012) Republished with permission of AAAS, from Vignieri S (2012) Magnetic sense. Neurosci Sci Signal 5(226):ec153; permission conveyed through Copyright Clearance Center, Inc
Viguier C (1882) Le sens d'orientation et ses organes chez les animaux et chez l'homme. Rev Phil France et de l'E′ tranger 14:1–36
Vilches-Troya J, Dunn RF, O'Leary DP (1984) Relationship of the vestibular hair cells to magnetic particles in the otolith of the guitarfish sacculus. J Comput Neurol 226(4):489–494. Copyright © 1984 Alan R. Liss, Inc
Vilstrup T (1951) Structure and function of the membranous sacs of the labyrinth in acanthias vulgaris. Ejnar Munksgaard, Copenhagen
von Baer KE (1837) Anatomische und zoologische Untersuchungen über das Walross (Trichenus rosmarus) und Vergleichung dieses Thiers mit andern See–Säugethieren. Mém de l'Acad Impér des Sciences de Saint–Pétersbourg, 6th sér. Sci Math Phys et Nat 4:96–236
von Schreiber JCD (1774) Die Säugethiere in Abbildungen nach der Natur. Wolfgang Walther, Erlangen, pp 1775–1855
Walcott C (1978) Annomalies in the earth's magnetic field increase the scatter of pigeon's vanishing bearings. In: Schmidt–König K, Keeton WTZ (eds) Animal migration, navigation, and homing. Springer, Berlin
Walcott C, Green RP (1974) Orientation of homing pigeons altered by a change in the direction of the applied magnetic field. Science 184:180–182
Walker MM (1998) On a wing and a vector: a model for magnetic field navigation by homing pigeons. J Theor Biol 192:341–349
Walker MM, Kirschvink JL, Chang SBR et al (1984) A candidate magnetic sense organ in the yellowfin tuna, *Thunnus albacares*. Science 224:751–753
Walker MM et al (1988) Republished with permission of The Company of Biologists Ltd, from Walker MM, Quinn TP, Kirschvink JL, Groot C (1988) Production of single-domain magnetite throughout life by sockeye salmon, Oncorhynchus nerka. J Exp Biol 140:51–63. Copyright (1988); permission conveyed through Copyright Clearance Center, Inc
Walker MM et al (1992) Republished with permission of The Company of Biologists Ltd, from Walker MM, Kirschvink JL, Ahmed G, Dizon AE (1992) Evidence that fin whales respond to the geomagnetic field during migration. J Exp Biol 171:67–78. Copyright (1992); permission conveyed through Copyright Clearance Center, Inc
Walker CE, Diebel CV, Haugh PM et al (1997) Structure and function of the vertebrate magnetic sense. Nature 390:371–376
Walker MM, Dennis TE, Kirschvink JL (2002) The magnetic sense and its use in long–distance navigation by animals. Curr Opin Neurobiol 12:735–744
Walker MM et al (2007) Reprinted from Walker MM, Diebel CE, Kirschvink JL (2007) Magnetoreception In: Hara TJ, Zielinski B (eds) Sensory systems neuroscience. Fish physiology

series, vol 25, pp 523. Elsevier Academic, Amsterdam, p. 369. Copyright (2007), with permission from Elsevier

Walsh MT, Murru FL (1987) Urogenital sinus calculi in a Sand Tiger Shark (*Odontaspis taurus*). J Wildl Dis 23(3):428–431

Wang Y, Kowalski PE, I T et al (1998) Otoconin–90, the mammalian otoconial matrix protein, contains two domains of homology to secretory phospholipase A2. Proc Natl Acad Sci USA 95:15345–15350

Wangkulangkul S, Thirakhupt K, Chantrapornsyl S (2000) Comparative study of eggshell morphology in wild and captive Olive ridley turtles *Lepidochelys olivacea* at Phuket, Thailand. In: Pilcher N, Ismail G (eds) Sea turtles of the Indo–Pacific: research, management and conservation. Asean Academic Press, London

Warren AA, Davey L (1992) Folded teeth in temnospondyls—a preliminary study, Alcheringa. Aust J Palaeontol 16:107–132

Warren A, Turner S (2006) Tooth histology patterns in early tetrapods and the presence of "dark dentine". Trans R Soc Edinb Earth Sci 96:113–130

Wei JD, Knittel I, Lang C et al (2011) Magnetic properties of single biogenic magnetite nanoparticles. J Nanopart Res 13(8):3345–3352

Weiner S, Wagner H (1998) The material bone: structure mechanical function relations. Annu Rev Mater Sci 28(1):271–298

Weiner S, Traub W, Wagner HD (1999) Lamellar bone: structure-function relations. J Struct Biol 126:241–255

Weisburd S (1984) Whales and dolphins use magnetic 'roads'. Sci News 126:389

Wesson JA, Ward MD (2007) Pathological biomineralization of kidney stones. Elements 3:415–421

Whitenack LB (2008) The biomechanics and evolution of shark teeth. PhD thesis, University of South Florida, Tampa, FL. Copyright © 2008, Whitenack LB. Reprinted with permission

Whitenack LB, Motta PJ (2010) Performance of shark teeth during puncture and draw: implications for the mechanics of cutting. Biol J Linn Soc 100:271–286. Copyright © 2010 The Linnean Society of London. Reprinted with permission

Whitenack LB et al (2010) Reprinted from Whitenack LB, Simkins Jr. DC, Motta PJ, Hirai M, Kumar A (2010) Young's modulus and hardness of shark tooth biomaterials. Arch Oral Biol 55(3):203–209. Copyright © 2010, with permission from Elsevier

Whitenack LB, Simkins DC, Motta PJ (2011) Biology meets engineering: the structural mechanics of fossil and extant shark teeth. J Morphol 272:169–179. Copyright © 2011 Wiley-Liss, Inc

Whitfield TT, Riley BB, Chiang MY, Phillips B (2002) Development of the zebrafish inner ear. Dev Dyn 223:427–458

Wicke B (1863) Chemisch–physiologische Notizen. Ann Chem Pharm 125:78–80

Wiley TR, Simpfendorfer CA, Faria VV et al (2008) Range, sexual dimorphism and bilateral asymmetry of rostral tooth counts in the smalltooth sawfish *Pristis pectinata* Latham (Chondrichthyes: Pristidae) of the southeastern United States. Zootaxa 1810:51–59

Wilson DE, Reeder DM (eds) (2005) Mammal species of the world. A taxonomic and geographic reference, 3rd edn. Johns Hopkins University Press, Baltimore

Wilson LE, Chin K, Jackson FD, Bray ES (2014) "I. Introduction to eggshells" Fossil eggshell: fragments from the past. Published on-line: http://www.ucmp.berkeley.edu/science/eggshell/eggshell1.php. Accessed 15 May 2014. Copyright © 2014, by the Regents of the University of California. Reprinted with permission

Wiltschko W, Wiltschko R (1972) Magnetic compass of european robins. Science 205:1027–1029

Wiltschko W, Wiltschko R (1981) Disorientation of inexperienced young pigeons after transportation in total darkness. Nature 291:433–434

Wiltschko W, Wiltschko R (1988) Magnetic orientation in birds. In: Johnston RF (ed) Current ornithology. Plenum Press, New York

Wiltschko R, Wiltschko W (1995) Magnetic orientation in animals. Zoophysiology. Springer, Berlin

Wiltschko W, Munro U, Wiltschko R et al (2002) Magnetite–based magnetoreception in birds: the effect of a biasing field and a pulse on migratory behaviour. J Exp Biol 205:3031–3037

Wiltschko W, Munro U, Ford H et al (2009) Avian orientation:the pulse effect is mediated by the magnetite receptors in the upper beak. Proc Biol Sci 276:2227–2232

Winkler JD (2006) Testing phylogenetic implications of eggshell characters in side–necked turtles (Testudines: Pleurodira). Zoology (Jena) 109(2):127–136

Winklhofer M, Holtkamp–Rötzler E, Hanzlik M et al (2001) Clusters of superparamagnetic magnetite particles in the upper–beak skin of homing pigeons:evidence of a magnetoreceptor? Eur J Mineral 13:659–669

Witzmann F (2009) Comparative histology of sculptured dermal bones in basal tetrapods, and the implications for the soft tissue dermis. Palaeodiversity 2:233–270

Wongdee K, Krishnamra N, Charoenphandhu N (2012) Endochondral bone growth, bone calcium accretion, and bone mineral density: how are they related? J Physiol Sci 62(4):299–307

Woo SL, Kwan MK, Lee TQ et al (1987) Perichondrial autograft for articular cartilage. Acta Orthop Scand 58:510–515

Woodall PF (1984) The structure and some functional aspects of the eggshell of the broad–shelled river tortoise *Chelodinia expansa* (Testudinata: Chelidae). Aust J Zool 32:7–14

Woodhouse CD, Rennie CJ (1991) Observations of vaginal calculi in dolphins. J Wildl Dis 27:421–427

Wroe S, Huber DR, Lowry M, McHenry C, Moreno K, Clausen P, Ferrara TL, Cunningham E, Dean MN, Summers AP (2008) Three-dimensional computer analysis of white shark jaw mechanics: how hard can a great white bite? J Zool 276:336–342. © 2008 The Authors. Journal compilation © 2008 The Zoological Society of London

Wu LQ, Dickman JD (2012) Neural correlates of a magnetic sense. Science 336:1054–1057

Wyeth RC (2010) Should animals navigating over short distances switch to a magnetic compass sense? Front Behav Neurosci 4:42–46

Xu Y, Zhang H, Yang H, Zhao X, Lovas S, Lundberg YW (2010) Expression, functional, and structural analysis of proteins critical for otoconia development. Dev Dyn 239:2659–2673. © 2010 Wiley-Liss, Inc

Yano A, Ogura M, Sato A et al (1997) Effect of modified magnetic field on the ocean migration of maturing chum salmon, *Oncorhynchus keta*. Mar Biol 129:523–530

Yaoi Y, Kikuyama S, Hayashi H et al (2001) Immunocytochemical localization of secretory phospholipase A(2)-like protein in the pituitary gland and surrounding tissue of the bullfrog, *Rana catesbeiana*. J Histochem Cytochem 49:631–637

Young JD (1950) The structure and some physical properties of the testudinian egg shell. Proc Zool Soc Lond 120:455–469

Young GC (2008) With kind permission from Springer Science+Business Media: Young GC (2008) Early evolution of the vertebrate eye—fossil evidence. Evol Educ Outreach 1(4):427–438. Copyright © 2008, Springer Science+Business Media, LLC

Yuki M, Sugimoto N, Takahashi K et al (2006) Enterolithiasis in a cat. J Fel Med Surg 8:349–352

Zangerl R (1981) Handbook of paleoichthyology. Chondrichthyes I: paleozoic elasmobranchii. Gustav Fischer Verlag, Stuttgart/New York

Zangerl R, Winter HF, Hansen MC (1993) Comparative microscopic dental anatomy in the Petalodontida (Chondrichthyes, Elasmobranchii). Fieldiana Geol Ser 26:1–43

Zaslansky P (2008) Dentin. In: Fratzl P (ed) Collagen: structure and mechanics. Springer, New York

Zeidel JD, Mathai JC, Campbell JD, Ruiz WG, Apodaca GL, Riordan J, Zeidel ML (2005) Selective permeability barrier to urea in shark rectal gland. Am J Physiol-Renal Physiol 289:F83–F89. ©The American Physiological Society (APS). Reprinted with permission

Zerbini AN, Cesar M, Santos O (1997) First record of the pygmy killer whale *Feresa attenuate* (Gray, 1874) for the Brazilian coast. Aquat Mamm 23(2):105–109

Zhang Z, Zhang YW, Gao H (2011) On optimal hierarchy of load–bearing biological materials. Proc R Soc B 278:519–525

Zhao X, Yang H, Yamoah EN, Lundberg YW (2007) Gene targeting reveals the role of Oc90 as the essential organizer of the otoconial organic matrix. Dev Biol 304:508–524

Zhao X, Jones SM, Yamoah EN, Lundberg YW (2008) Otoconin–90 deletion leads to imbalance but normal hearing: a comparison with other otoconia mutants. Neuroscience 153:289–299

Zioupos P (2005) *In vivo* fatigue microcracks in human bone: material properties of the surrounding bone matrix. Eur J Morphol 42(1/2):31–41

Zioupos P, Currey JD (1996) Pre–failure toughening mechanisms in the dentine of the narwhal tusk: microscopic examination of stress/strain induced microcracking. J Mater Sci Lett 15:991–994

Zioupos P, Currey JD, Casinos A et al (1997) Mechanical properties of the rostrum of the whale Mesoplodon densirostris, a remarkably dense bony tissue. J Zool Lond 241:725–737

Zustin et al (2010) Reprinted from Am J Pathol 177(3): Zustin J, Akpalo H, Gambarotti M et al (2010) Phenotypic diversity in chondromyxoid fibroma reveals differentiation pattern of tumor mimicking fetal cartilage canals development. Am J Pathol 177(3):1072–1078. Copyright (2010), with permission from American Society for Investigative Pathology. Published by Elsevier Inc

Zylberberg et al (1998) Reprinted from Zylberberg L, Traub W, de Buffrenil V, Allizard F, Arad T, Weiner S (1998) Rostrum of a toothed whale: ultrastructural study of a very dense bone. Bone 23(3):241–247. Copyright (1998), with permission from Elsevier

Part III
Marine Fishes as Source of Unique Biocomposites

Chapter 4
Fish Scales as Mineral-Based Composites

Abstract Structural elements called "scales" in fish have a structure and chemical content closer to teeth than to any other scale type. Scales exist in many shapes and sizes, and serve as protection (mechanical and anti-bacterial), camouflage, and plumage for marine fishes. In addition to protective properties, scales provide these animals with locomotive and, in the case of lateral lines, sensory abilities. A fish's locomotion is aided by the shape of scales, which help create a laminar flow of water around the animal. These properties are partially determined by the hierarchically composite-based structure of scales. Discussed in this chapter are numerous mineral-based composites and unique biological materials such as enamel, enameloid, dentines, cosmine, ganoine, hyaline and their derivatives. Also discussed is the diversity and structure of extinct and extant fish scales, scutes and denticles.

A scale (Greek *lepid*, Latin *squama*) is a rigid plate-like structural unit that grows out of a fish's skin to facilitate swimming under water and to provide protection. Scales show broad variety in structure and function. Scales are usually classified as part of integumentary system of organism (see for detailed review Sire et al. 2009). Because of their structural features, being closely examined, scales of fish are sources of information concerning the growth history and longevity of individual animal (Schneider et al. 2000). This information is of principal importance in fish identification as well as in their nomenclature and classification. They also may aid in determining the phylogeny, sexual dimorphism; and assist studies of growth and age. Past environmental conditions experienced by the fish are seen in its scales, allowing a measure of water pollution of the water body (see for review Dapar et al. 2012).

Teeth, scales, and hair are typical vertebrate appendages which, at first glance, appear to have little in common (Sharpe 2001) (see Fig. 4.1). Fish scales and hair are evolutionary and morphologically different in spite they are distributed over the body surface in an orderly pattern. Fish scales and teeth are morphologically different too, "but both are elements of the dermal skeleton and they are considered to be derived from a common ancestor," (Sharpe 2001; see also Huysseune and Sire 1998). Although teeth and hair are also very different from morphological and evolutionary view, they "both share a number of common developmental pathways, such as the Hedgehog, the Bone morphogenetic protein, and Wnt signalling pathways,

H. Ehrlich, *Biological Materials of Marine Origin*, Biologically-Inspired Systems 4,
DOI 10.1007/978-94-007-5730-1_4

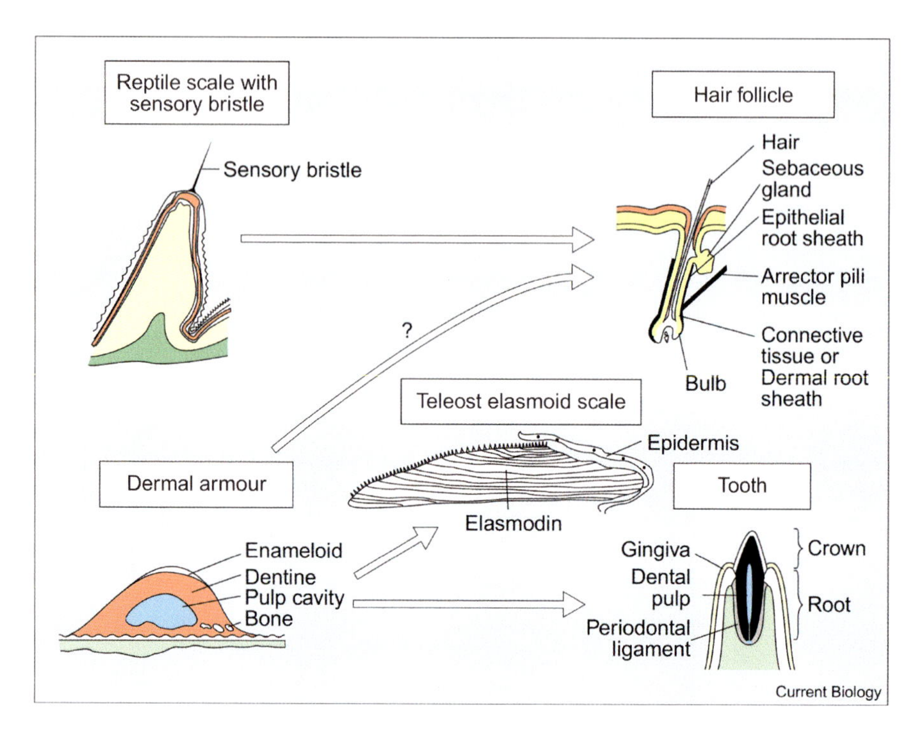

Fig. 4.1 Hypothetical evolutionary relationship between vertebrate ectodermal/dermal appendages (Reprinted from Sharpe (2001), Copyright (2001), with permission from Elsevier)

reflecting the similarities of their early development involving epithelial–mesenchymal interactions," (Sharpe 2001; see also Thesleff and Sharpe 1997). Corresponding confirmation has been reported in experiments with the rs-3 mutant in the teleost *Medaka* that is known as a model organism with a failure of the development of most scales. "The rs-3 locus has been shown to encode EDAR (Ectodysplasin A receptor) and thus mammalian hair loss and failure of fish scale development involve the same TNF pathway," (Sharpe 2001; see also Kondo et al. 2001).

In addition, there are differences in function and composition. For example, scales and hair are not homologous organs. Hairs like feathers are keratinous ectodermal structures localized in ectoderma. They probably evolved from keratinised epidermal scales in a common ancestor of mammals and reptiles. However, there is no keratin within fish scales. These observation lead to the conclusions formulated by Sharpe (2001) as follow:

- "The identification of a common requirement in the development of hair and fish scales thus reflects a common developmental mechanism for appendage formation. This involves epidermal–dermal interactions, and probably not a direct evolutionary link.

- The evolutionary link between fish scales and teeth is more substantial. Scales in teleost fish evolved from the dermal armour covering the body of ancient vertebrates.
- The structural and developmental similarities of fish dermal armour and mammalian teeth has led to the suggestion that teeth evolved by internalisation of dentine containing dermal armour 'odontodes' into the oral cavity.
- Teeth evolved from fish dermal armour," (Sharpe 2001).

Thus, each kind of fish scales possess both mineral and organic components, some of them are the same as in vertebrate teeth. Therefore, I find it necessary to introduce here briefly the corresponding compounds and biocomposites.

4.1 Enamel and Enameloid

Mature dental enamel is an example of highly mineralized hard tissue with respect to hierarchically structured hydroxyapatite crystals (up to 95 % by weight). Both chemistry and exceptionally organized structure of enamel make it the hardest substance in the human body with specific functional properties. According to Hu et al. (2007), dental enamel has no physiological means of repair outside of the remineralization and protective potential provided by saliva because of its acellularity (Hu et al. 2007). The ultrastructure of enamel "resembles a perfect pattern for knitting or croche," (Moradian-Oldak 2009). Hierarchical organization of mature enamel can be represented in the following way:

- Nanoscale: long fluoridated calcium hydroxyapatite crystals;
- Microscale: the crystals are aligned together in bundles to form 3 μm to 5 μm diameter prisms or rods.

The diameter of such enamel prism has approximately the same size as an enamel-making cell, or ameloblast. The extracellular environment between dentine and ameloblasts is that space where the formation of tooth enamel takes place initially, before the tooth erupts. The formation (amelogenesis) and development of enamel *in vivo* is a very complex process that includes gene expression, protein secretion, folding and assembly, and calcification. Margolis et al. (2006) defined enamel amelogenesis as "the result of highly orchestrated extracellular processes that regulate the nucleation, growth, and organization of forming mineral crystals," (Margolis et al. 2006). Although there are lot of open questions concerning understanding of the mechanism of enamel formation in different animal taxa as well as in human (see for detail Nanci 2003; Moradian–Oldak and Paine 2008), the common hypothesis is still based on crucial roles of protein-mineral and protein-protein interactions.

During collar enamel formation in actinopterigian fish (*Polypterus senegalus*), an amorphous fine enamel matrix without any presence of fibrillar collagen was recently found by Sasagawa and co-workers (2012) between the dentine and ameloblast layers. These authors characterized this type of enamel as follow:

"The fine structural features of collar enamel in Polypterus are similar to those of tooth enamel in Lepisosteus (gars), coelacanths, lungfish and amphibians. The enamel matrix shows intense immunoreactivity to the antibody and antiserum against mammalian amelogenins, and to the middle region- and C-terminal-specific anti-amelogenin antibodies. These findings suggest that the proteins in the enamel of Polypterus contain domains that closely resemble those of bovine and porcine amelogenins. The enamel matrix, which exhibits positive immunoreactivity to mammalian amelogenins, extends to the cap enameloid surface. This implies that amelogenin-like proteins are secreted by ameloblasts as a thin matrix layer that covers the cap enameloid after enameloid maturation," (Sasagawa et al. 2012).

Enameloid is an example of hypermineralized enamel-like tissue with unique pattern of mineralization (Sasagawa 2002; Sasagawa et al. 2009) that contains ectomesenchymal and ectodermal proteins. Scales of many fossil fish are the best location of the enameloid. It is assumed that the enameloid is an analogue of mammalian enamel, because the origin of enameloid is somewhat different from that of enamel (Sasagawa et al. 2006). According to Meinke (1986): "Enameloid is distinct from enamel in that enameloid develops prior to dentine and consists primarily of collagenous matrix that is deposited by odontoblasts," (Meinke 1986). While amelogenesis in mammals is a major subject of basic dental sciences, enameloid formation in bony fishes is paid little attention, so there are still many unclear aspects of enameloid formation.

The origin and evolution of both enamel and enameloid mineralization are well debated subjects from chemical, biochemical and genetic points of view. Enamel and enameloid were identified in early jawless vertebrates, about 500 million years ago. This suggests that enamel matrix proteins (EMPs) have at least the same age (Sire et al. 2007).

Three EMPs are secreted by ameloblasts during enamel formation: amelogenin (AMEL), ameloblastin (AMBN) and enamelin (ENAM). It was hypothesized, that "the full-length amelogenin uniquely regulates proper enamel formation through a process of cooperative mineralization, and not as a pre-formed matrix," (Margolis et al. 2006). Recently, two new genes, amelotin (AMTN) and odontogenic ameloblast associated (ODAM), were found to be expressed by ameloblasts during maturation, increasing the group of ameloblast-secreted proteins to five members (Sire et al. 2007). The evolutionary analysis of these five genes indicates that they are related: AMEL is derived from AMBN, AMTN and ODAM are sister genes, and all are derived from ENAM. Kawasaki et al. (2004) reported that they found genes for three major enamel ECM proteins (AMEL, AMBN, and ENAM) and five dentin bone ECM proteins:

- "dentine sialophosphoprotein (DSPP),
- dentine matrix acidic phosphoprotein 1 (DMP1),
- integrin-binding sialoprotein (IBSP),
- matrix extracellular, phosphoglycoprotein (MEPE),
- Osteopontin, or secreted phosphoprotein 1 (SPP1)," (Kawasaki et al. 2004).

These as well as some salivary proteins and milk caseins have a common ancestor that arise by gene duplication to form the secretory calcium binding phosphoprotein (SCPP) family (Kawasaki et al. 2004), some representatives of which are unusually acidic (Kawasaki 2011).

The difference between chemical content of fish teeth and scales' enameloid might be significant. For example, presence of iron in some fish teeth enamelloid has been reported (see for review Suga 1984; Motta 1987; Suga et al. 1989, 1991).

Interestingly, that strong definition of enameloid is still absent, partially because of diversity of this specific structure. The history of enameloid terminology was discussed in the work by Baume (1980). Below are some examples of specific enamelloids.

Coronoin is enameloid of elasmobranch fish (Bendix–Almgreen 1983). The inner dental epithelium (IDE) cells play crucial role in development of this type of enameloid. The detailed analysis of coronoin is given by Bendix-Almgreen and Bang (1997). Scleroblasts during their 'movement' inwards in the dental papilla, are responsible for deposition of collagen-rich pre-coronoin that is exclusively mesenchymatically derived. The change in shape of the IDE cells (they becomes elongated in the superficial direction) is performed synchronously with the phase of pre-coronoin formation. These specialized cells secrete enamelin and some proteins which participate in the degradation of the collagens, when coronoin forms out of the pre-coronoin. The IDE cells are apparently also responsible for the supply of the calcium constituents and even seem to participate in the removal of degraded collagens. During the first steps of calcification, biogenic HAP crystallite rudiments are laid down in alignment with the partially degraded pre-coronoin collagen fibres. Here, the specific properties of the recent coronoin:

1. absence of collagen remains (similar to acrodin and enamel);
2. the mineral phase is represented by the carbonate-fluorapatite crystallites;
3. the crystallites are oriented epitaxial to the original collagen fibre-bundles of the pre-coronoin (Bendix–Almgreen 1983).

In contrast to dentin, elasmobranch enameloid showed sharp diffraction peaks which indicated a high crystallinity of the enameloid. The lattice parameters of enameloid were close to those of the geological fluoroapatite single crystal. The inorganic part of shark teeth consisted of fluoroapatite with a fluoride content in the enameloid of 3.1 wt.% (Enax et al. 2012). The enameloid in fish teeth is 4–5 times harder than the dentine as has been recently evaluated by nanoindentation (Chen et al. 2011).

Adameloid is the term of elasmobranch enameloid proposed by Sasagawa (2002): "the origin of ectodermal enamel is older than that of enameloid. It is therefore likely that the enameloid in elasmobranchs is not a direct precursor of the ectodermal enamel in terrestrial vertebrates, but that enameloid is an analogous structure to enamel," (Sasagawa 2002; see also Smith 1995). Correspondingly, Sasagawa introduced the specific term for this enameloid of the elasmobranchs as "*adameloid*".

Acrodin is enameloid of teleost fish. This term was proposed by Ørvig (1978) for fossil actinopterygian fishes *Nephrotus* and *Colobodus*.

Prelomin is enameloid of holocefalans (members of the Chondrichthyes) (Ørvig 1967). This unique whitlockite-based highly mineralized tissue arises only from mesenchymal cells (Ishiyama et al. 1984, 1991). Also the ability of the mesenchymal cells themselves to be a player in biomineralization of the enamels is an interesting fact. It was reported, that "the initial crystals in prelomin are formed in the tubular saccules limited by a unit membrane and derived from mesenchymal cells," (Sasagawa 2002; see also Ishiyama et al. 1991).

4.2 Dentine and Dentine-Based Composite

Dentine is formed by two simultaneous processes, in which the odontoblasts are instrumental—the formation of the collagenous matrix, and mineral crystal formation in this matrix (Linde 1995). In other words, dentine is formed centripetally by the activity of odontoblasts and may be recognized by calcospheritic mineralization and the presence of odontoblast lacunae and tubules, which house the odontoblastic body and processes, respectively (see for detail Green 2009). The earliest known vertebrate that preserves tissue with these traits is *Anatolepis* from the late Cambrian, approximately 510 million years ago (Young et al. 1996). Dentin, which is less mineralized and less brittle than enamel, is necessary for the support of enamel. However, human dentine is a hard and brittle mineralized tissue. For example, it is difficult to prepare thin specimens (under 200 nm) of dentine for transmission electron microscope observation without demineralizing them (Miura et al. 2012). Tooth dentine is excellent described in humans (see for review Zaslansky 2008).

Probably the best source concerning the special terminology of the paleohistology of early fish dentine is the paper of Ørvig (1951), where corresponding literature on dentine structure has been reviewed. There are also numerous reports (see Baume 1980) on classification of specific types of dentine in marine and aquatic vertebrates like:

- **orthodentine** as the most common dentine that consists of two distinct layers; one with parallel tubes contained within an unmineralized "*predentine*" layer at the dentine-pulp junction, and a second with a mineralized matrix that is composed of intertubular (and sometimes peritubular) dentine (Green 2009). The two main features of orthodentine are as follow: «it is acellular (cell bodies of the odontoblasts are localized outside of the tissue matrix) and generally tubular,» (Sire et al. 2009). Additionally, "Each odontoblast has an elongate cell process that extends into the dentine, surrounded by concentric lamellae of matrix forming a series of parallel tubules. *Atubular orthodentine* was described in first-generation teeth of non-tetrapods, a feature that has been related to the small size of these teeth," (Sire et al. 2009; see also Sire et al. 2002);
- **durodentine** in devonian *Laccognathus panderi* fish (Schmidt 1948, 1959);

- **vasodentine** in teleost fish, a tissue containing capillary canals instead of dentineal tubules (Hay 1912; Herold 1970);
- **plicidentine** (infolded dentine) in sarcopterygians and the actinopterygian *Lepisosteus* (Maxwell et al. 2011),
- **mesodentine** in acanthodian fish scales with *Nostolepis*-type histological structure (Valiukevicius and Burrow 2005). Mesodentine "is characterized by odontoblasts (odontocytes) trapped within the matrix, comparable with cellular bone. Also similar to osteocytes these odontocytes demonstrate a reticulate (i.e. non-polarized) branching pattern of cell processes. Like orthodentine, mesodentine is tubular," (Sire et al. 2009);
- **semidentine** has been reported as unique structure of the extinct group Placodermi. "Similar to mesodentine, semidentine possesses embedded odontocytes within the matrix. However, the odontocyte cell processes are strongly polarized, thus resembling a structural intermediate between the tubular appearance of orthodentine and mesodentine," (Sire et al. 2009);
- **petrodentine** in dipnoan fish (Smith 1984, 1985);
- **elasmodine (isopedine)**, a plywood-like collagenous dentine (Witzmann 2011).

Probably the hardest type of dentine is petrodentine. This hypermineralized tissue is a functionally convergent structure with acrodin and enameloid in actinopterygian and chondrichthyan marginal teeth. In marked contrast to the above two structures, petrodentine forms entirely out of contact with any epithelium (Smith 1985), and is produced by mesenchyme cells in the pulpal region. It was proposed that petrodentine is laid down by "*petroblasts*" which are modified dentinoblasts (Denison 1974). It has an extremely intimate anatomical relationship with the neighbouring "standard" dentine, often called **osteodentine**. Petrodentine forms a blade, supported on both sides by dentine, but the dentine layer on one side is usually rather thin (Currey and Abeysekera 2003). According to the modern definition (Reisz et al. 2004), petrodentine is a hypermineralized tissue type, distinct from trabecular and circumpulpal dentine and bone, that approaches the hardness of enamel.

Dentine can be associated with some layers, which contain ions of metals other than calcium. For example, the distal part of the durodentine tip of teleostean teeth is occasionally provided with a very thin cap coloured by iron-oxid as reported by Schmidt (1969). In mammals, reptiles and amphibians the iron-oxid coloration of the tooth tip belongs exclusively to enamel. It is therefore concluded that the coloured cap of the teleostean teeth is to be considered as a very thin enamel layer.

Hyaloine has been identified as hypermineralized tissue that lacks on collagen and that covers the surface of the post-cranial scutes in armored catfish (Teleostei: Siluriformes) (Sire 1993). According to (Sire et al. 2009):

"Each scute initiates skeletogenesis via the osteogenic pathway: a bone primordium forms deep within the dermis and the presumptive scute grows by centrifugal ossification. Hyaloine matrix is deposited relatively late during development, once the bony scute surface has come into close proximity with the basal surface of the epidermis. Although outwardly similar to ganoine, hyaloine differs in that the superficial-most boundary of this tissue is always separated from the basalmost epidermal layer cells by a narrow (several microns) mesenchyme-filled space. Demineralized hyaloine has the appearance of a thin, stratified fibrillar material, suggestive of periodic deposition. To date the role of the epidermal cells in the formation of hyaloine remains elusive. However, hyaloine is structurally and spatially comparable with ganoine, and is similarly deposited in close proximity with the well-differentiated basalmost epidermal cells," (Sire et al. 2009; see also Sire 1993; Sire et al. 2002; Sire and Huysseune 2003).

Cosmine This biological material is "a unique combination of dentine, enameloid and, at least in the functional sense, some true bone with the pore-canal sensory system, and is found only in certain early fishes," (Thomson 1975). Cosmine is one of the few morphological and histological vertebrate structures that have no homologue among extant forms (Ørvig 1969). For example, dentine layers can be located in the form of concentric lamellar fashion at the margins of the cosmine as in the case of the fish *Ectosteorhachis nitidus* (Thomson 1975). In some species, "the superficial covering of the cosmine is an enameloid layer that is partially penetrated by the tips of the dentine tubules and, of course, is also perforated by the pore-cavity openings," (Thomson 1975; see also Borgen 1992). According to common suggestion, cosmine has both strengthening and protective function. Cosmine, at least in *Ectosteorhachis* and probably also in other osteolepid fishes, and is functionally associated in resorption (Borgen 1989) regeneration with a basal layer of diffuse true bony tissue. This is in addition to the dentine and enameloid that enclose the pore-canal and pulp-cavity systems (Thomson 1975).

Here, the characterization of cosmine made by Thomson (1975):

"A unique feature of all cosmine is that although it is functionally a full constituent of the dermal skeleton, it is topographically and to a great extent developmentally independent of the underlying constituents. The cosmine often forms in large sheets that extend over a large number of otherwise separate dermal elements, covering the sutures between them. In most osteolepid Rhipidistia, for example, the whole of the skull is covered with a shiny cosmine surface broken only by the tiny openings of the sensory pores, and no sign of the sutures between the dermal bone is visible. This obviously creates problems when it comes to growth in area of the units of the dermal skeleton. Thus, a second unique feature of cosmine in Dipnoi and Rhipidistia is that it is subject to periodic total resorption and redeposition, releasing the sutural regions for growth to occur and then covering them up again. Because cosmine may constitute up to 10 % of the calcified tissue in the body, the total resorption and redeposition of the cosmine represents an event of major biological significance to the animal," (Thomson 1975; see also Meinke 1984; Mondéjar–Fernández and Clément 2012). The origin of cosmine in fish seems to be ancient. For example, cosmine with locations in both dermal bones and scales has been reported in one of the oldest lungfish from the Lower Devonian *Uranolophus wyomingensis* (Meinke 1986) (Fig. 4.2).

Ganoine has been suggested to be either enamel proper or enameloid. Correspondingly, it must be "homologous with the highly mineralized layer coating the crown of vertebrate teeth" (Zylberberg et al. 1997; see for more information

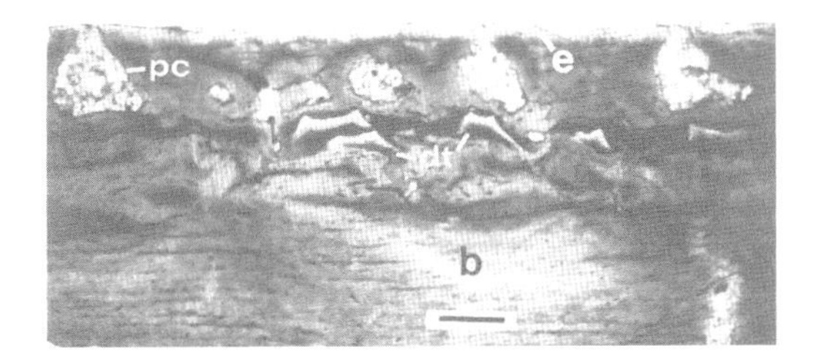

Fig. 4.2 Photograph of thin ground section, showing cosmine morphology of Low Devonian lungfish *Uranolophus wyomingensis* (FMNH 5089) overlying denticles (*dt*). *pc* pore-canals, *e* enameloid, *b* bone. Scale =100 μm (Reprinted from Meinke (1986) with permission of John Wiley and Sons. Copyright © 1986 Wiley-Liss, Inc)

Richter and Smith 1995) Experiments carried out on regenerating scales of *Lepisosteidae* (Sire 1994) as well as of *Polypteridae* (Sire et al. 1987) confirmed ectodermal origin of ganoine in both living primitive actinopterygians even on ultrastructural level. Similarity with tooth enamel was also confirmed with respect to mineralization features (Ørvig 1967; Sire 1995) as well as "the basal epidermal cells show a differentiation pattern similar to that described for the ameloblasts during mammalian tooth morphogenesis," (Zylberberg et al. 1997; see also Sire et al. 1987; Sire 1994). These ameloblasts synthesize the ganoine matrix that is a noncollagenous. Resorption and replacement of ganoin by dentine has been reported in a fossil polypterid (Daget et al. 2001). "Unlike enamel, ganoine is multilayered and, as evidenced by modern taxa, always localized deep to an epithelium" (Vickaryous and Sire 2009).

The formation of ganoine is excellently described by Sire (1994) in *Lepisosteus oculatus* (Holostei) in following way:

"Thus, nonregenerated scales of this fish are composed of a thick, avascular bony plate capped by ganoine that is covered either by the epidermis or by dermal elements. The ganoine surface is separated from the covering soft tissues by an unmineralized layer, the ganoine membrane. During the first 2 months of regeneration, the bony plate forms. It differs from the bony plate of nonregenerated scales only by its large, woven-fibered central region and by the presence of numerous vascular canals. Shortly before ganoine deposition, the osteoblasts cease their activity and an epithelial sheet comes to contact them and spreads on the bony surface. This epithelial sheet is connected to the epidermis only by a short epithelial bridge and is composed of two layers: the inner ganoine epithelium (IGE), in contact with the bone surface and composed of juxtaposed columnar cells that synthesize the ganoine matrix, preganoine; the outer ganoine epithelium (OGE), composed of elongated cells, the surface of which is separated from the overlying dermal space by a basal lamina. Isolated patches of preganoine are deposited by the IGE cells in the upper part of the osteoid matrix of the scale. The interpenetrated preganoine and osteoid matrices constitute an anchorage zone between ganoine and bone. Preganoine patches fuse and a continuous layer of preganoine is progressively synthesized by the IGE cells. Preganoine progressively mineralizes to become ganoine," (Sire 1994).

4.3 Fish Scales, Scutes and Denticles: Diversity and Structure

Fish scales are diverse according to shape and to content of main chemical components (see for review Williamson 1849, 1851; Goodrich 1907; Kerr 1952; Sire 1990; Sire et al. 2009). All of them are intriguing examples of naturally occurring calcium mineral-containing biocomposites. The first important studies which have been carried out on fish scale structure are those of Mandl (1839) and Agassiz (1833–1844). Mandl (1839) was the first to demonstrate the heterogeneity of scale organization in Teleosts (Fig. 4.3), against Agassiz's opinion, and he described how their structure is stratified with layers of parallel fibers whose directions shifted from one layer to the other (Meunier 1984). Williamson (1851), Baudelot (1873) and Hofer (1889) all confirmed the fibrillary nature of the basal plate and the regular change of the fibers' orientations, without giving, however, a thorough analysis of these changing fibrillar orientations. Mandl (1839) misinterpreted "*corpuscles*" as meaning cells. Later Williamson (1851) and especially Baudelot (1873) demonstrated, in their fine contributions, that those structures were mineralized concretions.

According to Sire and Akimenko (2004), in "fish literature, the term 'scale' is often used as a generalised term for all the hard, generally flattened, skeletal elements found in the skin of aquatic vertebrates. These include the scales of:

- chondrichthyans (**placoid scales**),
- the scales of basal actinopterygians (**ganoid scales**),

Fig. 4.3 Heterogenous character of Permian fossil *Elonichthys punctatus* fish scale. *Arrow* indicates the direction of scale growth. Abbreviations: *o* odontodes in a single layer, *d* dentine, *en* enamel, *i* isopedine (Reprinted from Dias et al. (2010) with permission. © Copyright 2010 by Unisinos)

- the bony scales of some actinopterygian taxa (**dermal bony scales and scutes**),
- the scales of basal sarcopterygian taxa and most actinopterygian species (**elasmoid scales**)," (Sire and Akimenko 2004).

The appendages listed above have a different structure in spite all of them represents derivatives of a common ancestral type. From the evolutionary point of view, ganoid and cosmoid scales, the two types of rhombic scales within the osteichthyans, can be traced back to a primitive scale similar to the scales of *Lophosteus* (Schultze 1977). The primitive rhombic scale did not have a peg-and-socket articulation, it is composed of lamellar bone superposed by many layers of spongy bone + dentine. This kind of superposition of layers of spongy bone + dentine (+enamel) has been retained in the cosmoid scale. In contrast, in the ganoid scale the growth of the dentine has become restricted to the lateral surface, the growth of ganoin to the outer surface, and the growth of bone to the inner surface of the scale. Interestingly, the scales of *Andreolepis* have a position between the primitive rhombic scale and the ganoid scale (Schultze 1977).

Cosmoid Scales The rhombic scale of basal sarcopterygians is the characteristic example of so called cosmoid scale. For example, such Devonian (about 410–415 Ma) fish as *Meemannia eos, Psarolepis romeri*, *Achoania jarvikii*, and *Styloichthys change* are the oldest known examples of cosmine-based scales (see for review Vickaryous and Sire 2009). Also animals like both Actinistia (coelacanths) and Dipnomorpha (lungfish) possess cosmoid-like scales (Meinke 1984). On histological level, there are some similarities with the ganoid scale: "a shiny superficial tissue comparable with enamel or enameloid, overlying a stacked sequence of dentine and lamellar bone," (Vickaryous and Sire 2009; see also Goodrich 1907; Meinke 1984). However, Vickaryous and Sire (2009), reported that "cosmoid scales (and cosmine) are uniquely characterized by an intrinsic, interconnected canal system, with numerous flask-shaped cavities and superficial pores. The cosmoid scale (and cosmine tissue) is extinct, and is no longer found in living species," (Vickaryous and Sire 2009).

The cosmine-free cosmoid-like scales of coelacanth are thinner than true cosmoid scales. Smith et al. (1972) described the structure of this scale as follow:

"The greater part of the scale of Latimeria chalumnae is composed of layers of unmineralized isopedine surmounted by the exposed portion of the scale which is pigmented and ornamented by a series of denticles of tubular dentine tipped with enameloid. Between these two parts is a thin ridged bone-like layer. In the electron micrographs the isopedine was seen to consist of layers of densely packed collagen fibres; the orientation of which was uniform in each layer but varied markedly from layer to layer. Only a few cells were found between the fibre bundles. The X-rays revealed numerous concentric annulae and, lying approximately at right angles to these, a further series of ridges radiating from the centre of the scales. It is suggested that the basal unmineralized isopedine and the ridged layer of bone-like tissue covering it represents a highly modified cosmoid scale on which the denticles and pigmented layer have become superimposed," (Smith et al. 1972).

Ganoid scales are examples of very hard, diamond-shaped, structures in which a layer of ganoin lies over the cosmine layer and under the enamel in contrast to cosmoid scales. Fish species, which possess ganoid scales, are related to reedfishes (family *Polypteridae*), gars (family *Lepisosteidae*) and bichirs. The schematic view of ganoid scale is represented in the (Fig. 4.4). According description made by Zylberberg and co-workers (1997):

"A non-regenerated ganoid scale of a polypterid is composed of a thick basal plate made of cellular bone covered by two layers of dental tissues, dentine and ganoine, constituting a so-called odontocomplex. The bony plate is anchored into the dermis by large Sharpey's fibers arising from the deep surface of the scale. The dentine is reduced and the ganoine is stratified, hypermineralized, and covered by the epidermis," (Zylberberg et al. 1997; see also Ørvig 1968).

According to opinion made by Vickaryous and Sire (2009), "ganoid scales are hypothesized to have given rise to elasmoid scales," (Vickaryous and Sire 2009).

From a biomaterials point of view, bony ganoid squamation as the plesiomorphic type in actinopterygians was replaced during evolution by weak and more flexible elasmoid scales. Gemballa and Bartsch (2002) provided a detailed comparative study on mechanical properties of the integument of the extant "ganoid" fishes (*Polypteridae* and *Lepisosteidae*) and "nonganoid" fishes, for closely related "lower" actinopterygians (*Amia calva*, *Acipenser ruthenus*,), and for "lower" sarcopterygians (*Neoceratodus forsteri*, *Latimeria chalumnae*). The aim of this work was to develop a functional understanding of the ganoid integument as a whole and especially its mechanical significance in dependence of morphology of ganoid scales. These researchers measured such parameters as "body curvatures during steady undulatory locomotion, sharp turns, prey-strikes, and fast starts, values of lateral strain on the convex and on the concave side of the fish body," (Gemballa and Bartsch 2002). Their results can be summarized as follow:

- "the ganoid squamation forms a protective coat, but at the same time it permits extreme body curvatures;
- this is reflected in characteristic morphological features of the ganoid scales, such as an anterior process, concave anterior margin, and peg-and-socket articulation;

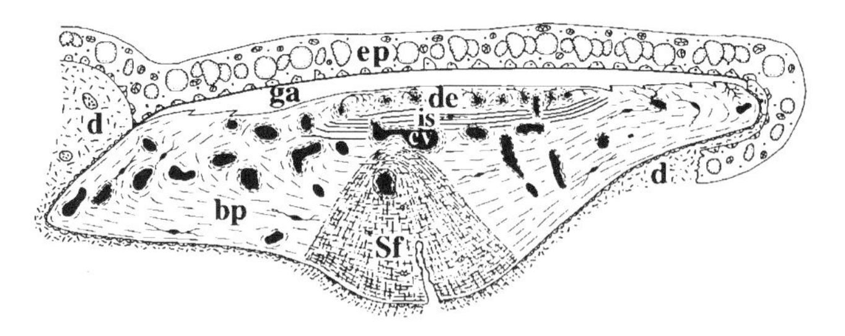

Fig. 4.4 Schematic view with regard to location of ganoine on the non-regenerated ganoid scale of *Calamoichthys calabaricus*. Abbreviations: *bp* osseous basal plate, *cv* vascular canal, *d* dermis, *de* dentine, *ep* epidermis, *ga* ganoine, *is* isopedine, *sf* Sharpey's fibers (Reprinted from Zylberberg et al. (1997) with permission of John Wiley and Sons. Copyright © 1986 Wiley-Liss, Inc)

- the anterior processes and anterior concave margin, together with the attached stratum compactum, guide movements in a horizontal plane during bending;
- ganoid scale rows, fibers of collagen layers of the stratum compactum, and the lateral myoseptal structures follow the same oblique orientation, which is needed to achieve extreme body curvatures;
- ganoid scales do not especially limit body curvature during steady undulatory locomotion;
- they do not act as torsion-resisting devices, but may be able to damp torsion together with the stratum compactum and internal body pressure," (Gemballa and Bartsch 2002).

Placoid Scales This type of scales, also called *denticles*, are similar in structure to teeth and has been observed on extinct and extant cartilaginous fish including sharks (Orvig 1951). The term placoid scales, as the dermal elements covering the body in chondrichthyans have been proposed by Agassiz (1833–1844) and Williamson (1849). Placoid forms of the scale are also known as examples of microsquamation in chondrichthyans. Intriguingly, some elements called "scales" have a structure closer to teeth than to any other scale type. They possess interesting features during their development and growth. For example, placoid scales do not increase in size as the fish grows: new scales are added between the older scales. Spines, which are diverse in form, shape and size, as well as a flattened rectangular base plate which is embedded in the fish body, are the main structural components of the typical placoid scale. Especially the spines of sharks give many species characteristic rough texture (Fig. 4.5).

The term "odontode" have been recently proposed to replace "placoid scale" (Sire and Huysseune 2003). "Indeed, odontodes, which were present in some early vertebrates, 500 million years ago, and which are considered the likely ancestors of all the elements of the dermal skeleton (including teeth) in living vertebrates, have a tooth-like structure," (Sire and Akimenko 2004; see also Reif 1982; Reif and

Fig. 4.5 SEM image: example of the placoid scales of a White Shark (Photo: Sue Lindsay ©Australian Museum)

Richter 2001). Morphology of odontodes in their nearly unchanged form can be observed only in Chondrichthyans. "In contrast, in the osteichthyan lineage, the odontodes have been progressively modified into various types of 'scales', including ganoid scales, dermal bony scales and elasmoid scales. Given their 'dental' structure, the 'placoid scales' must be distinguished from the other scale types," (Sire and Akimenko 2004; see also Sire 2001; Sire and Huysseune 2003). The importance of placoid scale morphology and structure for biomechanics and fish locomotion is discussed in the next chapter.

Thelodont scales Thelodonts are extinct jawless vertebrates, which characteristically possess a dermal skeleton composed of thousands of microscopic scales that were commonly dispersed after death and the decay of supporting soft tissues. Thus, the fossil record of these animals is dominated by jumbled collections of scales that are recovered by acid dissolution of limestones (e.g. Turner 1984).

The scales of Ordovican period origin belonging to loganiid thelodonts are the most distributed structures localized in the Harding Sandstone, Colorado. These scales are conical (Figs. 4.6 and 4.7) and the point of the cone occurs at a shallow angle to the rounded and flared base (Sansom et al. 1996; Turner 1984; Märss 2011).

Sire and co-workers (2009) proposed following classification of the thelodont scale-types are known based on their structure:

- "achanolepid,
- apalolepid,
- katoporid,
- loganiid,
- and thelodontid," (Sire et al. 2009; see also Märss et al. 2007).

Following biological materials can be found as structural elements of thelodont scales:

- orthodentine;
- mesodentine (known in loganiid- and katoporid-type scales);
- aspidin (see for review Sire et al. 2009).

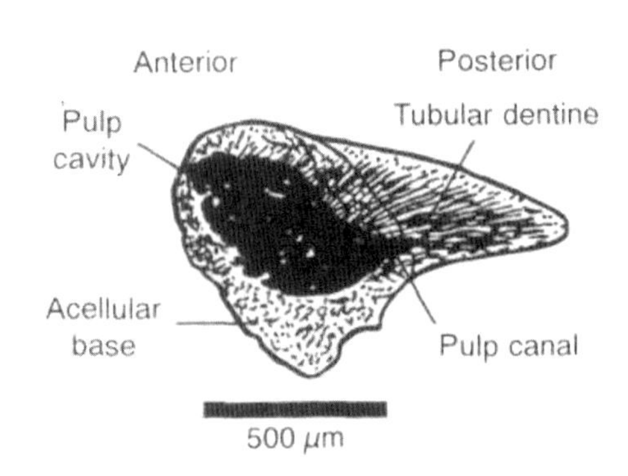

Fig. 4.6 Schematic view of theleodont scale (Reprinted by permission from Macmillan Publishers Ltd: Nature (Sansom et al. 1996), copyright (1996))

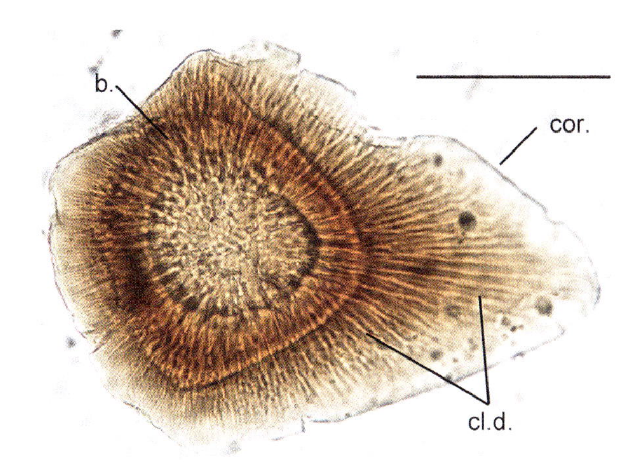

Fig. 4.7 Microstructure of Silurian *Thelodus laevis* (Pander), specimen TUG 1025–1052 scales with a smooth crown surface (Adapted from Märss (2011) with permission from the Estonian Academy Publishers)

The thin layer of monocrystalline enameloid is located on the surface of each scale. Some scales (pore-scales) of the Thelodontian *Phlebolepis elegans* are penetrated by one or more channels which extend on the inner side of the scales in the shape of tubes. The position and number of these channels is quite arbitrary. The channels of these pore-scales are interpreted as branch channels of the lateral organ (Gross 1967).

Elasmoid Scales The adjective "*elasmoid*" has been coined by Bertin (1944). These scales represent commonest type of scale, they are thin and flexible as well as possess typical lamellar structure and imbricated fitting in the tegument. This is the difference to the rhomboid scale arrangement of "lower" bony fishes (see Goodrich 1907). This definition allows the classification of structures as elasmoid scales of the living Coelacanthidae (*Latimeria*), Dipnoi (*Lepidosiren, Neoceratodus, Protopterus*), and Amiidae (*Amid*) (Meunier 1983). Structural features of elasmoid scales has been described in several species belonging to the sarcopterygian and the actinopterygian lineages (see Meunier 1983, 1984; Sire 1987; Huysseune and Sire 1998). Most of the 26,000 Teleostei species possess elasmoid scales, which forms in the fish dermis without the presence of a cartilaginous initium (see for review Francillon-Vieillot et al. 1990; Zylberberg et al. 1992). Sire and Akimenko (2004) reported that the elasmoid scales were mostly found to be invariably composed of three tissues:

(1) "the basal plate, a thick layer of incompletely mineralised tissue composed of elasmodin (previously called isopedin), itself consisting of several layers of type I and V collagen fibrils organised into a plywood-like structure (Meunier 1983; Schultze 1996);
(2) the external layer, a thin layer of well-mineralised tissue composed of a network of interwoven collagen fibrils;
(3) the limiting layer, a hyper-mineralised tissue devoid of collagen fibrils and deposited at the scale surface in the region close to the epidermis. The structure and organisation of this upper layer is the most variable amongst the various species," (Sire and Akimenko 2004).

As reviewed by Sharif et al. (2011), the collagen matrix of elasmoid scales is mineralised on the external (episquamal) side with hydroxyapatite crystals. The episquamal side of the scale possesses concentric ridges (circuli) and grooves (radii) radiating from the central focus to the edges of the scale. Each radius is covered by a dermal space with blood vessels and cells, which are embedded within a loose matrix. Scleroblasts synthesize and shape the scale matrix during ontogeny and regeneration. The external layer is synthesised first, followed by the elasmodine layer. The collagens of the elasmodine layer mineralize slowly from the external layer (Sharif et al. 2011).

As reported above, ganoid scales observed in basal taxa like actinopterygian clade, are thickly juxtaposed rhomboid structures. According to Meunier (2011):

> "these scales have evolved into imbricated thin and flexible elasmoid scales in various, more recent taxa. This evolutionary process has contributed to a lightening of the dermal skeleton, and has improved the efficiency of swimming. Similar specializations can be pointed out in the sarcopterygian clade, e.g. the thick cosmoid scales of extinct osteolepids and the elasmoid scales of extant dipnoids and coelacanths. In Neoceratodus and Latimeria, elasmoid scales present an extreme specialization. The basal plate is composed of an unmineralized network of elasmodine (isopedine), and the collagenous fibres are organized into a double twisted plywood," (Meunier 2011; see also Giraud et al. 1978).

The observations demonstrated that highly specialized structure isopedine is evolved from a bony tissue. For example, it was reported that "the fibrous basal plate of a Latimeria scale is homologous to the osseous basal plate of a cosmoid scale," (Meunier 1980).

The **limiting layer** develops in close proximity to the basalmost epidermal cells similarly to hyaloine and ganoine. It has been demonstrated that epidermal cells play a role in the formation of the limiting layer (Sire 1988). The main organic structural component of this well-mineralized layer is represented by collagenous tissue (exclusive of Sharpey's fibers). The limiting layer is localized superficially on the posterior field of teleost elasmoid scales (Sire et al. 2009). Its deposition is periodic because the limiting layer is superficially separated from basal most epidermal cells by a narrow mesenchymal space similar to hyaloine. The odontogenic pathway is the key way for initiation of skeletogenesis using elasmoid scales: "a scale papilla is formed in the dermis, immediately adjacent to the basal layer cells of the epidermis. Each presumptive elasmoid scale begins as a discrete accumulation of woven-fibered matrix, then is underpinned by multiple lamellae of collagen fibrils organized into a plywood-like arrangement" (Sire et al. 2009; see also Sire and Huysseune 2003; Sire and Akimenko 2004).

Leptoid Scales These scales have been observed on higher order bony fish and appear in two forms, cycloid and ctenoid structures. Both scales add concentric layers during their growth. Localization and orientation of leptoid scales seems to be ideal to reduce drag as well as to allow a smoother flow of water over the fish body.

Cycloid Scales These very common type of scales with a smooth outer edge can be observed on fish with soft fin rays, such as carp and salmon.

Ctenoid Scales Crappie and bass are representatives of fish with spiny fin rays and ctenoid sales, which **have** a toothed outer edge. The characteristic rough texture of these scales is determined by tiny teeth called *ctenii*, which are localized on their posterior edges (Fig. 4.8). Ctenoid scales lack on bone, however contain collagenous (deeper) layer as well as a surface layer containing hydroxyapatite and calcium carbonate. The distribution of both cycloid and ctenoid scales can be different within one fish species. Thus, in some flatfishes, ctenoid scales can be found in males and cycloid scales in females. Ctenii, radii, circuli, lateral line canal and other structures associated with scales (Figs. 4.9 and 4.10) have been used authentically for fish classification (see for review Patterson et al. 2002; Dapar et al. 2012).

There are three types of ctenoid scales based on their morphological and structural features. For example, in "true" ctenoid scales, the spines on the scale are distinct structures. In spinoid scales, the scale bears spines that are continuous with the scale itself, however, in crenate scales, the margin of the scale bears projections and indentations.

In contrast to ganoid and cosmoid scales, cycloid and ctenoid scales overlap, making them more flexible. During their growth, characteristic bands of uneven seasonal growth called annuli (singular annulus) became well visible. These annuli can be used to determine the age the fish. It is worth noting that both ctenoid and cycloid scales may be present on a single fish, such as the flounder. The shape of individual scales may vary a great deal on a single fish, as illustrated in Patterson et al.'s recent atlas of 48 fish species (Patterson et al. 2002).

Scutes and Dermal Plates As reviewed by Sire et al. (2009), "among some early chondrosteans (e.g. Saurichthys; early Triassic to early Jurassic, 250–200 Ma), elements of the integumentary skeleton resemble lepisosteoid scales: a bony base capped by a superficial region of multi-layered ganoine," (Sire et al. 2009). Lack of ganoine is known in bony plates and scutes of such living Acipenseriformes like paddlefish and sturgeons and related fossil taxa. These structures are composed

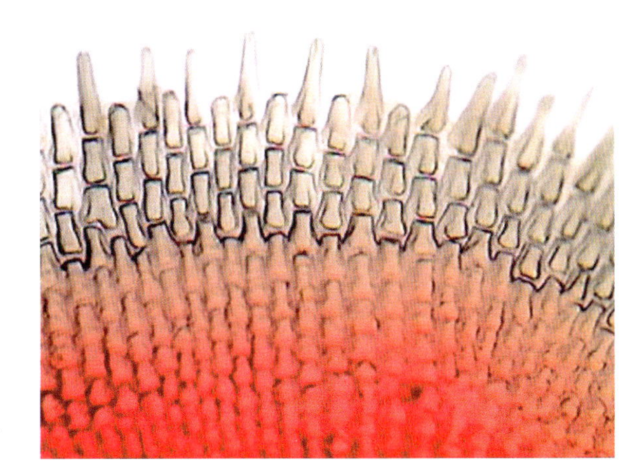

Fig. 4.8 Ctenii on the surface of ctenoid scale (Copyright © 2003 by Michael W. Davidson and The Florida State University. All Rights Reserved)

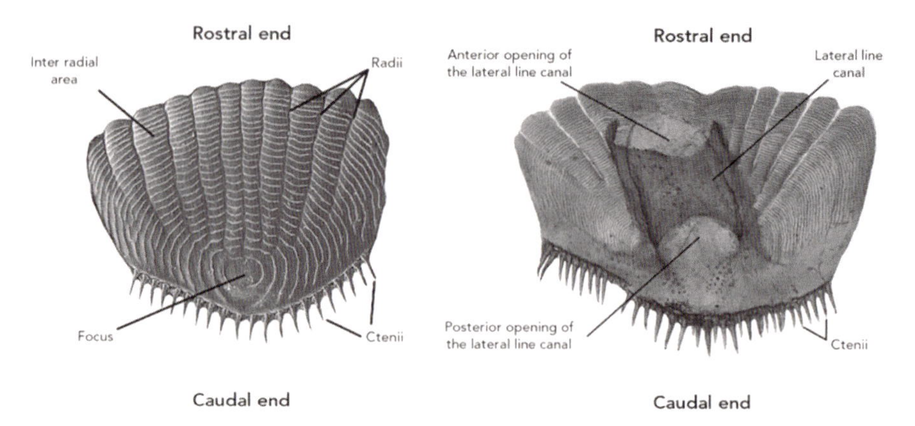

Fig. 4.9 Rostral and caudal ends of *Tripterygiid* scales (Adapted from Jawad (2005) with permission)

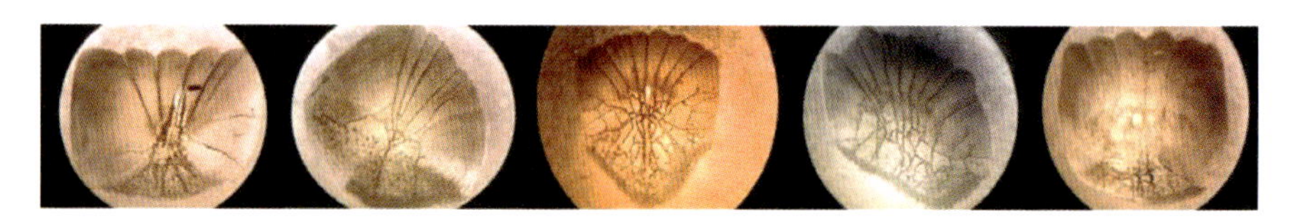

Fig. 4.10 Diversity of ctenoid fish scale shapes commonly found in male and female *Parupeneus indicus*. From *left to right*: rectangular, cycloid, triangular, oblong and square (Adapted from Dapar et al. (2012) with permission)

exclusively of parallel-fibered, cellular bone (Bemis et al. 1997). Special plates of cell-rich bone replaced elasmoid-type scales in various teleost species. Scutes with a layer of hyaloine and usually ornamented with dermal denticles are characteristic for armoured catfish (loricariids, callichthyids, and doradids) (Sire and Huysseune 1996; Sire 2001). Dermal plates, which are composed exclusively of bone, are known for representatives of taxa like syngnathiforms, tetraodontiforms and gasterosteiforms.

So called osteogenic primordium is located deep in the dermis and represents the base where ossification of dermal plates and scutes principally begin. This phenomenon is described by Sire and co-workers (2009) in a following way:

"As the bony plate begins to deposit osteoid and woven-fibered bone, pre-existing collagen fibers from the surrounding dermis are integrated into the matrix, in addition to various bone matrix components deposited by the osteoblasts. With continued ossification, this primordium becomes progressively surrounded by parallel-fibered bone, and eventually by a region made of plywood organized lamellar bone, in which are incorporated Sharpey's fibers," (Sire et al. 2009).

4.4 Conclusion

Scales originated as a natural consequence of biological evolution. Many animals and some plants grow them as a form of external protection, and they occur in nature in many shapes and sizes. The arrangement of scales on an animal's body, called its squamation, varies from species to species (Landreneau 2011). Different scale located biocomposites has been developed during the long history of animal evolution. For example, "the scale-like elements of *pteraspidomorphs* (Ordovician to Devonian; ~480–360 Ma) are characterized by a stratified combination of superficial enameloid overlying tubular dentine (orthodentine), set on a basal plate of acellular bone (aspidin)," (Vickaryous and Sire 2009). However, the integumentary elements of anapsids, which were distributed Silurian to Early Devonian (~443–400 Ma), contain only aspidin. In contrast, such biological materials as cellular bone, mesodentine and orthodentine represented the main structural components of integumentary elements in osteostracans (mid Silurian to Carboniferous; ~430–300 Ma). These data are used also for diagnostic aims. Today, it is well established (Donoghue et al. 2006; Sire et al. 2009) that the ability of vertebrates to mineralize the integument in such different mineralized tissue-types as bone, dentine and enamel/enamelod and their derivatives dates back at least to the Early Ordovician.

Independent of chemical composition, most types of scales are oriented in a certain direction. The reasons for this directionality are many. Fish benefit from improved laminar flow from scales directing the flow of water around their bodies. Both the shape and orientation vary depending on the flow of water, habits of the fish, and other environmental factors. Scale orientations also influence flexibility. While the segmented nature of scales provides more flexibility than rigid armor, certain movements are restricted based on orientation of the scale. For example, highly concave deformations may be restricted due to scales being flattened against each other or bunched together, creating restricting layers (Landreneau 2011).

The next chapter is dedicated to material and mechanical properties of fish scales, as well as to the unique armored constructs developed by extinct and recent fish taxa.

References

Agassiz L (1833–1844) Recherches sur les poissonsfossiles. Imprimerie Petitpierre, Neuchatel (Suisse)

Baudelot ME (1873) Recherches sur la structure et le developpement des ecailles des poisons osseux. Arch Zool Exp Gen 2:87–244; 429–480

Baume LJ (1980) The biology of pulp and dentine: a historic, terminologic–taxonomic, histologic–biochemical, embryonic and clinical survey. In: Myers HM (ed) Monographs in oral science, vol 8. Karger, Basel

Bemis WE, Findeis EK, Grande L (1997) An overview of *Acipenseriformes*. Environ Biol Fish 48:25–71

Bendix–Almgreen SE (1983) *Carcharodon megalodon* from the upper Miocene of Denmark, with comments on elasmobranch tooth enameloid: coronoiin. Bull Geol Soc Denmark 32:1–32

Bendix-Almgreen SE, Bang BS (1997) Aspects of enameloid ontogeny in fossils and recent selachians with comments on enameloids and their occurrence during early lower vertebrate phylogeny. In: The first European workshop on Vertebrate Palaeontology, Geological Museum, Copenhagen University. Dansk geologisk Forening, on-line series no. 1, pp 1–5
Bertin L (1944) Modifications proposees dans la nomenclature des ecailles et des nageoires. Bull Soc Zool Fr 69:198–202
Borgen UJ (1989) Cosmine resorption structures on three osteolepid jaws and their biological significance. Lethaia 22:413–424
Borgen UJ (1992) The function of the cosmine pore canal system. In: Mark–Kurik E (ed) Fossil fishes as living animals. Academy of Sciences of Estonia, Tallinn, pp 141–150
Chen PY, Schirer J, Simpson A et al (2011) Predation versus protection: fish teeth and scales evaluated by nanoindentation. J Mater Res 26:1–12
Currey JD, Abeysekera RM (2003) The microhardness and fracture surface of the petrodentine of *Lepidosiren* (Dipnoi), and of other mineralised tissues. Arch Oral Biol 48:439–447
Daget J, Gayet M, Jeunier FJ et al (2001) Major discoveries on the dermal skeleton of fossil and recent polypteriformes: a review. Fish Fish 2(2):113–124
Dapar MLG, Torres MAJ, Fabricante PC et al (2012) Scale morphology of the Indian goatfish, *Parupeneus indicus* (Shaw, 1803) (Perciformes: Mullidae). Adv Environ Biol 6(4):1426–1432
Denison RH (1974) The structure and evolution of teeth in lungfishes. Fieldiana (Geology) 33:31–58
Dias EV, Vega CS, Canhete MVU (2010) Microstructure of paleoniscid fish scales from Irati Formation, Permian (Cisuralian) of Paraná Basin. Braz J Geosci 6(2):69–75
Donoghue P, Sansom I, Downs J (2006) Early evolution of vertebrate skeletal tissues and cellular interactions, and the canalization of skeletal development. J Exp Zool B Mol Dev Evol 306B:278–294
Enax J, Prymak O, Raabe D et al (2012) Structure, composition, and mechanical properties of shark teeth. J Struct Biol 178(3):290–299
Francillon-Vieillot H, de Buffrénil V, Castanet J, Géraudie J, Meunier FJ, Sire JY, Zylberberg L, de Ricqlès A (1990) Microstructure and mineralization of vertebrate skeletal tissues. In: Carter JG (ed) Skeletal biomineralization: patterns, processes and evolutionary trends, vol 1. Van Nostrand Reinhold, New York, pp 471–530
Gemballa S, Bartsch P (2002) Architecture of the integument in lower teleostomes: functional morphology and evolutionary implications. J Morphol 253:290–309. Copyright © 2002 Wiley-Liss, Inc. Reprinted with permission from John Wiley and Sons
Giraud MM, Castanet J, Meunier FJ et al (1978) The fibrous structure of coelacanth scales: a twisted "plywood". Tissue Cell 10:671–686
Goodrich ES (1907) On the scales of fish, living and extinct, and their importance in classification. Proc Zool Soc London 77(4):751–773
Green JL (2009) Enamel–reduction and orthodentine in *Dicynodontia* (Therapsida) and *Xenarthra* (Mammalia): an evaluation of the potential ecological signal revealed by dental microwear. ProQuest dissertations and theses, North Carolina State University, North Carolina, Raleigh
Gross W (1967) Über Thelodontier–Schuppen. Palaeontographica Abt A 127:1–67
Hay OP (1912) On an important species of Edestus: with description of a new species *Edestus mints*. Proc US Natl Mus 42:31–38
Herold RC (1970) Vasodentine and mantle dentine in teleost fish teeth: a comparative microradiographic analysis. Arch Oral Biol 15(1):71–85
Hofer B (1889) Ueber den Bau und die Entwicklung der Cycloid und Ctenoidschuppen. Sitzber Ges Morph Phys Munchen 6:103–118
Hu JC, Chun YH, Al Hazzazzi T et al (2007) Enamel formation and amelogenesis imperfecta. Cell Tissue Organ 186(1):78–85
Huysseune A, Sire JY (1998) Evolution of patterns and processes in teeth and tooth–related tissues in non–mammalian vertebrates. Eur J Oral Sci 106:437–481
Ishiyama M, Sasagawa I, Akai J (1984) The inorganic content of pleromin in tooth plates of the living holocephalan, *Chimaera phantasma*, consists of a crystalline calcium phosphate known as β–$Ca_3(PO_4)_2$ (Whitlockite). Arch Histol Jpn 47:89–94

Ishiyama M, Yoshie S, Teraki Y et al (1991) Ultrastructure of pleromin a highly mineralized tissue comprising crystalline calcium phosphate known as whitlockite, in holocephalian tooth plates. In: Suga S, Nakahara H (eds) Mechanisms and phylogeny of mineralization in biological systems. Springer, Berlin
Jawad LA (2005) Comparative scale morphology and squamation patterns in triplefins (Pisces: Teleostei: Perciformes: Tripterygiidae). Tuhinga 16:137–167
Kawasaki K (2011) The SCPP gene family and the complexity of hard tissues in vertebrates. Cell Tissue Organ 194(2–4):108–112
Kawasaki K, Suzuki T, Weiss KM (2004) Genetic basis for the evolution of vertebrate mineralized tissue. Proc Natl Acad Sci U S A 101(31):11356–11361. Copyright (2004) National Academy of Sciences, U.S.A. Reprinted with permission
Kerr T (1952) The scales of primitive living actinopterygians. Proc Zool Soc London 122:55–78
Kondo S, Kuwahara Y, Kondo M et al (2001) The medaka rs–3 locus required for scale development encodes ectodysplasin–a receptor. Curr Biol 11:1202–1206
Landreneau EB (2011) Scales and scale–like structures. Dissertation, Texas A&M University, College Station
Linde A (1995) Dentin mineralization and the role of odontoblasts in calcium transport. Connect Tissue Res 33:163–170
Mandl L (1839) Recherches sur la structure intime des ecailles des poissons. Ann Sci Nat 2:337–371
Margolis HC, Beniash E, Fowler CE (2006) Role of macromolecular assembly of enamel matrix proteins in enamel formation. J Dent Res 85(9):775–793. Copyright © 2006 by International & American Associations for Dental Research. Reprinted by Permission of SAGE Publications
Märss T (2011) A unique Late Silurian Thelodus squamation from Saaremaa (Estonia) and its ontogenetic development. Estonian J Earth Sci 60:137–146
Märss T, Turner S, Karatajute–Talimaa V (2007) Handbook of paleoichthyology, vol 1B, 'Agnatha' II Thelodonti. Verlag Dr Friedrich Pfeil, Munich
Maxwell EE, Caldwell MW, Lamoureux DO (2011) The structure and phylogenetic distribution of amniote plicidentine. J Vertebr Paleontol 31:553–561
Meinke DK (1984) A review of cosmine: its structure, development, and relationships to the other forms of the dermal skeleton in osteichthyans. J Vertebr Paleontol 4:45–470
Meinke DK (1986) Morphology and evolution of the dermal skeleton in lungfishes. J Morphol 190:133–149. Copyright © 1986 Wiley-Liss, Inc. Reprinted with permission
Meunier FJ (1980) Les relations isopedine – tissu osseux dans le post-temporal et les ecailJes de la ligne laterale de *Latimeria chalumnae* (Smith). Zoologica Scripta 9:307–317. Copyright © 2008, John Wiley and Sons. Reprinted with permission
Meunier FJ (1983) Les tissus osseux des Ostéichthyens. Structure, genèse, croissance et évolution. Archives et Documentations, Micro–Edition, Institut d'Ethnologie, Museum National d'Histoire Naturelle, Paris, SN, 82–600–328
Meunier FJ (1984) Spatial organization and mineralization of the basal plate of elasmoid scales in osteichthyans. Am Zool 24:953–964
Meunier FJ (2011) Reprinted from: Meunier FJ (2011) The Osteichtyes, from the Paleozoic to the extant time, through histology and palaeohistology of bony tissues. C R Palevol 10:347–355, Copyright (2011), with permission from Elsevier
Miura J, Kubo M, Nagashima T, Takeshige F (2012) Ultra-structural observation of human enamel and dentin by ultra-high-voltage electron tomography and the focus ion beam technique. J Electron Microsc (Tokyo) 61(5):335–341
Mondéjar–Fernández J, Clément G (2012) Squamation and scale microstructure evolution in the Porolepiformes (Sarcopterygii, Dipnomorpha) based on *Heimenia ensis* from the Devonian of Spitsbergen. J Vertebr Paleontol 32:267–284
Moradian–Oldak J (2009) The regeneration of tooth enamel. Dimens Dent Hyg 7(8):12–15
Moradian–Oldak J, Paine ML (2008) Mammalian: enamel formation. In: Astrid S, Sigel H, Sigel RKO (eds) Metal ions in life sciences. Wiley, Chichester

Motta PJ (1987) A quantitative analysis of ferric iron in butterflyfish teeth (Chaetodontidae, Perciformes) and the relationship to feeding ecology. Can J Zool 65(1):106–112
Nanci A (2003) Enamel: composition, formation and structure. In: Nanci A (ed) Ten Cate's oral histology: development, structure, and function harcourt health. Elsevier Mosby, St. Louis
Ørvig T (1951) Histologic studies of Placoderms and fossil Elasmobranchs. I: the endoskeleton, with remarks on the hard tissues of lower vertebrates in general. Ark Zool 2(2):321–454
Ørvig T (1967) Phylogeny of tooth tissues: evolution of some calcified tissues in early vertebrates. In: Miles AEW (ed) Structural and chemical organization of teeth, vol I. Academic, New York
Ørvig T (1968) The dermal skeleton; general considerations. In: Ørvig T (ed) Current problems of lower vertebrate phylogeny. Proceedings of the 4th Nobel symposium. Almqvist and Wiskell, Stockholm
Ørvig T (1969) Cosmine and cosmine growth. Lethaia 2:241–260
Ørvig T (1978) Microstructure and growth of the dermal skeleton in fossil actinopterygian fishes: Nephrotus and Colobodus, with remarks on the dentition in other forms. Zool Scr 7:297–326
Patterson RT, Wright C, Chang AS et al (2002) Brit. Columbia fish–scale atlas. Palaeontol Electron 4(1):88
Reif WE (1982) Evolution of dermal skeleton and dentition in vertebrates: the odontode regulation theory. In: Hecht MK, Wallace B, Prauce GT (eds) Evolutionary biology, vol 15. Plenum Press, New York, pp 287–368
Reif WE, Richter M (2001) Revisiting the lepidomorial and odontode regulation theories of dermo–skeletal morphogenesis. Neues Jahr Geol Paläont 219:285–304
Reisz RR, Krupinina NI, Smith MM (2004) Dental histology in *Ichnomylax karatajae* sp. Nov., an early Devonian dipnoan from the Taymyr Peninsula, Siberia, with a discussion on petrodentine. J Vertebr Paleontol 24(1):18–25
Richter M, Smith MM (1995) A microstructural study of the ganoine tissue of selected lower vertebrates. Zool J Linnean Soc 114:173–212
Sansom IJ, Smith MM, Smith MP (1996) Scales of thelodont and shark-like fishes from the Ordovician of Colorado. Nature 379:628–630
Sasagawa I (2002) Reprinted from: Sasagawa I (2002) Mineralization patterns in elasmobranch fish. Microsc Res Tech 59:396–407. Copyright © 2002 Wiley-Liss, Inc., with permission from John Wiley and Sons)
Sasagawa I, Ishiyama M, Akai J (2006) Cellular influence in the formation of enameloid during odontogenesis in bony fishes. Mater Sci Eng C 26:630–634
Sasagawa I, Ishiyama M, Yokosuka H et al (2009) Tooth enamel and enameloid in actinopterygian fish. Front Mater Sci Chin 3:174–182
Sasagawa et al (2012) With kind permission from Springer Science + Business Media: Sasagawa I, Yokosuka H, Ishiyama M et al (2012) Fine structural and immunohistochemical detection of collar enamel in the teeth of *Polypterus senegalus*, an actinopterygian fish. Cell Tissue Res 347(2):369–381. Copyright © 2012, Springer-Verlag
Schmidt WJ (1948) Polarisationsoptische untersuchung schmelzartiger aussenschichten des zahnbeins von fischen. III. Das durodentin von myliobatis. Cell Tissue Res 34(2):165–178
Schmidt WJ (1959) Durodentin bei einem devonischen fisch (Laccognathus Panderi Gross). Cell Tissue Res 49:493–514
Schmidt WJ (1969) Die schmelznatur der eisenoxidhaltigen kappe auf teleostier–zähnen. Cell Tissue Res 93:447–450
Schneider JC, Laarman PW, Gowing H (2000) Age and growth methods and state averages: Chapter 9. In: Schneider JC (ed) Manual of fisheries survey methods II: with periodic updates, Fisheries special report 25. Michigan Department of Natural Resources, Ann Arbor
Schultze HP (1977) Ausgangsform und entwicklung der rhombischen schuppen der Osteichthyes (Pisces). Palöont Z 51:152–168
Schultze HP (1996) The scales of Mesozoic actinopterygians. In: Arratia G, Viohl G (eds) Mesozoic fishes systematic and palaeoecology. Verlag Dr. F. Pfeil, München
Sharif F, de Vrieze E, Metz JR et al (2011) Matrix metalloproteinases in osteoclasts of ontogenetic and regenerating zebrafish scales. Bone 48:704–712

Sharpe TP (2001) Reprinted from: Sharpe TP (2001) Fish scale development: hair today, teeth and scales yesterday? Curr Biol 11(18):R751–R752, Copyright (2001), with permission from Elsevier

Sire J-Y (1987) Structure, formation et régénération des écailles d'un poisson téléostéen, *Hemichromis bimaculatus* (Perciforme, Cichlidé). Dissertation, Université Paris, Paris

Sire J-Y (1988) Evidence that mineralized spherules are involved in the formation of the superficial layer of the elasmoid scale in the cichlids *Hemichromis bimaculatus* and *Cichlasoma octofasciatum* (Pisces, Teleostei): an epidermal active participation? Cell Tissue Res 253:165–171

Sire J-Y (1990) From ganoid to elasmoid scales in the actinopterygian fishes. Neth J Zool 40:75–92

Sire J-Y (1993) Development and fine structure of the bony scutes in *Corydoras arcuatus* (Siluriformes, Callichthyidae). J Morphol 215:225–244

Sire J-Y (1994) Light and TEM study of nonregenerated and experimentally regenerated scales of *Lepisosteus oculatus* (holostei) with particular attention to ganoine formation. Anat Rec 240:189–207. Copyright © 1994 Wiley-Liss, Inc. Reprinted with permission from John Wiley and Sons

Sire J-Y (1995) Ganoine formation in the scales of primitive actinopterygian fishes, lepisosteids and polypterids. Connect Tissue Res 32:535–544

Sire J-Y (2001) Teeth outside the mouth in teleost fish: how to benefit from a developmental accident. Evol Dev 3:104–108

Sire J-Y, Akimenko M-A (2004) Scale development in fish: a review, with description of sonic hedgehog (shh) expression in the zebrafish (*Danio rerio*). Int J Dev Biol 48:233–247. Copyright (c) 2004, UBC Press. Reproduced with permission

Sire J-Y, Huysseune A (1996) Structure and development of the odontodes in an armoured catfish, *Corydoras aeneus* (Callichthyidae). Acta Zool (Stockh) 77:51–72

Sire J-Y, Huysseune A (2003) Formation of skeletal and dental tissues in fish: a comparative and evolutionary approach. Biol Rev 78:219–249

Sire J-Y, Géraudie J, Meunier FJ et al (1987) On the origin of ganoine: histological and ultrastructural data on the experimental regeneration of the scales of *Calamoichthys calabaricus* (Osteichthyes, Brachyopterygii, Polypteridae). Am J Anat 180:391–402

Sire J-Y, Davit-Béal T, Delgado S et al (2002) The first generation teeth in non-mammalian lineages: evidence for a conserved ancestral character? Microsc Res Tech 59:408–434

Sire J-Y, Davit–Beal T, Delgado S et al (2007) The origin and evolution of enamel mineralization genes. Cell Tissue Organ 186:25–48

Sire JY et al (2009) Reprinted from: Sire JY et al (2009) Origin and evolution of the integumentary skeleton in non-tetrapod vertebrates. J Anat 214:409–440, Copyright © 2009 The Authors. Journal compilation © 2009 Anatomical Society of Great Britain and Ireland, with permission from John Wiley and Sons

Smith MM (1984) Petrodentine in extant and fossil dipnoan dentitions: microstructure, histogenesis and growth. Proc Linnean Soc NSW 107:367–407

Smith MM (1985) The patterns of histogenesis and growth of tooth plates in larval stages of extant lungfish. J Anat 140:627–643

Smith MM (1995) Heterochrony in the evolution of enamel in vertebrates. In: McNamara KJ (ed) Evolutionary change and heterochrony. Wiley, Chichester

Smith MM, Hobdell MH, Miller WA (1972) The structure of the scales of *Latimeria chalumnae*. J Zool 167:501–509. Copyright © 2009, John Wiley and Sons. Reprinted with permission from John Wiley and Sons.)

Suga S (1984) The role of fluoride and iron in mineralization of fish enameloid. In: Fearnhead RW, Suga S (eds) Tooth enamel. Elsevier Science Publishers BV, Amsterdam

Suga S, Wada K, Taki Y et al (1989) Iron concentration in teeth of tetra-odontiform fish and its phylogenetic significance. J Dent Res 68:1115–1123

Suga S, Taki Y, Wada K et al (1991) Evolution of fluoride and iron concentrations in the enameloid of fish teeth. In: Suga S, Nakahara H (eds) Mechanisms and phylogeny of mineralization in biological systems. Springer, Tokyo

Thesleff I, Sharpe P (1997) Signalling networks regulating dental development. Mechanisms Dev 67:111–123

Thomson KS (1975) On the biology of cosmine. Bulletin of the peabody museum of natural history, vol 40. Peabody Museum of Natural History, Yale University: New Haven. Courtesy of the Peabody Museum of Natural History, Yale University, New Haven, CT

Turner S (1984) Studies of Palaeozoic Thelodonti (Craniata: Agnatha), 2 vols. Unpublished PhD thesis, University of Newcastle-upon-Tyne, UK

Valiukevicius J, Burrow CJ (2005) Diversity of tissues in acanthodians with Nostolepis–type histological structure. Acta Palaeontol Pol 50:635–649

Vickaryous MK, Sire J-Y (2009) The integumentary skeleton of tetrapods: origin, evolution, and development. J Anat 214:441–464. © 2009 The Authors. Journal compilation © 2009 Anatomical Society of Great Britain and Ireland. Reprinted with permission from John Wiley and Sons

Williamson WC (1849) On the microscopic structure of scales and teeth of some ganoid and placoid fish. Phil Trans R Soc Lond Ser B Biol Sci 139:435–475

Williamson WC (1851) Investigations into the structure and development of the scales and bones of fishes. Philos Trans R Soc Lond 141:643–702

Witzmann F (2011) Morphological and histological changes of dermal scales during the fish–to–tetrapod transition. Acta Zool (Stockholm) 92:281–302

Young GC, Karatajute–Talimaa VN, Smith MM (1996) A possible Late Cambrian vertebrate from Australia. Nature 383:810–812

Zaslansky P (2008) Dentin. In: Fratzl P (ed) Collagen: structure and mechanics. Springer, New York

Zylberberg L, Géraudie J, Meunier FJ et al (1992) Biomineralisation in the integumental skeleton of the living lower vertebrates. In: Hall BK (ed) Bone – bone metabolism and mineralisation, vol 4. CRC Press Inc., Boca Raton, pp 171–224

Zylberberg L, Sire J-Y, Nanci A (1997) Immunodetection of amelogenin-like proteins in the ganoine of experimentally regenerating scales of *Calamoichthys calabaricus*, a primitive actinopterygian fish. Anat Rec 249:86–95. Copyright © 1997 Wiley-Liss, Inc. Reprinted with permission from John Wiley and Sons

Chapter 5
Materials Design Principles of Fish Scales and Armor

Abstract Models for new engineering designs are based on high-performance natural materials and systems including biological materials. Fish scales, which serve in the animal for many functional roles simultaneously, possess high biomimetic potential for bioinspiration and development of different armor-based constructs, superoleophobic as well as self cleaning surfaces. The diversity of fish scale shapes, their mechanical properties and broad variety of biomimetic applications are discussed in this chapter.

Fish scales first evolved millions of years ago as a form of flexible armor. Humans adapted scales for many uses, both artistic and practical. Scales continue to be utilized in a variety of formats today, including military science and engineering. Also representing them in Computer Graphics proves an interesting challenge (Landreneau 2011). Moreover, the principles of fish scale organization (Fig. 5.1) are used in modern optics. It was shown that metamaterial of bilayered fish-scale origin can function as a broad bandpass filter in the terahertz modus. The filter is relatively insensitive to incidence angles up to 45° (Zhang et al. 2012a, b).

From an evolutionary point of view, fish scales provided a new type of protection for organisms. Instead of surrounding an animal in large, rigid segments, scales cover an organism in a network of small, armored plates. Being smaller, the scales form a flexible sheetlike surface that is both resistant to attack and flexible. Scales can overlap, forming an imbricated network which fully covers the animals without the weak points of exoskeleton joints. These advantages make scales a powerful evolutionary advantage, one that persists in many animals today (Landreneau 2011).

Now, the popular source of bioinspiration for both bionics and biomimetics is the shark skin and its dermal denticles (placoid scales). It seems that we could also find alternatives for very impressive examples of fish scale protective properties in more ancient fish taxa. Scales contain tough materials that fossilize well, meaning that scaled animals will have a far more robust fossil record than similar, unscaled animals. In fact, a great number of primitive animals are identified by their scales alone, as the scales are all that remains of their bodies after fossilization. That species can be differentiated by their scales also emphasizes a large diversity in the structure and composition of scales (see for review Timms 1905; Creaser 1926; Sire et al. 2009).

H. Ehrlich, *Biological Materials of Marine Origin*, Biologically-Inspired Systems 4,
DOI 10.1007/978-94-007-5730-1_5

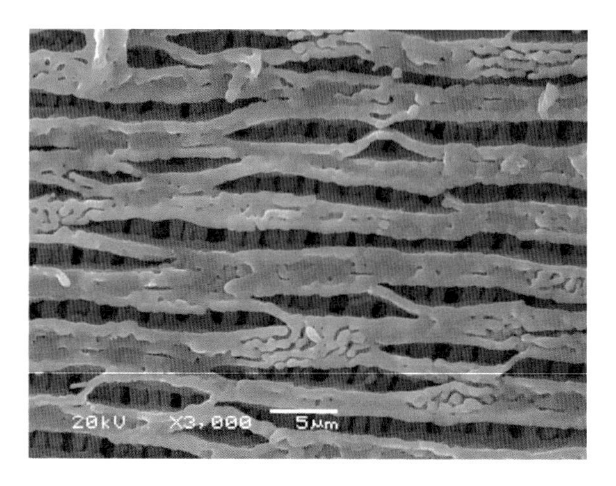

Fig. 5.1 SEM image of *P. major* fish scales: the fibrillary layer after heating at 1,473 K show unique ordered organization. It contains individual strips of calcium phosphate and interconnecting bridges (Reprinted from Ikoma et al. (2003) with permission from Elsevier)

As scales form a natural layer of armor for fish and reptilians, it is unsurprising that they prove useful in designing military armor since ancient times in human communities. Armor covered in scales bends to conform to the body of a soldier, deflecting attacks without sacrificing mobility. The selfsimilar, tileable nature of scales also simplifies manufacture of the armor, as scales can be mass produced and fitted to any underlying shape. The individual scales are produced separately, then typically woven together and attached to an underlying armor (Landreneau 2011). While segmented scale armor is inherently more flexible than a solid plate of the same thickness, both engineers, the Nature in the case of fish armor, and human in the case of military armor, used the same physical principles.

The use of scales in protective armor dates back at least to the ancient Persians. Herodotus also makes mention of Persian scaled armor, describing colored tunics covered in iron fish scales (Laufer 1914; Herodotus 1954). Persian Elamite artwork as early as the third millennium BCE depicts people wearing armor covered in scales. Scale armor remained prominent throughout the reign of the Roman Empire, where two types of fish scale mimicking lorica's has been developed. The first one, *Lorica Squamata*, was a type of scaled armor used in Rome as equipment for soldiers (Fig. 5.2). Manufacture of such scale armors was cheap. Their protective properties for the tasks in the battlefield were acceptable. Even after more advanced armors were invented, this version was used for a long time. Here, the description of this construct according to http://www.history-of-armor.com/LoricaSquamata.html: the "squamae", or individual scales, were either brass or iron, or other alternating metals on the same shirt. The multiple layers gave satisfactory protection due to the multidirectional overlapping scales. These were wired together in horizontal rows which were in special manner laced to the backing. Correspondingly between 4 and 12 holes, which were necessary for wiring of each scale to the next in the row, can be observed on examples of this kind of lorica.

The armor consisted of innumerable metal scales arranged in overlapping horizontal rows (Fig. 5.3). The leather was the commonest strong backing material to that metal scales has been attached. Scaled armors were produced individual,

Fig. 5.2 Soldiers in lorica squamata. Fragment of the mosaic from Villa del Casale, Piazza Armerina – Enna, Italy, datable at about IV sentury bC (Parco archeo-logico della Villa romana del Casale-Piazza Armerina-IT. Reprinted with permission by ESTRELA S.A.S. Copyright © ESTRELA S.A.S.)

Fig. 5.3 Lorica squamata (Copyright © State of Lower Austria – Archaeological Park Carnuntum, Bad Deutsch-Altenburg (photo: N. Gail))

therefore broad variety of scales with respect to their size, thickness, and shapes are to be found up today. For examples, there are scales with bottoms which are rounded, or flat, or pointed. It is thought that the Romans developed lamellar armor, that didn't require a backing. In this type of scaled armor the main principle to manufacture the construct was wiring each scale to each of its neighbours, making it more resilient (see for review Robinson 1975).

The second type of lorica, *Lorica Plumata* (feathered armor; ribbed scale armor) was a hybrid mail covered in very small scales. It was used by officers as well as tribunes and above, but typically generals due to the high cost of production and maintenance (see for review Price 1983; Goldsworthy 1996). The plumata armor had scales about 7 mm wide by 10 mm tall. The normal armor's scales were two or three times larger than this. They could be squared off at the bottom, pointed or rounded. They could be made of brass sheet, iron sheet or a mixture of the two, sometimes so as to make a pattern of the scales, sometimes just as a repair using what came to hand (Fig. 5.4).

The third type of military lorica is the *Lorica Segmentata*. This replaced the older lorica hamata (mail shirt). Lorica segmentata (Fig. 5.5) is a type of Roman ferrous plate armour articulated on leather straps. It was first used at the end of the first century BC, and continued in service with the Roman army until the middle of the

Fig. 5.4 Fragments of lorica plumata (Image courtesy of Andrew Iingard)

Fig. 5.5 Front view of the commander's "updated" Newstead Lorica Segmentata assembled by Joe Piela of Lonely Mountain Forge. This version is based on the plans of M.C. Bishop in his JRMES monograph no. 1 "Lorica Segmentata vol 1 (2002) (http://www.legionxxiv.org/newsteadenlrg/)

third century AD (see for review Bishop 2002). In contrast to other types of plate armour, the pieces are not raised/dished, but rather curved. In this construct, a wide piece of metal pipe can be used as an anvil. The main idea was to have the abdomen protected by a series of overlapping hoops which hung from vertical leather straps. The plates which form the shoulder assembly are also held together by straps. Plate overlap was akin to roof shingles, with everything pointing down.

There are a couple guidelines in Internet today on how to adjust all the pieces to construct lorica segmentata using modern materials and tools. Intriguingly, however, is the point that we have a problem of explaining the existence and development of *lorica segmentata-like armors* known in fossil fish specimens like Placodermi (see Fig. 5.6). Placoderms, with their characteristic armor of bony plates, were the most successful and diverse group of fish during the Devonian. They also have an excellent fossil record because their dermal bones were generally robust and easily preserved (see for review Young 2010).

Another example of *lorica segmentata-like armors* we can observe in a modern fish is the three-spine stickleback *Gasterosteus aculeatus* (Fig. 5.7). This small fish

Fig. 5.6 The Devonian Placoderms (armored fish) possess broad variety of plate and scale armored dermal skeletons visually similar to Lorica Segmentata (Adapted from Young 2010. Modified with permission from the Annual Review of Earth and Planetary Sciences, Volume 38 © 2010, pgs 523–550 by Annual Reviews, http://www.annualreviews.org)

Fig. 5.7 Oceanic sticklebacks with about 30 armored plates covering the fish body from head to tail (Image courtesy www.sakhalin.ru)

habituates in lakes, streams and oceans, across the northern hemisphere. Subspecies from freshwater niches differ from their ocean-dwelling relatives with respect to the amount of protective armor that covers fish bodies. In contrast to freshwater sticklebacks with only several plates that sit closer to the front of the body, oceanic specimens possess about 30 armored plates extending from head to tail. "The dermal plates of the three-spine stickleback are anisotropic in shape (oval to rectangular), conformal to the body of the fish, porous, and composed of acellular lamellar bone," (Song et al. 2010). Recently, the fish was investigated by Christine Ortiz and co-workers at MIT as model system for better understanding of material design principles in naturally occurring armors. These researchers used micro-computed tomography (μCT) techniques for quantitative comparative studies on armor structures of both marine and freshwater *G. aculeatus*. As reported by Song (2011).

"The convolution of plate geometry in conjunction with plate-to-plate overlap allows a relatively constant armor thickness to be maintained throughout the assembly, promoting spatially homogeneous protection and thereby avoiding weakness at the armor unit interconnections. Plate-to-plate junctures act to register and join the plates while permitting compliance in sliding and rotation in selected directions. SEM and μCT revealed a porous, sandwich-like cross-section of lateral plates beneficial for bending stiffness and strength at minimum weight," (Song et al. 2010)

The armor assembly of marine *G. aculeatus* possesses a significant amount of porosity: "the larger pores occupy the centrally thickened region of the lateral plates, while the smaller pores are located throughout the plate," (Song et al. 2010). The internal porosity lateral to the sandwich-like structure of the *G. aculeatus* provides plates with special microarchitecture that provides stiffness and strength in bending at minimum weight. Moreover, this internal microstructure plays important role against the penetration of predator's teeth. According to Song et al. (2010).

"a penetrating tooth will first indent and bend the textured outer surface layer of the plate which would provide a greater resilience to penetration events compared to a fully dense internal structure. This structure would also provide greatly different dissipative deformation mechanisms during more aggressive penetration loading, where indentation will be accommodated by more distributed deformation events compared to a dense structure," (Song et al. 2010).

This special naturally designed structure observed in three-spine stickleback is used as bioinspired model for development of novel and highly effective lightweight armor systems (Song 2011).

Armor made of scales and plates remained popular long after the decline of the Roman Empire, finally falling out of fashion during the Middle Ages. It was cheap to produce, could be made by unskilled personnel, and was easy to repair in the field. It is thought that it was at least as common as mail armour. Although scale armour declined in popularity with the advent of modern warfare, using scales to provide flexible protection persists. There are no doubts that inflexible and heavy body armors limit soldier's speed and agility on the battlefield. Therefore, for example the US Army laboratories and affiliated companies showed great interest in designing more flexible body armor systems.

The overlapping scales motive on the skin of teleost fish is an example of "multilayer system that alternates hard and soft layers in an arrangement reminiscent of the design of bulletproof glass," (Zhu et al. 2012b). It was analysed that "in addition, overlapping scales ensures compliance and breathability, two highly desirable properties for personal armors. A biomimetic design at the individual scale level could therefore be combined with a clever arrangement of the scales at the macroscale to yield a hierarchical protective system with attractive properties," (Zhu et al. 2012b). One example of modern scale armor was Dragon Skin (Fig. 5.8), a ballistic vest produced by the Pinnacle Armor corporation (USA). However, in 2008, US Army rejected Pinnacle Armor's Dragon Skin system.

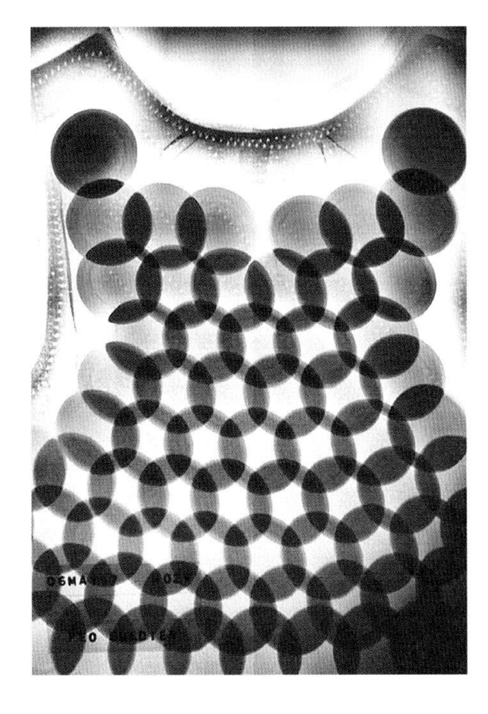

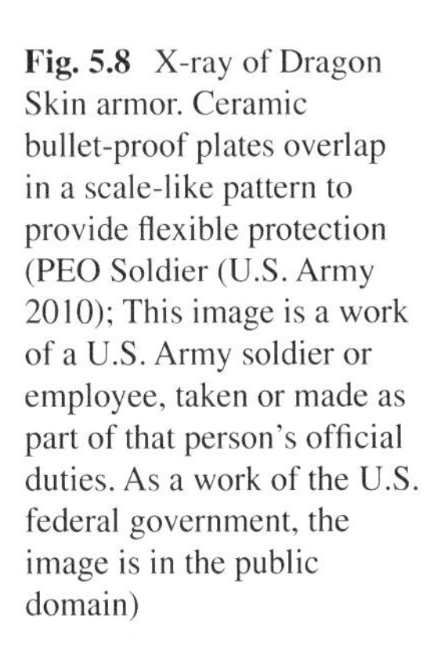

Fig. 5.8 X-ray of Dragon Skin armor. Ceramic bullet-proof plates overlap in a scale-like pattern to provide flexible protection (PEO Soldier (U.S. Army 2010); This image is a work of a U.S. Army soldier or employee, taken or made as part of that person's official duties. As a work of the U.S. federal government, the image is in the public domain)

According to information on the website of the Coorstek Company, "CoorsTek produces ceramic armor components used as the ballistic armor strikeface of lightweight composite armor systems. CeraShield™ ceramics, working in conjunction with an appropriate backing system, can defeat various threats including armor-piercing rounds" (see for details Ceramic Armor/Military Defense & Security: published on-line http://www.coorstek.com/products/ceramic-armor.asp?gclid=CMbp vtPX7rECFQFAzQodoxIAEQ).

Since 2009, US Army funded engineers at MIT (*Institute for Soldier Nanotechnologies*, Director Prof. Christine Ortiz), which are studying materials properties and natural design principles of fish scales with the goal to use the knowledge in development of novel lighter and more flexible armor for military use. Besides sticklebacks, scientists used "dinosaur eel", the *Polypterus senegalus*, which can be found in the muddy freshwater shallows in Africa. The body of this species is covered with multiple scale layers, each of which is about 100 μm thick (see for details Bruet et al. 2008). Characteristic properties of ancient as well as modern wish species with elasmoid scales like light weight and flexibility (Vernerey and Barthelat 2010), transparency and breathability (Zhu et al. 2012a, b) are of interest for materials scientists today. Additionally, studies on resistance to penetration using perforation tests with a sharp needle (tip radius = 25 μm) indicated that a single 200–300 μm thick fish scale of striped bass (*Morone saxatilis*) shows a high resistance to penetration which is superior to artificial polymers like polycarbonate and polystyrene. Here, the results of such kind of experiments reported by Zhu and co-authors (2012a, b):

"Under puncture, the fish scale undergoes a sequence of two distinct failure events: First, the outer bony layer cracks following a well-defined cross-like pattern which generates four 'flaps' of bony material. The deflection of the flaps by the needle is resisted by the collagen layer, which in biaxial tension acts as a retaining membrane. Remarkably this second stage of the penetration process is highly stable, so that an additional 50 % penetration force is required to eventually puncture the collagen layer. The combination of a hard layer that can fail in a controlled fashion with a soft and extensible backing layer is, probably, the key to the resistance to penetration of individual scales," (Zhu et al. 2012b). It was shown that "while the bony layer is brittle, the collagen layer can undergo large deformations, eventually failing by extensive fiber pullout," (Zhu et al. 2011).

5.1 Biomechanics of Fish Scales

From the structure-function relationship point of view, the fish scale is constructed during evolution from hierarchically organized biological materials in the multilayered form (Fig. 5.9.) that determined both excellent protective and hydrodynamics properties. As biocomposite-based structures, fish scales provide a protective layer resisting penetration by pathogenic microorganisms and providing a physical barrier against attack from predators, as well as abrasive and wear-resistance against

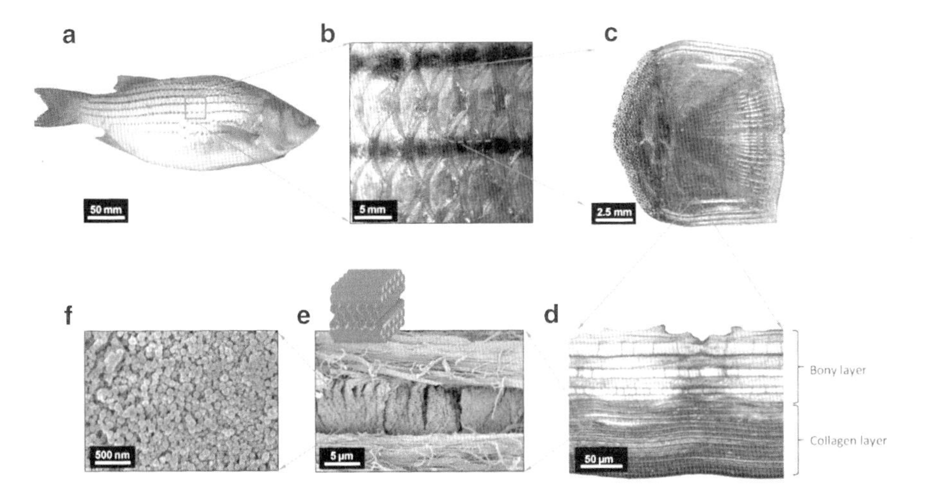

Fig. 5.9 The fish scale of striped bass, *M. saxatilis* and its hierarchical structure. (**a**) Whole fish, (**b**) staggered multiple scales, (**c**) an individual scale, (**d**) cross-section of a scale, (**e**) cross-ply collagen structure, (**f**) collagen fibrils (Zhu et al. 2012b. Copyright © 2012 WILEY-VCH Verlag GmbH & Co. KGaA, Weinheim. Reprinted with permission from John Wiley and Sons)

rocky surfaces in fast-flowing streams. At the same time, the ideal fish scale must possess properties which correspond to the optimal hydrodynamic conditions for swimming. I mean drug reduction properties, superoleophobicity in water and self cleaning. There are no doubts that fish scales as structurally highly diverse and complex biocomposites require multi-scale (Ortiz and Boyce 2008; Bruet et al. 2008; Song et al. 2011) and interdisciplinary physics (Wang et al. 2012) research to connect the structure to properties and ultimately to function.

Because mechanical characterization of fish scales is crucial to understand naturally occurring design principles for engineering of composites and protective systems, selected modern approaches are used to obtain results in detail.

According to the modern strategies, micromechanical models based on finite element and artificially constructed macroscale prototypes are employed to measure mechanical response to blunt and penetrating indentation loading. For example, in the recent Master Thesis by Ashley Browning (MIT) entitled "Mechanics of Composite Elasmoid Fish Scale Assemblies and their Bioinspired Analogues," (Browning et al. 2013), "deformation mechanisms of fish scale bending, scale rotation, tissue shear, and tissue constraint were found to govern the ability of the composite to protect the underlying substrate. These deformation mechanisms, the resistance to deformation, and the resulting energy absorption can all be tailored by structural parameters including architectural arrangement (angle of the scales, degree of scale overlap), composition (volume fraction of the scales), morphometry, and material properties (tissue modulus and scale modulus)," (Browning et al. 2013).

There is still lack on information concerning the contribution and synergies of each length of scale in spite the hierarchical organization of the fish skin probably plays a crucial role in its overall mechanical performance (Fratzl and Weinkamer 2007). For example, how neighbouring scales interact to minimize skin deflection and, finally, to prevent penetration is not known (Vernerey and Barthelat 2010). Little is also known about the mechanical properties of extinct fish dermal armor and scales, as well as the specific *mechanical roles* (Bruet 2008) of the mathematical properties for "material form variations (e.g. gradients) both within and between various layers, the number of layers, the layer and junction thickness, structures and geometries, the constitutive laws of each layer, and the relationship of these parameters to larger scale biomechanical performance and environmental stresses," (Bruet 2008).

Recently, Vernerey and Barthelat (2010) proposed one idealized model of fish scale structure. Scale deformation is described in terms of homogeneous curvature **j**. During bending, the support shape is "described as an arc circle of radius **R** = **1**/**j** (Fig. 5.10), which results in the rotation and deformation of scales and the development of contact forces between adjacent scales. The elastic energy stored in this deformed configuration determines the bending stiffness of the fish-structure. To further simplify the system, one may take advantage of two distinct features of fish scale structures: (i) the fish scale structure is made of a periodic pattern and (ii) during uniform bending deformation, every scale undergoes the same deformation," (Vernerey and Barthelat 2010).

The free-body diagram of a fishscale subjected to several applied forces and moments arising from scale-scale and dermis-scale interactions are depicted in Fig. 5.10. In particular, the authors consider:

- "The force exerted by the left scale applied at the point of contact. This force is comprised of a normal force $\mathbf{f}^{L}_{n}$ and a tangential force $\mathbf{f}^{L}_{t}$ resulting from friction.
- The force exerted by the right scale applied at the right extremity of the principal scale. This force can also be decomposed into a normal $\mathbf{f}^{R}_{n}$ and tangential $\mathbf{f}^{R}_{t}$ component, with respect to the right scale. Because of the periodicity argument, the magnitude of these forces is equal and opposite to $\mathbf{f}^{L}_{n}$ and $\mathbf{f}^{L}_{t}$.
- A moment $\mathbf{m}^{D}$ resisting scale rotation around its support," (Vernerey and Barthelat 2010).

Generally, results obtained by Vernerey and Barthelat (2010) indicate that:

- "Strain-stiffening characteristic increases with increasing scale density and decreasing scale-dermis attachment rotational stiffness (relative to a scale's bending stiffness);
- The contact force (relative to macroscopic moment) between scales increases exponentially with a measure of scale density;
- The average macroscopic bending stiffness increases in a nonlinear fashion with the ratio $\mathbf{K}^{D}/\mathbf{EI}$ (angular stiffness to scale stiffness);
- Finally, shear deformation of the scale tends to decrease both the average stiffness and the strain-stiffening characteristic of the fish scale response," (Vernerey and Barthelat 2010).

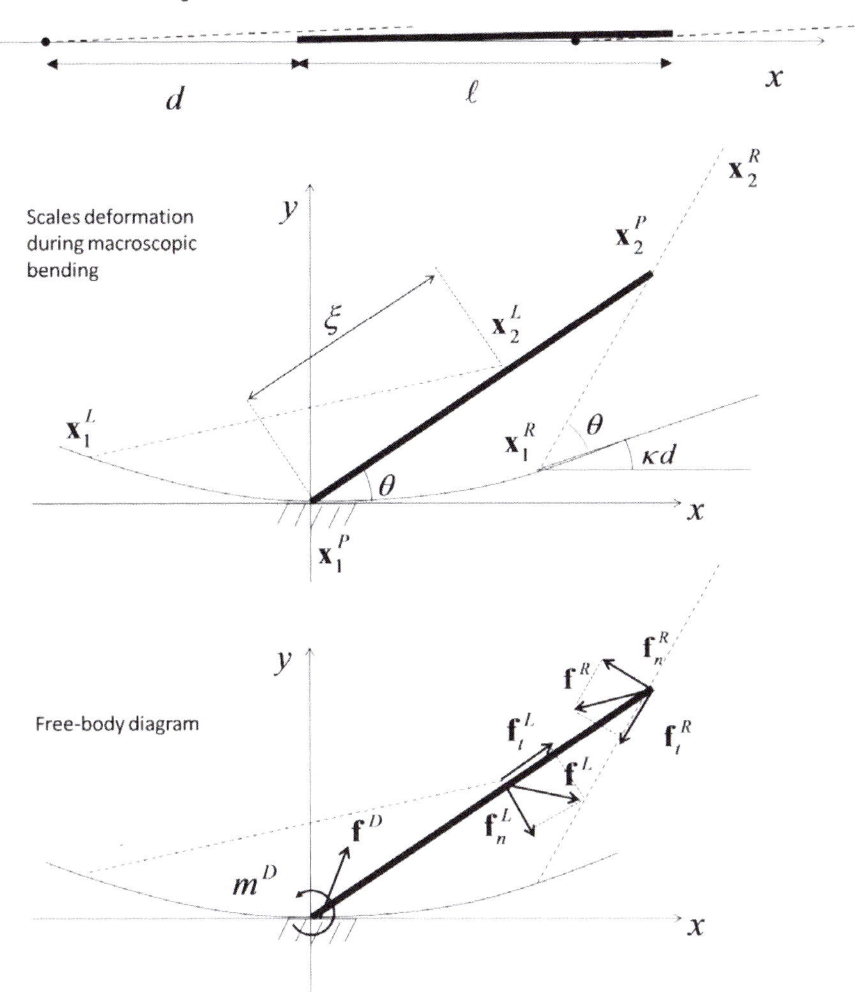

Fig. 5.10 Free body diagram of a primary fish scale (*bold line*) during macroscopic bending. *Dotted lines* represent left and right scales (Reprinted from Vernerey and Barthelat (2010), Copyright (2010), with permission from Elsevier)

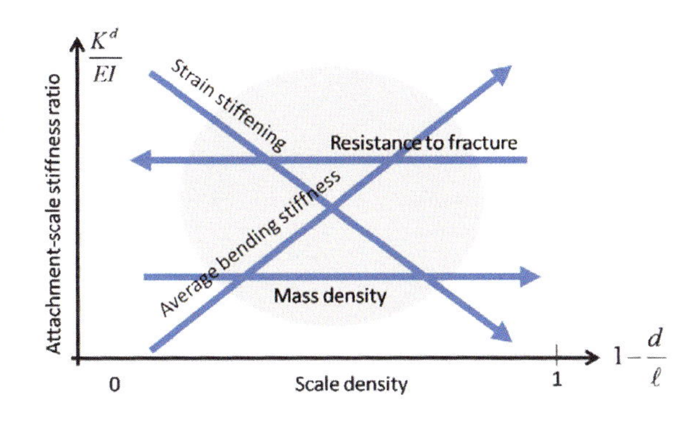

Fig. 5.11 Trends in fishscale response in terms of its underlying microstructure (Reprinted from Vernerey and Barthelat (2010), Copyright (2010), with permission from Elsevier)

Figure 5.11 summarizes these trends. The authors suggest that "according to its environment and size, a species of fish may emphasize certain functions over others. For instance, in cases where locomotion speed is emphasized, the structure will be 'designed' in terms of the strain-stiffening (the role of external tendons) and being light-weight. On the other hand, if protection against predator is critical, the design will favour a higher resistance to fracture and average bending stiffness. This flexibility in choice may explain the large diversity of scale structures encountered in nature, from large to small scale, high to low density, all of them within region of acceptable design depicted by the grey region," (Vernerey and Barthelat 2010) in Fig. 5.11.

As reviewed by Browning et al. (2013), tensile testing of elasmoid fish scales reports an effective scale elastic modulus is about 1 GPa, but varies among species and along the length of the body of the fish. Testing of scales in a hydrated state (more closely resembling in vivo conditions) yields a reduced elastic modulus 0.1–0.8 GPa and tensile strength to range 22–65 MPa. The elastic properties of fish scales are in the range of common engineering polymers (e.g., polycarbonate, acrylonitrile butadiene styrene). The density of elasmoid fish scales ranges between 1.92 and 2.43 g/cm^3 based on measurements and available literature (Ikom et al. 2003; Torres et al. 2008).

The mechanical properties of really existing elasmoid fish scales have been described by Ikom et al. (2003) on example of sea bream, *Pagrus major* and recently by Garrano et al. (2012) on carp (*Cyprinus carpio*). The scale of sea beam shows an orthogonal plywood structure of 1–2 μm thick stratified lamellae. These formations are constructing of closely packed collagen fibers of about 80 nm in diameter. The mineral phase in the scale (Fig. 5.12) was identified as "calcium-deficient hydroxy-

Fig. 5.12 The scale of *P. major* under SEM investigation. The sliding of the collagen lamella structure and pulling out and breakage of individual collagen fibers are visible (Reprinted from Ikom et al. (2003), Copyright (2003), with permission from Elsevier)

apatite containing a small amount of sodium and magnesium ions, as well as carbonate anions in phosphate sites of the apatite lattice," (Ikom et al. 2003). The tensile strength of this fish scale reaches 90 MPa. This value is high because of the hierarchically ordered structure of "mineralized collagen fibers, in association with long, narrow platelike apatite crystals that are aligned along the crystallographic c-axis parallel to the collagen fibers," (Ikom et al. 2003). Functional advantages based on mechanical anisotropy of collagen-mineral-containing composite matrices are well known (Bigi et al. 1996). The tensile stress–strain curve obtained by Ikom et al. (2003) was initially linear with a corresponding Youngs modulus (stress/strain) of 2.2 GPa. In comparison with data reported for red deer (50 %, 6.1 GPa) and axis deer (80 %, 31.6 GPa) (Currey and Brear 1990; Mann 2001), this value confirms the relatively low stiffness of the fish scales due to their small mineral content (46 %). As reported in the discussed work, "at high stress values, the tensile stress–strain curve showed considerable plastic yielding before fracture. Corresponding SEM images of the fracture surface indicated that sliding of the collagen lamellae and pulling out of individual collagen fibers, 2–3 μm in thickness, were responsible for the plasticity close to the yield point (Fig. 5.12). Demineralization of the fish scales considerably reduced the average tensile strength and Young's modulus to values of 36 MPa and 0.53 GPa, respectively, although the fracture behavior was essentially the same as for the mineralized tissue," (Ikom et al. 2003).

Definitively, model presented by Vernerey and Barthelat (2010) is very much idealized and can, probably, be used to assess general trends in the mechanics of fish scale structures, probably only with respect to elasmoid scales. The situation with, for example, multi-layered ganoid scale seems to be more complex. Thus, in ganoid *Polypterus seneglus* scale, "each material layer was found to have significantly different mechanical properties compared to others (except for bone), as compared to the isopedine layer," (Bruet 2008). These nanocomposite structured layers are of different thickness (from outer to inner): ganoine (ca. 10 μm), dentin (ca. 46 μm), isopedine (ca. 45 μm) and the basal bone plate (ca 300 μm). The plastic and elastic properties through the four layers were investigated spatially using high resolution nanomechanical methods. The obtained data showed that both "indentation modulus and hardness decrease with distance from the outer to the inner surfaces of the scale, from 62 to 17 GPa and 4.5 to 0.54 GPa, respectively," (Bruet 2008).

The study on the ganoid armor scales of *P. senegalus* has elucidated the biomechanical advantages of ganoine–dentin–isopedine–bone- based multilayers in penetration resistance (Bruet et al. 2008), but at the added weight of the higher density external layers (Song et al. 2010) as well as at the energetic "cost" of additional biomineralization (Vermeij 2006). Recently, it was reported by Wang et al. (2009) that "the elastic-plastic anisotropy of the outmost ganoine layer of P. senegalus scale enhances the load-dependent penetration resistance of the multilayered armor compared with the isotropic ganoine layer by (i) retaining the effective indentation modulus and hardness properties, (ii) enhancing the transmission of stress and dissipation to the underlying dentin layer, (iii) lowering the ganoine/dentin interfacial stresses and hence reducing any propensity toward delamination, (iv) retaining the suppression of catastrophic radial surface cracking, and favoring localized circumferential cracking, and (v) providing discrete structural pathways (interprism) for circumferential cracks to propagate normal to the surface for easy arrest by the underlying dentin layer, hence containing damage locally. These results indicate the potential to use anisotropy of the individual layers as a means for design optimization of hierarchically structured material systems for dissipative armor," (Wang et al. 2009).

The next intriguing topic on fish scale biomechanics is related to poorly investigated interlocking mechanisms between scales as well as between interfaces of different layers. Bruet (2008) observed the interfaces between layers in *P. senegalus* scales to be exceedingly strong and tough, e.g. the ganoin-dentin junction was able to arrest microcracks. Localized fractures within the brittle ganoin layer and detachment of fragments of the ganoine layer of the scale was seen, while the ganoin-dentin junction remains wholly intact; perhaps as a sacrificial mechanism. "The corrugated junction between the layers is expected to lead to spatially heterogenous stresses and a higher net interfacial compression which could serve to prevent delamination," (Bruet 2008).

Since 2008, the giant scales of *Arapaima gigas* a fish from the Amazonian region with a rather prehistorical aspect became the status of subject of scientific investigations with regard to materials science and biomineralization. It is known for its large tough scales which are used in handcrafts and souvenirs (Torres et al. 2008). Arapaima is a giant fish measuring over 2 m in length and weighing in at over 100 kg. Their scales of varying degrees of mineralization serve as a dermal armor to natural predators, including piranhas. These scales are elasmoid and consist of an elasmodine basal plate of a twisted plywood arrangement of collagen fibers capped with different hypermineralized tissues (Meunier and Brito 2004; Sire et al. 2009; Torres et al. 2008). Scales of *A. gigas* have a partially mineralized base and well mineralized limiting layer of hydroxyapatite. Recently, Marc Meyers and coworkers from UC San Diego (Lin et al. 2011) published a paper on laminate structure and mechanical properties of *A. gigas* scales. This work received a large mass media resonance because of the unique experiments that were carried out in the Lab using teeth of piranhas as the potential biological enemy of Arapaima. Piranhas use their famous razor-like teeth to trap skin and muscle in a guillotine-like bite, and then tear the prey into bits. Meyers and colleagues embedded Arapaima scales in a soft rubber surface (to mimic the underlying muscle on the fish), and attached piranha teeth to an industrial-strength hole punch. When the tooth was pushed into the scale, it partially penetrated the outer defences, but cracked before it could puncture the muscle. Here is the clue (Brown 2012): the heavily mineralised scales with hard enamel-like outer layer overlap each other like shingles. Their surface is wrinkled, too, to allow it to bend without cracking. Collagen fibers with plywood-like orientation are stacked underneath. This combination of a softer internal architecture and a strong outer layer determines unique mechanical properties of the construct. This lets the aquatic animal remain mobile while staying heavily armoured and alive.

In the work by Lin and co-workers (2011), following results have been reported: "The micro-indentation hardness of the external layer (550 MPa) of A. gigas scale is considerably higher than that of the internal layer (200 MPa), consistent with its higher degree of mineralization. Tensile testing of the scales carried out in the dry and wet conditions shows that the strength and stiffness are hydration dependent. As is the case of most biological materials, the elastic modulus of the scale is strain-rate dependent. The strain-rate dependence of the elastic modulus, as expressed by the Ramberg-Osgood equation, is equal to 0.26, approximately ten times higher than that of bone. This is attributed to the higher fraction of collagen in the scales and to the high degree of hydration (30 % H_2O). Deproteinization of the scale reveals the structure of the mineral component consisting of an interconnected network of platelets with a thickness of ~50 nm and diameter of ~500 nm," (Lin et al. 2011).

The hydration state of fish scale seems to play an important role for obtaining of experimental data. For example, it was reported about significant differences in mechanical properties of scales from carp (*Cyprinus carpio*) as a function of their anatomical position and hydration. Garrano et al. (2012) showed that "fully hydrated scales from the head exhibited an elastic modulus and strength nearly twice that of those properties for the tail. However, after dehydration there were no significant differences in the mechanical properties as a function of anatomical position. Considering all three regions of evaluation, dehydration had the largest influence on changes in the elastic modulus, and scales from the tail underwent the most significant changes in properties with moisture loss," (Garrano et al. 2012).

Unfortunately, with exception of works by Ikom et al. (2003) and Song (2011), there are no detailed studies on mechanical properties of scales from marine fish including specimens with very large scales like Tarpon (*Megalops atlanticus*) (Figs. 5.13 and 5.14). Tarpons are known as one of the great saltwater game fishes. It regularly grows to 1.8 m and 45.4 kg or larger. Recently, I received numerous specimens of these scales from Adam Summers. Corresponding experiments are in progress in our Lab now.

Fig. 5.13 Armored head of Tarpon (*Megalops atlanticus*) (Courtesy of Brett Colvin)

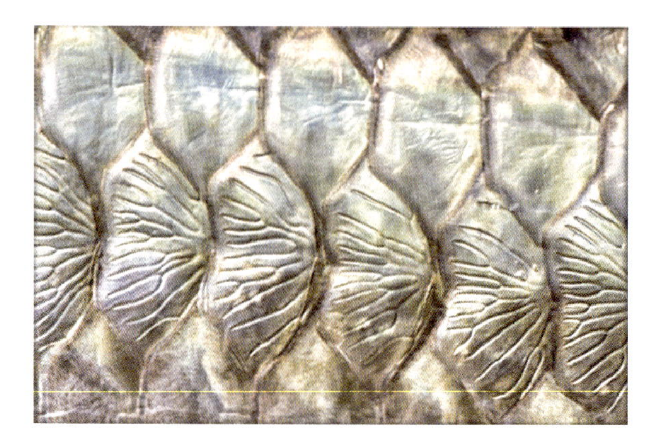

Fig. 5.14 Hierarchically structured scales of Tarpon (*M. atlanticus*) are the intriguing specimen for investigations on their material properties (Image courtesy of Brett Colvin)

5.2 Fish Swimming and the Surface Shape of Fish Scale

The biological function of fish scale is to represent an intermediate trade-off between body protection and maximum mobility (swimming). Fish as aquatic animals "have evolved form and structures to minimize drag, with the primary decrease coming from a streamlined body shape to reduce flow separation," (Lang et al. 2008). A great deal of research has examined the links between fish body, fish skin and fish scale design and swimming performance (see for review the "*Fish Biomechanics*" by Shadwick and Lauder 2006; Lauder 2011; Oeffner and Lauder 2012). Much of this work has had a bio-engineering focus (Blake 1983, 2004; Colgate and Lynch 2004), attempting to understand how specific structures work.

In general, fish interact with their surrounding fluid medium trough their scales (Sudo et al. 2002), which possess the mucous layer and specifically structured and oriented riblets (Fig. 5.15). Thus, such scales with micro-grooved surfaces (Cui and Fu 2012) might seem compatible with good (or even excellent) hydrodynamics. "The mucus secreted by fish causes a reduction in drag as they move through water, and also protects the fish from abrasion, by making the fish slide across objects rather than scrape; and disease, by making the surface of the fish difficult for microscopic organisms to adhere to," (Dean and Bhushan 2010; see also Shephard 1994).

According to definition by Oeffner and Lauder (2012), "riblets are fine rib-like surface geometries with sharp surface ridges that can be aligned either parallel or perpendicular to the flow direction and might reduce drag," (Oeffner and Lauder 2012). The diversity of forms of riblets are well described and discussed in the literature (see for review Lee and Lee 2001; Sudo et al. 2002; Lang et al. 2008; Dean and Bhushan 2010; Cui and Fu 2012) (Fig. 5.15). The nature of fluid flow over an effective fish skin surface is an important point in understanding of the mechanism of fish skin drag reduction. Three types of drag are described for the swimming animal. These include:

(i) form drag due to a difference in pressure around the body,
(ii) drag-dueto- lift, and
(iii) skin friction due to boundary layer formation (Bushnell and Moore 1991).

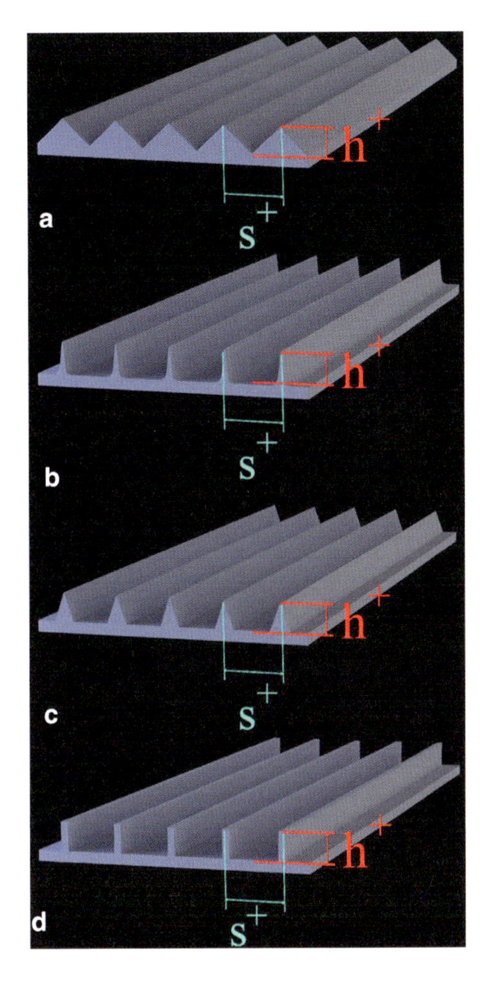

Fig. 5.15 The configuration of various riblets (**a**) Sawtooth riblets. (**b**) Scalloped riblets. (**c**) Spaced-V riblets. (**d**) Blade riblets (Designed by Alexei Rusakov after inspiration according to Sudo et al. 2002)

As reviewed by Dean and Bhushan (2010), "fluid drag comes in several forms, the most basic of which are pressure drag and friction drag. Pressure or form drag is the drag associated with the energy required to move fluid out from in front of an object in the flow, and then back in place behind the object. Much of the drag associated with walking through water is pressure drag, as the water directly in front of a body must be moved out and around the body before the body can move forward. The magnitude of pressure drag can be reduced by creating streamlined shapes. Drag is caused by the interactions between the fluid and a surface parallel to the flow, as well as the attraction between molecules of the fluid (Fig. 5.16). Friction drag is similar to the motion of a deck of cards sliding across a table. The frictional interactions between the table and the bottom card, as well as between each successive card, mimic the viscous interactions between molecules of fluid. Moving away from the surface of an object in a fluid flow, each fluid layer has higher velocity until a layer is reached where the fluid has velocity equal to the mean flow. Fluids of higher

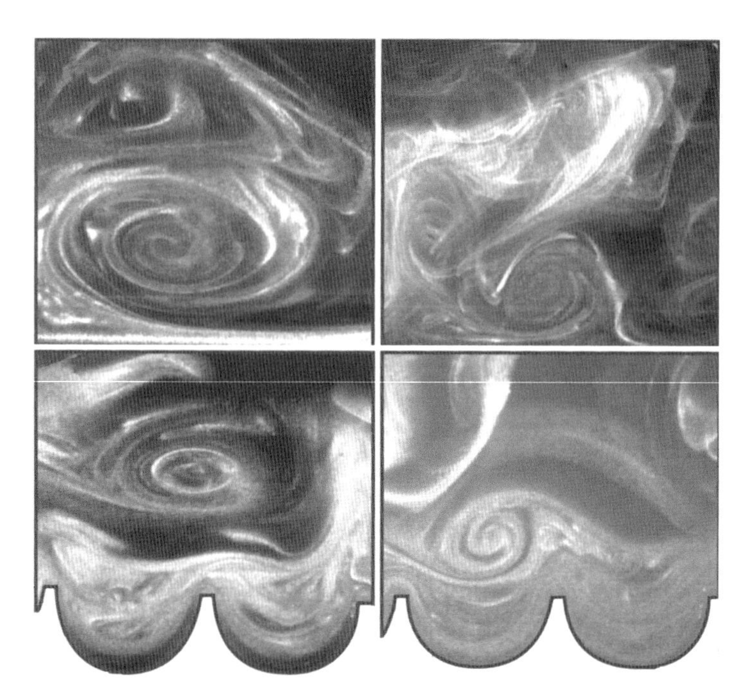

Fig. 5.16 Schematic view of the turbulent-flow of streamwise vortices in a vertical cross section over a flat plate: (*left up*) drag decreasing case (V = 3 ms^{-1}) and (*right up*) drag increasing case (V = 5 ms^{-1}). Riblet surfaces: (*left bottom*) drag decreasing case (V = 3 ms^{-1}) and (*right bottom*) drag increasing case (V = 5 ms^{-1}) (Adapted from Dean and Bhushan (2010) by permission of the Royal Society)

viscosity—the attraction between molecules—have higher apparent friction between fluid layers, which increases the thickness of the fluid layer distorted by an object in a fluid flow. For this reason, more viscous fluids have relatively higher drag than less viscous fluids. A similar increase in drag occurs as fluid velocity increases. The drag on an object is, in fact, a measure of the energy required to transfer momentum between the fluid and the object to create a velocity gradient in the fluid layer between the object and undisturbed fluid away from the object's surface," (Dean and Bhushan 2010).

The skin of sharks is covered by specific (0.2–0.5 mm) bony placoid scales, which can vary in size and shape and represent an example of riblet-containing surfaces which are regularly spaced (30–100 μm) (see for review Lang et al. 2008; Dean and Bhushan 2010). Their base is located in the deeper collagenous layer termed the *stratum compactum*, with the crown exposed to the water. The angle between collagenous fibers, to which the base of each scale is attached, to the longitudinal axis of the shark body is of approximately 50–60°. The role of this type of scales also with respect to bending flexibility, and elastic energy storage during swimming of sharks was studied previously (Motta 1977; Wainwright et al. 1978; Hebrank 1980).

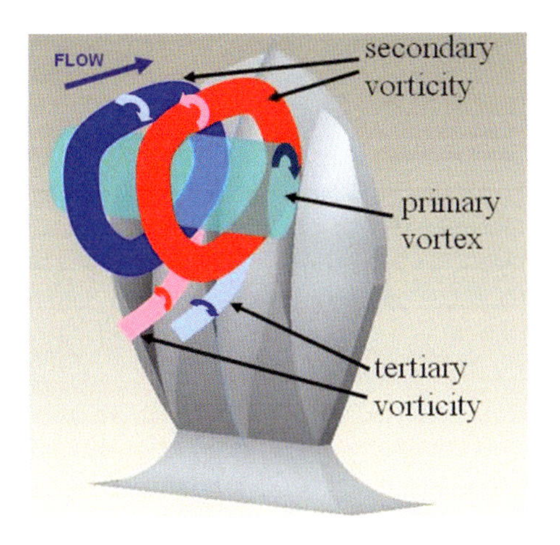

Fig. 5.17 Bristled shark skin denticles on the dorsal side of one scale can form vorticity as represented in the scheme. Positive vorticity is represented in *red*, and the negative – in *blue* colouring (Reprinted from Lang et al. (2008). © IOP Publishing. Reproduced with permission. All rights reserved)

Interestingly, sharks can even bristle their scales at higher swimming speeds. The phenomenon of the scale erection may be determined by the change the helical orientation of the collagen bundles to which the scales are attached. Probably it accomplished through increased overall skin tension at higher swimming speeds, as the subcutaneous pressure has been demonstrated to increase from 7 to 14 Pa in a resting lemon shark *Negaprion brevirostris* to 200 kPa (29 psi) during fast swimming (Wainwright et al. 1978). In their *Water Tunnel Experiments* with bristle shark skin, Lang et al. (2008), showed that "there exist counterrotating pairs of vortices forming between the three ribs, and the ribs appear to promote the formation of the secondary vorticity (Fig. 5.17) while additionally preventing the flow from passing around the sides of the tips, thereby also increasing the strength of the primary cavity vortex. The turbulent cavity vortex flow field on average also resembled that found under laminar conditions (Fig. 5.17), just embedded slightly deeper into the cavity and of slightly smaller size but higher strength. Experiments modeling an extreme angle of bristling for shortfin mako (*Isurus oxyrinchus*) denticles confirmed the formation of embedded vortices within the inter-denticular cavities. It was postulated that the unique microgeometry of bristled shark skin can have the potential to result in more than one means of controlling the boundary layer to decrease overall drag," (Lang et al. 2008).

Direct replication of creatural scarf skins to form biomimetic surfaces with relatively vivid morphology was proposed as new attempt for the bio-replicated forming technology of an animal body. Taking shark skins as the replication templates, and the micro-embossing and micro-molding as the material forming methods, the micro-replicating technology of the outward morphology on shark skins was demonstrated (Han and Zhang 2008). The preliminary analysis on replication precision indicated that the bio-replicated forming technology can replicate the outward morphology of the shark scales with good precision, which validates the application of the bio-replicated forming technology in the direct morphology replication of the firm creatural scarfskins.

Shark skin has been accepted as the main source of bioinspiration to develop and to design the "*biomimetic skin*". There are numerous reports on corresponding experimental and comparative works (Bechert et al. 1985, 1997; Benjanuvatra et al. 2002; Mollendorf et al. 2004; Han et al. 2008).

The recent report by Oeffner and Lauder (2012) contains experimental results that suggest "that one important effect of the shark skin denticles is to enhance thrust, and not simply to reduce drag. The overwhelming emphasis of the existing literature on shark skin has been on drag reduction, but denticles alter vortex location. This is particularly true for the tail surface where flow separation and vortex formation have been demonstrated, and could increase thrust," (Oeffner and Lauder 2012; see also Wilga and Lauder 2002, 2004).

Not only the fish skin riblets-like surface, but also the surface structure of lateral plates plays an important role for swimming. For example, "the exterior surface topography of the lateral plates of three-spine stickleback may serve a hydrodynamic role during swimming, e.g., viscous drag reduction of turbulent boundary layers and/or reduction of surface shear stress and skin friction," (Song et al. 2010).

5.2.1 *Superoleophobicity of Fish Scale Surfaces*

The next phenomenon that is also crucial for swimming behaviour in fish is related to the nanolevel and surface chemistry of fish scale. For carps, sectorlike scales are covered by oriented micropapillae with nanostructures, presenting not only drag-reducing function but superoleophilicity in air and *superoleophobicity* in water (Liu and Jiang 2011; Bhushan 2011, 2012). "Superoleophobic surfaces—those that display contact angles greater than 150° with organic liquids and have appreciably lower surface tensions than that of water—are extremely rare. Calculations suggest that creating such a surface would require a surface energy lower than that of any known material," (Tuteja et al. 2007). The superoleophobic fish surface originates from the water-phase micro/nano hierarchical structures. Oil droplets on fish scales in water showed antiwetting behaviour. This phenomenon was used as bioinspiration for developing of a superoleophobic and low-adhesive interface on a solid substrate with multiscale structures, via oil/water/solid three-phase systems (see for review Liu et al. 2012; Zhang et al. 2012a). For example, as reported by Cheng et al. (2012), an underwater pH-responsive superoleophobic surface successfully demonstrated a reversible switch of oil-adhesion on a nanostructured poly (acrylic acid) (PAA) surface by changing the environmental pH values. At low pH, intramolecular hydrogen bonding of PAA is formed, and results in high oil-adhesion. As for high pH, the oil droplets can easily roll off due to the intermolecular hydrogen bonding between PAA and surrounding water. Bioinspired by the surface structures of fish scales, Liu et al. (2009) have elaborated micro/nanostructure silicon surfaces exhibiting superoleophobic properties; but only if the surface was immerged in water.

5.2.2 Selfcleaning of Fish Scales and Biomimetic Applications

The ribbed texture of the scales of a shark provided the inspiration for the design of multifunctional coatings with not only drag-reducing, but also self-cleaning functions. Self-cleaning surfaces are of scientific interest especially for antifouling coatings (see for review Liu and Jiang 2012). Numerous marine invertebrates are responsible for biofouling on underwater surfaces of both commercial and naval vessels. Correspondingly, modern surface-active and non-toxic antifouling technologies using biomimetic approaches are in trend today (Salta et al. 2010; Callow and Callow 2011; Banerjee et al. 2011). Surface roughness can affect the hydrodynamic performance of antifouling coatings and influence the settlement behaviour of fouling larvae, which makes it an important parameter in the evaluation of novel coatings. Surface roughness is a parameter that originates from marine engineering, but has been used extensively by marine scientists to characterise novel coatings and to investigate microtopographies that might inhibit settlement behaviour (Howell and Behrends 2006).

Recently, mimicking the surface structures of shark skin, hierarchical engineered anti-fouling microtopographies on the polymer basis surfaces were patented (Brennan et al. 2006) and fabricated (Genzer and Marmur 2008). For example, a polydimethyl siloxane elastomer (PDMSe) material termed as Sharklet AF™ was inspired by surface microtopography based on the skin of sharks. As reported by Chung et al. (2007):

"The Sharklet AF™ PDMSe was tested against smooth PDMSe for biofilm formation of *Staphylococcus aureus* over the course of 21 days. The smooth surface exhibited early-stage biofilm colonies at 7 days and mature biofilms at 14 days, while the topographical surface did not show evidence of early biofilm colonization until day 21. At 14 days, the mean value of percent area coverage of *S. aureus* on the smooth surface was 54 % compared to 7 % for the Sharklet AF™ surface ($p<0.01$). These results suggest that surface modification of indwelling medical devices and exposed sterile surfaces with the Sharklet AF™ engineered topography may be an effective solution in disrupting biofilm formation of *S. aureus*," (Chung et al. 2007).

Sharklet Technologies are patented and include antifouling surfaces against the settlement of zoospores of the ship fouling alga *Ulva* (Schumacher et al. 2007) as well as several medical tools and devices including a Sharklet Urinary Catheter to help reduce hospital-acquired infections (see for review http://www.sharklet.com/technology/).

The fabrication of the bionic shark-skin coatings with life-sized scale-like microstructure is the trend today. The shark skin structure has inspired the Speedo company to develop technology called FAST SKIN®. This technology is used in the manufacturing of swimwear in order to minimise the water resistance (http://www.speedo.de/de/swimwear_products/performance/fastskinfsii/index.html).

According to the website of Sym Tech Company (USA), "Sharkskin Surface Protection is the next generation protection coating for automotive surfaces. Once applied, Sharkskin's nano-sized silicon dioxide molecules re-assemble themselves on the vehicle surface creating a negatively charged, highly repellant surface. The treated surface repels water, dirt and other pollutants. It is UV stable and increases resistance to scratches. The surface is also easier to clean and keeps the vehicle looking like new," (http://sym-tech.ca/programs-products/sharkskin/).

Recently, Han and Wang (2011) proposed a novel method for fabrication of the biomimetic shark skin coating. Using hot embossing technology, the direct micro replication of the microstructure on shark skin was investigated for the first time. Modeled after the shark skin sample, the negative structure was directly replicated and printed on PMMA flat plate in the hot embossing process, which relied on the bionic shark-skin coating made of silica gel that was fabricated in the end. The preliminary experiment results indicate that this method is a high precision, high throughput, high efficiency and low cost way to fabricate bionic microstructure in micron and submicron scale with good repeatability and availability.

Modern naval industry is still looking for effective artificially created shark skin-like materials which could help ships sailing smoothly. A synthetic shark skin of elastic silicone that makes it harder for barnacles to gain a foothold has been created (Kesel and Liedert 2007a, b) and patented (Hochschule Bremen 2007) in Bremen, Germany by Ralph Liedert of the University of Applied Sciences (Fig. 5.18).

An intriguing project entitled the "HAUT"-program was funded by the German Research Foundation (DFG). The research topics are surfaces which can adapt to the surrounding atmospheric conditions at temperatures over 600 °C similar to a living skin (in German "Haut"). The project entitled "*A shark skin for high temperature application*" was dedicated to the designing of the surface of selected substrates by using the natural oxidation behaviour of the metal oxides-containing material at temperatures up to 1,000 °C.

Fig. 5.18 Shark skin antifouling coating produced according to patent EP 06 018 001.5 for sell in Germany (Courtesy Fa. VOSSCHEMIE)

5.3 Conclusion

Today, ultimately we aim at the design and manufacture of synthetic and hierarchical material systems or structures in which the organization is designed and controlled on length scales ranging from nano to micro, even all the way to macro. Millions of years of fish evolution and natural selection have yielded a hierarchically constructed biocomposite – a fish scale that possess numerous optimal properties with regard to structure-function relationship; namely on nano-, micro- and macro levels. Both fish scale as individual construct, and fish scale- made armor, are currently popular subjects for bioinspiration. Armored fish are, probably, the key to the successful evolution of fish. Military science as well as Civil Engineering shows interest in fish scale and armor investigations. "Knowledge of the structure–property–function relationships of the dermal scales of armoured fish could enable pathways to improved bioinspired human body armour, and may provide clues to the evolutionary origins of mineralized tissues" (Bruet et al. 2008). To achieve success in this challenging topic, modern science uses a multiscale experimental and computational approach.

References

Banerjee I, Pangule RC, Kane RS (2011) Antifouling coatings: recent developments in the design of surfaces that prevent fouling by proteins, bacteria, and marine organisms. Adv Mater 23(6):690–718

Bechert DW, Hoppe G, Reif WE (1985) On the drag reduction of the shark skin. In: AIAA shear flow control conference, vol AIAA 85–0546. AIAA, Boulder, pp 1–18

Bechert DW, Bruse M, Hage W et al (1997) Experiments on drag reducing surfaces and their optimization with an adjustable geometry. J Fluid Mech 338:59–87

Benjanuvatra N, Dawson G, Blanksby B (2002) Comparison of buoyancy, passive and net active drag forces between Fastskin TM and standard swimsuits. J Sci Med Sport 5:115–123

Bhushan B (2011) Biomimetics inspired surfaces for drag reduction and oleophobicity/philicity. Beilstein J Nanotechnol 2:66–84

Bhushan B (2012) Bioinspired structured surfaces. Langmuir 28(3):1698–1714

Bigi A, Gandolfi M, Koch MHJ et al (1996) X–ray diffraction study of in vitro calcification of tendon collagen. Biomaterials 17:1195–1201

Bishop MC (2002) Lorica segmentata vol I. A handbook of articulated roman plate armour, JRMES monograph 1. The Armatura Press, Duns, pp vii–120

Blake RW (1983) Fish locomotion. Cambridge University Press, Cambridge

Blake RW (2004) Fish functional design and swimming performance. J Fish Biol 65(1193):1222

Brennan AB, Baney RH, Carman ML, Estes TG, Feinberg AW, Wilson LH, Schumacher JF (2006) Surface topography for non–toxic bioadhesion control. US patent no 7,143,709, 5 Dec 2006

Brown M (2012) Piranha-proof fish offer body armour inspiration. http://www.wired.co.uk/news/archive/2012-02/13/piranha-proof-fish. Accessed 14 June 2012. Published on-line 13 Feb 2012. Copyright © 2012, Condé Nast UK. Reprinted with permission

Browning et al (2013) Reprinted from Browning A, Ortiz C, Boyce MC (2013) Mechanics of composite elasmoid fish scale assemblies and their bioinspired analogues. J Mech Behav Biomed Mater 19:75–86. Copyright (2013), with permission from Elsevier

Bruet BJF (2008) Multiscale structural and mechanical design of mineralized biocomposites. Dissertation, Massachusetts Institute of Technology, Cambridge

Bruet et al (2008) Reprinted by permission from Macmillan Publishers Ltd: Bruet BJF, Song J, Boyce MC and Ortiz C (2008) Materials design principles of ancient fish armour. Nat Mater 7:748–756. Copyright © 2008, Rights Managed by Nature Publishing Group

Bushnell D, Moore K (1991) Drag reduction in nature. Annu Rev Fluid Mech 23:65–79

Callow JA, Callow ME (2011) Trends in the development of environmentally friendly fouling–resistant marine coatings. Nat Commun 2:244

Ceramic Armor/Military Defense & Security (published on-line http://www.coorstek.com/markets/aerospace_defense/armor_protection.php), Accessed 15 May 2014. Copyright © 2014 CoorsTek, Inc. Reprinted with permission

Cheng Q, Li M, Yang F et al (2012) An underwater pH–responsive superoleophobic surface with reversibly switchable oil–adhesion. Soft Matter 8:6740–6743

Chung et al (2007) Reprinted with permission from Chung KK, Schumacher JF, Sampson EM et al (2007) Impact of engineered surface microtopography on biofilm formation of *Staphylococcus aureus*. Biointerphases 2(2):89–94. Copyright 2007, AIP Publishing LLC

Colgate JE, Lynch KM (2004) Mechanics and control of swimming: a review. IEEE J Ocean Eng 29(660):673

Creaser CW (1926) The structure and growth of scales of fishes in relation to the interpretation of their life history with special reference to the sunfish *Eupomotis gibbosus*. Misc Publ Univ Michigan Mus Zool 17:1–82

Cui J, Fu Y (2012) A numerical study on pressure drop in microchannel flow with different bionic micro–grooved surfaces. Bionic Eng 9:99–109

Currey JD, Brear K (1990) Hardness, Young's modulus and yield stress in mammalian mineralized tissues. J Mater Sci Mater Med 1:14–20

Dean B, Bhushan B (2010) Shark-skin surfaces for fluid-drag reduction in turbulent flow: a review. Phil Trans R Soc A 368(1929):4775–4806, by permission of the Royal Society

Fratzl P, Weinkamer R (2007) Nature's hierarchical materials. Progr Mater Sci 52(8):1263–1334

Garrano et al (2012) Reprinted from Garrano MCA, La Rosa G, Zhang D, Niu LN, Tay FR, Majd H, Arola D (2012) On the mechanical behavior of scales from *Cyprinus carpio*. J Mech Behav Biomed Mater 7:17–29. Copyright (2012), with permission from Elsevier

Genzer J, Marmur A (2008) Biological and synthetic self–cleaning surfaces. MRS Bull 33:742–746

Goldsworthy AK (1996) The Roman army at war 100 BC–200AD. Clarendon, Oxford

Han X, Wang J (2011) A novel method for fabrication of the biomimetic shark–skin coating. Adv Mater Res 239–242:3014–3017

Han X, Zhang DY (2008) Study on the micro–replication of shark skin. Sci China Ser E Tech Sci 51:890–896

Han X, Zhang D, Li X, Li Y (2008) Bio–replicated forming of the biomimetic drag–reducing surfaces in large area based on shark skin. Chinese Sci Bull 53:1587–1592

Hebrank MR (1980) Mechanical properties and locomotor functions of eel skin. Biol Bull 158:58–68

Herodotus (1954) The histories. Penguin, New York

Hochschule Bremen (2007) Europäische patentanmeldung "antifouling coating" EP 06 018 001.5

Howell D, Behrends B (2006) A review of surface roughness in antifouling coatings illustrating the importance of cutoff length. Biofouling 22(5–6):401–410

Humer F (2006) Legionsadler und Druidenstab. Archäologischer Park Carnunthum, Petronell-Carnuntum, 327 pp

Ikom et al (2003) Reprinted from Ikom T, Kobayashi H, Tanakaa J, Walshb D, Mann S (2003) Microstructure, mechanical, and biomimetic properties of fish scales from *Pagrus major*. J Struct Biol 142(3):327–333. Copyright (2003), with permission from Elsevier

Kesel AB, Liedert R (2007a) Antifouling coating. EP 06 018 001.5. 2007

Kesel AB, Liedert R (2007b) Learning from nature: non–toxic biofouling control by shark skin effect. In: SEB 2007 Abstracts. CBP Part A 146(4):130

Landreneau EB (2011) Scales and scale–like structures. Dissertation, Texas A&M University, College Station

Lang AW, Motta P, Hidalgo P, Westcott M (2008) Bristled shark skin: a microgeometry for boundary layer control? Bioinsp Biomim 3:046005 (9pp). doi:10.1088/1748-3182/3/4/046005. Copyright © 2008 IOP Publishing Ltd. Reprinted with permission

Lauder GV (2011) Swimming hydrodynamics: ten questions and the technical approaches needed to resolve them. Exp Fluids 51:23–35

Laufer B (1914) Chinese clay figures, vol 13. Field Museum of Natural History, Chicago

Lee S–J, Lee S–H (2001) Flow field analysis of a turbulent boundary layer over a riblet surface. Exp Fluids 30:153–166

Lin et al (2011) Reprinted from Lin YS, Wei CT, Olevsky EA, Meyers MA (2011) Mechanical properties and the laminate structure of *Arapaima gigas* scales. J Mech Behav Biomed Mater 4(7):1145–1156. Copyright (2011), with permission from Elsevier

Liu K, Jiang L (2011) Bio–inspired design of multiscale structures for function integration. Nano Today 6:155–175

Liu K, Jiang L (2012) Bio–Inspired self–cleaning surfaces. Annu Rev Mater Res 42:231–263

Liu MJ, Wang ST, Wei ZX et al (2009) Bioinspired design of a superoleophobic and low adhesive water/solid interface. Adv Mater 21:665–669

Liu X, Gao J, Xue Z et al (2012) Bioinspired oil strider floating at the oil/water interface supported by huge superoleophobic force. ACS Nano 6(6):5614–5620

Mann S (2001) Biomineralization. Principles and concepts in bioinorganic materials chemistry. Oxford University Press, New York

Meunier F, Brito P (2004) Histology and morphology of the scales in some extinct and extant teleosts. Cybium 28:225–235

Mollendorf J, Termin A, Openheim E et al (2004) Effect of swim suit design on passive drag. Med Sci Sports Exerc 36:1029–1035

Motta P (1977) Anatomy and functional morphology of dermal collagen fibers in sharks. Copeia 1977:454–464

Oeffner and Lauder (2012) Republished with permission of The Company of Biologists Ltd, from Oeffner J, Lauder GV (2012) The hydrodynamic function of shark skin and two biomimetic applications. J Exp Biol 215:785–795. Copyright (2012); permission conveyed through Copyright Clearance Center, Inc

Ortiz C, Boyce MC (2008) Materials science–bioinspired structural materials. Science 319:1053–1054

Price P (1983) An interesting find of *Lorica Plumata* from the Roman Fortress at UK. In: Bishop MC (ed) Roman military equipment: proceedings of a seminar held in the department of ancient history and classical archaeology at the University of Sheffield, Sheffield

Robinson HR (1975) The armour of imperial Rome. Lionel Leventhal Limited, London

Salta M, Wharton JA, Stoodley P et al (2010) Designing biomimetic antifouling surfaces. Philos Trans A Math Phys Eng Sci 368(1929):4729–4754

Schumacher JF, Carman ML, Estes TG et al (2007) Engineered antifouling microtopographies–effect of feature size, geometry, and roughness on settlement of zoospores of the green alga Ulva. Biofouling 23(1–2):55–62

Shadwick RE, Lauder GV (2006) Fish biomechanics. Elsevier Academic Press, San Diego

Shephard KL (1994) Functions for fish mucus. Rev Fish Biol Fish 4:401–429

Sire JY, Donoghue PCJ, Vickaryous M (2009) Origin and evolution of the integumentary skeleton in non–tetrapod vertebrates. J Anat 214:409–440

Song J (2011) Multiscale materials design of natural exoskeletons: fish armor. PhD dissertation, Materials Science and Engineering, Massachusetts Institute of Technology, Cambridge, 282 p

Song et al (2010) Reprinted from Song J, Reichert S, Kallai I et al (2010) Quantitative microstructural studies of the armor of the marine threespine stickleback (*Gasterosteus aculeatus*). J Struct Biol 171(3):318–331. Copyright (2010) with permission from Elsevier

Song J, Ortiz C, Boyce MC (2011) Threat–protection mechanics of an armored fish. J Mech Behav Biomed Mater 4(5):699–712

Sudo S, Tsuyuki K, Ito Y et al (2002) A study on the surface shape of fish scales. JSME Int J Ser C 45:1100–1105

Timms HWM (1905) The development, structure and morphology of the scales in some teleostean fish. Quart J Micr Soc 49:39–69

Torres FG, Troncoso OP, Nakamatsu J et al (2008) Characterization of the nanocomposite laminate structure occurring in fish scales from *Arapaima gigas*. Mater Sci Eng C 28:1276–1283

Tuteja et al (2007) From Tuteja A, Choi W, Ma M et al (2007) Designing superoleophobic surfaces. Science 318(5856):1618–22. Reprinted with permission from AAAS

Vermeij GJ (2006) Nature: an economic history. Princeton University Press, Princeton

Vernerey FJ, Barthelat F (2010) Reprinted from Vernerey FJ, Barthelat F (2010) On the mechanics of fishscale structures. Int J Solids Struct 47(17):2268–2275. Copyright © 2010 Elsevier Ltd, with permission from Elsevier

Wainwright SA, Vosburgh F, Hebrank JH (1978) Shark skin: a function in locomotion. Science 202:747–749

Wang LF, Song JH, Ortiz C et al (2009) Anisotropic design of a multilayered biological exoskeleton. J Mater Res 24(12):3477–3494. © Cambridge University Press 2009. Reproduced with permission

Wang X, Li J, Lee JD, Eskandarian A (2012) On the multiscale modeling of multiple physics. In: Li S, Gao X (eds) Handbook of micromechanics and nanomechanics. Pan Stanford Publishing Pte Ltd, Singapore

Wilga CD, Lauder GV (2002) Function of the heterocercal tail in sharks: quantitative wake dynamics during steady horizontal swimming and vertical maneuvering. J Exp Biol 205:2365–2374

Wilga CD, Lauder GV (2004) Hydrodynamic function of the shark's tail. Nature 430:850

Young GC (2010) Placoderms (armored fish): dominant vertebrates of the Devonian period. Annu Rev Earth Planet Sci 38:523–50

Zhang L, Zhang Z, Wang P (2012a) Smart surfaces with switchable superoleophilicity and superoleophobicity in aqueous media: toward controllable oil/water separation. NPG Asia Mater 4:e8

Zhang X, Gu J, Cao W et al (2012b) Bilayer–fish–scale ultrabroad terahertz bandpass filter. Optics Lett 37:906–908

Zhu et al (2011) With kind permission from Springer Science+Business Media: Zhu D, Vernerey F, Barthelat F (2011) The Mechanical Performance of Teleost Fish Scales. In: Mechanics of biological systems and materials, vol 2: conference proceedings of the Society for Experimental Mechanics series. Springer. Copyright © 2011, Springer Science+Business Media, LLC

Zhu D, Barthelat F, Vernerey FJ (2012a) The intricate multiscale mechanical response of natural fish–scale composites. In: Li S, Gao X (eds) Handbook of micromechanics and nanomechanics. Pan Stanford Publishing Pte Ltd, Singapore

Zhu D, Ortega CF, Motamedi R et al (2012b) Structure and mechanical performance of a "modern" fish scale. Adv Eng Mater 14:B185–B194. Copyright © 2012 WILEY-VCH Verlag GmbH & Co. KGaA, Weinheim. Reprinted with permission from John Wiley and Sons

Chapter 6
Fish Skin: From Clothing to Tissue Engineering

Abstract The epidermis and the dermis are two basic layers of fish skin. The specific material properties of fish skin have been known since ancient times. Some human cultures developed unique techniques to prepare fish leather from fish skin and used this leather for clothing, shoes and homes. Another fish skin-based biomaterial with incredible sturdiness is known as shagreen. Additionally, some modern approaches regarding the practical applications of fish scales in tissue engineering as well as novel cultivation techniques of fish skin cell lines are described in this chapter.

According to Rakers et al. (2011), "the term 'fish skin' stands for the skin from all aquatic, non-tetrapod vertebrates with or without scales. It exhibits enormous vital functions; including chemical and physical protection, sensory activity, or hormone metabolism," (Rakers et al. 2011). From the biomaterials point of view, skin, with its complex structure and close association with various tissue elements, is a fibre reinforced, pliant composite (Wainwright et al. 1976) and shows variations in its mechanical properties depending on the region of the animal body which it covers (Muthiah et al. 1967). Depending on mechanical attributes, skins have been classified as active or passive structures (Craig et al. 1987).

Definitively, we can accept animal skin, including that of fish origin, as composite-based biological material, which is the source of the leather – a man-made material. Salmon species, sharks and rays are the main fish groups, which has been used for practical applications starting in ancient times and continue to see use today. As discussed in the previous chapter, scientists are using shark skin as a bioinspired construct for developing new antifouling coatings for naval industry. However, as reported previously (http://www.rojeleather.com) before the skins of shark and skate found application for leather goods, the fishermen used their exotic skins as waste. Sometimes they has been used as the substitute material for sandpaper, or even in boat building.

H. Ehrlich, *Biological Materials of Marine Origin*, Biologically-Inspired Systems 4,
DOI 10.1007/978-94-007-5730-1_6

6.1 Fish Skin Clothing and Leather

The earliest records of wide scale use of fish skins appear to have been used by the Japanese Ainu culture in the fifth century (Price 2005). Inuit and some southern maritime cultures have used fish skins, especially salmon, since the Middle Ages for their clothing accessories such as jackets and outer lining of shoes. For example, broad variety of fish skin-made leather goods is known from the Nanai culture. French Jesuit geographers travelling on the Ussury and the Amur in 1709 described Nanai people for the first time. They were also known as *fish-skin tartars* or Yupi Tartars (Du Halde 1735). Their settlements were distributed in the region of the Ussury River and on the Amur above the mouth of the Dondon River (which falls into the Amur between today's Khabarovsk and Komsomolsk-on-Amur in Russia). These people produced salmon fish skins for numerous goals (Fig. 6.1). Thus, in the Museum of Ethnology in St. Petersburg an old fish-skin door cover from Sakhalin Island in Siberia may be seen.

Recently, numerous unique examples of Nanai Cottage may be seen in Siberian museums (Figs. 6.2 and 6.3). The Nanai and the Hezhen are the same people, with different names in Russia and in China. Here, I take the liberty to insert detailed description of Hezhen traditions as follow:

Fish have an overwhelming influence on the Hezhen society. Age is calculated via counting the times they ate fish, they wear fish skin garments, and fish skin is used for housing material. Fishing forms the basis of their culture, and is prevalent in their social interactions. Amongst these, perhaps the most striking aspect is the fish skin clothing.

Fig. 6.1 Fish skin technology by Nanai people, also known as *fish-skin tartars* (Image courtesy www.sakhalin.ru)

Fig. 6.2 The "olochi" – fish skin-made shoes (Image courtesy www.sakhalin.ru)

Fig. 6.3 The "Amiri" clothing for women is also handmade from the salmon fish skin (Image courtesy www.sakhalin.ru)

The patterns from the fish scales lend fish skin it natural beauty, and this carefully processed hide was commonly worn in the early days of Hezhen culture. This processing involves skinning, drying, tanning, cutting, sewing, and embellishing the skin. These old methods will soon be lost, as this labor-intensive protocol is difficult for artists.

These days, the most common fish skin used is salmon from the Heilongjiang River. With over a dozen separate steps, tanning fish skin is an important industrial art form. Bamboo, which does not damage the skin, is used to separate the skin from the animal. The skin is then air dried, where it hardens, and scraped with a knife. This produces a tanned leather product. Needles used to sew this leather are derived from fish bone; and the thread made from either animal sinews or thin strips of fish skin. Due to the small size of fish, the symmetry of the joints and natural colors need to be accounted for as the garment is assembled- each piece of clothing is pieced together from a patchwork of hides. Dozens of fish are needed to make a full fish skin suit. Unlike cotton and silk, the skin is thick and strength is required to pierce the hide with the bone needle (http://en.showchina.org/Features/15/3/200907/t368164.htm).

Yet it has not been until relatively recent times that Western culture has utilized more than a few species of fish to make leather (see for review Price 2005). Sharks, for example, have been used for their strong, rough lenticular abrasive surface, that is ideal for military boots, and non-slip sword handles. Eels on the other hand have extremely smooth, thin, skin, but it too is very strong and consequently used in the high-end shoe and purse/wallet markets. During World War II, Norwegians covered their feet with salmon skins. In the Western Fjords of Iceland, people made shoes from the Atlantic wolfish and even measured the length of mountain-roads with their endurance as they counted how many pair of shoes they had to use to get from a to b. These were not durable shoes, and as a result people mostly used them as slippers and to wear in the hay fields when harvesting grass (see for details http://www.huld.is/).

In the last decade, however, the skin of farmed fish, Tilapia and salmon, has stormed into the global designer fashion industry for clothing and small items such as belts, purses, wallets, and bikinis. It is well established that the fish skin can be processed and transformed into a quality raw material of unique and peculiar aspect after tanning, due to its resistance and drawing on its surface, mainly the skins of fish with scales.

According to Ingram and Dixon (1994), the skin of fish is considered an exotic and innovative leather, and is being accepted in different clothing industry segments. It has been verified that in the fish skin commercialization and industrialization there are problems associated with its small size and apparent fragility. However, because of the need to know and test the quality of this raw material, some works have been developed to test its resistance by physical-mechanical interrogations. The after-tanning exotic drawing of the skins of fish with scales compensates its reduced size. The original drawing of these skins, which can hardly be imitated by impression plates on other leathers, hinders the falsification of this type of product, mainly if the lamellas of insertion of the scale are more elongated (Price 2005).

The physical-mechanical tests confirm that the skins of fish present variable resistance in function of a series of factors such as: the species of fish and composition of collagen fibers, the fish dimensions, technique of tanning employed, region of the skin and orientation or direction of the leather (longitudinally and transversally to the length of the fish), among others.

As for the species, the resistance of the skin is related to its architectural histology, e.g., the disposition, orientation and composition of the collagen fibers. According to Junqueira et al. (1983), the structural arrangement of the collagen fibers in the compact dermis, as well as the thickness of its stratus, allows for the skin to present great resistance to different traction forces. Therefore, the skin of some species of fish can be used commercially in the confection of leather devices. As for the size of the fish, it is directly related to the thickness of the skin: as the fish grow, the skin proportionally thickens due to the increasing amount of collagen fibers, which will react with the tanning agents and give the characteristic resistance to the leather.

According to Craig et al. (1987) it has been verified in the skins of some species the distribution of collagen fibers in accordance with their size. The parameters that indicate the traction (load of force, tension of traction and elasticity) can be correlated with the amount and orientation of collagen fibers. The dermis thickness is mainly determined by the collagen fibers' ratio in the skin (Fujikura et al. 1988).

For the transformation of the skin into leather, an unputrefiable product, it is necessary to tan it. In this way, the fibrous nature is kept after being previously separated, for the removal of the interfibrillar material makes it easy for chemical products to act. After this preparation of the skin, they are treated with tanning substances, which transform them into leathers or processed skins (tanned), preserved from autolitic processes or microbial attacks (Hoinacki 1989). Thus, the skin is transformed into an unputrefiable material with characteristic softness, elasticity, flexibility, and traction resistance. In sum, it has specific physical-mechanical qualities due to the tanning process. During the tanning process, the skin goes under modifications due to the use of chemical products that react with the collagen fibers, giving the leather a greater resistance associated to the disposal and orientation of collagen fibers (Price 2005).

The skins from fish are now considered the principal residual by-product of the aquaculture business. As reported by Price (2005), several renowned designers are now marketing corresponding lines of a pink salmon shoes, for example Christian Dior (U$ 800), or, John Galliano (U$ 800). Stingray clutch for U$ 1,180 and a small evening stingray purse for U$ 1,620 were proposed by Bottega Veneta and Givenchy, respectively. Also, wide selection of fish-skin wallets, boots, and other products are producing in the USA by the companies like Upscale Leather, and Ocean Leather.

The haute couture fashion has embraced fish skin leather because of not only its novelty; but also praises for its softness, strength, and versatility. Part of the reason for not using fish leather has been that up until recently few tanneries had managed to get rid of the "fish" smell completely. With modern tanning procedures, this is no longer an issue. It is established that the fish skins are as strong as crocodile leather and have the durability and strength of a manmade material (Price 2005).

The high-end market also aligns itself with the current politically correct marketing strategy as "*environmentally friendly alternative exotic leather*" because they use an otherwise wasted product. The fish leather market trend is a spin-off from the dramatic increase in fish farming around the world of Atlantic salmon (marine fish source) and Tilapia (fresh-water fish source).

The national study concerning the ecological use of natural materials stimulated approaches to use salmon skin leather for clothes. Finally, designer Claudia Escobar (SKINI London) represented a line of perfectly crafted bikinis made entirely of salmon skin. Price level – $US355 and more (Price 2005).

From a biological materials point of view, the methods of fish skin preparation, finishing and tanning are still of interest. I take the liberty to recommend for reading the texts of some patents, which have been published since the beginning of the twentieth century. Thus, Allen (1920) patented a method for the preparation of shark-skins and the like for tanning. Knudsen (1923) and Adrien (1929) patented the process for tanning fish skins and the process for the treatment of skins of animals

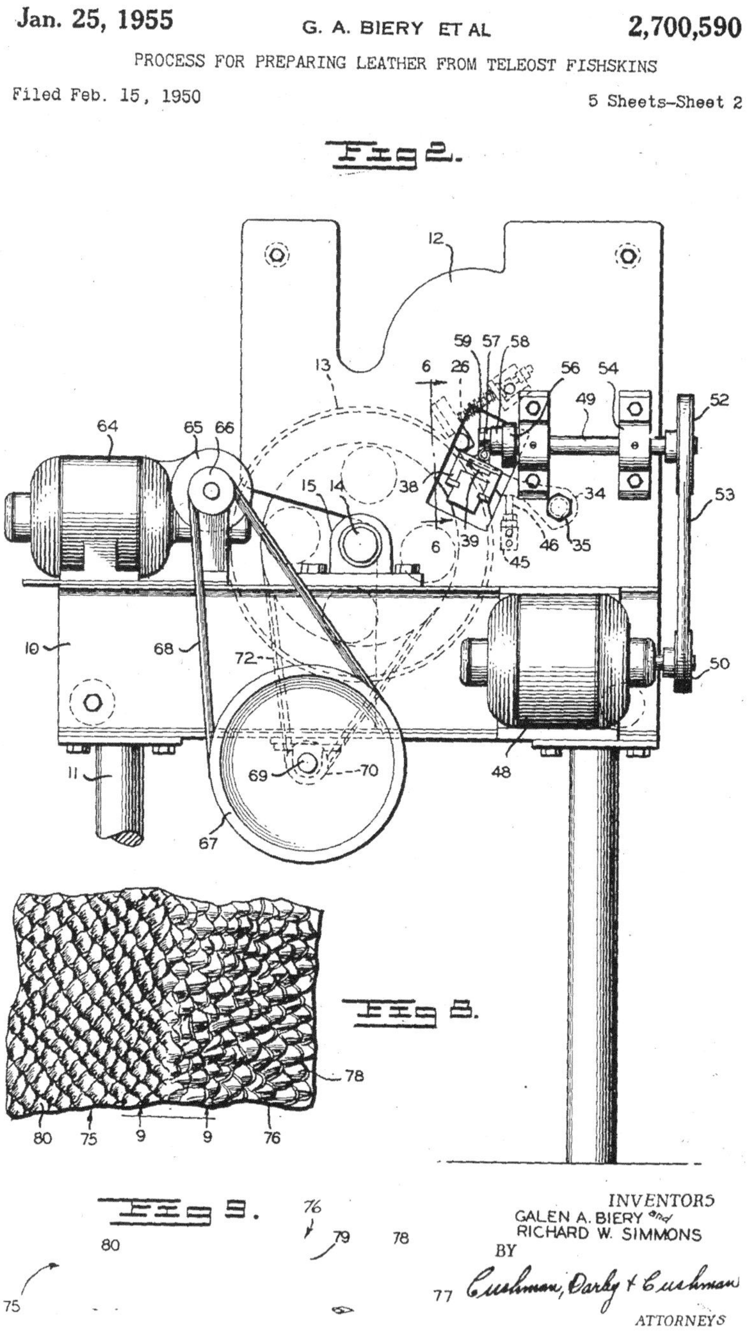

Jan. 25, 1955 G. A. BIERY ET AL 2,700,590

PROCESS FOR PREPARING LEATHER FROM TELEOST FISHSKINS

Filed Feb. 15, 1950 5 Sheets-Sheet 2

INVENTORS
GALEN A. BIERY and
RICHARD W. SIMMONS
BY
Cushman, Darby & Cushman
ATTORNEYS

Fig. 6.4 Image of the apparatus proposed by Biery and Simmons in their patent entitled "Process for preparing leather from teleost fish skins" (1955) (Adapted from the US Patent 2,700,590)

containing calcified formations, respectively. After patents by Erkel (1940) and Rose (1953), Biery and Simmons (1955) patented the process for preparing leather from teleost fishskins (Fig. 6.4). There are also more recent patents published. In the of making fish skin leather method patented by Chen (2006), a fish skin with fish scales thereon is washed, cleaned, and tanned. Then, the fish scales are removed from the tanned fish skin to expose a plurality of ridges underlying the fish scales and presenting fish skin grains. The resultant fish skin leather has a skin substrate surface with fish skin grains that include a plurality of ridges which project from the surface of the skin substrate. I suggest that the methodological principles for fish skin treatment described in the documents listed above, can be used in the modern laboratories for experimental works of fish skin properties and modification like mineralization and metallization in vitro.

6.2 Shagreen

It cannot be ruled out that ray skin has been used in ancient Egypt, and during the Chinese Han Dynasty (202 BC – AD 220). Later, during feudal times, Japanese samurai crafted their sword handles with biomaterial obtained from shark or ray skin named *shagreen*. It was known as material with incredible sturdiness.

The properties of fish skin as source suitable to manufacture shagreen is described as follow:

> Sharks and skates skins are usually covered with hard calcified placoid scales. Their size is chiefly dependent on the size and age of the specimen. These scales are worn down to give a textured surface of rounded pale protrusions, between which the dye shows when the skin is coloured from the other side (http://www.edtanner.co.uk/leather-facts/).

According to Margot Brunn, Conservator Provincial Museum of Alberta, Canada, "shagreen is prepared by scraping, stretching and drying. It is not a true leather but a rawhide (Fig. 6.5). It is processed into a usable material by grinding, filing and polishing the surface denticles in order to flatten their sharp points. The skin may also have been dyed, or have pearly white denticles interspaced with black varnish. A chemical process to decalcify and soften placoid scales for producing commercially tanned leather did not exist prior to the 1920s. Black filler in-between closely spaced denticles is common, as are natural, light-coloured skins with ground-down spikes, e.g., from the stingray. Copper salts and other dyes were used for contrasting denticles and skin tones in shades of green, red, brown and yellow," (Brunn 2001).

It is suggested that Europe began importing shagreen-covered tools since the seventeenth century. Jean-Claude Galluchat (d. 1774) was the first famous tanner and shagreen expert in Paris in eighteenth century. As reported in "A brief history of the Shagreen (Stingray skin)" (http://www.fauxshagreenfurniture.net/en/news/a-brief-history-of-the-shagreen-stingray-skin.html): "His name has been transformed

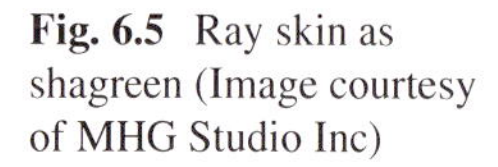

Fig. 6.5 Ray skin as shagreen (Image courtesy of MHG Studio Inc)

into galuchat, the French word for shagreen. Louis XV's most famous mistress, Madame de Pompadour, was the dominant patron of Monsieur Galluchat, and it was said that a week didn't go by when she didn't buy some new object, often in shagreen."

Nearly 1,000 shagreen made artefacts including candlesticks and frames, handbags and belts, vases and elaborate boxes, wallets and briefcases has been produced in London from 1899 to 1933 by English artisan John Paul Cooper. Shagreen saw its hey-day in the twentieth century when its use by makers such as Clément Rousseau, Jacques-Émile Ruhlman, Jean-Michel Frank, André Groult and Jules Leleu made it the height of chic. Caunes and Perfettini wrote a definitive book on the material in 1994 with the same publishers. The exhibition and catalogue were the consequences of the popularity and importance of this earlier work (see for review De Caunes and Perfettini 1995; Lefranc 2004; Perfettini 2005).

Throughout history, shagreen has been revered for its mesmerizing beauty, durability and functionality (Fig. 6.6). It was also frequently used in furniture and accessories during the Art Deco period.

There are some debates concerning shagreen as "environmentally correct material" because of animal protection policy. Probably, due to this situation such prominent companies that use shagreen as Ironies and R&Y Augousti, confirm that they use only sustainably-sourced remnants from the fishing industry.

Fig. 6.6 Metallic red stingray leather silver capped cuff (Photo by Karen Koenig © Karen Koenig. Reprinted with permission)

Today, the representatives of *Dasyatidae* family of rays are used as source for modern shagreen. For example, such species as *Dasyatis stephan* and *D. bleekeri* carry no restrictions including CITES. Their skins are a by-product of the fishing industry as well as an important foodstuff in South East Asia.

6.3 Fish Scales and Skin as Scaffolds for Tissue Engineering

Both epidermis and dermis are the basic layers localized within skin of vertebrates including the skin of all fish species. Rakers and co-workers (2011) described the organization of skin in the following way:

"The stratified squamous epithelium is composed of three strata with cell – cell junctions providing the only epithelial coherence. Undifferentiated epidermal progenitor cells emerging from the intermediate and probably from the basal layer are induced to proliferate and/or differentiate into various specialized cell types (mucus cells, club cells, sensory cells) on demand and subsequently recruited to the outermost epidermal layer. Dermal cells however, are mostly fibroblasts, interspersed with different chromatophores. The dermis further houses the scales, which are formed and anchored in the superficial region of the dermis, the stratum laxum. Underneath, the dermal stratum compactum is located, with a densely packed collagenous matrix," (Rakers et al. 2011). From materials science point of view, design principles, which can be observed within hierarchically organized fish scales as naturally occurring scaffolds may be useful in the broad variety of practical applications in biomedicine and modern technologies. Following features of fish scales have been reported as advantageous:

- "fish scales, which are usually considered as marine wastes, were acellularized, decalcified and fabricated into collagen scaffolds;
- these treatments did not affect the naturally three dimensional, highly centrally-oriented micropatterned structure of the material," (Lin et al. 2010).

- the fish scale is thermostable with respect to both high and very low temperature;
- the orthogonally organized ventral side plates of collagenous fibers embedded in proteoglycans can be useful as scaffolds for tissue engineering as well as cell culturing (Ravneet et al. 2009);
- "Fish-scale collagen has *[…]* the low probability of zoonotic infection," (Takagi and Kazuhiro 2007).

Keratoprostheses are examples of artificial corneas that are a promising alternative to obtaining tissue replacements for corneal transplantation. "Especially to patients, who are blind due to corneal defects, artificial corneas could potentially benefit and the demand is increasing," (Lin et al. 2010). Teleost fish scales have been recently proposed as biological model for the development of biomaterials for regeneration of the corneal stroma (Takagi and Kazuhiro 2007; Lin et al. 2010) (Fig. 6.7).

Both fish scale collagen (Nomura et al. 1996; Ikoma et al. 2003; Sastry et al. 2008; Ikoma and Tanaka 2012) and hydroxyapatite (Mondal et al. 2010) were reported as biomimetic materials for biomedicine (see also **Chap. 8** in this work).

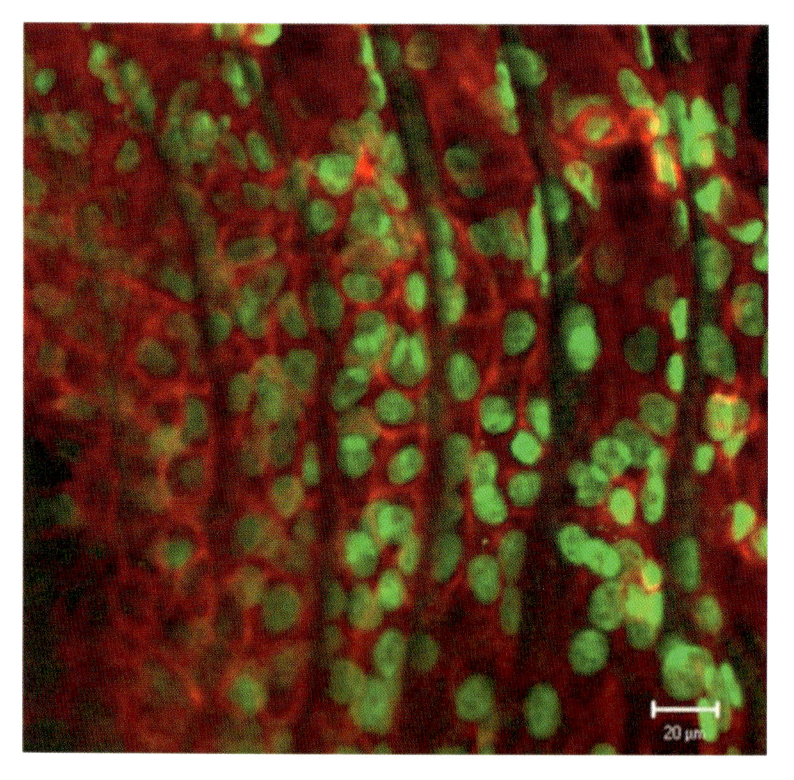

Fig. 6.7 The image shows the distribution of corneal cells after 7 days of cultivation using acellular fish scale scaffold (Adapted from Lin et al. 2010). The cells were correspondingly fixed and stained using Alexa Fluor 488-phalloidin (for visualization of F-actin, *red*) and Hoechst 33342 (to detect nuclei, *green*) (Lin et al. 2010; Copyright (c) 2010, Lin et al. Reprinted with permission)

A lateral face evaporation method was recently proposed for fabrication of collagen fibril membranes (CFMs) based on type I atelocollagen extracted from tilapia fish scales. In order to increase the mechanical property of these membranes they were crosslinked in gaseous glutaraldehyde for different durations. The density and thickness of the CFM obtained were 0.51 ± 0.04 mg/cm^3 and 50 ± 5 μm (Ikoma and Tanaka 2012).

Fish cell lines have been useful in many areas of research. Originally developed to support the growth of fish viruses for studies in aquatic animal viral diseases, fish cell lines have been isolated from numerous species as well as tissues of origin. As reviewed by Wagg and Lee (2005), Mauger et al. (2009) and Lakra et al. (2011), fish immunology, physiology, genetics and development, toxicology, ecotoxicology, endocrinology, disease control, biotechnology, aquaculture as well as biomedical research are the areas for application of wide variety of fish cell lines. The intriguing question about the possibility to cultivate fish skin cells can be positively answered today. Although, the establishment of the eurythermic line of fish cells in vitro was described by Wolf and Quimby in 1962, primary cell cultures of Indian carp (*Labeo rohita*) and rainbow trout (*Oncorhynchus mykiss*) epidermal cells were first reported by Lakra and Bhonde (1996) and Lamche et al. (1998), respectively. Recently, Rakers et al. (2011) reported the fabrication of a permanent skin cell culture from (*O. mykiss*). Thus, "the cells of the fish skin cell culture could be propagated over 60 passages so far. It is possible to integrate freshly harvested rainbow trout scales into this new fish skin cell culture. The epithelial cells derived from the scales survived in the artificial micro-environment of surrounding fibroblast-like cells. Also, antibody staining indicated that both cell types proliferated and started to build connections with the other cell type. It seems that it is possible to generate an 'artificial skin' with two different cell types. This could lead to the development of a three-dimensional test system, which might be a better in vitro representative of fish skin in vivo than individual skin cell lines," (Rakers et al. 2011).

One of the goals of such kind of studies is related to the establishment of a biosynthetic fish skin for application in aquatic robots that can emulate fish like autonomous underwater vehicles known as the "RoboTuna" (see the next chapter). The paper with a very long but intriguing title "*Tissue Engineering of Fish Skin: Behavior of Fish Cells on Poly(ethylene glycol terephthalate)/Poly(butylene terephthalate) Copolymers in Relation to the Composition of the Polymer Substrate as an Initial Step in Constructing a Robotic/Living Tissue Hybrid*" was published by Pouliot et al. in 2004. Researchers studied the attachment behaviour as well as the proliferation brown bullhead (*Ameiurus nebulosus*) (BB) and of chinook salmon (*Oncorhynchus tshawytscha*) embryo (CHSE-214) cells. These cells were placed on different compositions of a poly(ethylene glycol terephthalate) (PEGT) and poly(butylene terephthalate) (PBT) copolymer (Polyactive) films. It was shown that "when a 55 wt.% and a 300-Da molecular mass form of PEGT was used, maximum attachment and proliferation of CHSE-214 and BB cells was achieved. Histological studies and immunostaining indicate the presence of collagen and cytokeratins in the extracellular matrix formed after 14 days of culture. Porous scaffolds of PEGT/PBT copolymers were also used for three-dimensional tissue engineering of fish skin, using BB cells. Overall, the obtained results indicated that fish cells can attach, proliferate, and express fish skin components on dense and porous polymer scaffolds," (Pouliot et al. 2004).

6.4 Conclusion

Further study would serve not only to investigative dermatology as thought previously (Rackers et al. 2010), but also the biological materials science and biomedical applications. An interesting direction was recently discovered for cancer research. Evidence of melanoma in wild marine fish *Plectropomus leopardus* has been reported for the first time (Sweet et al. 2012). The authors studied "extensive melanosis and melanoma (skin cancer) in wild populations of an iconic, commercially-important marine fish, the coral trout P. leopardus. The syndrome observed has strong similarities to previous studies associated with UV induced melanomas in the well-established laboratory fish model Xiphophorus. Relatively high prevalence rates of this syndrome (15 %) were recorded at two offshore sites in the Great Barrier Reef Marine Park. In the absence of microbial pathogens and given the strong similarities to the UV-induced melanomas, it was concluded that the likely cause was environmental exposure to UV radiation," (Sweet et al. 2012).

Because of natural selection the fish scale and fish skin as mechanical systems evolved in marine fish are highly efficient for the corresponding mode of life and habitat in each fish species. Their amazing abilities inspire today innovative designs to improve the ways that man-made systems function in and interact with the aquatic niches, including those like underwater biomimetic robotics. Researchers mimic the marine fish (mostly tuna, sharks and rays) locomotion, shape, profile, and structure, in their development of machines or devices for underwater propulsion as well as manoeuvrability. This is the topic of the next chapter.

References

A brief history of the Shagreen (Stingray skin). Published on-line http://www.fauxshagreenfurniture.net/en/news/a-brief-history-of-the-shagreen-stingray-skin.html. Accessed 15 May 2014. Copyright by fauxshagreenfurniture.net. Reprinted with permission

Adrien GR (1929) Process for the treatment of skins of animals containing calcified formations. United States Patent 1,725,629

Allen R (1920) Treating of shark–skins and the like preparatory to tanning. United States Patent 1,338,531

Biery GA, Simmons RW (1955) Process for preparing leather from teleost fishskins. United States Patent 2700590

Brunn M (2001) Shagreen (http://cool.conservation-us.org/byform/mailing-lists/cdl/2001/0513.html) published on-line 17.04.2001; Accessed 15 May 2014. Copyright © 2001, Foundation of the American Institute for Conservation of Historic and Artistic Works (FAIC). Reprinted with permission

Chen HK (2006) Fish skin leather and method of making the same. United States Patent 7,150,763
Craig AS, Eikenberry EF, Parry DAD (1987) Ultrastructural organisation of skin: classification on the basis of mechanical role. Connect Tissue Res 16:213–223
De Caunes L, Perfettini, J (1995) Le galuchat, Catalogue exposition, Bibliothèque Forney, Hôtel de Sens 14 mars–13 mai 1995, Les éditions de l'amateur
Du Halde JB (1735) Description Beographique Historique dl'Empire de la Chine. vol 4, Paris
Erkel FC (1940) Preparing shark and other hides for tanning. United States Patent 2,222,656
Fujikura K, Kurabuchi S, Tabuchi M, Inoue S (1988) Morphology and distribution of the skin glands in *Xenopus laevis* and their response to experimental stimulation. Zool Sci Tokyo 5:415–430
Hoinacki E (1989) Peles e couro: origens, defeitos e industrialização 2°ed. Ver. E ampl. Porio Alegre: CFP de Aries Gráficas "Herique d'Avila Beriaso", 319p. http://www.worldfish.org/PPA/PDFs/SemiAnnual%20V%20English/5th%20s.a.%20eng_D.5.pdf
Ikoma T, Tanaka J (2012) Effect of glutaraldehyde on properties of membranes prepared from fish scale collagen. MRS proceedings. doi:dx.doi.org/10.1557/opl.2012.396
Ikoma T, Kobayashi H, Tanaka J et al (2003) Physical properties of type I collagen extracted from fish scales of *Pagrus major* and *Oreochromis niloticas*. Int J Biol Macromol 32:199–204
Ingram P, Dixon G (1994) Fish skin leather: an innovate product. J Soc Leather Technol Chem 79:103–106
Junquera LCU, Joazeiro PP, Montes GS (1983) É possivel o aproveitamento industrial da pele dos peixes de couro? Tecnicouro, Novo Hamburgo 5:4–6
Knudsen E (1923) Process for tanning fish skins. United States Patent 1,467,858
Lakra WS, Bhonde RR (1996) Development of primary cell culture from the caudal fin of an Indian major carp, *Labeo rohita* (Ham). Asian Fish Sci 9:149–152
Lakra WS, Swaminathan TR, Joy KP (2011) Development, characterization, conservation and storage of fish cell lines: a review. Fish Physiol Biochem 37:1–20
Lamche G, Meier W, Suter M et al (1998) Primary culture of dispersed skin epidermal cells of rainbow trout *Oncorhynchus mykiss walbaum*. Cell Mol Life Sci 54(9):1042–1051
Lefranc C (2004) Le galuchat, une gaine en peau de raie. Connaiss Arts 620:104–107
Lin CC, Ritch R, Lin SM et al (2010) A new fish scale-derived scaffold for corneal regeneration. Eur Cell Mater 19:50–57. Copyright (c) 2010, Lin et al. Reprinted with permission
Mauger PE, Labbé C, Bobe J et al (2009) Characterization of goldfish fin cells in culture: some evidence of an epithelial cell profile. Comp Biochem Physiol B Biochem Mol Biol 152(3):205–215
Mondal S, Mahata S, Kundu S et al (2010) Processing of natural resourced hydroxyapatite ceramics from fish scale. Adv Appl Ceram Struc Funct Bioceram 109:234–239
Muthiah PL, Ramanathan N, Nayudamma Y (1967) Mechanical properties of skins, hides and constituent fibres. Biorheology 4:185–191
Nomura Y, Sakai H, Ishii Y et al (1996) Preparation and some properties of type I collagen from fish scales. Biosci Biotechnol Biochem 60:2092–2094
Perfettini J (2005) Le Galuchat. Une materiau mysterieux, une technique oubliee. Editions H. Vial, Paris, p 70
Pouliot et al (2004): Reprinted from Pouliot R, Azhari R, Qanadilo HF et al (2004) Tissue engineering of fish skin: behavior of fish cells on poly(ethylene glycol terephthalate)/poly(butylene terephthalate) copolymers in relation to the composition of the polymer substrate as an initial step in constructing a robotic/living tissue hybrid. Tissue Eng 10(1–2):7–21, with permission. Copyright © 2004, Mary Ann Liebert, Inc. The publisher for this copyrighted material is Mary Ann Liebert, Inc. publishers.
Price WS (2005) Value-added options: background research for development of appropriate adaptations to artisanal fisheries in the Três Marias area of Brazil. Report prepared for World Fisheries Trust. In: World Fisheries Trust. http://www.worldfish.org/PPA/PDFs/Semi-Annual V English/5th s.a. eng_D.5.pdf. Accessed 10 Jan 2012
Rakers S, Gebert M, Uppalapati S et al (2010) "Fish matters": the relevance of fish skin biology to investigative dermatology. Exp Dermatol 19:313–324

Rakers et al (2011): Reprinted from Rakers S, Klinger M, Kruse C et al (2011) Pros and cons of fish skin cells in culture: long-term full skin and short-term scale cell culture from rainbow trout, *Oncorhynchus mykiss*. Eur J Cell Biol 90(12):1041–1051. Copyright (2011), with permission from Elsevier

Ravneet et al (2009) High resolution scanning electron microscope examination of the fish scale: inspiration for novel biomaterials. J Biomim Biomater Biomed Eng 4:13–20. doi:10.4028/www.scientific.net/JBBBE.4.13. © 2014 by Trans Tech Publications Inc. Reprinted with permission

Rose H (1953) Process of finishing fish skins. United States Patent 2,633,730

Sastry TP, Sankar S, Mohan R et al (2008) Preparation and partial characterization of collagen sheet from fish (*Lates calcarifer*) scales. Int J Biol Macromol 42:6–9

Sweet M et al (2012) Reprinted from Sweet M, Kirkham N, Bendall M et al (2012) Evidence of melanoma in wild marine fish populations. PLoS ONE 7(8):e41989. doi:10.1371/journal.pone.0041989. Copyright © Sweet et al. This is an open-access article distributed under the terms of the CC-BY-2.5 License

Takagi Y, Ura K (2007) Teleost fish scales: a unique biological model for the fabrication of materials for corneal stroma regeneration. J Nanosci Nanotechnol 7(3):757–762. Journal of nanoscience and nanotechnology by American Scientific Publishers. Reproduced with permission of American Scientific Publishers in the format Republish in a book via Copyright Clearance Center

Wagg SK, Lee LEJ (2005) A proteomics approach to identifying fish cell lines. Proteomics 5:4236–4244

Wainwright SA, Biggs WD, Currey JD, Gosline JM (1976) Mechanical design in organisms. Edward Arnold, London

Wolf K, Quimby MC (1962) Established eurythermic line of fish cells *in vitro*. Science 135:1065–1066

Chapter 7
Fish Fins and Rays as Inspiration for Materials Engineering and Robotics

Abstract Marine fish show unique properties: they can move in both water and air. These properties are determined by numerous factors like body and skin shape, fins and tails, rays and ray-like structures, and muscles. The diversity, structure and function of fish fins and rays, including an analysis of the specific biological materials they are made of, are discussed in this chapter. Special attention is payed to biomimetics and bioinspiration for fish robotics and devices. This chapter also aims to explore the possibilities for fields such as fish fin regeneration and tissue engineering.

According modern point of view, "the gradual mineralization of the vertebral elements, appearance of fin rays and new median fins, and transverse and then horizontal segmentation of the axial musculature are all features correlated with increases in swimming speed, manoeuvrability, and body size of early chordates and vertebrates," (Koob and Long 2000). Thus, most teleost fish are obligate axial swimmers (Nelson 1994) and swim by oscillating or laterally undulating their body and propulsive caudal fin. The role of muscles during swimming is briefly described by Ann Pabst as follow:

"Axial muscles, arranged in complexly folded myomeres, transmit contractile forces via the connective tissue 'fabrics' of myosepta, horizontal and vertical septa, and skin. These forces do work against a variably flexible beam, the vertebral column, to affect swimming movements. Myomeres and myosepta are connected directly to the dermis of the skin, as well as to the vertebral column. The function of the myomeres is to produce lateral bending in the axial skeleton. During steady swimming, the pattern of muscle activity is both unilateral and uniphasic—at any point along the body, muscles on only one side of the animal are active at a time, and there is only one bout of muscle activity per side, per locomotor cycle," (Ann Pabst 2000; see also Wainwright 1983; Westneat et al. 1993).

The structure and function of fish muscles, which are involved in swimming, is not the subject of this book. However, fish fins are crucial biocomposite-containing structures, which are the main players in swimming phenomena and therefore are under discussion in this chapter. The dynamic interaction between the fin and the water is determined due to existence of the forces created especially by highly deformable, ray-finned fins. Briefly, "the fin moves, pushes against the water, bends, stores and releases energy, and creates vortices and jets that are shed into the flow," (Tangorra et al. 2011). Fish locomotion is based on activity of two major classes of

H. Ehrlich, *Biological Materials of Marine Origin*, Biologically-Inspired Systems 4,
DOI 10.1007/978-94-007-5730-1_7

fins: median, which are along the body center line, and those that are paired (Lauder and Madden 2006). Their flexibility is important for thrust production and vectoring forces. Thus, fishes mostly possess two sets of paired fins, the *pelvics* and *pectorals*, and three median fins, the *dorsal*, *anal* and *caudal fin*. Recent work showed that the fish's locomotion mechanism is mainly controlled by its paired pectoral fins and caudal fin (Lauder and Drucker 2004; Westneat et al. 2004). Due to the special structure and unique material properties of their fins, fishes are efficient swimmers, and possess high manoeuvrability. Fish are "able to follow trajectories, can efficiently stabilize themselves in currents and surges, create fewer wakes, and have noiseless propulsion," (Sitorus et al. 2009).

I would like to direct your attention to the fact that fish can also fly! A few marine fish are not only swimmers, but possess the ability to fly and glade over the sea surface. So, flying-fish aerodynamics (Latimer-Needham 1951; Fish 1999) is of the same scientific importance as swimming fish hydrodynamics. Therefore, it's not surprising that there is growing interest in researching mechanical and control system for flying robotics, as well as for underwater vehicles using fish fins as sources for bioinspiration. "These ongoing research efforts are motivated by more pervasive applications of such vehicles, including: seabed oil and gas explorations, scientific deep ocean surveys, military purposes, ecological and water environmental studies," (Sitorus et al. 2009), and also for entertainment (see for review Sfakiotakis et al. 1999; Mittal 2004; Kato et al. 2005; Kodati 2006; Lauder et al. 2011).

7.1 Fish Fins and Rays: Diversity, Structure and Function

Over than 28,000 of fish species are related to ray-finned fishes, which are known for their diversity in locomotory styles (Lauder and Drucker 2004; Lauder 2006; Alben et al. 2007). Ray-finned fish due to both median (midline) and paired fins (Fig. 7.1) can control their body position and generate force during locomotion.

The main types of fins are as follows: dorsal fins, caudal fin, pectoral fins, pelvic fins, anal fin, adipose fin and finlets. Finlets in birchis, for example, are represented in the form of small fins located only on the dorsal surface.

According Lauder et al. (2012), "fish species vary in body stiffness, and in how the body is moved during swimming," (Lauder et al. 2012), by the manner in which these types of fins are used. Webb (1988) classifies all swimming vertebrae in four classes:

- **Class A** uses body and/or caudal fins (BCF) for periodic propulsion and is best suited for long-term swimming at relatively high speeds.
- **Class B** uses body and/or caudal fins for transient propulsion, well suited for quick starts and turns. The bodies of members of this class are flexible and have a large tail area.
- **Class C** uses median and/or paired fins (MPF) for slow swimming and precise manoeuvring, and has better efficiency at low speeds.
- **Class D** include those fish that swim only rarely (see also for review Sitorus et al. 2009).

Fig. 7.1 The skeleton of Snowy grouper (*Epinephelus niveatus*), showing the positions of the major fins and their internal skeletal supports (Adapted from Lauder et al. 2007). The pectoral and pelvic fins are paired, while the dorsal, anal and caudal fins are median (midline) fins. The dorsal and anal fins of ray-finned fishes have internal skeletal supports (pterygiophores), which support musculature that moves the fin rays (Republished with permission of Company of Biologists Ltd, from Lauder et al. (2007); permission conveyed through Copyright Clearance Center, Inc)

MPF swimming can be done by different appendages, like the pectoral, dorsal or anal fins, or combinations of these. Intriguingly, the marine boxfishes (Teleostei: *Ostraciidae*) are also related to MPF swimming species and have proven useful as a source for bioinspired designs (see for review Bartol et al. 2005, 2008). Here, the characteristic of swimming behavior of these animals made by Bartol et al. (2008):

- “Boxfishes exhibit a unique combination of high stability and manoeuvrability. Despite having a somewhat ungainly exterior, they exhibit some of the lowest amplitude recoil movements during swimming thus far detected in any fish;
- they also have the ability to spin around with a minimal turning radius, and hold precise control of their positions and orientations;
- boxfishes have a rigid carapace encasing a significant portion of their bodies, requiring them to swim predominantly using coordinated movements of their five fins rather than relying heavily on undulatory body movements like most fish;
- the carapace of one boxfish, the smooth trunkfish *Lactophrys triqueter*, plays an integral role in the hydrodynamic stability of swimming. The ventro-lateral keel of this species is broadly triangular in cross-section, and produces leading edge vortical flows that increase in magnitude when the carapace pitches and yaws at greater angles. The vortices are strongest at posterior regions of the carapace, and produce self-correcting trimming forces that probably help *L. triqueter* maintain smooth swimming trajectories in turbulent environments,” (Bartol et al. 2008; see also Bartol et al. 2003; Gordon et al. 2001; Hove et al. 2001; Walker 2000).

Rays swim by passing a running wave through their pectoral fins (*rajiform*). The bluegill sunfish makes a rowing motion with its pectoral fins (*labriform*), some fish pass running waves past their long dorsal of caudal fins (*amiiform*) and gymnotiform, while others oscillate these fins (*balistiform*) (Mattheijssens et al. 2012). Additionally, some marine fish species are observed that show swimming behaviour that can be defined as "*hydrodynamic parasitism*", or hitchhiking (Fish 2010).

Thus, "hitchhiking is a mechanism to reduce locomotor costs by direct physical attachment to another animal in motion. As the hitchhiker is passively towed along, it saves considerable amounts of energy that it would have to expend by muscular contraction to swim. For this behavior to be of benefit to the hitchhiker without being a detriment to the other animal, the hitchhiker must be much smaller in body size compared to the host animal," (Fish 2010; see also O'Toole 2002).

Typical examples are eight species of remoras or sharksuckers (*Echeneididae*) that rely upon hitchhiking. As reviewed by Frank Fish (2010), "these fishes attach onto much larger fish, turtles, and whales. Attachment onto large mobile hosts minimizes the remona's energy expenditure from locomotion. When attached to a moving object or in a current, remoras use the water flow for respiration, switching from active branchial ventilation of the gills to passive ram gill ventilation," (Fish 2010; see also Fertl and Landry 1999; Guerrero-Ruiz and Urbán 2000; Sazima and Grossman 2006; Steffensen and Lomholt 1983).

The term BCF imply under body and/or caudal fin locomotion that is characteristic for most fish species generating thrust by bending their bodies into a backward-moving propulsive wave that extends to the caudal fin. There are four undulatory BCF locomotion modes proposed by their amplitude envelope of the propulsive wave: *anguilliform*, *subcarangiform*, *carangiform* and *thunniform*.

They different types are specialized as follows:

Dorsal Fins As described by Suzuki et al. (2003), "in teleosts, the embryonic fin fold consists of a peridermis, an underlying epidermis and a small number of mesenchymal cells. Beginning from such a simple structure, the fin skeletons, including the proximal and distal radials and finrays, develop in the dorsal fin fold at the larval stage. In early larvae the mesenchymal cells grow between the epidermis and spinal cord to form a line of periodical condensations, which are proximal radial primordia, to produce chondrocytes. The prescleroblasts, which ossify the proximal radial cartilages, differentiate within the mesenchymal cells remaining between the cartilages. Then, mesenchymal condensations occur between the distal ends of the proximal radials, forming distal radial primordia, to produce chondrocytes. Simultaneously, condensations occur between the distal radial primordia and peridermis, which are finrays primordia, to produce prescleroblasts," (Suzuki et al. 2003).

In adult fish, dorsal fin possesses a unique structure and plays an important role in their swimming behaviour. According Standen and Lauder (2007), "recent kinematic and hydrodynamic studies on fish median fins have shown that dorsal fins actively produce jets with large lateral forces. As the location of dorsal fins above the fish's rolling axis, these lateral forces, if unchecked, would cause fish to roll," (Standen and Lauder 2007).

In some sharks the dorsal fin shows very specific structural properties. As reported in the paper by Lingham-Soliar (2005a) on examples of several shark species, "the transverse sections of the skin in the dorsal fin of the white shark,

Carcharodon carcharias, tiger shark, *Galeocerdo cuvier*, and spotted raggedtooth shark, *Carcharias taurus*, show large numbers of dermal fiber bundles, which extend from the body into the fin. The bundles are tightly grouped together in a staggered formation (not arranged in a straight line or in rows). This arrangement of dermal fibers gives tensile strength without impeding fiber movement. Tangential sections indicate that the fibers in all three species are strained and lie at angles in excess of 60°. The overall results indicate that the dorsal fin of *C. carcharias* functions as a dynamic stabilizer and that the dermal fibers are crucial to this role. The fibers work like riggings that stabilize a ship's mast. During fast swimming, when the problems of yaw and roll are greatest, hydrostatic pressure within the shark increases and the fibers around the body, including in the dorsal fin, become taut, thereby stiffening the fin. During slow swimming and feeding the hydrostatic pressure is reduced, the fibers are slackened, and the muscles are able to exert greater bending forces on the fin via the radials and ceratotrichia. In *C. carcharias* there is a trade-off for greater stiffness in the dorsal fin versus that of flexibility," (Lingham-Soliar 2005a).

Caudal Fin The role of caudal fin in the swimming and propulsion of some fish was excellently described by Bainbridge (1963). During steady, fairly fast, straight locomotion, forward progression is effected entirely by lateral movements of the body and the caudal fin. Dependent to a certain extent on the forward velocity, the paired pectoral and pelvic fins and the dorsal fin are usually closely pressed to the body. The median anal fin remains more or less extended during this type of swimming, but its propulsive significance cannot be very great. As the velocity gradually diminishes, the dorsal fin and then the paired fins start to rise from the body. The lateral propulsive movements of the body are in the form of a wave which travels, with increasing amplitude, from the front end of the body to the rear. The amplitude is at a maximum at the most posterior edge of the caudal fin. It was concluded (Bainbridge 1963) that the fish has active control over the speed, and the amount of bending and area of the caudal fin during transverse movement. The current though is that (Esposito et al. 2012) in many fish the caudal fin is used not exclusively for thrust production, but also functions to create forces and moments that control the orientation of the fish. These occur via complex conformational changes and motions. These shape changes and kinematic patterns are produced by fin rays within the caudal fin. The rays are moved by intrinsic caudal musculature that is distinct from the segmented body muscles (see Fig. 7.2).

The importance of caudal locomotion has been highlighted by George Lauder (1989) in his famous review paper entitled "*Caudal Fin Locomotion in Ray-finned Fishes: Historical and Functional Analyses*". It was suggested that "the perfection of caudal locomotion has probably been the single greatest achievement of the teleostean fishes," (Lauder 1989; see also Gosline 1971). The morphology of form and shape seen in caudal fins between different fish lineages is quite diverse. "The tail of Polypterus, the most primitive living actinopterygian (Fig. 7.2), is specialized compared to other lower ray-finned fishes in that it has a diphycercal morphology. Dissections of Polypterus reveal that the lateral myotomes thin in a posterior direction

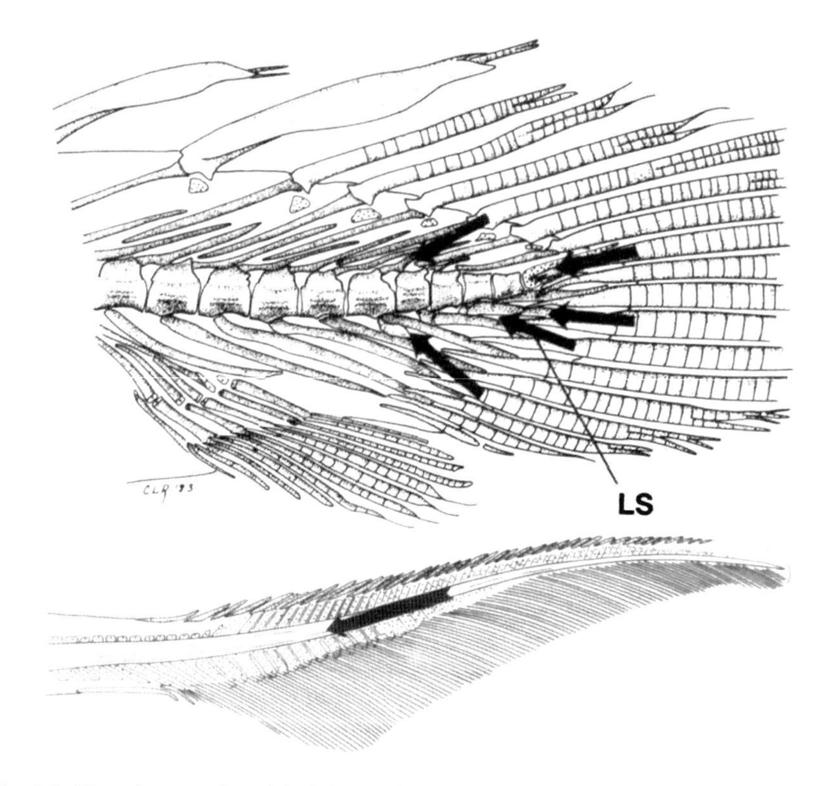

Fig. 7.2 Musculature and caudal skeleton of *Polypterus senegalus* (*above*). According Lauder (1989), "*black lines* indicate major muscles and their lines of action. Polypterus lacks intrinsic caudal musculature and shares with other primitive ray-finned fishes the condition of having the lateralis superficialis (LS) myotomal fibers attaching to the heads of the fin rays (*black arrows*). Caudal skeleton and musculature of *Acipenser stellatus* is represented below. It is well visible that the lateral body myotomes condense posteriorly to a series of long tendons that run along the notochord (*black arrow*)" (Reprinted from Lauder (1989) by permission of Oxford University Press)

and attach to the heads of the caudal fin rays. There is no differentiation of superficial and deep caudal musculature in Polypterus and no separation of distinct intrinsic muscles," (Lauder 1989).

No intrinsic caudal muscles are present in representatives of Chondrostei (Fig. 7.2), where numerous hypaxial fin rays are attached to small cartilages, which are located ventral to the notochord. Tendons of the posterior myotomes origin are attached to the connective tissue overlying the notochord and tail cartilages.

The principal trend in the evolution of fish is the change in tail structure to specific environmental pressures. Examples of this diversity include the morphologically "asymmetrical heterocercal tail shape seen in sharks and other basal ray-finned fishes, and the externally symmetrical homocercal tail of most teleost fishes," (Lauder et al. 2012). However, from bionic point of view, different morphologies of caudal fins in elasmobranch and bony fishes are of interest. Giant sharks in particular are a potentially rich source of bioinspiration. As an example, I would like to present the description of specific dynamic locomotory structure that is formed by the caudal peduncle and caudal fin of the giant shark *C. carcharias* as follow:

> "The caudal peduncle is a highly modified, dorsoventrally compressed and rigid structure that facilitates the oscillations of the caudal fin. Its stiffness appears to be principally achieved by a thick layer of adipose tissue that composes 28–37 % of its cross-sectional area, reinforced by cross-woven collagen fibers. The overlying layers of collagen fibers of the stratum compactum, oriented in steep left- and right-handed helices (~65° to the shark's long axis), prevent bowstringing of the perimysial fibers that lie just below the dermal layer. Perimysial fibers, muscles, and the notochord are restricted to the dorsal lobe of the caudal fin and comprise the bulk of its mass. Adipose tissue reinforces the leading edge of the dorsal lobe of the caudal fin and contributes to maintaining the ideal cross-sectional geometry required of an advanced hydrofoil.
>
> These dermal fibers of the stratum compactum in the dorsal lobe occur in numerous distinct layers. The layers are more complex than in other sharks and appear to reflect a hierarchical development in C. carcharias. The fiber layer comprises a number of thick fiber bundles along the height of the layer, and the layers get thicker deeper into the stratum compactum. Each of these layers alternates with a layer a single fiber-bundle deep, a formation thought to give stability to the stratum compactum and to enable freer movements of the fiber system. In tangential sections of the stratum compactum the fiber bundles in the dorsal lobe can be seen oriented, with respect to the long axis of the shark, at ~55–60° in left- and right-handed helices.
>
> Due to the backward sweep of the dorsal lobe (~55° to the shark's long axis), the right-handed fibers also parallel the lobe's long axis. In the dorsal lobe, ceratotrichia are present only along the leading edge (embedded within connective tissue), apparently as reinforcement. Stratum compactum fiber bundles of the ventral lobe, viewed in transverse section, lack the well-ordered distinctive layers of the dorsal lobe. Instead, they occur as irregularly arranged masses of tightly compacted fiber bundles of various sizes. In tangential sections the fiber bundles are oriented at angles of ~60°, generally in one direction, i.e., lacking the left- and right-handed helical pattern. Tensile load tests on the caudal fin indicate high passive resistance to bending by the skin. The shear modulus G showed that the skin's contribution to stiffness (average values from three specimens at radians 0.52 and 1.05) is 33.5 % for the dorsal lobe and 41.8 % for the ventral. The load tests also indicate greater bending stiffness of the ventral lobe compared to the dorsal. The helical fiber architecture near the surface of the caudal fin is analogous to strengthening of a thin cylinder in engineering. High fiber angles along the span of the dorsal lobe are considered ideal for resisting the bending stresses that the lobe is subjected to during the locomotory beat cycle. They are also ideal for storing strain energy during bending of the lobe and consequently may be of value in facilitating the recovery stroke. Thus, the complex fiber architecture of the caudal fin and caudal peduncle of C. carcharias provides considerable potential for an elastic mechanism in the animal's swimming motions, and consequently for energy conservation," (Lingham-Soliar 2005b).

Furthermore, it was additionally reported (Lingham-Soliar 2005c), that "the buoyancy may play a dominant role in larger white sharks by permitting slow swimming while minimizing energy demands needed to prevent sinking. In contrast, hydrodynamic lift is considered more important in smaller white sharks. Larger caudal fin spans and higher lift/drag ratio in smaller C. carcharias indicate greater potential for prolonged, intermediate swimming speeds and for feeding predominantly on fast-moving fish. Meanwhile, slow-swimming search patterns of larger individuals are advantageous in seeking predominantly large mammalian prey. Such data may provide some answers to the lifestyle and widespread habitat capabilities of this still largely mysterious animal," (Lingham-Soliar 2005c).

As recently reviewed by Esposito et al. (2012), bony fish such as bluegill sunfish (*Lepomis macrochirus*) are able to actively deform the surface of their fins through individual fin ray motions and/or by altering the stiffness of individual fin rays. The caudal fin of this fish has been demonstrated to generate locomotor forces in the lateral, lift, and thrust directions, indicating that the tail is being used for more than just propulsion. Fish are capable of altering the relative magnitudes of these forces and to vector water momentum in appropriate directions to execute maneuvers; this ability is a function of the design of the caudal fin with its individually controllable fin rays. George Lauder and co-workers (Esposito et al. 2012) "designed a robotic fish caudal fin with six individually moveable fin rays based on the tail of the bluegill sunfish. Previous fish robotic tail designs have loosely resembled the caudal fin of fishes, but have not incorporated key biomechanical components. These includes fin rays that can be controlled to generate complex tail conformations, and motion programs similar to those seen in the locomotor repertoire of live fishes. The researchers used this robotic caudal fin to test for the effects of fin ray stiffness, frequency and motion program on the generation of thrust and lift forces. Five different sets of fin rays were constructed to be from 150 to 2,000 times the stiffness of biological fin rays, appropriately scaled for the robotic caudal fin, which had linear dimensions approximately four times larger than those of adult bluegill sunfish. Five caudal fin motion programs were identified as kinematic features of swimming behaviours in live bluegill sunfish. It was shown that more compliant fin rays produced lower peak magnitude forces than the stiffer fin rays at the same frequency. Thrust and lift forces increased with increasing flapping frequency. Thrust was maximized by the 500× stiffness fin rays, and lift was maximized by the 1,000× stiffness fin rays," (Esposito et al. 2012).

Pectoral Fins The axial propulsion only near peak speeds is used by numerous fish species due to the pectoral fins (Fig. 7.3) as the primary propulsors (Wainwright et al. 2002). Here, some examples. As reviewed by Hale et al. (2006), "pectoral fin, or labriform, locomotion is particularly common among reef fishes. In this form of

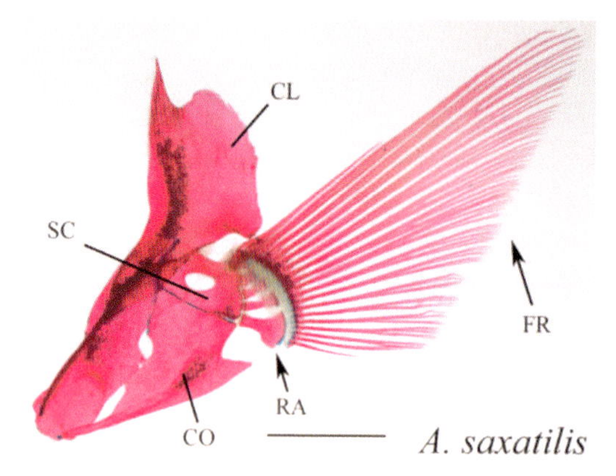

Fig. 7.3 Skeletal morphology of cleared and stained pectoral girdles and pectoral fin shapes of *Abudefduf saxatilis* (Adapted from Thorsen and Westneat 2005). Shown is a lateral view of the labroid species. Abbreviations: *CL* cleithrum, *CO* coracoid, *FR* fin rays, *RA* radials, *SC* scapula. Scale bar 1 cm (Reprinted from Thorsen and Westneat (2005) with permission of John Wiley and Sons. Copyrights © 2004 Wiley-Liss, Inc.)

swimming, the pectoral fins are generally actuated synchronously with one another in forward swimming. Speed can be increased gradually by increasing fin beat amplitude and/or frequency. Labriform swimmers change to body-caudal fin swimming near their peak speed," (Hale et al. 2006). The bluegill sunfish (*Lepomi machrochirus*) use their pectoral fin as a flapping propulsor for both manoeuvring and steady swimming. "This fin is a highly deformable and controllable surface, which can be made to flap, feather, and row, as well as to take on a variety of complex three-dimensional shapes," (Gottlieb et al. 2010).

The transmission of muscle force and motion to individual fin rays determine the thrust production as well as the common biomechanics of fin motion in fish. Following observations have been reported in this aspect:

- "Force transmission to the fin is determined by the contractile mechanics of the muscles and the structure of the fin skeleton;
- the osteological differences in fin shape, fin ray morphology, bony processes at the bases of the fin rays, and the fin radials have functional implications;
- fin shape has clear consequences for fin thrust mechanics, related to a trade-off between the efficiency of fast swimming and maneuverability in slowly swimming species;
- a rowing stroke is capable of producing stronger thrust transients for maneuvers when it is performed with a more rounded, distally broadened, paddle-shaped fin. In contrast, a more slender, tapering, wing-like fin is more efficient for swimming at higher sustained speeds using a dorsoventral flapping stroke;
- aspect ratio shows a strong correlation with swimming performance, with higher aspect ratio fins enabling higher critical swimming speeds than paddle-shaped fins;
- the labriform swimmers with elongated, wing-like fins and a steeper (more dorsoventral) stroke plane can achieve and maintain higher swimming speeds than can labriform swimmers; who have lower-aspect-ratio paddle-like fins and shallower stroke planes," (Thorsen and Westneat 2005; see also Walker and Westneat 1997, 2000, 2002).

Furthermore, it was reported that "the variation in bony processes at the bases of the fin rays is striking and has clear functional implications for flapping and rowing fins. Flappers have processes that are pronounced toward the leading edge and taper off toward the trailing edge. Rowers have pronounced processes along all rays, with special connections for the ABS and ABP that allow for control over a central pivot. Mechanical (force) advantage of fin rays can be expected to vary on a continuum between rowers and flappers; rowers should have a higher mechanical advantage, while higher performance swimmers should have a lower mechanical advantage with a concomitant increase in fin velocity advantage," (Thorsen and Westneat 2005).

Recently, a biorobotic pectoral fin was developed in the Lab of George Lauder. This model construct was used in investigations concerning the flexural rigidities and the fin's propulsive forces (see for details Tangorra et al. 2010; Gottlieb et al. 2010). Thus, briefly, "the design of the biorobotic fin was based on a detailed analysis of the pectoral fin of the bluegill sunfish (*L. macrochirus*). The biorobotic fin was made to execute the kinematics used by the biological fin during steady swimming, and to have structural properties that modeled those of the biological fin. This resulted in an engineered fin that had a similar interaction with the water as the biological fin and that created close approximations of the three-dimensional motions, flows, and forces produced by the sunfish during low speed, steady swimming. Experimental trials were conducted during which biorobotic fins of seven different stiffness configurations were flapped at frequencies from 0.5 to 2.0 Hz in flows with velocities that ranged from 0 to 270 mm/s. The results of the trials revealed that slight changes to the fin's mechanical properties or to the operating conditions can have significant impact on the direction, magnitude and time course of the propulsive forces. In general, the magnitude of the 2-D (thrust and lift) propulsive force scaled with fin ray stiffness, and increased as the fin's flapping speed increased or as the velocity of the flow decreased" (Tangorra et al. 2010).

Interestingly, searobins (*Triglidae*) and batfish (*Ogcocephalidae*) use their pectoral fins to walk on substrates (Ward 2002).

Anal Fin The anal fish fin is a single structure with location on the underside of the body just forward of the caudal fin. The function of anal fin is to stabilize the animal while it is swimming. It was reported (Standen and Lauder 2005) that "both dorsal and anal fins in bony fish *L. macrochirus* produce balancing torques during steady swimming. During maneuvers, fin area is maximized and mean lateral excursion of both fins is greater than during steady swimming, with large variation among maneuvers. Fin surface shape changes dramatically during maneuvers. At any given point in time the spanwise (base to tip) curvature along fin rays can differ between adjacent rays, suggesting that fish have a high level of control over fin surface shape. Also, during maneuvers the whole surface of both dorsal and anal fins can be bent without individual fin rays exhibiting significant curvature," (Standen and Lauder 2005). However, in case of brook trout (*Salvelinus fontinalis*) the relationship between anal and dorsal fins is more complex. As reported by Standen and Lauder (2007):

1. Anal fins produce lateral jets to the same side as dorsal fins, confirming the hypothesis that anal fins produce fluid jets that balance those produced by dorsal fins.
2. In contrast to previous work on sunfish *L. macrochirus*, neither dorsal nor anal fins produce significant thrust during steady swimming; flow leaves the dorsal and anal fins in the form of a shear layer that rolls up into vortices similar to those seen in steady swimming of eels.
3. Dorsal and anal fin lateral jets are more coincident in time than would be predicted from simple kinematic expectations; shape, heave and pitch differences between fins, and incident flow conditions may account for the differences in timing of jet shedding.
4. Relative force and torque magnitudes of the anal fin are larger than those of the dorsal fin; force differences may be due primarily to a larger span and a more squarely shaped trailing edge of the anal fin compared to the dorsal fin; torque differences are also strongly influenced by the location of each fin relative to the fish's centre of mass.
5. Flow is actively modified by dorsal and anal fins resulting in complex flow patterns surrounding the caudal fin. The caudal fin does not encounter free-stream flow, but rather moves through incident flow greatly altered by the action of dorsal and anal fins.

6. Trout anal fin function differs from dorsal fin function; although dorsal and anal fins appear to cooperate functionally, there are complex interactions between other fins and free stream perturbations that require independent dorsal and anal fin motion and torque production to maintain control of body position," (Standen and Lauder 2007).

Propulsion in some fish species is based on function of the long anal fins that are moved in an undulating manner (Snyder et al. 2007). For example, Amazonian black ghost knifefish (*Apteronotus albifrons*) possess characteristic elongated ribbon-like anal fin, the structure that is used due to its primary propulsive mechanism. As described by Shirgaonkar et al. (2008), "knifefish continuously emit a weak electric field, which is perturbed by objects that enter the field and distort the field because their electrical properties differ from the surrounding fluid. These perturbations are detected by thousands of electroreceptors on the surface of the body. MacIver and coworkers have shown that the knifefish is able to both sense and move omnidirectionally," (Shirgaonkar et al. 2008; see also Snyder et al. 2007). Furthermore, "they swim forward by sending undulations along the ribbon fin in the anterior to posterior direction. They can easily swim backward by reversing the direction of the traveling waves, and can move vertically by simultaneously sending traveling waves from the head and tail toward the center of the fin. This cancels longitudinal forces and amplifies the vertical force," (Curet et al. 2011a). These fish showed different maneuvers including rapid body rolls (MacIver et al. 2001), and even omnidirectional moving over short time intervals (Snyder et al. 2007).

The ability of this fish to switch movement direction rapidly (in $\approx$100 ms) (MacIver et al. 2001), inspired numerous experiments with the goal of developing a similar robotic device (Curet et al. 2011a, b) (Fig. 7.4).

In addition to swimming, there are other roles for the anal fin in fish. The anal fin in some fish like Japanese medaka (*Oryzias latipes*), is a typical sexual secondary

Fig. 7.4 The south american knifefish (*Apteronotus albifrons*) (**a**, Image courtesy of Per Erik Sviland) and its bioinspired robotic implementation (**b**, Reprinted from Curet et al. (2011b). © IOP Publishing. Reproduced with permission. All rights reserved)

characteristic. Moreover, experimental results showed that anal fins of this species may be a "sensitive bio-indicator for screening environmental estrogenic chemicals," (Hayashi et al. 2007).

Pelvic Fins The paired pelvic (or ventral) fins are located forward from the anal fin, and provide additional stability during swimming. In some species, pelvic fins are modified into thread-like long fins and used as a tactile organ. For example, pelvic fins by catfish (*Corydoras*) are used to hold their eggs during spawning. Fish like flying gurnards (*Dactylopteridae*) use these fins to walk on substrates (Macesic and Kajiura 2010).

In contrast to the fins discussed above, paired pelvic fin function remains poorly investigated. The first study about detailed three-dimensional movements of pelvic fins in fish was carried out by Standen (2008). The author also evaluated hypotheses for how these fins might function. Thus, "during slow-speed swimming, *[rainbow]* trout *[Onchorynchus mykiss]* moved their pelvic fins in contralateral oscillations. Each individual pelvic fin moved in two major oscillations that overlapped during the fin beat cycle (left and right fins act 180° out of phase): (1) oscillations relative to the transverse plane (initiated first); and (2) oscillations relative to the sagittal plane (initiated 120° after the transverse plane oscillations). Both the direction and timing of these oscillations suggested that pelvic fin motion is, at least partially, the result of active muscle use, and not due to body and water motion alone," (Standen 2008). "Based on kinematic analysis it appears that pelvic fin oscillation produced a series of forces. During the stroke phases when the fin was actively pushing against flow, one can assume the fin produced hydrodynamic force. Altogether, during slow-speed steady swimming in trout, pelvic fins have a complex, active motion that appears to have both a dynamic-powered and a static-trim force producing function" (Standen 2008).

Some fish species use pelvic fins for so called "*bentic locomotion*". One of the forms of bentic locomotion is "*punting*": "while keeping the rest of the body motionless, the skate's pelvic fins are planted into the substrate and then retracted caudally, which thrusts the body forward," (Macesic and Kajiura 2010; see also Koester and Spirito 2003). The benthic locomotion has been observed only within three families of Elasmobranchii:

- "The epaulette and bamboo sharks (*Hemiscyllidae*), and horn sharks (*Heterodontidae*), use their flexible pectoral and pelvic fins to walk and station-hold on the substrate," (Macesic and Kajiura 2010; see also Pridmore 1995; Compagno 1999; Goto et al. 1999; Wilga and Lauder 2001).
- "Within the batoids (rays and skates), only members of the family Rajidae, the skates, are reported to use their specialized bilobed pelvic fins, termed crura, to walk (each fin alternately) and punt (both fins synchronously) on substrates," (Macesic and Kajiura 2010; see also Lucifora and Vassallo 2002; Koester and Spirito 2003).

As reported by Macesic and Kajiura (2010), "only the clearnose skate, *Raja eglanteria*, and the lesser electric ray, *Narcine brasiliensis*, performed "true punting", in which only the pelvic fins were engaged. The skate punted significantly faster than the other species. Examination of the pelvic fin musculature revealed more specialized muscles in the true punters than in the augmented punters. This concordance of musculature with punting ability provides predictive power regarding the punting kinematics of other elasmobranchs based upon gross muscular examinations," (Macesic and Kajiura 2010).

7.1.1 *Fish Wings: Fins of Flying Fish*

The Exocoetidae (flying fish) are first appearing in the Eocene (Müller 1985). Their specific aerodynamic designs affect the maximum distance travelled in flight, which is considered to be a major factor in the evolution of these marine vertebrates as to decrease the energetic expenditure of locomotion, or as a response to predation (Fish 2010). Correspondingly, flying fish has been classified into two distinct aerodynamic designs due to distribution of their wing area (Breder 1930) as well as to lift, speed and stability, based on the wing loading and aspect ratio of the fins:

- the monoplane type (*Exocoetus*) (Fig. 7.5) possess a single set of long narrow main wings (pectoral fins) (Crossland 1911);
- the biplane type (*Cypselurus*) has under wings (pelvic fins) staggered far back from the main wings.

Fig. 7.5 Museum specimen of flying fish (*Exonautes rondeletii*) originally collected by Prof. Grimpe in 1929. Property of Senckenberg Museum, Dresden (Image courtesy Andre Ehrlich)

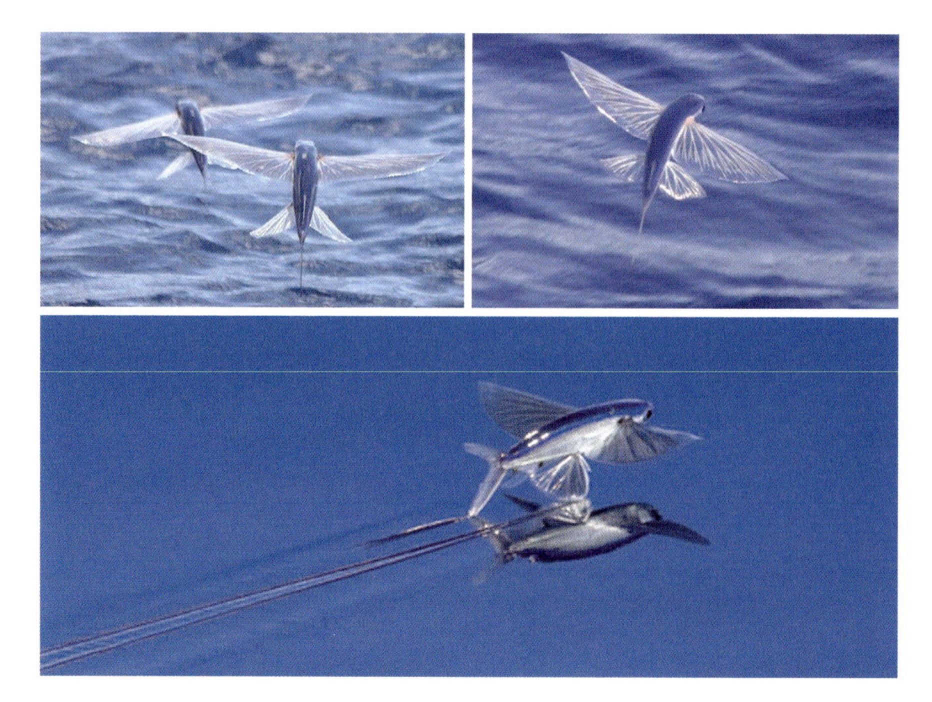

Fig. 7.6 Flying fish motion (Image courtesy www.society.ezinemark.com and Geoff Jones)

The wing-like process of the fin rays (lepidotrichia) could help in stiffening the fins when spread through it. Each lepidotrichium gave support to the previous one. There are differences in glide distances between different families of flying fish (Fig. 7.6). Thus, these distances are greater for *Cypselurus* species compared to representatives of *Exocoetus*. Here, some interesting data as follow:

- "typical flights of 15–92 m for *Cypselurus* were reported, although it was claimed that when flying with the wind these fish could traverse 400 m in a single glide!" (Fish 1990);
- "the glide path of flying fish is regarded as relatively flat *[…]*. Maximum height of the glide is 6–7 m above the water," (Fish 1990);
- "at the end of the glide when speed and altitude are decreasing, flying fish can either fold up their wings and fall back into the sea or drop their tail into the water and reaccelerate for another flight. This capacity for successive flights greatly increases the possibilities for air time," (Fish 1991–1992).

Flying behavior of Exocoetidae is the object of some modern studies. As recently reported by Park and Choi (2010), "the aerodynamic performance of flying fish is comparable to those of various bird wings, and the flying fish has some morphological characteristics in common with the aerodynamically designed modern aircrafts. The maximum lift coefficient of flying fish is measured at a = 30–35° where the flying fish is observed to emerge from the sea. The lift-to-drag ratio is largest at a = –5~0°, indicating that the best gliding performance can be achieved when the flying fish glides nearly parallel to the sea surface. As the lateral dihedral angle of the pectoral fins decreases, the lift coefficient slightly increases. In addition to the enlarged pectoral fins, the large pelvic fins have an important role in enhancing the lift-to-drag ratio and longitudinal static stability. The enhancement of the lift-to-drag ratio from the pelvic fin is attributed to the jet-like flow existing between the pectoral and pelvic fins. For both solid and water surfaces, the drag coefficient decreases and thus the lift-to-drag ratio increases as a result of the ground effect, indicating that the flying fish obtains substantial advantages by gliding close to the sea surface. The ground effect is more pronounced for the water surface, which has a slip boundary condition," (Park and Choi 2010).

7.2 Fish Fin Spines and Rays

According Alben et al. (2007), "since fins provide the structural interface between the fin muscles and the fluid environment, understanding the mechanics of fin function is an essential component of a complete analysis of how locomotor forces are transmitted by fish to the aquatic environment," (Alben et al. 2007). Therefore, fish fins as specifically designed and macrostructured formations are the most intriguing subjects for developing fish-bioinspired robotic devices. However, at the microlevel of their structural organization other biocomposite-based constructs – *spines* and *rays* – are of great interest to biological materials scientists as well as experts in bionics and biomimetics. In Osteichthyes, most fins may have rays or spines. "Fin ray structure in ray-finned fishes (Actinopterygii) largely defines fin function. Fin rays convert the muscle activity at the base of the fin to shape changes throughout the external fin web," (Taft and Taft 2012). Despite their critical functional significance, very little is known about the relationship between form and function in this key vertebrate structure. Such feature as segmentation of rays is the main difference that separates them from spines, which are usually sharp and stiff. Additionally, rays appear generally as flexible structures which may be branched. A fish fin may contain only soft rays, only spiny rays, or a combination of both.

Since the nineteenth century (Krukenberg 1885; Harrison 1893), the structure of fish fin rays was under investigations. This topic has continued to attract the attention of researchers in early twentieth century (Goodrich 1904) and continues to this day. The diversity of dermal fin rays is excellently described in the classic paper by Edwin Goodrich entitled "*On the Dermal Fin-rays of Fishes-Living and Extinct*" (Goodrich 1904). He mentioned that real dermal rays are absent in *Amphioxus* and the Cyclostomes. The ray-like structures figured and described by various authors in the larva of *Amphioxus* (Willey 1890) are elongated epidermal cells. In the Cyclostomes, on the other hand, the fins are supported by delicate cartilaginous rays, prolongations of the neural and haemal arches of the axial endoskeleton (Goodrich 1904).

Fish fin rays can be defined according to their origin and structural properties into *dermatotrichia*, *laterotrichia*, *ceratotrichia*, *camptotrichia*, *lepidotrichia*, and *actinotrichia*. Here, very brief description of these biological materials-containing structures.

Dermatotrichia. The dermal fin rays of elasmobranchs and bony fishes are known collectively as dermatotrichia (Lagler et al. 1977).

Laterotrichium a small unpaired bone rod above the first pelvic ray probably derived from a dermatotrichia in some lower Teleostei like *Esocidae*, *Cyprinidae*, *Salmonidae*, *Osmeridae*, *Cobitidae*, *Siluridae* and *Balitoridae*.

Ceratotrichia (Goodrich 1904) are large dermal fibers which are located bilaterally in sharks and are homologous with the fibrous rays called actinotrichia, which are localized at the distal ends of bony rays in teleosts. As cited in paper by Tsuchiya and Nomura (1953), the horny fibre-like Ceratotrichia, embedded in the shark fin are the only part of special human diet in China in the form of soup. Ceratotrichia showed features similar to both collagen and elastin that led Krükenberg to name "elastoidin" to the main protein within ceratotrichia (Krukenberg 1885).

Camptotrichia – has been found as specific rays which support the fin membranes in Crossopterygii and Dipnoi. It was suggested that camptotrichia may be homologous to the ceratotrichia of Elasmobranchii. As described in Geraudie and Meunier (1982):

"*Camptotrichia* constitute the only skeleton of the Dipnoan fins. Their fine structure is indeed quite different from the two other *Osteichthyan dermotrichia*, namely the actinotrichia and the lepidotrichia found simultaneously in the fin margin of both Actinopterygians and Crossopterygians fins. Although different from lepidotrichial bone, the fine structure of camptotrichia is more reminiscent of it than of the elastoidin-formed actinotrichia," (Geraudie and Meunier 1982).

In dipnoan species, *Protopterus* and *Neoceratodus*, the camptotrichia have "a dual structure: only the superficial region facing the stratified epidermis is mineralized while the deep one is made of a dense unmineralized network of collagen fibrils forming a permanent pre-osseous tissue. In the living lungfish *Neoceratodus*, the camptotrichia is made of cellular bone," (Geraudie and Meunier 1984).

Lepidotrichia are well described in the paper of Johanson et al. (2005) as "dermal elements located at the distal margin of osteichthyan fins. In sarcopterygians and actinopterygians, the term has been used to denote the most distal bony hemisegments and also the more proximal, scale-covered segments which overlie endochondral bones of the fin. In certain sarcopterygian fishes, including the *Rhizodontida*, these more proximal, basal segments are very long, extending at least half the length of the fin. The basal segments have a subcircular cross section, rather than the crescentic cross section of the distal lepidotrichial hemisegments, which lack a scale cover and form short, generally regular, elements. In rhizodonts and other sarcopterygians, e.g. *Eusthenopteron*, the basal elements are the first to appear during fin development, followed by the endochondral bones and then the distal lepidotrichia," (Johanson et al. 2005).

Two parallel and symmetrical bony demirays that form jointed segments within the fin are the main structural elements of lepidotrichia (Fig. 7.7). The demirays

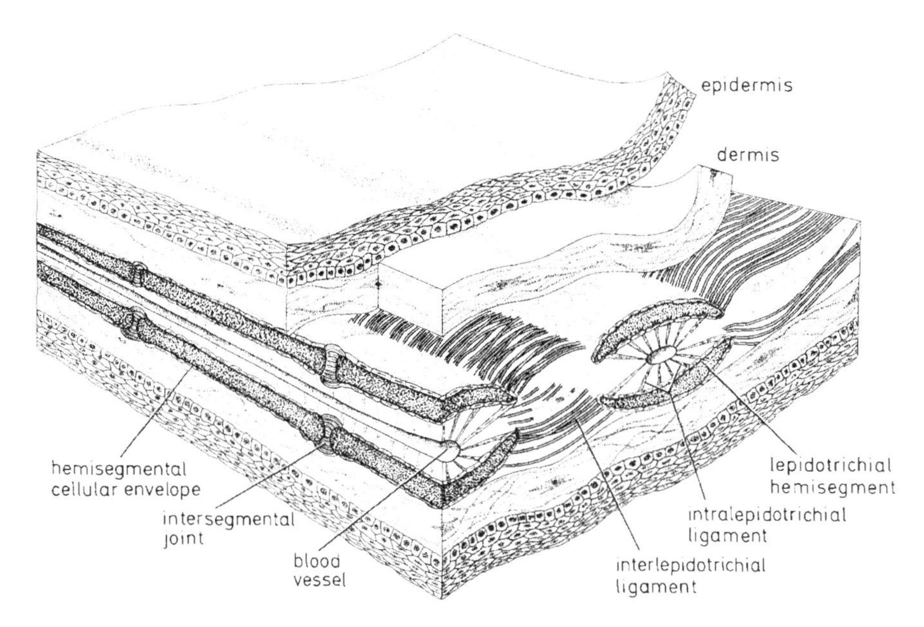

Fig. 7.7 Schematic 3D drawing of a section for a characteristic adult fin (Adapted from Becerra et al. 1983). Note that the two apposed, segmented hemirays are joined by ligaments. Each hemiray consists of skeletal material known as lepidotrichia. The whole structure is surrounded by a multilayered epidermis (With kind permission from Springer Science+Business Media: Becerra et al. (1983) © 1983, Springer)

biomineralize in a proximodistal direction within the extracellular collagenous network of the basal lamella belonging to the epidermal-dermal interface of the fin (Géraudie and Landis 1982).

According to description of Geraudie and Landis (1982), "needle- and plate-like particles of a solid mineral phase appear to be associated with the collagen fibrils and with a fine, granular, interfibrillar material central to the demirays. Cellular processes and membrane-bound vesicles are absent from the regions of calcification. During fin growth, the bony, acellular lepidotrichia are separated from the epidermal-dermal interface by infiltrating mesenchymal cells in proximal fin regions; in distal areas, the lepidotrichia remain within the basal lamella. Considerations of embryologic and structural features of fin components fail to support the hypothesis that individual segments of lepidotrichia are modified scales in all fish," (Géraudie and Landis 1982).

Recently, in experiments with developing zebrafish fins, it was shown that "the transcription factor Evx1 is expressed in the joints between individual lepidotrichia segments and at the distal tips of the lepidotrichia. It is also expressed in the apical growth zone in regenerating fins. This is a very specific phenotype as both lepidotrichia hemisegment separations and lepidotrichia bifurcations still form normally in Evx1 mutant fins, as do joints in the more proximal endoskeletal radials," (Schulte et al. 2011).

Actinotrichia are located in brush-fashioned groups at the distal tip of each ray, where they provide a flexible support to the fin edge being non-calcified fusiform spicules (Montes et al. 1982; Becerra et al. 1996). According to Géraudie and Landis (1982), the actinotrichia are "rods of elastoidin that occupy the distal margin of the fin and precede the differentiation of lepidotrichia. Once the lepidotrichia form, actinotrichia lie preferentially between their demirays. In some instances, structural interactions are suggested between actinotrichia and lepidotrichia," (Géraudie and Landis 1982). Actinotrichia serving as a scaffold for the migration into the fin fold of the mesenchymal cells that will form the future fin connective tissue (Wood 1982; Wood and Thorogood 1984). Actinotrichia are localized within a loose connective tissue with blood vessels and nerves, and surrounded by a stratified epidermis (Becerra et al. 1983). They are synthesized by the so-called *actinotrichia forming cells* (AFC), whose exact identity and origin (epidermal or mesenchymal) is still unknown (Durán et al. 2011). It is suggested that actinotrichia fulfil following functions:

- a morphogenetic role in lepidotrichia morphogenesis, possibly by inducing differentiation of scleroblasts or *lepidotrichia forming cells* (LFC) (Santamaría and Becerra 1991);
- actinotrichia might be responsible for maintaining the structural integrity of the early fin fold (Dane and Tucker 1985);
- loss of actinotrichia during evolution may have led to loss of lepidotrichia and contributed to the fin-to-limb transition (Zhang et al. 2010).

From a biological materials point of view, both the chemistry and structural biology of actinotrichia as unique fibrillar collagen-containing composites are very interesting (Fig. 7.8). Collagen-related aspects of these structures are well described in the work of Duran et al. (2011) as follow:

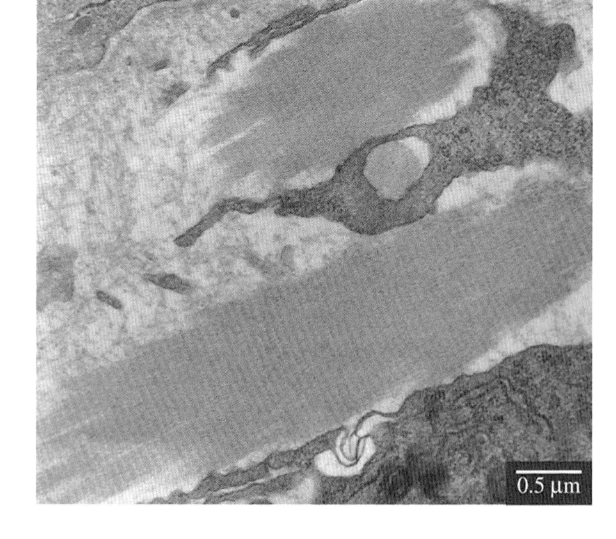

Fig. 7.8 TEM image (Adapted from Böckelmann and Bechara 2009). Observe the longitudinal section of the actinotrichium and the regular transversal striation characteristic of collagen proteins (Reprinted from Böckelmann and Bechara (2009) with permission)

"Actinotrichia are composed not by a bundle of discretely separated collagen fibrils, but rather of hyperpolymerized collagen molecules. Given the size of the actinotrichia, such a degree of polymerization would require careful post-translational processing during collagen biosynthesis and collaboration with other molecules to reach such a large size. One good candidate for this task is Lysyl hydroxylase 1 (lh1 or plod1), which has been described in cells surrounding the actinotrichia during fin development. Lysyl hydroxylases are essential for collagen biosynthesis, catalyzing the addition of hydroxyl groups to lysine residues. These hydroxylysine residues serve as attachment sites for carbohydrate chains and participate in the formation of intermolecular cross-links. In the teleostean actinotrichia, the presence of, at the very least, a collagen fraction in them was confirmed additionally by electron microscopy studies that showed the typical banding pattern of collagen in actinotrichia longitudinal sections. Though recent immunohistochemical studies have suggested actinotrichia are composed of Collagen type II, which would interact with other collagens, such as Collagen IX," (Durán et al. 2011; see also Gross and Dumsha 1958; Huang et al. 2009; Montes et al. 1982; Ramachandran 1962; Sastry and Ramachandran 1965; Schneider and Granato 2007).

Recently, the non-collagen fraction of actinotrichia has been termed as *actinodins*, which are correspondingly encoded by actinodin genes (And). These genes only present in fish lineages and, significantly, they seem to be implicated in fin/limb evolution (Zhang et al. 2010).

Recently van den Boogaart et al. (2012) demonstrated that actinotrichia contribute to increase the mass of water accelerated backward during swimming. The amount, dimensions, orientation and growth of actinotrichia were measured at various locations along the finfold in several developmental stages of common carp (*Cyprinus carpio*) and zebrafish (*Danio rerio*). Actinotrichia morphology correlated with expected lateral forces exerted on the water during swimming. The authors proposed the analytical model that predicts the extent of camber from the oblique arrangement of the actinotrichia and curvature of the body. Camber of the finfold during swimming was measured from high-speed video recordings and used to evaluate the model predictions. Based on structural requirements for swimming and strain limits for collagen, the model also predicts optimal orientations of actinotrichia (van den Boogaart et al. 2012).

7.3 Chemistry of Fish Fin: Elastoidin

Protein of the shark fin origin with properties of both collagen and elastin was termed *elastoidin* by Krukenberg (1885). He identified significant amount of sulfur within elastoidin fibers, which did not yield gelatin when boiled with water. Unlike elastin, the elastoidin fibers were resistant to enzymatic treatment with pancreatic trypsin (Kemp 1977). In some shark species, elastoidin fibers may be several millimeters wide and up to 30 cm long (Damodaran et al. 1956). They show longitudinal striations, however, are shiny and transparent, and may be brownish or yellowish, in contrast to the white ordinary connective tissue surrounding them. They are spindle-shaped, tapering to fine points at either end. Cross sections show peripheral grooves and layers of concentric lamellae (Kemp 1977). Some authors (Damodaran et al. 1956; Sastry and Ramachandran 1965; Kimura and Kubota 1966; Chandross 1982) suggested that elastoidin may be responsible for the abnormal size of ceratotrichia fibrils. It was found in elasmobranch and teleost fish species (Garrault 1936) as well as in *Latimeria* (Geraudie and Meunier 1980). According to Tsuchiya and Nomura (1953), scientist with the name Kuo-Hao Lin in 1926 analysed the yellow fish fiber in the tissue of the shark fin, and called it "substance A". The author found its nitrogen distribution and its amino acid content. These authors also proposed to change the definition "elastoidin" into "*selachin*".

Initial studies into the amino acid composition (Damodaran et al. 1956; Kimura and Kubota 1969), physico-chemical properties (Kimura and Kubota 1966), the ultrastructure (McGavin and Pyper 1964; Woodhead-Galloway and Knight 1977; Chandross and Bear 1979), as well as X-ray diffraction patterns (Ratho and Misra 1970; Hukins et al. 1976) of elastoidin have been performed. Some studies (Ramachandran 1962; Sastry and Ramachandran 1965) showed that elastoidin was "a mix of collagenous and non-collagenous proteins, but neither the particular collagen types nor the identity of the non-collagenous proteins could be determined," (Durán et al. 2011). Although the hydroxyproline content of elastoidin fibers is similar to that of collagen from teleost fish (Damodaran et al. 1956), they differed from other collagens in containing of relatively high amount of tyrosine (7.15 %) as well as cysteine (0.35 %). Gross and Dumsha (1958) reported that elastoidin fibrils contained 0.77 % carbohydrate.

Studies on molecular packing in elastoidin spicules of the spurhound *Squalus acanthias* has been carried out by Woodhead-Galloway et al. (1978). Here, the results obtained:

"Low-angle X-ray diffraction showed that, despite the well-defined regular axially projected structure, there is no long-range lateral order in the packing of molecules in native (undried) or dried elastoidin spicules from the fin rays of this fish. The equatorial intensity distribution of the X-ray diffraction pattern from native elastoidin indicates a molecular diameter of 1.1 nm and a packing fraction for the structure projected on to a plane perpendicular to the spicule (fibril) axis of 0.31 (the value for tendon is much higher, around 0.6). Density measurements support this interpretation. When the spicule dries, the packing fraction increases to 0.43, but there is still no long-range order in the structure. It was observed that the X-ray diffraction patterns provide no convincing evidence for any microfibrils or subfibrils in elastoidin. Gel electrophoresis shows that the three chains in the elastoidin molecule are identical. The low packing fraction for collagen molecules in elastoidin explains the difference in appearance between electron micrographs of negatively stained elastoidin and tendon collagen. In elastoidin, but not in tendon collagen, an appreciable proportion of the stain is able to penetrate between the collagen molecules," (Woodhead-Galloway et al. 1978).

Using controlled pepsin digestion, a new type of collagen was isolated from elastoidin of great blue shark (*Prionace glauca*). According Kimura and co-workers, "the collagen alpha chain of elastoidin, designated alpha 1(E), was very similar in electrophoretic and chromatographic behavior and amino acid composition to shark skin alpha 1(I) chain, but they were genetically-distinct on the basis of CNBr-peptide maps. The collagen molecule of elastoidin was shown to be an [alpha 1(E)]3 homotrimer," (Kimura et al. 1986).

The elastoidin fibres are responsible for some stiffness, probably, because they are packed tightly along their length on either side of the cartilaginous radials of the shark fin and they extend to the edge of the fin (Alexander 1974). The mechanical properties and fracture behaviour of dry and native elastoidin have been studied as a function of strain rate, and the plastic set behaviour of the dry elastoidin is found to be sensitive to strain rate (Arumugam and Sanjeevi 1987). The results are correlated with the scanning electron microscopy done on the fractured ends of dry and native elastoidin. Broken ends of dry elastoidin, fractured at a strain rate of 10.0 min^{-1}, appear blunt. Under the same conditions, the native specimen's ends appear sharp. The tensile properties and mode of fracture of shark elastoidin has been studied by Rajaram et al. (1981):

"Elastoidin fibres were stronger than tendon in the dry state, whereas the opposite was observed for fibres tested in the wet state. However, elastoidin was stiffer than tendon whether dry or wet. Scanning electron micrographs of the cross-sections and fractured surfaces revealed that elastoidin fibres consisted of fibrils with varying diameter arranged in a lamellar fashion. From the nature of the fractured surfaces, it could be deduced that the primary failure mechanism for elastoidin was probably through the structure fissuring," (Rajaram et al. 1981).

It is probable that elastoidin plays an important role in fish fin regeneration. As suggested by Mari-Beffa et al. (1989), "during teleostean fin regeneration the actinotrichia are immersed in the blastema, maintaining their apical position. In this epimorphic event the latter fact might be achieved by either a cellular carriage or a continuous turn-over of these hyperpolimerized fibrils. A 3H-proline pulse and radioautographic chase experiment of the isolated actinotrichia found a turn-over of collagen within the structure," (Mari-Beffa et al. 1989).

Recently, it was shown that "two zebrafish proteins actinodin 1 and 2 (And1 and And2), are essential structural components of elastoidin. The presence of actinodin sequences in several teleost fishes and in the elephant shark (*Callorhinchus milii*, which occupies a basal phylogenetic position), but not in tetrapods, suggests that these genes were lost during tetrapod species evolution. Double gene knockdown of And1 and And2 in zebrafish embryos results in the absence of actinotrichia and impaired fin folds. Gene expression profiles in embryos lacking and1 and and2 function are consistent with pectoral fin truncation, and may offer a potential explanation for the polydactyly observed in early tetrapod fossils. It was proposed that the loss of both actinodins and actinotrichia during evolution may have led to the loss of lepidotrichia, and may have contributed to the fin-to-limb transition," (Zhang et al. 2010).

7.4 Fin Regeneration and Fin Cell Culture

Contrarily to mammals, adult teleost fish have the ability to regenerate their appendages (Nakatani et al. 2007). Correspondingly, it is not surprising that studies on the role of stem cells in tissue regeneration of fish are currently fashionable (see for review Kawakami 2010).

As reviewed by Akimenko et al. (2003), a group of undifferentiated and proliferative cells termed as blastema, is the key player in regeneration (see for details Santamaría and Becerra 1991). A series of regulation mechanisms takes place between the cells that are involved in the regeneration of corresponding part of the fin. The simple structure of the Teleosts fin at the cellular level determines its fast regeneration. Moreover, each fin ray exists as independent regeneration component. These features make it an appropriate model to determine molecular mechanisms of regeneration including different multiple pathways.

The fins of actinopterygian can regenerate following amputation (Fig. 7.9). Especially the ray as the structural unit of fish fins, might regenerate independent of this appendage (see for review Nabrit 1929; Géraudie and Singer 1992; Becerra et al. 1996; Marí-Beffa et al. 1996).

This phenomenon is well described in the work of (Murciano et al. 2007) as follow:

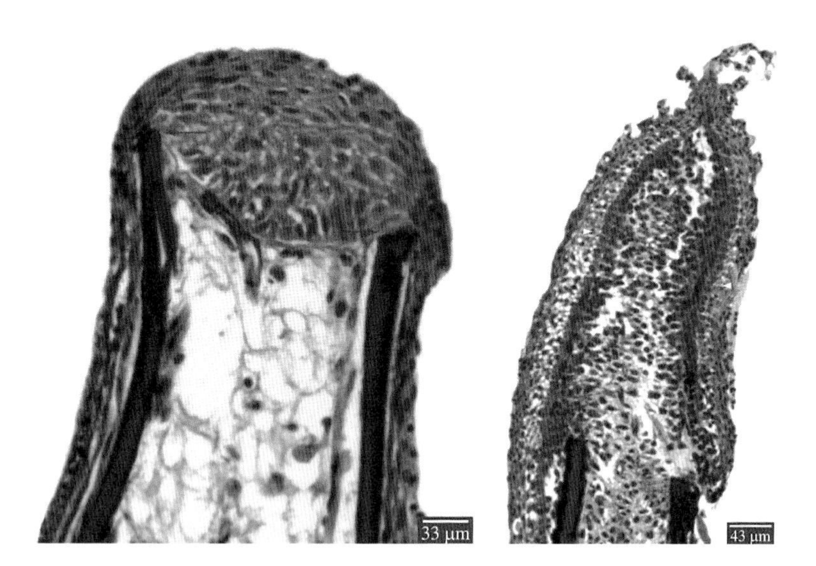

Fig. 7.9 *Left*: longitudinal section through the distal end of a regenerating tail fin of a carp fish (*Cyprinus carpio*) of control group, 24 h after amputation stained with picrosirius-hematoxylin. Observe that the regenerating epidermis has already fully covered the amputated region of the tail fin, forming the epidermal cap. *Right*: longitudinal section of the tail fin of the fish (*C. carpio*) of control group after 4 days of regeneration stained with picrosirius-hematoxylin (Reprinted from Böckelmann and Bechara (2009) with permission)

"Each fin ray is formed by two apposed contralateral hemirays. A hemiray may autonomously regenerate and segmentate in a position-independent manner. This is observed when heterotopically grafted into an interray space, after amputation following extirpation of the contralateral hemiray or when simply ablated. During this process, a proliferating hemiblastema is formed. This hemiblastema shows a pattern of gene expression for a domain similar to half ray blastema. It was suggested that there are contralateral interactions between hemiblastema of each ray, and that hemiblastema may vary its morphogenesis, always differentiating as their host region. These non-autonomous, position-dependent interactions control coordinated bifurcations, segment joints, and ray length independently," (Murciano et al. 2007).

Fin regeneration is, probably, dependent on the hedgehog and bone morphogenetic signaling pathways. It was shown, for example, that the hedgehog signalling pathway is involved in both growth of the fin regenerate, and formation and patterning of the dermal bones composing the fin rays (Avaron et al. 2006). Also Bone Morphogenic Protein signalling is required for both the growth of the regenerate, and for the differentiation of the bone-secreting cells (Smith et al. 2006).

While much progress has been made in regeneration research based on such model organisms as zebrafish and medaka, other fish species also seem to be very appropriate for these studies. For example, the plesiomorphic and derived characters found in Polypteriformes make these fish attractive subjects for performing evolutionary and developmental comparisons. In the recently published work (Cuervo et al. 2012), the authors evaluated the ability of the *Polypterus* fin to regenerate upon a quasi-complete amputation. They observed that *Polypterus* regenerates its fins with remarkable accuracy, only comparable to the regeneration observed in amphibian urodeles. Thus, "pluridisciplinary approaches led to the notion that fin regeneration is an intricate phenomenon involving epithelial-mesenchymal and reciprocal exchanges throughout the process, as well as interactions between ray and interray tissue," (Akimenko et al. 2003).

I suggest that progress in fish fin cell culture research might be important for studies on regeneration as well as on tissue engineering. However, medically important topic is still in its infancy. Development of cell lines from both freshwater and marine fish for both identifying the pathogenesis and for vaccine production in case of viral diseases, is well known imperative, and seems to be of commercial importance (Lakra et al. 2010; Han et al. 2011). According Mauger et al. (2006), "it is essential to recover the somatic cells from the fish body without sacrificing the animal, and fin explants are good candidates for this purpose. They are easy to sample and they have natural regenerative capacities. This prevents long-term disabling of the fish and it should provide a good proliferation of cells from the fin in a culture system," (Mauger et al. 2006; see also Akimenko et al. 2003; Prasanna et al. 2000).

Several papers have described methods for fin cells culture, mostly of freshwater fish (Mothersill et al. 1995) (Fig. 7.10), mostly to study a karyotype of the cells (*Esox lucius*, *Carassius auratus*, *Apogon imberbis*, and *Sparus aurata* Alvarez et al. 1991; Wang et al. 2003), as well as a cell proliferation and ageing (*C. auratus*; Shima et al. 1980). Selected cells cultured from fin explants of medaka (*Oryzias latipes*) and bighead carp (*Aristichthys nobilis*) were showed to be useful for cloning

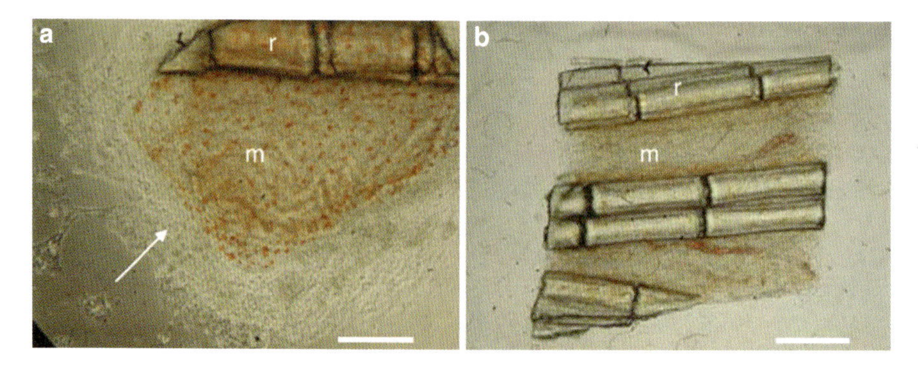

Fig. 7.10 Imagery of explants from goldfish anal fin after a 72 h cultivation at 20 °C (Adapted from Mauger et al. 2006). (**a**) Adhering explant surrounded by outgrowing cells (*arrow*)=cell-donor explant. (**b**) Loosely attached explant without outgrowing cells. *r* bony ray, *m* inter-ray membrane. *Scale bar*=0.25 mm (Reprinted from Mauger et al. (2006), Copyright (2006), with permission from Elsevier)

(Liu et al. 2002; Ju et al. 2003). Also, the new continuous cell line (KCF-1) from caudal fin of koi (*Cyprinus carpio koi*), that consists of short fibroblast-like cells, was recently developed (Dong et al. 2012).

Unfortunately, relatively speaking only a few fin cell lines were developed in marine fish. For example, from fin tissues of sheepshead (*Archosargus probatocephalus*) (Gregory et al. 1980), of gilt-head seabream (*Sparus aurata*) (Bejar et al. 1997), of a grouper (*Epinephelus coioides*) (Chi et al. 1999), of a flounder (*Paralichthys olivaceus*) (Kang et al. 2003), of a tropical grouper (*Epinephelus awoara*) (Lai et al. 2003), from caudal fin of brakish water fish *Etroplus suratensis* (Swaminathan et al. 2010), from the a tail fin (Imajoh et al. 2007) as well as from dorsal fin of red sea bream (*Pagrus major*) (Ku et al. 2010) have all been developed.

Of course, the idea to use fish fin cell lines in tissue engineering with the challenging task to develop corresponding fish-inspired robotic devices, seems to be very speculative today. Indeed, it does beg the question: what is the state of the art in fish robotics research today?

7.5 Robotic Fish-Like Devices

The diversity of form and design in fishes is determined by tremendous evolutionary success of these aquatic vertebrates (Fish 1992). Principally, the design specializations shown by fish are all still elaborations of features generic to the lower vertebrate classes (Katz 2002). During last decades, “biologists and engineers have made considerable progress in understanding how animals like fish moving underwater use their muscles to power movement. They have described body and appendage motion during propulsion, and conduct experimental and computational analyses of fluid movement and attendant forces,” (Lauder et al. 2007).

Nowadays, underwater surveillance, salvage, and military missions are research areas for the use of unmanned undersea vehicles (UUVs). The principal difference between underwater behaviour of these artificial constructs and fish is as follow:

> "The maneuverability and control of UUVs during such operations is an obvious concern. Typically, UUVs have rigid bodies, are driven using propellers, and produce their maneuvering forces with rigid control surfaces that are effective only when the flow of the water past the UUV exceeds a minimum velocity.
>
> In contrast, fish, which are remarkable in their ability to maneuver and to control their body position, have, with few exceptions, flexible bodies, and use flexible fins that are actively controlled to make the appropriate movement and assume the appropriate shape for generating the forces required by a particular situation (e.g., maneuvers, hovering, high-speed stability, and braking)," (Tangorra et al. 2007b).

Thus, to achieve success in the development of robotic constructs, principally we need not only to be inspired by the form and shape of selected fish, but use the corresponding biological or artificial materials for designing of robotic functional segments like fins or tails. "Robotics engineers are able to combine the study and the findings of biology and engineering, while a group of researchers is actively exploring the lightweight or micro-robotic fish with smart materials for actuation and locomotion," (Low 2009). In numerous studies, the robotic fins were constructed to characterize specific features of the fish fins with respect to investigate "how the fins' kinematic patterns, their spatially varying mechanical properties, and the fluid's rate of flow affected the magnitude and direction of each fin's propulsive forces," (Tangorra et al. 2011). Today, the design of smart materials to assist in studies on stiffness and in the construction of robotic fish-like devices is the interesting challenging task (Lauder et al. 2011), its realization is dependent on our knowledge in biological materials of fish origin.

7.5.1 Fish and Designing of Smart Materials

Due to their higher energy conversion efficiency and quieter operation compared to traditional motors, special material actuators have been explored for propulsion. Diverse smart material actuators include piezoelectric actuators, shape memory alloy actuators, and electroactive polymer actuators (see for review Tan et al. 2006; Aureli et al. 2010; Lauder et al. 2011). Ionic polymer-metal composites (IPMCs), including ionic polymeric gel muscles (Shahinpoor 1992) are a particularly "promising class of actuation materials for underwater robots, since they produce large bending motions under low voltages (1–2 V), work well in water and other fluids, and are flexible and biocompatible," (Tan et al. 2006; see also Kim et al. 2005). The electrical potential can determine geometry of electroactive polymer actuators. Elasticity, damage tolerance and large actuation strains make them functionally similar to biological muscles (Bar-Cohen 2001).

Tangorra and co-authors (2007) reported about development of "conducting polymer actuators based on polypyrrole are for use in biorobotic fins that are designed to create and control forces like the pectoral fin of the bluegill sunfish (*L. macrochirus*). *[...]* However, the speed of these large polymer films was slow, and must be increased if the fin's shape is to be modulated synchronously with the fin's flapping motion. Free standing linear conducting polymer films can generate large stresses and strains, but there are many engineering obstacles that must be resolved in order to create linear polymer actuators that generate simultaneously the forces, displacements, and actuation rates required by the fin. [*It was demonstrated*] two approaches that are being used to solve the engineering challenges involved in utilizing conducting polymer linear actuators: the manufacture of long, uniform ribbons of polymer and gold film, and the parallel actuation of multiple conducting polymer films," (Tangorra et al. 2007a).

George Lauder and co-workers have developed a computer controlled robotic flapping foil device that allows measurement of the effect of different flexible foil motion programs and materials on swimming speed. This construct and details of self-propelled swimming speed measurement are described in Lauder et al. (2007). These researches used so called LabView program that takes input from linear encoders on the flapping foil robotic apparatus to tune the flow tank speed so that the mean position is constant over a flapping cycle (Lauder et al. 2011; Oeffner and Lauder 2012).

As analysed by Lauder and co-authors (2011), "at present, most robotic fish models of whole fish or fish fins use a single rigid or uniformly flexible membrane to transmit force to the water. Even in the more complex robotic fish models with jointed and individually actuated fin rays, the surface conformation of the propulsor cannot be easily altered. The advent of smart materials that allow surface conformational changes in a controlled way will greatly enhance our ability to design robotic fish-like devices with performance that is closer to real animals. Thus, despite some recent significant advances in understanding the material composition and function of fish, there is still only the most general understanding of the materials that make up a fish body and fins and how these materials function during natural behaviors such as swimming," (Lauder et al. 2011; see also Shadwick and Lauder 2006; Summers and Long 2006).

7.5.2 *Fish Biorobotics*

Recently, numerous excellent works on a robot resembling real fish have been published (see for review Lauder et al. 2007; Low 2009; Kopman and Porfiri 2011; Garnier 2011; Polverino et al. 2012). Moreover, there are experiments which include a combination of both the living fish and a robot. "The integration of biomimetic robots in a fish school may enable a better understanding of collective behaviour, offering a new experimental method to test group feedback in response to behavioural modulations of its 'engineered' member," (Marras and Porfiri 2012). Recently, Marras and Porfiri (2012) analysed "a robotic fish and individual golden shiners (*Notemigonus crysoleucas*) swimming together in a water tunnel at different flow velocities. It was found that biomimetic locomotion is a determinant of fish preference as fish are more attracted towards the robot when its tail is beating rather than when it is statically immersed in the water as a 'dummy'. At specific conditions, the fish hold station behind the robot, which may be due to the hydrodynamic advantage obtained by swimming in the robot's wake. This work makes a compelling case for the need of biomimetic locomotion in promoting robot-animal interactions, and it strengthens the hypothesis that biomimetic robots can be used to study and modulate collective animal behaviour," (Marras and Porfiri 2012).

Interest in fish biorobotics is not new, and there is a long history, dating back to early experimental work using models (Houssay 1912; Breder 1926; Gray 1953). As reviewed by Alexander (1983) and Lauder et al. (2007), these investigators constructed mechanical models that allowed them to investigate power output, undulatory wave formation, and the function of the tail during fish locomotion. This work greatly increased our early understanding of how fish generate propulsive forces. Since the 1980s, biomimetics of cetacean or fish has principally focused on BCF mode with a great number of constructs, such as RoboTuna, RoboPike and VCUUV (MIT, USA) (Triantafyllou and Triantafyllou 1995; Triantafyllou et al. 2002), SPC-I/II/III (BUAA, China) (Liang et al. 2005), and Essex-G8/9 (Essex, UK) (Liu and Hu 2006). Here, I take the liberty to represent briefly only few bioinspired fish robotic systems. For more detailed information, I recommend the review by Low (2009).

(a) **Robotic manta ray** (**RoMan-II**). Manta rays (*Manta birostris*) are the largest species of ray's family (Chondrichthyes), which use median paired fins to swim. Manta rays are recognized from their diamond-shaped flat body. There are numerous examples of bioinspiration for swimming behaviour of these fish (see for recent review Fish et al. 2012). Similar to cownose, eagle and bat ray, manta ray is uses a flapping pectoral fin to swim (Rosenberger 2001). The development of the improved manta ray robot (RoMan-II), its flapping motion control, and gliding motion with buoyancy control are described by Zhou and Low (2010). The designed fish robot (Fig. 7.11) achieves an average velocity at approximately

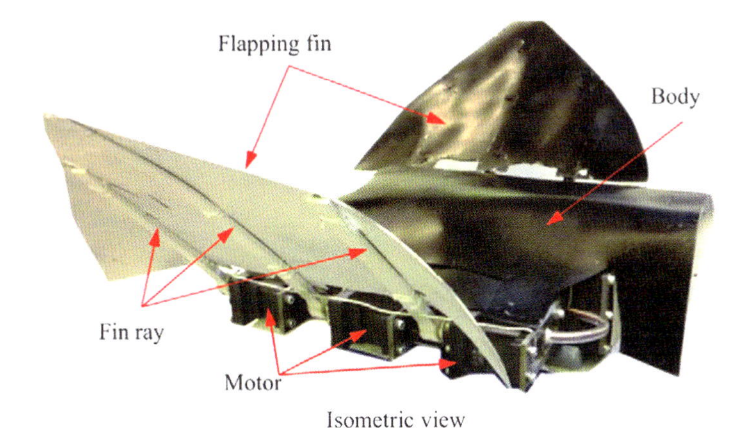

Fig. 7.11 Isometric view of the manta ray robot RoMan-II (Reprinted from Zhou and Low (2010), with permission from Elsevier. Copyright © 2010 Jilin University. Published by Elsevier Ltd All rights reserved)

one body length per second. The elastic fin ray is made of Polypropylene. Its Young's modulus is 1,350 MPa and density is 946 kg•m^{-3}. As the allowable payload capacity of the developed robot is 4 kg, the vehicle is able to carry various sensors or underwater communication equipment. The power unit can also be accommodated in the flat fish body. The electrical power is provided by six cell 3,300 mAh battery. The manta ray robot can perform the forward/backward swimming, pivot turning, and gliding with the control function. The initial testing showed that the operation time can last about 6 h for swimming in still water and 90 h for pure gliding (Zhou and Low 2010).

(b) **Aqua_Ray**. German company Festo AG proposed a remote-controlled fish with a water-hydraulic drive unit with the name "Aqua_ray" (Fig. 7.12) that development was bioinspired by the movement patterns of the manta ray. In this construct, hollow elastomer tubes with integrated woven aramide fibres serve as actuators. These tubes are similar to the fluidic muscle which can be filled with water or air, its diameter increases and it contracts longitudinally, giving rise to smoothly flowing elastic movement. The fluidic muscle from Festo, in combination with the Fin Ray Effect®, represents Aqua_ray's central propulsion and control unit.

(c) **"Cownose ray-I" robotic** fish mimic the Cownose ray (*Rhinoptera bonasus*) that belongs to the Myliobatiformes (Fig. 7.13), cartilaginous fish characterized by a dorsoventrally compressed rhombic profile and an elongated tail. Its pectoral fins are greatly enlarged and fused to the cranium, forming large, highly modified, wing-like structures. In order to gracefully propel itself, the Cownose ray moves

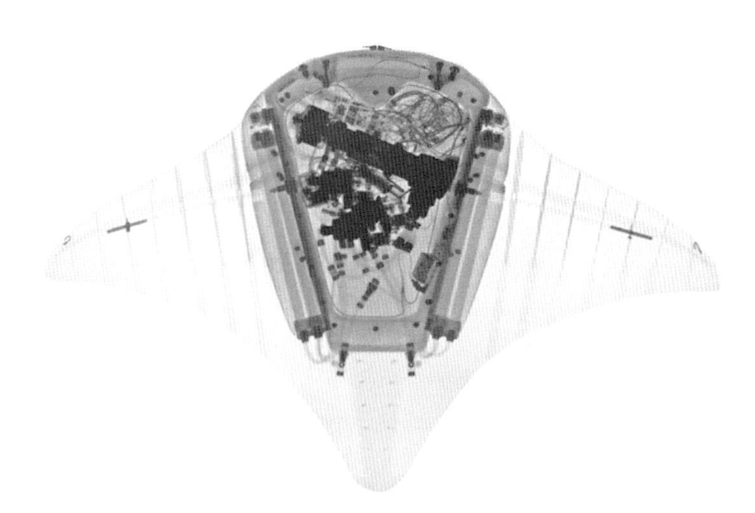

Fig. 7.12 X-ray image of the Aqua_ray (Image courtesy of Walter Fogel, Angelbachtal, Germany and Festo AG). Technical data: overall length: 61.5 cm; Dry chamber: 41.0 cm; overall width: 96.0 cm; dry chamber: 31.0 cm; height: 14.5 cm; weight: approx. 10 kg; materials: wings and tail: CURV®, water-jet carved; torso: fibreglass-reinforced plastic; skin: polyamide with elastan content; power rating: 24 V, 10 Ah; maximum speed: approx. 1.8 km/h; (See also: www.festo.com)

its fins up and down in a plane that is roughly perpendicular to the main axis of the body. The robotic fish, (Yang et al. 2008) (Fig. 7.13) based on the principle of "likeness in shape", is dorsoventrally flattened without a tail, including a main body and two pectoral fins. The body is made of aluminum, while the triangular pectoral fins are made of elastic silica gel. The length (distance from snout to posterior disc margin) of the robotic fish is 300 mm, and the span (distance between the two fin-tips when the pectoral fins are flat) is 500 mm.

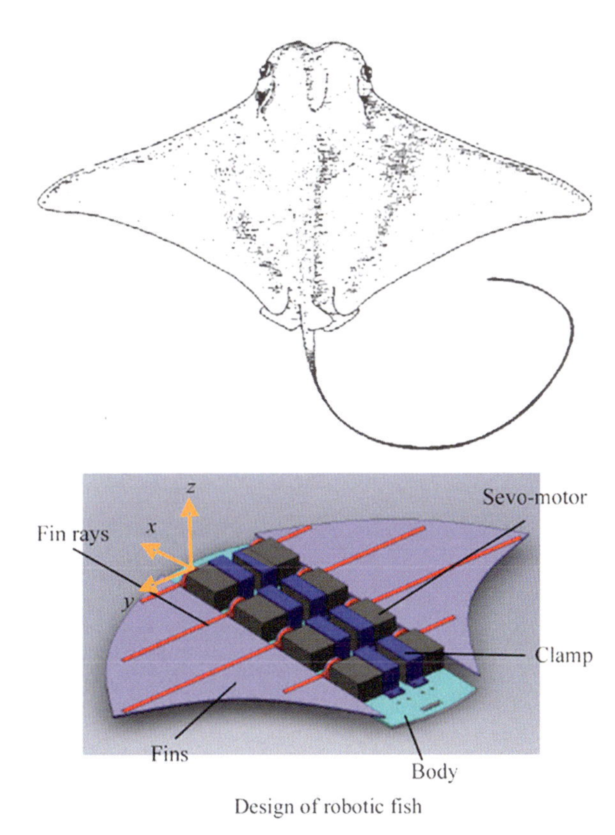

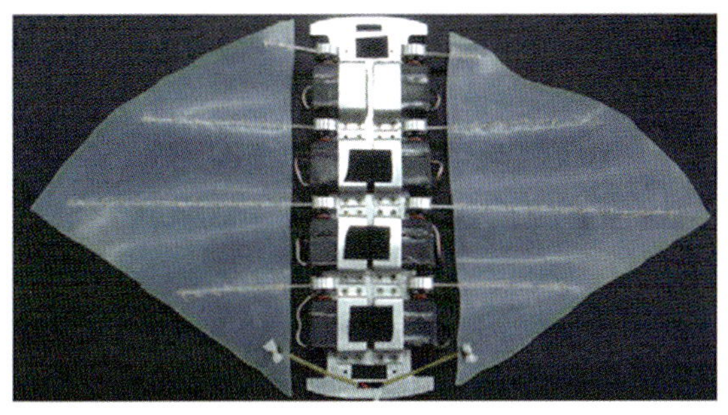

Fig. 7.13 "Cownose ray-I" robotic fish mimics the cownose ray (*Rhinoptera bonasus*) (Reprinted from Yang et al. (2009), Copyright (2009), with permission from Elsevier)

After the robotic fish is adjusted to neutral buoyancy, the total weight is about 1 kg. In the robotic fish, the propulsive waves along the pectoral fins are generated by controlling the oscillation of the fin rays. When the rigid fin rays are driven by eight eudipleural servo-motors producing a sine wave, a propulsion wave with a wave number of less than 0.5 is propagated along the fins. In the water, the robotic fish is capable of freely surging forward and backward, as well as rapidly swerving without the gyration radius (Yang et al. 2009).

(d) **SHOAL fish**. According to report by Jacob Aron (http://www.newscientist.com/article/dn21836-robotic-fish-shoal-sniffs-out-pollution-in-harbours.html), "the SHOAL fish are one and a half metres long, comparable to the size and shape of a tuna, but their neon-yellow plastic shell means they are unlikely to be mistaken for the real thing. A range of onboard chemical sensors detect lead, copper and other pollutants, along with measuring water salinity. They are driven by a dual-hinged tail capable of making tight turns that would be impossible with a propeller-driven robot.

They are also less noisy, reducing the impact on marine life. The robots are battery powered and capable of running for 8 h between charges. At the moment the researchers have to recover them by boat, but their plan is that the fish will return to a charging station by themselves," (Aron 2012).

SHOAL robots are currently being tested in the Port of Gijon in Spain. SHOAL is a collaboration between BMT Group, University of Essex, Tyndall National Institute, University of Strathclyde, Thales Safare, and the Port Authority of Gijon. To visit the SHOAL project website please go to: www.roboshoal.com

(e) **A Lamprey-Based Undulatory Robot**. This lamprey bioinspired underwater robot (Fig. 7.14) has been designed for the specific mission of mine hunting in the littoral and sub-littoral zones where it will systematically search an area to identify and locate mines and obstacles (Wilbur et al. 2002). The choice to use this animal as model for designing of underwater vessel is based on the swimming behaviour of lamprey (Currie 1999). As reported by Wilbur and co-workers,

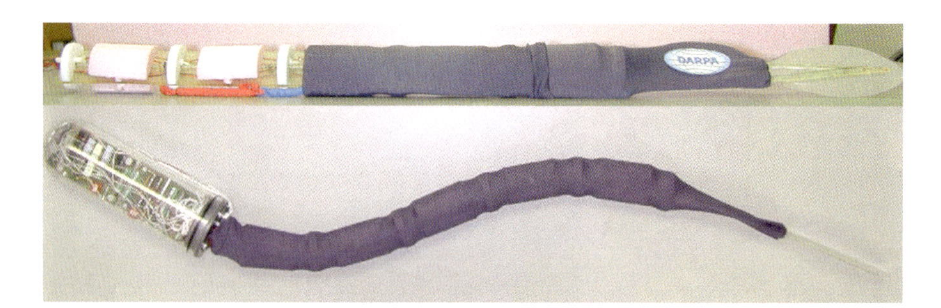

Fig. 7.14 The Undulatory Robot from nose to tail (average length 36 in.), tail of fiberglass shimstock, notochord of polyurethane, watertight electronics bay. Notochord comprised of teflon vertebrae (*white*), close-cell foam buoyancy elements between vertebrae, lycra skin, and lead shot encased in latex (Wilbur et al. (2002), figure 14.5, © 2002 Massachusetts Institute of Technology, by permission of The MIT Press)

"Lampreys swim by rhythmic lateral undulations of the body axis. The lamprey swims forward by propagating lateral axial undulations that increase in amplitude from nose to tail. Similar waves travelling from tail to nose can propel the lamprey backward. Rhythmic alteration of muscle activity on either side of the body axis produces a propulsive wave. Lamprey swimming is uncomplicated by pectoral, pelvic, or anal fins. Endurance rather than high speed is characteristic of pure anguilliform mode," (Wilbur et al. 2002). Thus, "the lamprey was selected as an animal model because of its ability to be free to move in three dimensions, its highly efficient and maneuverable form of locomotion, and its ability to navigate in currents," (Wilbur et al. 2002).

(f) **RoboGnilos** is an example of a motor-driven fin actuator. As explained by researchers, It was inspired by amiiform fish (Gymnarchus niloticus), "which generally swim by undulations of a long flexible dorsal fin," (Hu et al. 2009). "RoboGnilos consists of five parts: the base, fin rays, a membrane surface, drivers and the controller (see Fig. 7.15). The base is the mounted support of the entire device, on which mounting slots and lock holes of all joints are arranged evenly. Note that the driver's output shaft is directly connected to the swing axis. The fin rays are the backbone of the bionic undulating fin, on which there are a mounting plate, a rectangular slot, and a setting hole," (Hu et al. 2009).

(g) **Flying Fish Robot**. It's not a surprise that researches working on bioinspired fish robotics have not forgotten to mimic flying fishes. Such a robot (Fig. 7.16) was designed in Franklin W. Olin College of Engineering (Needham, MA, USA) (Project ENGR 2330: Intro to Mechanical Prototyping). The fish is made mostly aluminium sheet metal, with the tail being made of robust spring steel. The robot uses two single-motor Tamiya gearboxes powered and controlled via a wired remote control. The final prototype is fully functional and exhibits a nice biomimetic motion (see for details http://raphaelcherney.com/projects/2012.html).

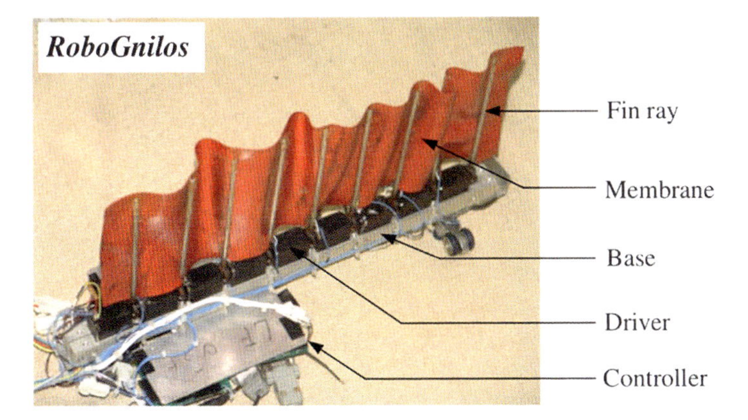

Fig. 7.15 The RoboGnilos, with nine-fin rays connected by a membrane surface, is nothing else as biomimetic undulating fin (Reprinted from Hu et al. (2009), Copyright (2009), with permission from Elsevier)

Fig. 7.16 Aluminium-made flying fish robot (Image courtesy of Raphael Cherney)

I am sure that developing of fish robotics is still a current popular topic in the literature and will be progressing in the future. For example, novel magnetic fish-robot has been recently reported (Kim et al. 2012). Researchers present "a biologically inspired fish-robot driven by a single flexible magnetic actuator with a rotating magnetic field in a three-axis Helmholtz coil. *[…]* The proposed robot can swim and perform a variety of maneuvers with the addition of pectoral fins and control of the magnetic torque direction. The robot's dynamic actuation correlates with the magnetic actuator and the rotating magnetic field. The proposed robot is also equipped with new features, such as a total of 6° of freedom, a new control method that stabilizes posture, three-dimensional swimming, a new velocity control, and new turning abilities," (Kim et al. 2012).

7.6 Conclusion

In 1958 R. B. Clark and J. B. Cowey published a paper in which they presented a simple geometric model, based on the idea of a fibre-reinforced cylinder, to explain the mechanism underlying shape changes in ribbon worms and flatworms (Clark and Cowey 1958). While their results may have been of interest to only a few biologists at that time, the essential idea of this paper, that a structure composed of inextensible fibres could accommodate large extensibility, has endured and its application has become widespread. It was first used in numerous biomechanical case studies and, more recently, in modern biomimetics and mechanical engineering. The basic model that was developed is now entrenched as a design principle in biomechanics. A fish is a cylinder only from principal point of view. As discussed above, numerous hierarchically designed biological structures like scales, skin, fins and fin rays are responsible for evolutionary success of fish as "living cylinders". Fish can swim, fly and walk on some surfaces. Unfortunately, the main structural element of fish rays – the elasmoidin – is so poorly investigated even today that we have no clear chemical definition of this structure; except that this is a hyperpolymerized fibrillar collagen with very special molecular packing that contains some non-collagenous proteins. I have no doubts that studies of highly specific and "evolutionary approved for use" swimming mechanisms of fish reported here can inspire a efficient underwater vehicle thruster design and mechanism. The achievements in fundamental science and engineering reached by Adam Summers, Frank Fish and George Lauder and their teams are amazing. However, by mimicking of the design principles of fish, we still use different artificial polymeric and metal-containing materials, foils and membranes. In this case, bioinspiration has a long way to go to discover basics of the biological materials that compose fish fins and rays.

References

Akimenko MA, Marí-Beffa M, Becerra J et al (2003) Old questions, new tools, and some answers to the mystery of fin regeneration. Dev Dyn 226(2):190–201. Copyright © 2003 Wiley-Liss, Inc. Reproduced with permission

Alben S, Madden PG, Lauder GV (2007) The mechanics of active fin-shape control in ray-finned fishes. J R Soc Interface 4(13):243–256. Copyright © 2007, The Royal Society

Alexander R McN (1974) Functional design in fishes. B.I. Publications, Bombay

Alexander RM (1983) The history of fish mechanics. In: Webb PW, Weihs D (eds) Fish biomechanics. Praeger, New York

Alvarez MC, Otis J, Amores A et al (1991) Short-term cell culture technique for obtaining chromosomes in marine and freshwater fish. J Fish Biol 39:817–824

Ann Pabst D (2000) To bend a dolphin: convergence of force transmission designs in cetaceans and scombrid fishes. Am Zool 40(1):146–155 by permission of Oxford University Press

Aron J (2012) Robotic fish shoal sniffs out pollution in harbours. New scientist: environment. Published on-line http://www.newscientist.com/article/dn21836-robotic-fish-shoal-sniffs-out-pollution-in-harbours.html. Accessed 15 May 2014. © Copyright Reed Business Information Ltd

Arumugam V, Sanjeevi R (1987) Effect of strain rate on the mode of fracture in elastoidin. J Mater Sci 22:2691–2694

Aureli M, Kopman V, Porfiri M (2010) Free-locomotion of underwater vehicles actuated by ionic polymer metal composites. IEEE/ASME Trans Mechatron 15:603–614

Avaron F, Hoffman L, Guay D et al (2006) Characterization of two new zebrafish members of the hedgehog family: a typical expression of the zebrafish Indian hedgehog gene in skeletal elements of both endochondral and dermal origins. Dev Dyn 235:478–489

Bainbridge R (1963) Caudal fin and body movement in the propulsion of some fish. J Exp Biol 40:23–56

Bar-Cohen Y (2001) Electroactive polymer (EAP) actuators as artificial muscles – reality, potential and challenges, vol PM98. SPIE Press, Bellingham

Bartol IK, Gharib M, Weihs D et al (2003) Hydrodynamic stability of swimming in ostraciid fishes: role of the carapace in the smooth trunkfish *Lactophrys triqueter* (Teleostei: Ostraciidae). J Exp Biol 206:725–744

Bartol IK, Gharib M, Webb PW et al (2005) Body-induced vortical flows: a common mechanism for self-corrective trimming control in boxfishes. J Exp Biol 208:327–344
Bartol IK, Gordon MS, Webb P et al (2008) Evidence of self-correcting spiral flows in swimming boxfishes. Bioinspir Biomim 3:014001. doi:10.1088/1748-3182/3/1/014001. © IOP Publishing. Reproduced with permission. All rights reserved
Becerra J, Montes GS, Bexiga SR et al (1983) Structure of the tail fin in teleosts. Cell Tissue Res 230:127–137
Becerra J, Junqueira LC, Bechara IJ et al (1996) Regeneration of fin rays in teleosts: a histochemical, radioautographic, and ultrastructural study. Arch Histol Cytol 59:15–35
Bejar J, Borrego JJ, Alvarez MC (1997) A continuous cell line from the cultured marine fish gilt-head seabream (*Sparus aurata*). Aquaculture 150:143–153
Böckelmann PK, Bechara IJ (2009) The regeneration of the tail fin actinotrichia of carp (*Cyprinus carpio*, Linnaeus, 1758) under the action of naproxen. Braz J Biol 69(4):1165–1172
Böckelmann PK, Bechara IJ (2010) Influence of indomethacin on the regenerative process of the tail fin of teleost: morphometric and ultrastructural analysis. Braz J Biol 70(3)(suppl)):889–897
Breder CM (1926) The locomotion of fishes. Zoological 4:159–256
Breder CM Jr (1930) On the structural specialization of flying fishes from the standpoint of aerodynamics. Copeia 1930:114–121
Chandross RJ (1982) Structure and packing of dry elastoidin: a collagen phase change. Coll Relat Res 2(4):331–348
Chandross RJ, Bear RS (1979) Comparison of mammalian collagen and elasmobranch elastoidin fiber structures, based on electron density profiles. J Mol Biol 130(3):215–219
Chi SC, Hu WW, Lo BJ (1999) Establishment and characterization of a continuous cell line (GF-1) derived from grouper. Epinephelus coioides: a cell line susceptible to grouper nervous necrosis virus (GNNV). J Fish Dis 22:173–182
Clark RB, Cowey JB (1958) Factors controlling the change of shape of certain nemertean and turbellarian worms. J Exp Biol 35:731–748
Compagno LJV (1999) Checklist of living elasmobranchs. In: Hamlett WC (ed) Sharks, skates, and rays: the biology of elasmobranch fishes. John Hopkins University Press, Maryland
Crossland C (1911) The flight of Exocoetus. Nat Lond 86:279–280
Cuervo R, Hernández-Martínez R, Chimal-Monroy J et al (2012) Full regeneration of the tribasal Polypterus fin. PNAS 109(10):3838–3843
Curet OM, Patankar NA, Lauder GV et al (2011a) Aquatic manoeuvering with counter-propagating waves: a novel locomotive strategy. J R Soc Interface 8(60):1041–1050
Curet OM, Patankar NA, Lauder GV et al (2011b) Mechanical properties of a bio-inspired robotic knifefish with an undulatory propulsor. Bioinspir Biomim 6(2):026004. doi:10.1088/1748-3182/6/2/026004. Copyright © 2011 IOP Publishing Ltd. Reprinted with permission
Currie S (1999) Reverse engineering of lamprey-based undulatory AUV. In: Biomimetic underwater robot program progress report Y01Q1-3
Damodaran M, Sivaraman C, Dhavalikar RS (1956) Amino acid composition of elastoidin. Biochem J 62(4):621–625
Dane PJ, Tucker JB (1985) Modulation of epidermal cell shaping and extracellular matrix during caudal fin morphogenesis in the zebra fish *Brachydanio rerio*. J Embryol Exp Morphol 87:145–161
Dong C, Weng S, Li W et al (2012) Characterization of a new cell line from caudal fin of koi, *Cyprinus carpio koi*, and first isolation of cyprinid herpesvirus 3 in China. Virus Res 161(2):140–149
Durán et al (2011) Reprinted from Durán I, Marí-Beffa M, Santamaría JA et al (2011) Actinotrichia collagens and their role in fin formation. Dev Biol 354(1):160–72. Copyright © 2011, with permission from Elsevier
Esposito et al (2012) Republished with permission of Company of Biologists Ltd., from Esposito CJ, Tangorra JL, Flammang BE et al (2012) A robotic fish caudal fin: effects of stiffness and motor program on locomotor performance. J Exp Biol 215(Pt 1):56–67, copyright (2012) permission conveyed through Copyright Clearance Center, Inc

Fertl D, Landry AM Jr (1999) Sharksucker (*Echeneis naucrates*) on a bottlenose dolphin (*Tursiops truncatus*) and a review of other cetacean-remora associations. Mar Mamm Sci 15:859–863
Fish FE (1990) Wing design and scaling of flying fish with regard to flight performance. J Zool 221:391–403. Copyright © 2010 Wiley-Liss, Inc. Reproduced with permission
Fish FE (1992) On a fin and a prayer. Scholars 3(1):4–7
Fish FE (1999) Energetics of swimming and flying in formation. Comm Theor Biol 5:283–304
Fish FE (2010) Reproduced with permission of Science Publishers from Fish FE (2010) Chapter 4. Swimming strategies for energy economy. In: Domenici P, Kapoor BG (eds) Fish locomotion an eco-ethological perspective, pp 90–122 Science Publishers, Enfield, p 102. Permission conveyed through Copyright Clearance Center, Inc
Fish FE, Haj-Hariri H, Smits AJ et al (2012) Biomimetic swimmer inspired by the manta ray, Chapter 17. In: Bar-Cohen Y (ed) Biomimetics: nature-based innovation. CRC Press, Boca Rotan
Garnier S (2011) From ants to robots and back: how robotics can contribute to the study of collective animal behavior. In: Meng Y, Jin Y (eds) Bio-inspired self-organizing robotic systems. Springer, Berlin/Heidelberg, pp 105–120
Garrault AF (1936) Développment des fibres d'elastoidine (actinotrichia) chez les salmonides. Arch Anat Microsc Morphol Esp 32:105–137
Géraudie J, Landis WJ (1982) The fine structure of the developing pelvic fin dermal skeleton in the trout Salmo gairdneri. Am J Anat 163(2):141–156. Copyright © 1982 Wiley-Liss, Inc. Reproduced with permission
Geraudie J, Meunier FJ (1980) Elastoidin actinotrichia in Coelacanth fins: a comparison with teleosts. Tissue Cell 12(4):637–645
Geraudie J, Meunier FJ (1982) Comparative fine structure of the *Osteichthyan dermotrichia*. Anat Rec 202(3):325–328. Copyright © 1982 Wiley-Liss, Inc. Reproduced with permission
Geraudie J, Meunier FJ (1984) Reproduced from Geraudie J, Meunier FJ (1984) Structure and comparative morphology of camptotrichia of lungfish fins. Tissue Cell 16(2):217–236. Copyright © 1984, with permission from Elsevier
Géraudie J, Singer M (1992) The fish fin regenerated. In: Taban CH, Boilly B (eds) Keys for regeneration. Karger, Basel
Goodrich ES (1904) On the dermal fin-rays of fishes living and extinct. Q J Microsc Sci 47:465–522
Gordon MS, Hove JR, Webb PW et al (2001) Boxfishes as unusually well controlled autonomous underwater vehicles. Physiol Biochem Zool 73:663–671
Gosline WA (1971) Functional morphology and classification of teleostean fishes. University of Hawaii Press, Honolulu
Goto T, Nishida K, Nakaya K (1999) Internal morphology and function of paired fins in the epaulette shark, *Hemiscyllium ocellatum*. Ichthyol Res 46:281–287
Gottlieb JR et al (2010) Reprinted from Gottlieb JR, Tangorra JL, Esposito CJ et al (2010) A biologically derived pectoral fin for yaw turn manoeuvres. Appl Bionics Biomech 7(1):41–55. Copyright (2010) with permission from IOS Press
Gray J (1953) How animals move. Cambridge University Press, Cambridge
Gregory PE, Howard-Peebles PN, Ellender RD, Martin BJ (1980) Analysis of a marine fish cell line from a male sheepshead. J Hered 71(3):209–211
Gross J, Dumsha B (1958) Elastoidin: a two component member of the collagen class. Biochim Biophys Acta 28(2):268–270
Guerrero-Ruis M, Urbán JR (2000) First report of remoras on killer whales (*Orcinus orca*) in the Gulf of California, Mexico. Aquat Mamm 26(2):148–150
Hale ME et al (2006) Republished with permission of Company of Biologists Ltd., from Hale ME, Day RD, Thorsen DH et al (2006) Pectoral fin coordination and gait transitions in steadily swimming juvenile reef fishes. J Exp Biol 209:3708–3718, copyright (2006) permission conveyed through Copyright Clearance Center, Inc
Han JE, Choresca CH Jr, Koo OJ et al (2011) Establishment of glass catfish (*Kryptopterus bicirrhis*) fin-derived cells. Cell Biol Int Rep (2010) 18(1):e00008. doi:10.1042/CBR20110002

Harrison RJ (1893) Ueber die Entwicklung der nicht knoerpelig vorgebildeten Skeletteile in den Flossen der Teleostier. Arch Mikrosk Anat 42:248–278

Hayashi H, Nishimoto A, Oshima N et al (2007) Expression of the estrogen receptor alpha gene in the anal fin of Japanese medaka, *Oryzias latipes*, by environmental concentrations of bisphenol A. J Toxicol Sci 32(1):91–96. Copyright © 2007 The Japanese Society of Toxicology

Houssay F (1912) Forme, puissance et stabilité des poissons. Herman, Paris

Hove JR, O'Bryan LM, Gordon MS et al (2001) Boxfishes (Teleostei: Ostraciidae) as a model system for fishes swimming with many fins: kinematics. J Exp Biol 204:1459–1471

Hu T et al (2009) Reprinted from Hu T, Shen L, Lin L, Xu H (2009) Biological inspirations, kinematics modeling, mechanism design and experiments on an undulating robotic fin inspired by *Gymnarchus niloticus*. Mech Mach Theory 44(3):633–645. Copyright (2009), with permission from Elsevier

Huang CC, Wang TCB-H, Lin BH et al (2009) Collagen IX is required for the integrity of collagen II fibrils and the regulation of vascular plexus formation in zebrafish caudal fins. Dev Biol 332:360–370

Hukins DW, Woodhead-Galloway J, Knight DP (1976) Molecular tilting in dried elastoidin and its implications for the structures of other collagen fibrils. Biochem Biophys Res Commun 73(4):1049–1055

Imajoh M, Ikawa T, Oshima S (2007) Characterization of a new fibroblast cell line from a tail fin of red sea bream, *Pagrus major*, and phylogenetic relationships of a recent RSIV isolate in Japan. Virus Res 126(1–2):45–52

Johanson Z, Burrow C, Warren A et al (2005) Homology of fin lepidotrichia in osteichthyan fishes. Lethaia 38:27–36. Copyright © 2007, John Wiley and Sons. Reproduced with permission

Ju B, Pristyazhnyuk I, Ladygina T et al (2003) Development and gene expression of nuclear transplants generated by transplantation of cultured cell nuclei into non-enucleated eggs in the medaka *Oryzias latipes*. Dev Growth Differ 45:167–174

Kang MS, Oh MJ, Kim YJ, Jung J et al (2003) Establishment and characterization of two cell lines derived from flounder, *Paralichthys olivaceus* (Temminck & Schlegel). J Fish Dis 26:657–665

Kato N, Liu H, Morikawa H (2005) Biology-inspired precision maneuvering of underwater vehicles. Int J Offshore Polar Eng 15:81–87

Katz SL (2002) Design of heterothermic muscle in fish. J Exp Biol 205:2251–2266

Kawakami A (2010) Stem cell system in tissue regeneration in fish. Dev Growth Differ 52(1):77–87

Kemp NE (1977) Banding pattern and fibrillogenesis of ceratotrichia in shark fins. J Morphol 154:187–204

Kim B, Kim DH, Jung J et al (2005) A biomimetic undulatory tadpole robot using ionic polymer-metal composite actuators. Smart Mater Struct 14:1579–1585

Kim SH, Shin K, Hashi S et al (2012) Magnetic fish-robot based on multi-motion control of a flexible magnetic actuator. Bioinspir Biomim 7(3):036007. doi:10.1088/1748-3182/7/3/036007. Copyright © 2012 IOP Publishing. Reprinted with permission

Kimura S, Kubota M (1966) Studies on elastoidin. I. Some chemical and physical properties of elastoidin and its components. J Biochem 60(6):615–621

Kimura S, Kubota M (1969) Tyrosine derivatives in a structural protein, elastoidin. J Biochem 65(1):141–143

Kimura S et al (1986) Reproduced from Kimura S, Uematsu Y, Miyauchi Y (1986) Shark (*Prionace glauca*) elastoidin: characterization of its collagen as [alpha 1(E)]3 homotrimers. Comp Biochem Physiol B 84(3):305–308. Copyright © 1986, with permission from Elsevier

Kodati P (2006) Biomimetic micro underwater vehicle with ostraciiform locomotion: system design, analysis and experiments. Master's thesis, University of Delaware, Newark

Koester DM, Spirito CP (2003) Punting: an unusual mode of locomotion in the little skate, *Leucoraja erinacea* (Chondrichthyes: Rajidae). Copeia 3:553–561

Koob TJ, Long JH Jr (2000) The vertebrate body axis: evolution and mechanical function. Am Zool 40(1):1–018, by permission of Oxford University Press

Kopman V, Porfiri M (2011) A miniature and low-cost robotic fish for ethorobotics research and engineering education I: bioinspired design ASME dynamic systems and control conference, Arlington, pp 209–216

Krukenberg C (1885) Über die chemische beschaffenheit der sog. hornfäden von mustelus und über die zusammensetzung der keratinösen hüllen um die eier von *Scyllium stellate*. Mittheilungen Zool Stat Neapel 6:286–296
Ku CC, Lu CH, Wang CS (2010) Establishment and characterization of a fibroblast cell line derived from the dorsal fin of red sea bream, *Pagrus major* (Temminck & Schlegel). J Fish Dis 33(3):187–196
Lagler KF, Bardach JE, Miller RR et al (1977) Ichthyology, 2nd edn. Wiley, New York
Lai YS, John JA, Lin CH (2003) Establishment of cell lines from a tropical grouper, *Epinephelus awoara* (Temminck & Schlegel), and their susceptibility to grouper irido- and nodaviruses. J Fish Dis 26(1):31–42
Lakra WS, Swaminathan TR, Rathore G et al (2010) Development and characterization of three new diploid cell lines from *Labeo rohita* (Ham.). Biotechnol Prog 26(4):1008–1013
Latimer-Needham CH (1951) Flying-fish aerodynamics. Flight 26:535–536
Lauder GV (1989) Caudal fin locomotion in ray-finned fishes: historical and functional analyses. Am Zool 29(1):85–102, by permission of Oxford University Press
Lauder GV (2006) Locomotion. In: Evans DH, Claiborne JB (eds) The physiology of fishes, 3rd edn. CRC Press, Boca Raton, pp 3–46
Lauder GV, Drucker EG (2004) Morphology and experimental hydrodynamics of fish fin control surfaces. IEEE J Ocean Eng 29:556–571
Lauder GV, Madden PGA (2006) Learning from fish: kinematics and experimental hydrodynamics for roboticists. Int J Autom Comput 4:325–335
Lauder GV et al (2007) Republished with permission of Company of Biologists Ltd., from Lauder GV, Anderson EJ, Tangorra J, Madden PG (2007) Fish biorobotics: kinematics and hydrodynamics of self-propulsion. J Exp Biol 210(Pt 16):2767–2780. Copyright (2007) permission conveyed through Copyright Clearance Center, Inc
Lauder GV, Madden PGA, Tangorra JL et al (2011) Bioinspiration from fish for smart material design and function. Smart Mater Struct 20:094014. doi:10.1088/0964-1726/20/9/094014. Copyright © 2011 IOP Publishing. Reprinted with permission. All rights reserved
Lauder GV, Flammang B, Alben S (2012) Passive robotic models of propulsion by the bodies and caudal fins of fish. Integr Comp Biol 52(5):576–587. doi:10.1093/icb/ics096 by permission of Oxford University Press
Liang J, Zou D, Wang S, Wang Y (2005) Trial voyage of SPC-II fish robot. J Beijing Univ Aeronaut Astronaut 31(7):709–713
Lingham-Soliar T (2005a) Dorsal fin in the white shark, *Carcharodon carcharias*: a dynamic stabilizer for fast swimming. J Morphol 263(1):1–11. Copyright © 2004 Wiley-Liss, Inc. Reprinted with permission
Lingham-Soliar T (2005b) Caudal fin in the white shark, *Carcharodon carcharias* (Lamnidae): a dynamic propeller for fast, efficient swimming. J Morphol 264:233–252. Copyright © 2005 Wiley-Liss, Inc. Reprinted with permission
Lingham-Soliar T (2005c) With kind permission from Springer Science+Business Media: Lingham-Soliar T (2005c) Caudal fin allometry in the white shark *Carcharodon carcharias*: implications for locomotory performance and ecology. Naturwissenschaften 92(5):231–236. Copyright © 2005, Springer-Verlag
Liu J, Hu H (2006) Biologically inspired behaviour design for autonomous robotic fish. Int J Autom Comput 3(4):336–347
Liu TM, Yu XM, Ye YZ et al (2002) Factors affecting the efficiency of somatic cell nuclear transplantation in the fish embryo. J Exp Zool 293:719–725
Low (2009) Reproduced from Low KH (2009) Modelling and parametric study of modular undulating fin rays for fish robots. Mech Mach Theory 44(3):615–632, Copyright © 2009, with permission from Elsevier
Lucifora LO, Vassallo AI (2002) Walking in skates (Chondrichthyes. Rajidae): anatomy, behaviour and analogies to tetrapod locomotion. Biol J Linn Soc 77:35–41
Macesic LJ, Kajiura SM (2010) Comparative punting kinematics and pelvic fin musculature of benthic batoids. J Morphol 271(10):1219–1228. Copyright © 2010 Wiley-Liss, Inc. Reproduced with permission

MacIver MA, Sharabash NM, Nelson ME (2001) Prey-capture behavior in gymnotid electric fish: motion analysis and effects of water conductivity. J Exp Biol 204:543–557
Marí-Beffa M et al (1989) With kind permission from Springer Science+Business Media: Marí-Beffa M, Carmona MC, Becerra J (1989) Elastoidin turn-over during tail fin regeneration in teleosts. A morphometric and radioautographic study. Anat Embryol (Berl) 180(5):465–70. Copyright © 1989, Springer-Verlag
Marí-Beffa M, Mateos I, Palmqvist P et al (1996) Cell to cell interactions during teleosts fin regeneration. Int J Dev Biol Suppl 1:179S–180S
Marras S, Porfiri M (2012) Fish and robots swimming together: attraction towards the robot demands biomimetic locomotion. J R Soc Interface 9(73):1856–1868. doi:10.1098/rsif.2012.0084 by permission of the Royal Society
Mattheijssens J, Marcel JP, Bosschaerts W et al (2012) Oscillating foils for ship propulsion. In: 9th National Congress on Theoretical and Applied Mechanics, Brussels, 9-10-11 May 2012
Mauger et al (2006) Reproduced from Mauger P-E, Le Bail P-Y, Labbé C (2006) Cryobanking of fish somatic cells: optimizations of fin explant culture and fin cell cryopreservation. Comp Biochem Physiol B: Biochem Mol Biol 144(1):29–37. Copyright 2006, with permission from Elsevier
McGavin S, Pyper AS (1964) An electron microscope study of elastoidin. Biochim Biophys Acta 79:600–605
Mittal R (2004) Computational modeling in biohydrodynamics: trends, challenges, and recent advances. IEEE J Ocean Eng 29:595–604
Montes GS, Becerra J, Toledo OM et al (1982) Fine structure and histochemistry of the tail fin ray in teleosts. Histochemistry 75:363–376
Mothersill C, Lyng F, Lyons M et al (1995) Growth and differentiation of epidermal cells from the rainbow trout established as explants and maintained in various media. J Fish Biol 46:1011–1025
Müller AH (1985) Lehrbuch der Paläozoologie. Teil 1. 2. Auflage. Band 3, VEB Custav Fischer Verlag Jena, 655 pp
Murciano C et al (2007) Reproduced from Murciano C, Pérez-Claros J, Smith A et al (2007) Position dependence of hemiray morphogenesis during tail fin regeneration in *Danio rerio*. Dev Biol 312(1):272–283. Copyright © 2007, with permission from Elsevier
Nabrit SM (1929) The role of the rays in the regeneration in the tail-fins of fishes (in Fundulus and goldfish). Biol Bull 56:235–266
Nakatani Y, Kawakami A, Kudo A (2007) Cellular and molecular processes of regeneration, with special emphasis on fish fins. Dev Growth Differ 49(2):145–154
Nelson JS (1994) Fishes of the world, 3rd edn. Wiley, New York
O'Toole B (2002) Phylogeny of the species of the superfamily Echeneoidea (Perciformes: Carangoidei: Echeneidae, Rachycentridae, and Coryphaenidae), with an interpretation of echeneid hitchhiking behaviour. Can J Zool 80:596–623
Oeffner J, Lauder GV (2012) The hydrodynamic function of shark skin and two biomimetic applications. J Exp Biol 215:785–795
Park H, Choi H (2010) Republished with permission of Company of Biologists Ltd., from Park H, Choi H (2010) Aerodynamic characteristics of flying fish in gliding flight. J Exp Biol 213:3269–3279. Copyright (2010) permission conveyed through Copyright Clearance Center, Inc
Polverino G, Abaid N, Kopman V, Macrì S, Porfiri M (2012) Zebrafish response to robotic fish: preference experiments on isolated individuals and small shoals. Bioinspir Biomim 7(3):036019. doi:10.1088/1748-3182/7/3/036019
Prasanna I, Lakra WS, Ogale SN et al (2000) Cell culture from fin explant of endangered golden masheer, *Tor putitora* (Hamilton). Curr Sci 79:93–95
Pridmore PA (1995) Submerged walking in the epaulette shark *Hemiscillium ocellatum* (Hemiscyllidae) and its implications for locomotion in rhipidistian fishes and early tetrapods. Zool Anal Complex Syst 98:278–297
Rajaram A et al (1981) With kind permission from Springer Science+Business Media: Rajaram A, Sanjeevi R, Ramanathan N (1981) The tensile properties and mode of fracture of elastoidin. J Biosci 3(3):303–309. Copyright © 1981, Indian Academy of Sciences

Ramachandran LK (1962) Elastoidin–a mixture of three proteins. Biochem Biophys Res Commun 6:443–448
Ratho T, Misra T (1970) Estimation of parameters of elastoidin by low-angle X-ray method. Colloid Polym Sci 239:574–577
Rosenberger LJ (2001) Pectoral fin locomotion in batoid fishes: *Undulation versus* oscillation. J Exp Biol 204:379–394
Santamaría JA, Becerra J (1991) Tail fin regeneration in teleosts: cell-extracellular matrix interaction in blastemal differentiation. J Anat 176:9–21
Sastry LV, Ramachandran LK (1965) The protein components of elastoidin. Biochim Biophys Acta 97:281–287
Sazima I, Grossman A (2006) Turtle riders: remoras on marine turtles in southwest Atlantic. Neotropical Ichthyol 4:123–126
Schneider VA, Granato M (2007) Genomic structure and embryonic expression of zebrafish lysyl hydroxylase 1 and lysyl hydroxylase 2. Matrix Biol 26:12–19
Schulte CJ, Allen C, England SJ et al (2011) Evx1 is required for joint formation in zebrafish fin dermoskeleton. Dev Dyn 240(5):1240–1248. Copyright © 2011 Wiley-Liss, Inc. Reproduced with permission
Sfakiotakis M, Lane DM, Davies JBC (1999) Review of fish swimming modes for aquatic locomotion. IEEE J Ocean Eng 24:237–252
Shadwick RE, Lauder GV (eds) (2006) Fish physiology vol 23: fish biomechanics. Academic, San Diego
Shahinpoor M (1992) Conceptual design, kinematics and dynamics of swimming robotic structures using ionic polymeric gel muscles. Smart Mater Struct 1:91–94
Shima A, Nikaido O, Shinohara S et al (1980) Continued in vitro growth of fibroblast-like cells (RBCF-1) derived from the caudal fin of the fish, *Carassius auratus*. Exp Gerontol 15:305–314
Shirgaonkar AA et al (2008) Republished with permission of Company of Biologists Ltd., from Shirgaonkar AA, Curet OM, Patankar NA et al (2008) The hydrodynamics of ribbon-fin propulsion during impulsive motion. J Exp Biol 211:3490–3503. Copyright (2008) permission conveyed through Copyright Clearance Center, Inc
Sitorus PE et al (2009) Reprinted from Sitorus PE, Nazaruddin YY, Leksono E et al (2009) Design and implementation of paired pectoral fins locomotion of labriform fish applied to a fish robot. J Bionic Eng 6(1):37–45. Copyright (2009), with permission from Elsevier)
Smith A, Avaron F, Guay D et al (2006) Inhibition of BMP signaling during zebrafish fin regeneration disrupts fin growth and scleroblast differentiation and function. Dev Biol 299:438–454
Snyder JB, Nelson ME, Burdick JW et al (2007) Omnidirectional sensory and motor volumes in an electric fish. PLoS Biol 5:2671–2683
Standen EM (2008) Republished with permission of Company of Biologists Ltd., from Standen EM (2008) Pelvic fin locomotor function in fishes: three-dimensional kinematics in rainbow trout (*Oncorhynchus mykiss*). J Exp Biol 211:2931–2942. copyright (2008) permission conveyed through Copyright Clearance Center, Inc
Standen EM, Lauder GV (2005) Republished with permission of Company of Biologists Ltd., from Standen EM, Lauder GV (2005) Dorsal and anal fin function in bluegill sunfish *Lepomis macrochirus*: three-dimensional kinematics during propulsion and maneuvering. J Exp Biol 208(Pt 14):2753–2763. Copyright (2005) permission conveyed through Copyright Clearance Center, Inc
Standen EM, Lauder GV (2007) Republished with permission of Company of Biologists Ltd., from Standen EM and Lauder GV (2007) Hydrodynamic function of dorsal and anal fins in brook trout (*Salvelinus fontinalis*). J Exp Biol 210(Pt 2):325–339, copyright (2007) permission conveyed through Copyright Clearance Center, Inc
Steffensen JF, Lomholt JP (1983) Energetic cost of active branchial ventilation in the sharksucker, *Echeneis naucrates*. J Exp Biol 103:185–192
Summers A, Long J (2006) Skin and bones, sinew and gristle: the mechanical behavior of fish skeletal tissues. In: Shadwick RE, Lauder GV (eds) Fish physiology vol 23: fish biomechanics. Academic, San Diego

Suzuki T, Haga Y, Takeuchi T et al (2003) Differentiation of chondrocytes and scleroblasts during dorsal fin skeletogenesis in flounder larvae. Dev Growth Differ 45(5–6):435–448. Copyright © 2004, John Wiley and Sons. Reprinted with permission

Swaminathan TR, Lakra WS, Gopalakrishnan A et al (2010) Development and characterization of a new epithelial cell line PSF from caudal fin of Green chromide, *Etroplus suratensis* (Bloch, 1790). In Vitro Cell Dev Biol Anim 46(8):647–656

Taft NK, Taft BN (2012) Republished with permission of Company of Biologists Ltd., from Taft NK, Taft BN (2012) Functional implications of morphological specializations among the pectoral fin rays of the benthic longhorn sculpin. J Exp Biol 215(Pt 15):2703–2710. Copyright (2012) permission conveyed through Copyright Clearance Center, Inc

Tan X, Kim D, Usher N et al (2006) An autonomous robotic fish for mobile sensing. In: Proceedings of the 2006 IEEE/RSJ international conference on intelligent robots and systems October 9–15, 2006, Beijing, pp 5424–5429. doi:10.1109/IROS.2006.282110. © 2006 IEEE. Reprinted, with permission

Tangorra J, Anquetil P, Fofonoff T et al (2007a) The application of conducting polymers to a biorobotic fin propulsor. Bioinspir Biomim 2:S6–S17. Copyright © 2007 IOP Publishing. Reprinted with permission. All rights reserved

Tangorra JL, Davidson SN, Hunter I et al (2007b) The development of a biologically inspired propulsor for unmanned underwater vehicles. IEEE J Ocean Eng 32(3):533–550. © 2007 IEEE. Reprinted, with permission

Tangorra JL et al (2010) Republished with permission of Company of Biologists Ltd., from Tangorra JL, Lauder GV, Hunter IW, Mittal R, Madden PG, Bozkurttas M (2010) The effect of fin ray flexural rigidity on the propulsive forces generated by a biorobotic fish pectoral fin. J Exp Biol 213(Pt 23):4043–4054. Copyright (2010); permission conveyed through Copyright Clearance Center, Inc

Tangorra J, Phelan C, Esposito C et al (2011) Use of biorobotic models of highly deformable fins for studying the mechanics and control of fin forces in fishes. Integr Comp Biol 51(1):176–189, by permission of Oxford University Press

Thorsen DH, Westneat MW (2005) Diversity of pectoral fin structure and function in fishes with labriform propulsion. J Morphol 263(2):133–150. Copyright © 2004 Wiley-Liss, Inc. Reprinted with permission

Triantafyllou MS, Triantafyllou GS (1995) An efficient swimming machine. Sci Am 272(3):62–70

Triantafyllou MS, Techet AH, Zhu Q et al (2002) Vorticity control in fish-like propulsion and maneuvering. Integr Comp Biol 42:1026–1031

Tsuchiya Y, Nomura T (1953) Chemical nature of the shark fin fiber. Tohoku J Agric Res 4(1):43–53

van den Boogaart JGM, Muller M, Osse JWM (2012) Structure and function of the median finfold in larval teleosts. J Exp Biol 215:2359–2368

Wainwright SA (1983) To bend a fish. In: Webb PW, Weihs D (eds) Fish biomechanics. Praeger, New York

Wainwright PC, Bellwood DR, Westneat MW (2002) Ecomorphology of locomotion in labrid fishes. Environ Biol Fish 65:47–62

Walker JA (2000) Does a rigid body limit maneuverability? J Exp Biol 203:3391–3396

Walker JA, Westneat MW (1997) Motor patterns of Labriform locomotion: kinematic and electromyographic analysis of pectoral fin swimming in the labrid fish *Gomphosus varius*. J Exp Biol 200:1881–1893

Walker JA, Westneat MW (2000) Mechanical performance of aquatic rowing and flying. Proc R Soc Lond B 267:1875–1881

Walker JA, Westneat MW (2002) Performance limits of labriform propulsion and correlates with fin shape and motion. J Exp Biol 205:177–187

Wang G, LaPatra S, Zeng L et al (2003) Establishment, growth, cryopreservation and species of origin identification of three cell lines from white sturgeon, *Acipenser transmontanus*. Methods Cell Sci 25:211–220

Ward AB (2002) Kinematics of the pectoral fins in batfishes (Ogcocephalidae) during aquatic walking. Integr Comp Biol 42:1331–1331

Webb PW (1988) Steady swimming kinematics of tiger musky, an esociform accelerator, and rainbow trout, a generalist cruiser. J Exp Biol 138:51–69

Westneat MW, Hoese W, Pell CA et al (1993) The horizontal septum: mechanics of force transfer in locomotion of scombrid fishes (Scombridae, Perciformes). J Morphol 217:183–204

Westneat M, Thorsen D, Walker J et al (2004) Structure, function, and neural control of pectoral fins in fishes. IEEE J Ocean Eng 29:674–677

Wilbur C, Vorus W, Cao Y, Currie S (2002) A lamprey-based undulatory vehicle. In: Ayers J, Davis JL, Rudolph A (eds) Neurotechnology for biomimetic robots. MIT Press, Cambridge, MA, pp 285–296. Published by The MIT Press

Wilga CD, Lauder GV (2001) Functional morphology of the pectoral fins in bamboo sharks. *Chiloscyllium plagiosum*: benthic vs. pelagic station-holding. J Morphol 249:195–209

Willey A (1890) On the development of the atrial chamber of amphioxus. Proc R Soc Lond 48:80–89

Wood A (1982) Early pectoral fin development and morphogenesis of the apical ectodermal ridge in the killifis *Aphyosemion scheeli*. Anat Rec 204:349–356

Wood A, Thorogood P (1984) An analysis of *in vivo* cell migration during teleost fin morphogenesis. J Cell Sci 66:205–222

Woodhead-Galloway J, Knight DP (1977) Some observations on the fine structure of elastoidin. Proc R Soc Lond B Biol Sci 195(1120):355–364

Woodhead-Galloway J et al (1978) Reprinted from Woodhead-Galloway J, Hukins DWL, Knight DP, Machin PA, Weiss JB (1978) Molecular packing in elastoidin spicules. J Mol Biol 118(4):567–578. Copyright © 1978, with permission from Elsevier

Yang SB, Han XY, Zhang DB et al (2008) Design and development of a kind of new pectoral oscillation robot fish. Robotics 30:508–515

Yang SB, Qiu J, Han XY (2009) Kinematics modeling and experiments of pectoral oscillation propulsion robotic fish. J Bionic Eng 6:174–179

Zhang J et al (2010) Reprinted by permission from Macmillan Publishers Ltd: Zhang J, Wagh P, Guay D et al (2010) Loss of fish actinotrichia proteins and the fin-to-limb transition. Nature 466(7303):234–237. Copyright © 2010, Rights Managed by Nature Publishing Group

Zhou C, Low KH (2010) Better endurance and load capacity: an improved design of manta ray robot (RoMan-II). J Bionic Eng 7(Suppl):S137–S144

Part IV
Marine Biopolymers of Vertebrate Origin

Chapter 8
Marine Collagens

Abstract There are three main categories of marine collagens: collagens of invertebrate origin (sponges, jellyfish, molluscs), fish collagens, and marine mammal collagens. Marine fish collagens isolated from skin, meat, scales, fins and waste materials are of particular interest from an industrial point of view. Different types of fish collagen-based biomaterials (gels, scaffolds, sponges, films, membranes, composites) and their biomedical application are discussed in this chapter.

Collagens are structural proteins, definitively, of fundamental evolutionary significance in both marine invertebrates and in vertebrate taxa. All of them possess a characteristic structural element in the form of rod-like triple-helical segment, however differ otherwise in their size, and functional roles. "As a family of proteins with unique structural features, marine invertebrate collagens have been a focus of structure–function correlation studies; as well as studies interrelating successive levels of structural organization, from the amino acid sequence to the anatomically defined fibril," (Ehrlich 2010). Numerous review papers (e.g., Gross et al. 1956; Engel 1997; Garrone 1999; Exposito et al. 2002) and books (e.g., Garrone 1978; Bairati and Garrone 1985; Ehrlich 2010) are dedicated entirely to collagens of marine invertebrates like sponges, corals, worms, molluscs, echinoderms, and crustaceans. Using a gentle, alkali-based slow-etching method we have recently isolated fibrillar collagens from *Hyalonema sieboldi* and *Monorhaphis chuni* glass sponges. The isolated organic fraction is revealed to be dominated by a hydroxylated fibrillar collagen that contains an unusual [Gly-3Hyp-4Hyp] motif (Ehrlich et al. 2010). Hydroxylated collagen appears to form the basis for the specific optical and extraordinary mechanical properties of hexactinellid specular structures.

The self-assembly properties of collagen and its templating activity for silicification processes are useful for inspiring current ideas about the development of hierarchical silica-based architectures. Macroscopic bundles of silica nanostructures result from the kinetic cross-coupling of two molecular processes: a dynamic supramolecular self-assembly, and a stabilizing silica mineralization. The feedback interactions between template growth and inorganic deposition are driven non-enzymatically by means of hydrogen bonding. We speculate that the hydroxylated glass sponge collagen may change the nature of silica in aqueous solution by converting the distribution of oligomers to an arrangement favourable for the mineralization. We suggest that

H. Ehrlich, *Biological Materials of Marine Origin*, Biologically-Inspired Systems 4,
DOI 10.1007/978-94-007-5730-1_8

the "[Gly-3Hyp-4Hyp] motif is predisposed for silica precipitation, and provides a novel template for biosilicification in nature," (Ehrlich et al. 2010).

Intriguingly, the deep-sea glass sponges habituate in aquatic niches with temperatures between −1.5 and 4 °C. Therefore, we can hypothesize that hydroxylated collagen could play an important role as a scaffold and/or as a template for biosilicification at temperatures near 0 °C; i.e., in extremely cold aquatic environments. However, the mechanism of this "*psychrophilic biosilicification*" is still unknown. It is well accepted that antifreeze proteins (AFPs) enable organisms to avoid freezing under extreme conditions. These proteins have been reported in bacteria, insects, fish, and other organisms that need to survive in cold temperatures (see for review Venkatesh and Dayananda 2008). AFPs protect the organisms by arresting the growth of ice crystals in their bodies. Interestingly, there are no reports on possible role of these proteins in biomineralization phenomena, including both collagen-based biosilicification and calcification. While the amino-acid composition and thermal stability of the skin collagen of the Antarctic ice-fish is well known (Rigby 1968), there is a lack of knowledge regarding the role of this collagen in biomineralazation. From this point of view, different species of Antarctic ice fishes seems to be appropriate model organisms to study collagen-based calcification under extremely cold conditions.

If we understand the principles of collagen chemistry, as well as the structural organization underlying survival at these freezing temperatures, then it is almost certain that we will understand how to create and establish the principles of Extreme Biomimetics using collagenous templates (Ehrlich 2012).

Thus, we can define three main categories of marine collagens: collagens of invertebrate origin, fish collagens, and marine mammal collagens. Marine fish collagens isolated from skin, meat, scales, fins and waste materials are especially of interest from industrial point of view (see also Chap. 12.1 in this work).

8.1 Isolation and Properties of Fish Collagens

Type I, type II, type V/XI and type XVIII genes encoding collagens have been identified in fish (see for review Guellec et al. 2004). "The number of published sequences is still limited, but additional sequences are available in databases. The comparison between fish and tetrapod collagen sequences indicates that the main characteristics of collagens were conserved during vertebrate evolution. In skin and intramuscular connective tissue of fish, type I and type V collagens have been identified as the major and minor collagen respectively," (Guellec et al. 2004). As reviewed by Eastoe (1957), amino acid analyses of fish collagens have been reported by Beveridge and Lucas (1944), who used mainly gravimetric methods, for isinglass from the swim bladder of hake (Urophyci); by Block, Horwith and Bolling in 1949 for the scales of herring (Clupea) and by Neuman in 1949, who used microbiological- assay techniques, for halibut skin and for gelatin prepared from the scales of an unspecified fish. Amino acid composition of fish collagens has been also studied by Piez and Gross (1960). Fish collagens and gelatins resemble those of mammals (Eastoe 1957) in that they contain the same amino acids in, broadly speaking, similar proportions. A rather wider range of composition is observed in fish than in mammals, which is not surprising in view of the greater evolutionary age of fishes. Hydroxyproline, in particular, is notably low, although showing considerable variation among fish species (Takahashi and Yokoyama 1954). Increased amounts of the aliphatic hydroxyl amino acids serine and threonine were present in all the fish proteins examined, compared with the amounts in mammalian collagen. The hydroxylysine content was substantially increased in two of the fish materials. The increase of hydroxyl groups in the molecule, owing to the larger numbers of aliphatic hydroxy amino acid residues in fish, was approximately equal to the loss of groups from the diminished number of hydroxyproline residues. This balancing effect results in the collagen molecule containing approximately the same proportion of hydroxyl groups in fish and mammals (Eastoe 1957).

Kawaguchi (1985) investigated the differences in the chemical composition between shark dentine and skin with respect to their hydroxylation and phosphorylation. The author reported that "The hydroxylation of prolyl and lysyl residues occurred more in the dentine alpha chains than in the skin chains. Among the four alpha chains, the phosphate content was the highest in the alpha 2 chain of the dentine collagen. These differences in hydroxylation and phosphorylation have been observed among alpha chains in mammalian mineralized and unmineralized tissues. The preferential dimerization to form alpha 1-alpha 2, characteristic of shark-skin collagen, was not observed in the dentine collagen," (Kawaguchi 1985).

As recently reviewed in Hayashi et al. (2012), "glycine is the most abundant amino acid [*in fish*] collagens and accounted for more than 30 % of all amino acids. Further, the degree of hydroxylation of proline was calculated to be 40–48 %, which was also similar level to that of the mammalian (about 45 %). The linear relationship between collagen stability and hydroxyproline content was demonstrated by that the degree of hydroxylation of proline in the fish collagen peptides. This was calculated to be about 35 %," (Hayashi et al. 2012).

The physico-chemical properties of marine fish collagens as well as their molecular forms (Kimura and Ohno 1987), subunits (Kimura et al. 1987), and self-assembly (Nomura et al. 1997) have all been thoroughly investigated. The high solubility of fish skin collagen upon heating has been known for a long time (Gustavson 1942). In the comparative study by Rose et al. (1988), it was found that the hydration of warm-water fish collagen is greater than that of cold-water fish collagen (halibut). Although the intrinsic viscosities of warm-water fish (bigeye-tuna, carp and catfish) collagens are almost the same, the hydrated volume of bigeye-tuna collagen is approximately 1.5 and 3 times those of freshwater fishes, such as carp and catfish collagens respectively. It was calculated that the denaturation temperatures of halibut, bigeye-tuna, carp and catfish collagens are 17, 31, 32 and 26–30 °C respectively (Rose et al. 1988).

As an alternative for the mammalian collagens, nowadays, especially fish collagen has gained increasing interest. As recently reviewed by Simpson et al. (2012), it can be generally isolated and manufactured from by-products generated during processing of both freshwater and marine fish; including deep-sea species (Wang et al. 2007b). The potential raw materials include skin (Nagai et al. 2002), bone

(Nagai and Suzuki 2000), scale (Ikoma et al. 2003; Nagai et al. 2004), fins (Nagai 2004), wing muscles in skate (Mizuta et al. 2002), the backbone (Zelechowska et al. 2010), swimbladder (Fernandes et al. 2008), shark placoid-scale dentine (Kawaguchi 1985) and so on. Numerous technological approaches concerning collagen isolation have been reported (Jongjareonrak et al. 2005; Nalinanon et al. 2007; Skierka and Sadowska 2007). For example, "to increase the extraction yield, pepsins from mammalian and fish origins which specifically cleave at telopeptide region, have been used successfully without the changes in molecular properties. Fish collagen can be used in food, biomedical, and pharmaceutical applications," (Benjakul et al. 2012; see also Pati et al. 2012).

Fish Skin Collagen The dermis of teleost fish can be divided into two regions (for reviews see Guellec et al. 2004). More detailed:

- "the superficial region (stratum laxum) is composed of a loose collagenous matrix housing numerous fibroblasts, nerves, some pigment cells, and scales (with a role in protection and hydrodynamics: Burdak 1979; Sire 1986). As the skin is an accessory respiratory surface, the upper layer of the dermis is most strongly vascularised, particularly around the scales. The deep region (stratum compactum) is characterised by a dense, plywood-like organisation of the collagen matrix, in which some fibrocytes are interspersed. This region is penetrated in some places by vertical bundles of collagen fibrils. In naked species, such as a number of catfishes, the dermis is principally composed of densely packed collagen fibres and is thicker (0.6–2.0 mm) than in scaled species (e.g., less than 0.2 mm in zebrafish)," (Guellec et al. 2004).
- "Initially, the subepidermal region of the fish skin, characterised by a plywood-like organisation of the collagen fibrils, was called 'subepidermal collagenous lamella' and even renamed 'basement lamella'. This collagenous stroma constitutes the first deposited matrix of the future dermis, and in order to avoid confusion with the basement membrane, it was suggested that 'collagenous dermal stroma' for the larval and early juvenile skin; and 'stratum compactum' instead of 'basement lamella' for the adult skin" (Guellec et al. 2004; see also Fujii 1968; Nadol et al. 1969).

There are numerous patents on the isolation of collagen from the fish skin (see as examples Higheberger 1961; Allard et al. 1995), as well as scientific papers. Thus, collagens were isolated from the skin of numerous marine fish species: lamprey (Kimura et al. 1981), skate (*Raja kenojei*) (Wang et al. 2007a) and different sharks (Kimura et al. 1981; Nomura et al. 1995; Yoshimura et al. 2000; Nomura 2004; Kittiphattanabawon et al. 2010), hake (*Merluccius hubbsi*) (Ciarlo et al. 1997); threadfin bream (*Nemipterus* spp.) (Nalinanon et al. 2008), brown backed toadfish (*Lagocephalus gloveri*) (Senaratne et al. 2006), Baltic cod (*Gadus morhua*) (Sadowska et al. 2003; Sadowska and Ko1odziejska 2005; Skierka and Sadowska 2007), arabesque greenling (*Pleurogrammus azonus*) (Nalinanon et al. 2010), black drum (*Pogonias cromis*) and sheepshead seabream (*Archosargus probatocephalus*) (Ogawa et al. 2003), brownstripe red snapper (*Lutjanus vitta*) (Jongjareonrak et al.

2005), bigeye snapper (Priacanthus tayenus) (Kittiphattanabawon et al. 2005), ocellate puffer fish (*Takifugu rubripes*) (Nagai et al. 2002), redfish (*Sebastes mentella*) (Wang et al. 2007b), walleye pollack (*Theragra chalcogramma*) (Yan et al. 2008), yellowfin tuna (*Thunnus albacares*) (Woo et al. 2008), hoki (*Macruronus novaezelandiae*) and ling (*Genypterus blacodes*) (Hofman and Newberry 2011).

Methods of collagen isolation from fish skin differ in details. However, there are some steps which are common for all methods used. Here, as example, I give briefly description of isolation of collagen from skin of brown backed toadfish (*Lagocephalus gloveri*) according to Senaratne et al (2006):

"All the preparative steps were carried out at 4 °C with continuous stirring. Skin pieces were treated with 0.1 N NaOH to remove non-collagenous proteins at a solid to solution ratio of 1:10 (w/v) for 3 days. After this, they were washed with distilled water till the washed water became neutral or slightly basic pH. The alkali solution was changed every day. Then, the sample was defatted with 10 % butyl alcohol at a solid to solvent ratio of 1:10 (w/v) for 24 h, washed with ample amounts of distilled water, and lyophilized. The lyophilized matter was extracted with 0.5 M acetic acid for 3 days. Then, 10 % of pepsin (1:10,000 units) according to the lyophilized weight was added and hydrolysed for 48 h. The viscous solution was centrifuged at 12,000 g for 1 h at 4 °C and supernatant was collected. The supernatant was salted out by adding NaCl to a final concentration of 0.7 M. This was followed by precipitation by addition of NaCl to a final concentration of 2.3 M in 0.05 M Tris–HCl (pH 7.5). The resultant precipitate was separated by centrifugation at 12,000 g for 1 h at 4 °C. The precipitate was then dissolved in 0.5 M acetic acid, dialysed against 0.1 M acetic acid, distilled water, and lyophilized," (Senaratne et al. 2006).

From industrial point of view collagen isolated from the skins of freshwater aquaculture cultivated fish species like carp, tilapia, and rainbow trout (Tabarestani et al. 2012) are of great interest.

Fish Scale Collagen The collagen fibrils-containing basal plate of Amiidae (*Amia*) and Sarcopterygii (*Leptdosiren*, *Latimeria*, *Protopterus*, *Neoceratodus*) looks like plywood, "a system of superimposed layers of parallel fibers or fibrils the directions of which rotate with a regular angle in two successive layers," (Meunie 1984; see also Bigi et al. 2001). As characterized by Meunier (1984), "the double twisted plywood is constituted of two imbricate systems, the odd and the even, where the rotation of the fibrillar directions is right-handed in Sarcopterygii and left handed in Amiidae and numerous primitive Teleostei. The orthogonal plywood, with its two main orthogonal fibrillar directions, characterizes the evolved Teleostei and some more primitive ones," (Meunie 1984). Thus, "the basal plate is a stacking of collagenous fibred layers. In one layer, fibrils are parallel to each other. The diameter of fibrils (30–190 nm) is greater than in bone (20–50 nm) or osseous layer (20–30 nm)," (Meunie 1984; see also Meunier 1983; Zylberberg et al. 1988). Recently Youn and Shin (2009) studied the supramolecular organization of collagen fibrils in fish scales of red seabream (*Pagrus major*). These authors reported, "a collagen fiber consists of helical substructures of collagen fibrils wrapped with incrustation. *[…]* Freshly growing edge region of fish scale, embedded into fish skin, showed rarely patched and one directionally arranged collagen fibers, in which specifically triple helical assemblies of collagen fibrils were found. On the contrary, the relatively aged region of the rostral field close to the scalar focus displayed randomly directed and densely packed collagen fibers, in which loosened- and deteriorated-helical assemblies of collagen fibrils were mostly found," (Youn and Shin 2009).

Interestingly, comparisons between the stable isotope composition of carbon in collagen excised from juvenile (freshwater) and adult (marine) portions of scales from Atlantic salmon *Salmo salar* demonstrated that c. 75 % of carbon analysed in the 'juvenile' portion of the scale derives from older collagen. Scale collagen analyses were effectively restricted to the last season of growth (Hutchinson and Trueman 2006).

Isolation and characterisation of fish scale collagen is also well described (see for review Toshiyuki et al. 2003; Nagai et al. 2004; Zhang et al. 2011; Mori et al. 2013). In the past, the first step for extracting collagen from fish scales was to remove the fat with an organic solvent, such as acetone. The second step was removal of the hydroxyapatite by acidity solution; and the third step was filtration to gain the crude collagen (Wu and Chai 2007). Recently, the novel extraction of collagen from fish scales with papain under ultrasonic pretreatment was reported (Jiang et al. 2012). The results showed that the optimum conditions for collagen extraction from fish scales were: ultrasonic pretreatment time 4 min, ratio of papain to fish scales 4 %, temperature 60 °C and extraction time 5 h. Under the optimum conditions the extraction rate of collagen reached 90.7 %.

Unfortunately, little attention was paid to the other proteinaceous components of fish scales like *ichtylepidin* (Burley and Solomons 1955). Mörner (1898) has shown that the scales of many species of fish contain, in addition to mineral matter and collagen, a peculiar albuminoid. To this albuminoid he gave the name *ichtylepidin*. Mörner prepared his ichthylepidin in the following way: "the clean scales were digested at room temperature with a large excess of 5 % hydrochloric acid, 0.05 % caustic potash, and 0.01 % acetic acid. Each digestion extended over several days. This treatment removed soluble proteins, most of the guanin, the chrondroitin-sulphuric acid, and the inorganic matter. The residual scales were then digested with 0.1 % hydrochloric acid at 40 ° C. The residue thus freed from collagen was washed with alcohol and ether, and dried. The substance so obtained (pure ichthylepidin) was insoluble in boiling water, in cold dilute acids, and in alkalies; but it was soluble in hot solutions both of dilute acids and alkalies, and in the cold concentrated solutions of the same. It gave a strong Millon's reaction and contained much loosely combined sulphur (as shown by the blackening of the substance when boiled with an alkaline solution of lead acetate)," (Green and Tower 1901; see also Mörner 1898 [Ger]). By the two latter reactions the presence of ichthylepidin may, according to Mörner, be determined in fish scales (Green and Tower 1902). The relation between content of ichtylepidin to collagen is fish scales are reported as 24–76 % (Green and Tower 1902).

Ichtylepidin isolated from carp scales contains 50,87 % C; 6,56 % H; 15,69 % N; 1,02 % S, and 26,8 % O (Abderhalden and Voitinovic 1907). As reported by Kalyani (1997), Winter in 1954 made a complete analysis of the Pilchard scale ichthylepidin

(*Sardina ocellata*), tested the stability and effect of proteinases on it and emphasized the occurrence of chondroitin sulfate. The preparation of ichthylepidin, its nature and amino acid composition are discussed elsewhere (Seshaiya et al. 1963). The most characteristic feature of ichthylepidin in contrast to collagen is that it is not converted to gelatin upon boiling with water. The content of neutral sugars, hexosamine and uronic acid in fish scales was measured by Kalyani (1997) as following (g per 100 g dry weight): glucose 0.9–1.0; galactose 0.6; hexosamine 0.47–0.61 and uronic acid 0.2.

Thus, collagens of fish origin (both marine and fresh water) are produced today in large scales. Mostly these collagens are sources for gelatine production, however there are some special fields of application of the collagens in pharmacy and biomedicine. For example, Chen et al. (2012) extracted a glycoprotein from the cartilage of blue shark (*Prionace glauca*) and identified it as shark type II collagen (SCII). The authors "aim to confirm the effects of oral tolerance of SCII on inflammatory and immune responses to the ankle joint of rheumatoid-arthritis rats induced by Complete Freund's Adjuvant (CFA). [*The obtained*] results suggested that appropriate dose of SCII can not only ameliorate symptoms but also modify the disease process of Complete-Freunds-Adjuvant-induced arthritis," (Chen et al. 2012).

Fish Swimbladder Collagen The swimbladder of teleost fishes is a gas-filled sac which serves primarily to make the fish neutrally buoyant in sea water, but occasionally assumes other functions (see for review Davenport 2005; Tibbetts et al. 2007). The gas contained in the swimbladder is largely oxygen, at a pressure very close to the external hydrostatic pressure. The difference in gas partial pressure between the gaseous contents of the swimbladder and the blood and tissue fluids is large in fish living at any considerable depth. The hydrostatic pressure increases about 1 atm with each 10 m depth, while the partial pressures of gases in sea water and body fluids are relatively independent of depth and together give a pressure of only about 1 atm. The difference in partial pressure of oxygen alone across the wall of the swimbladder of a fish living at 3,000 m depth is close to 300 atm (Wittenberg et al. 1980). Collagen is located within the membranous walls of fish swimbladder and can be extracted using acetic acid (2.5 pH), and precipitated by the addition of NaCl up to 3.0 M (see for details Fernandes et al. 2008).

Product known as *isinglass* is also of the fish swimbladder origin. Here, the description of isinglass by Hickman et al. (2000):

"It is widely used commercially to clarify alcoholic beverages by aggregation of the yeast and other insoluble particles. It is derived from swim bladders of tropical fish by solubilisation in organic acids and consists predominantly of the protein collagen. The low content of intermolecular cross-links allows ready dissolution of swim bladder compared to bovine hide; which is cross-linked by a high proportion of stable bonds and requires enzymic digestion to solubilise. Isinglass is no longer effective as a clarifying agent if thermally denatured hence the collagenous triple helical structure must be maintained. Thermal denaturation of isinglass occurs at 29 °C, compared to 40–41 °C for mammalian collagens, primarily due to the lower hydroxyproline content. Studies on the mechanism of action of isinglass have shown that higher molecular weight aggregates that increase the length of the collagen molecules (trimers, tetramers, etc.) increase efficiency and that their surface charges are important in the clarification process," (Hickman et al. 2000).

8.2 Fish Collagen as a Biomaterial

Collagen is not only easily available and highly versatile, but also possesses a major advantage in being biocompatible and biodegradable. The use of collagens as biomaterials in biomedicine is determined by their possible degradation by human collagenases. (see for review Parenteau-Bareil et al. 2010; Hayashi et al. 2012). Additionally, "type I collagen is a suitable material for implantation since only a small amount of people possess humoral immunity against it, and a simple serologic test can verify if a patient is susceptible to an allergic reaction in response to this collagen-based biomaterial," (Parenteau-Bareil et al. 2010; see also Eaglstein et al. 1999).

Two basic techniques are used for preparation of collagen-based biomaterials. "The first one is a decellularized collagen matrix preserving the original tissue shape and ECM structure, while the other relies on extraction, purification and polymerization of collagen and its diverse components to form a functional scaffold," (Parenteau-Bareil et al. 2010). According to Gilbert et al. (2006) three main methods are used for tissue decellularization:

"**Physical methods** include snap freezing that disrupt cells by forming ice crystals, high pressure that burst cells and agitation, that induce cell lysis and are used most often in combination with chemical methods to facilitate penetration of active molecules in the tissue.

Chemical methods of decellularization include a variety of reagents that can be used to remove the cellular content of ECM. These substances range from acid to alkaline treatments, as well as chelating agents such as EDTA, ionic or nonionic detergents and solutions of extreme osmolarity.

Enzymatic treatments such as trypsin, which specifically cleaves proteins and nucleases that remove DNA and RNA are also commonly used to produce acellular scaffold. However, none of these methods can produce an ECM completely free of cellular debris and a combination of techniques is often required to obtain a material free of any cell remnant," (Parenteau-Bareil et al. 2010).

There are several types of collagen-based biomaterials of fish origin: gels, scaffolds, sponges, films, membranes etc.

Collagen Gels In contrast to marine chitosan gels, which are well investigated with respect to tissue engineering, fish collagen gels possess some limiting factors. For example, the low denaturation temperature (Td) of most fish collagens with the exception of shark collagen (Nomura et al. 2000a, b), renders these structural biopolymers difficult to handle. One of the crucial properties of the shark collagen solution with Td of about 30 °C (Nomura et al. 1995), is that the dissolution of the fibrillar gel obtained from this type of collagen takes place at 37 °C (Nomura et al. 2000a). Correspondingly, this gel could not be practically used at the actual physical temperature of human medical application. Similar situation is known with unstable chum salmon collagen with the T_d of approximately 19 °C (Matsui et al. 1991).

The development of bio-inspired collagen fibrillar gel from salmon, that served only as a scaffold for cell culture has been demonstrated in several works (Yunoki et al. 2003, 2004; Nagai et al. 2007, 2008a). However, recently a chemical cross-linking method for improving the thermal stability of collagen gel derived from chum salmon (*Oncorhynchus keta*) using water-soluble carbodiimide during in vitro collagen fibrillogenesis, was reported (see for review Shen et al. 2008). The obtaining salmon collagen fibrillar gel has a denaturation temperature of 55 °C and showed good biological properties in numerous in vitro studies. It was reported also about the use of cross-linking agents like 1-ethyl-3-(3-dimethylaminopropyl)-carbodiimide (EDC) during fibril formation. This led to increases of the denaturation temperature in the fish collagen gels up to 47 °C (Kawaguchi et al. 2011).

To overcome the obstacle with low level of T_d in fish collagens, a development of the chitosan/marine-originated collagen composite has been recently reported (Wang et al. 2010). In detail: "The chitosan gel including N-3-carboxypropanoil-6-O-(carboxymethyl) chitosan of 3 mol%, 6-O-(carboxymethyl) chitosan of 62 mol% and 6-O-(carboxymethyl) chitin of 35 mol% was prepared and compounded with the salmon atelocollagen (SA) gel at different mixture ratios. The SA gel first, within 2 weeks, and then chitosan in the composite gel was slowly absorbed after implantation, followed by soft tissue formation. It is expected that this composite gel will be available as a carrier for tissue filler and drug delivery systems," (Wang et al. 2010).

Collagen Scaffolds For the scaffold manufacturing, collagens originated from marine products are indispensable due to severe infection problems from animal source; such as bovine spongiform encephalopathy, avian and swine influenzas, and tooth-and-mouth disease in bovine, pig, and buffalo that occur all over the world (Hayashi et al. 2012). Three scaffold-design parameters are accepted as influencing tissue regeneration: (i) modification of scaffold surfaces to enhance cell interaction, (ii) controlled release of growth factors from scaffolds, and (iii) scaffold mass transport (Hollister 2009).

Collagen scaffolds with excellent biocompatibility which were derived from fresh water fish have been reported recently by Pati et al. (2012):

"The fish collagen scaffolds exhibited considerable cell viability and were comparable with that of bovine collagen. SEM and fluorescence microscopic analysis revealed significant proliferation rate of cells on the scaffolds, and within 5 days the cells were fully confluent. These findings indicated that fish collagen scaffolds derived from fresh water origin were highly biocompatible in nature," (Pati et al. 2012). Marine fish collagens are also known as good sources for designing tissue scaffolds.

Some years ago, Lin et al. (2010) proposed a new scaffold based on acellularized and decalcified fish scales for corneal regeneration. Rapid rabbit corneal cells proliferation and migration at different time periods on this scaffold have been reported. "Collectively, the authors demonstrated the superior cellular conductivity of the newly developed material. The highly centrally-oriented micropatterned structure of the scaffold was beneficial for efficient nutrient and oxygen supply to the cells cultured in the three-dimensional matrices, and therefore it is useful for high-density cell seeding and spreading," (Lin et al. 2010).

Recently, Terada and co-authors reported in article "Construction and characterization of a tissue-engineered oral mucosa equivalent based on a chitosan-fish scale collagen composite,":

"This study was designed to assess the *in vitro* biocompatibility of a chitosan-collagen composite scaffold constructed by blending commercial chitosan and tilapia scale collagen with oral mucosa keratinocytes. *[…]* These findings demonstrated that these hybrid scaffolds have a potential application for epithelial tissue engineering, and provides a new potential therapeutic device for oral mucosa regenerative medicine," (Terada et al. 2012).

The animal implantation studies have demonstrated (Kawase et al. 2010) that, "after osteogenic processing, cultured human periosteal sheets form osteoid tissue ectopically without the aid of conventional scaffolding materials. To improve the osteogenic activity of these periosteal sheets, we have tested the effects of including a scaffold made of salmon collagen-coated ePTFE mesh. Periosteal sheets were produced with minimal manipulation without enzymatic digestion. Outgrown cells penetrated into the coated mesh fiber networks to form complex multicellular layers and increased expression of alkaline phosphatase activity in response to the osteoinduction. *In vitro* mineralization was notably enhanced in the original tissue segment regions, but numerous micro-mineral deposits were also formed on the coated-fiber networks. When implanted subcutaneously into nude mice, periosteal sheets efficiently form osteoid around the mineral deposits. These findings suggest that the intricate three-dimensional mesh composed of collagen-coated fibers substantially augmented the osteogenic activity of human periosteal sheets both *in vitro* and *in vivo*," (Kawase et al. 2010).

One of the modern challenging tasks is to obtain restorable collagen/hydroxyapatite composites for applications in bone regeneration. In the report by Michael Meyer and co-workers (Hoyer et al. 2012), "established procedures for mineralization of bovine collagen were adapted to a new promising source of collagen from salmon skin took on the challenge of the low denaturation temperature. Therefore, in the first instance, variation of temperature, collagen concentration, and ionic strength was performed to reveal optimized parameters for fibrillation and simultaneous mineralization of salmon collagen. Porous scaffolds from mineralized salmon collagen were prepared by controlled freeze-drying and chemical cross-linking. The scaffolds exhibited interconnecting porosity, were sufficiently stable under cyclic compression, and showed elastic mechanical properties. Human mesenchymal stem cells were able to adhere to the scaffolds, cell number increased during cultivation, and osteogenic differentiation was demonstrated in terms of alkaline phosphatase activity," (Hoyer et al. 2012).

Collagen Sponges with microporous structures from tilapia were also manufactured. According to Sugiura et al. 2009, corresponding sponge-like constructs were "reconstituted collagen fibrils using freeze-drying and cross-linked by dehydrothermal treatment *[...]* or additional treatment with water-soluble carbodiimide. *[...]* The pellet implantation tests into the paravertebral muscle of rabbits demonstrated that tilapia collagen caused rare inflammatory responses at 1- and 4-week implantation, statistically similar to those of porcine collagen and a high-density polyethylene as a negative control," (Sugiura et al. 2009).

Kawaguchi et al. (2011) reported on the development of "crosslinked salmon derived atelocollagen (SC) sponge, which has a denaturation temperature of 47 °C (Fig. 8.1). Sixty-four knees of 32 mature rabbits were randomly divided into 4 groups after creating an osteochondral defect in the femoral trochlea. Defects in Groups I, II, and III were filled with the crosslinked SC sponge, the crosslinked porcine collagen (PC) sponge, and the non-crosslinked PC sponge, respectively. In Group IV, defects were left untreated as the control. At 12 weeks after implantation, the histological score showed that Group I was significantly greater than Groups III (P=0.0196) and IV (P=0.0021). In addition, gene expression of type-2 collagen, aggrecan, and SOX9 was the greatest in Group I at 12 weeks. The fundamental in vivo properties of the crosslinked SC sponge showed that this is a promising biomaterial, specifically as a scaffold for cartilage tissue engineering," (Kawaguchi et al. 2011).

Collagen Films Swim bladders of tropical fish species like Amarela (*Cynoscion acoupa*), Gurijuba (*Arius parkeri*) and Branca (*Cynoscion leiarchus*) has been used for extraction under acidic conditions (pH 2.5) and preparation of special collagen films using precipitation by the addition of NaCl (Fernandes et al. 2008). "Differential

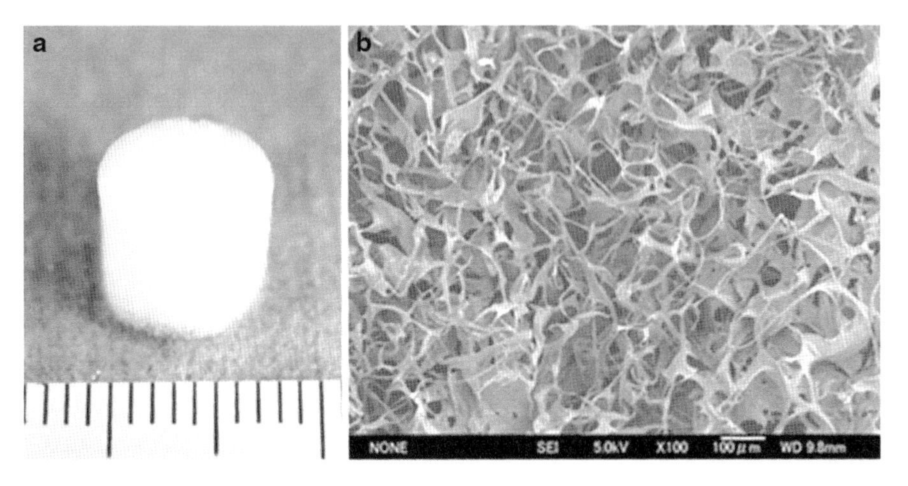

Fig. 8.1 SEM imagery of the collagenous cylindrical sponge constructs (**a**). Crosslinked salmon collagen sponge is represented on the image (**b**) (With kind permission from Springer Science and Business Media: Kawaguchi et al. (2011); Copyright (2011) Springer)

scanning calorimetry revealed high denaturation temperature peaks at temperatures ranging from 65.9 to 74.8 °C. The micrographs showed no fibrillar organization along the material, but rather a spongy structure, with cavity diameters relatively uniform at around 2 μm. The impedance spectroscopy presented a distributed relaxation process. A. parkeri's films showed piezoelectricity," (Fernandes et al. 2008).

Nagai et al. (2009) proposed balloon-expandable or self-expandable covered stents with a biodegradable salmon collagen (SC) film (Fig. 8.2). Thus, more detailed:

"Since conventional metallic stents are unable to inhibit smooth muscle overgrowth and extracellular matrix production leading to restenosis, much effort has been concentrated on finding new methods to prevent the outgrowth by covering the stent pores using cover films. The SC-covered stents were fabricated by placing a bare stent in a mixture of acidic SC solution and a fibrillogenesis-inducing buffer (pH 6.8) including a cross-linking agent (water-soluble carbodiimide), and subsequent incubation at 4 °C for 24 h and lyophilization. The stents obtained were completely

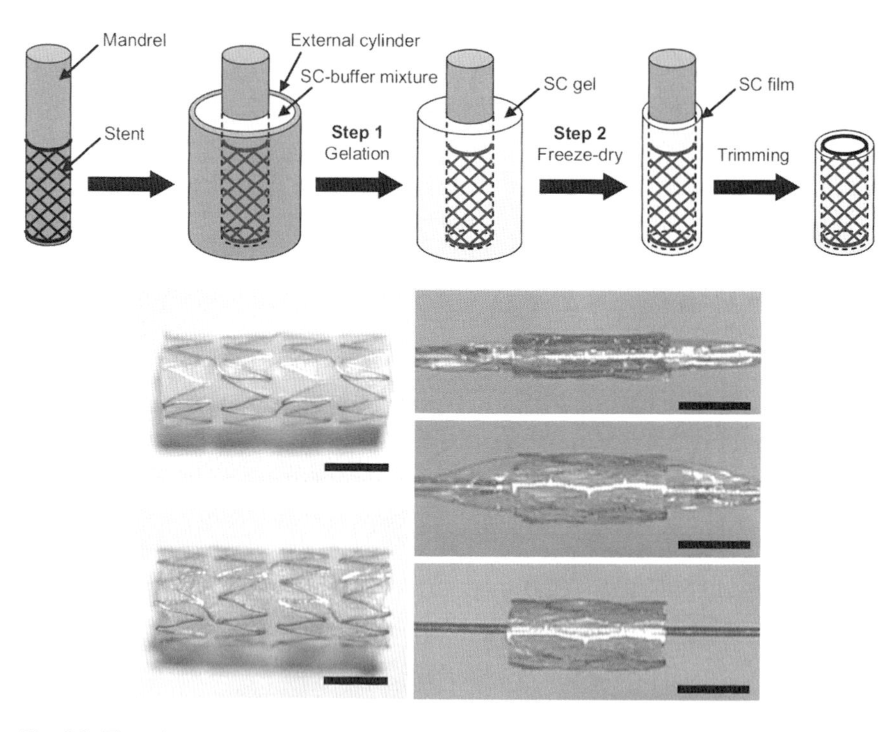

Fig. 8.2 The salmon collagen (SC)- covered stents have been produced as follow: "The stent is placed on a mandrel and incubated in an SC–buffer mixture at 4 °C for 24 h. Stents covered with the cross-linked SC gel were lyophilized, resulting in SC-covered stents. The gross appearance of the self-expandable SC-covered stent before and after immersion in water. Gross appearance of the balloon expandable SC-covered stent after mounting on a percutaneous transluminal angioplasty balloon catheter, after dilation by expanding the balloon, and after removal of the balloon catheter. Bars 5 mm," (With kind permission from Springer Science+Business Media: Nagai et al. (2009). Copyright (2009) Springer)

covered with an SC film with a nanofibrous structure (fibril diameter, about 70 nm). On immersion in water, the film is converted to a gel with slight swelling. There was no rupture of the SC cover after mounting on a balloon catheter or after expansion. Preliminary implantation was conducted by placing the balloon-expandable covered stents in the common carotid arteries of beagles. One month after implantation, angiography showed that all stented arteries were patent with no significant neointimal thickening. The authors suggested that SC is potentially useful as a cover material of endovascular stents to enhance patency," (Nagai et al. 2009).

Collagen Membranes Fish collagen membranes have also been designed as well as patented (see for review Andre et al. 2000). A bioresorbable collagen barrier membrane (Periocol®) of fish origin has been developed for guided tissue regeneration applications in human periodontal intrabony and furcation defects (Singh et al. 2011). Periocol® collagen membrane have been used as a sustained release chlorhexidine chip in chronic periodontitis patients and reported to resorb after 30 days (Divya and Nandakumar 2006). Recently, high mechanical stable collagen fibril membranes (CFMs) of type I atelocollagen extracted from tilapia scales were designed using a lateral face evaporation method, (see for details Xu et al. 2011). These membranes showed following features:

"The density and thickness of the CFM obtained were 0.51 ± 0.04 mg/cm^3 and 50 ± 5 μm. The collagen fibrils in the CFM had a similar periodic stripped pattern of 67 nm with native collagen fibrils. The CFM was crosslinked in gaseous glutaraldehyde for different duration in order to increase the mechanical property. The crosslinking degrees of the CFMs analyzed by free amino groups gradually increased to 70.3 % against the exposure duration until 6 hours, and reached a plateau. The denaturation temperatures of the CFMs with the crosslinking degrees at 20.4 % to 43 % were linearly increased from 49 °C to 75 °C. The tensile strength of the CFMs was slightly improved until the crosslinking degree at 33.3 %, at which point the tensile strength rapidly increased to be 68 MPa. It was suggested that a percolation phenomenon took place in the CFMs by crosslinking of collagen fibrils with polymerized GA molecules," (Xu et al. 2011).

Collagen Composites Fish collagen composites with chitosan are discussed above. However, Shen et al. (2008) reported about manufacture of DNA and salmon collagen (SC) composite materials for wound dressing. More detailed information is as follow:

"the sDNA/SC composites were prepared by incubating a mixture of an acidic SC solution, an sDNA solution, and a collagen fibrillogenesis inducing buffer (pH 6.8) containing a crosslinking agent (water-soluble carbodiimide) for gelation. A subsequent ventilation-drying process gave sDNA/SC films. The sDNA/SC films with various doses of sDNA (sDNA/SC weight ratios of 1:5, 1:10, and 1:20) were used for *in vitro* cell cultures to evaluate their growth potentials of normal human dermal fibroblasts (NHDF) and normal human epidermal keratinocytes (NHEK). It was found that NHDF proliferation was increased by sDNA conjugation, whereas NHEK proliferation was dose-dependently inhibited. In light of the *in vitro* results, the appropriate dose of sDNA for *in vivo* study was determined to be at a ratio of 1:10. For the implantation in full-thickness skin defects in the rat dorsal region, the sDNA/SC films were reinforced by incorporating them on a porous SC sponge, because the sDNA/SC films exhibited early contraction and inadequate morphologic stability when implanted *in vivo*. The regenerated tissue in the sDNA/SC sponge group showed similar morphology to native dermis. Meanwhile, the SC sponge group without sDNA showed epithelial overgrowth, indicating that additional sDNA could reduce epidermal overgrowth. Furthermore, blood capillary formation was significantly enhanced in the sDNA/SC sponge group when compared to the SC sponge group. In conclusion, the results suggest that the sDNA/SC composite could be a potential wound dressing for clinical applications," (Shen et al. 2008).

Several studies were dedicated to the inflammatory response of different fish collagen specimens. Previously, it was shown that collagen is responsible for the activation of professional phagocytes of the marine gilthead seabream (*Sparus aurata*) (Castillo-Briceño et al. 2009). Recently, attention was paid to fibroblasts which are involved in the initiation of tissue regeneration including wound repair. Therefore, Castillo-Briceño and co-workers (2011) used "SAF-1 cells (gilthead seabream fibroblasts) to identify the binding motifs in collagen by end-point and real-time cell adhesion assays using the collagen peptides and Toolkits. The authors identified the collagen motifs involved in the early magnesium-dependent adhesion of these cells. Furthermore, it was found that peptides containing the GFOGER and GLOGEN motifs (where O is hydroxyproline) present high affinity for SAF-1 adhesion, expressed as both cell number and surface covering. In cell suspensions, these motifs were also able to induce the expression of the genes encoding the proinflammatory molecules interleukin-1β and cyclooxygenase-2. These data suggest that specific collagen motifs are involved in the regulation of the inflammatory and healing responses of teleost fish," (Castillo-Briceño et al. 2011).

Fish collagen has also been developed in other physical forms like sheet (Sankar et al. 2008), or vascular graft (Nagai et al. 2008b). A very interesting form of collagen in a biomaterial – the cornea from tuna fish – was reported by Parravicini et al. (2012). Among available biomaterials, the cornea is almost completely devoid of cells and is composed only of collagen fibers oriented in an orderly pattern, which contributes to low antigenicity. *Thunnus thynnus*, the Atlantic bluefin tuna, is a fish with large eyes that can withstand pressures of approximately 10 MPa. Eyes from freshly caught Atlantic bluefin tuna were harvested and preserved in a fixative solution. Sterilized samples of corneal stroma were embedded in paraffin and stained with hematoxylin and eosin, and the histologic features were studied. Physical and mechanical resistance tests were performed in comparison with bovine pericardial strips and porcine mitral valves. Corneal material was implanted subcutaneously in seven rats, to evaluate in vivo calcification rates. Mitral valves made from tuna corneal leaflets were implanted in nine sheep. It was found that the corneal tissue consisted only of parallel collagen fibers without evidence of vascular or neural structures. In tensile strength, the tuna corneal specimens were substantially similar to bovine pericardium. After 23 days, the rat-implanted samples showed no calcium or calcium salt deposition. Hydrodynamic and fatigue testing of valve prototypes yielded acceptable functional and long-term behavioral results. In the sheep, valvular performance was stable during the 180-day follow-up period, with no instrumental sign of calcification at the end of observation. The researchers conclude that low antigenicity and favorable physical properties qualify tuna cornea as a potential material for durable bioimplantation (Parravicini et al. 2012).

Fish collagen-based biomaterials can be used also for non-biomedical applications. For example, Moura et al. (2012) developed Poly (glutaraldehyde)-stabilized fibrillar collagen from fish scale for heavy metal sorption. As described by these authors, "fibrillar collagen on scales of Corvina fish (*Micropogonias furnieri*) was crosslinked and used as a new adsorbent for sorption of Cr (VI) from aqueous solutions. Characterization has suggested that the crosslinked collagen of Corvina scale has higher denaturation temperature in relation to the raw scales. In addition, electrostatic interactions between collagen positive charges and chromate negative charges constitute the majority of the interactions. Solution microcalorimetry experiments have indicated that water swelling of the crosslinked scales is slightly exothermic and increased with increasing temperature. Sorption of Cr (VI) by crosslinked scales increases with increasing initial Cr (VI) concentration in solution and decreases with temperature increasing. The kinetic data of Cr (VI) sorption on crosslinked scales were best fitted to a multilinear exponential model. The values of Cr (VI) diffusion constants increase with both temperature and initial Cr (VI) concentration in solution. The maximum sorption capacity of the new adsorbent for Cr (VI) was found to be at 39 mg/g and is higher than some commercial adsorbent samples," (Moura et al. 2012).

8.3 Conclusion

Collagens of fish origin are truly impressive alternative sources for collagen-based biomaterials of animal origin. Unfortunately, marine mammals are also used currently as a source for collagen. Of course, there is scientific interest on whale and dolphin collagens (see for review Ludowieg et al. 1973; Nagai et al. 2008c). However, "the Whale and Dolphin Conservation Society said in their reports that thousands of approved patents for products or processes could contain whale ingredients. Japan was selling sperm whale myoglobin and chondroitin for treating osteoarthritis to researchers worldwide. Japanese researchers were also using whale collagen for beauty treatments and as an anti-inflammatory. Norway was examining the use of whale oil for pharmaceutical and health supplements," (Darby 2010).

I am absolutely sure that our scientific community can exclude marine mammal collagens as any kind of biomaterial with practical application, and replace them with ones derived from a more abundant marine species.

References

Abderhalden E, Voitinovic I (1907) Weitere Beiträge zur Kenntnis der Zusammensetzung der Proteine. Hoppe Seylers Z Physiol Chem 52(3–4):368–374

Allard R, Devictor P, Huc A et al (1995) Unpigmented fish skin, particularly from flat fish, as a novel industrial source of collagen, extraction method, collagen and biomaterial thereby obtained. US Patent 5,420,248

Andre V, Abdul Malak N et al (2000) Use of collagen of aquatic origin for the production of supports for tissue engineering, and supports and biomaterials obtained. US Patent 6,541,023

Bairati A, Garrone R (1985) Biology of invertebrate and lower vertebrate collagens. Plenum, New York

Benjakul S, Nalinanon S, Shahidi F (2012) Fish collagen. In: Simpson BK (ed) Food biochemistry and food processing, 2nd edn. Wiley-Blackwell, Oxford. Copyright © 2012 John Wiley & Sons, Inc. Reproduced with permission

Beveridge JMR, Lucas CC (1944) Amino acids of isinglass. J Biol Chem 155:547–556

Bigi A, Burghammer M, Falconi R et al (2001) Twisted plywood pattern of collagen fibrils in teleost scales: an X–ray diffraction investigation. J Struct Biol 136:137–143

Burdak VD (1979) Morphologie fonctionnelle du tégument écailleux des poissons. La Pensée Scientifique, Kiev (en russe). traduction française in Cybium (Paris), 1986, 10 (3) supplément:147 p

Burley RW, Solomons CC (1955) The action of fluorodinitrobenzene on ichthylepidin. Acta Bioenerg 18(1):137–138

Castillo–Briceño P, Sepulcre MP, Chaves–Pozo E et al (2009) Collagen regulates the activation of professional phagocytes of the teleost fish gilthead seabream. Mol Immunol 46:1409–1415

Castillo-Briceño P, Bihan D, Nilges M et al (2011) A role for specific collagen motifs during wound healing and inflammatory response of fibroblasts in the teleost fish gilthead seabream. Mol Immunol 48(6–7):826–834. © 2010 Elsevier Ltd. All rights reserved. (Published under a Creative Commons license)

Chen L, Bao B, Wang N et al (2012) Oral administration of shark type II collagen suppresses complete Freund's adjuvant-induced rheumatoid arthritis in rats. Pharmaceuticals (Basel) 5(4):339–352. © 2010 by the authors; licensee Molecular Diversity Preservation International, Basel, Switzerland. This article is an open-access article distributed under the terms and conditions of the Creative Commons Attribution license (http://creativecommons.org/licenses/by/3.0/)

Ciarlo AS, Paredi ME, Fraga AN (1997) Isolation of soluble collagen from hake skin (*Merluccius hubbsi*). J Aqua Food Prod Technol 6:65–77

Darby A (2010) Australian firm sells sperm whale extract. Published on-line http://www.smh.com.au/environment/whale-watch/australian-firm-sells-sperm-whale-extract-20100615-ydbm.html (16.07.2010). Accessed 15 May 2014. Copyright © 2014 Fairfax Media. Reprinted with permission

Davenport J (2005) Swimbladder volume and body density in an armoured benthic fish, the streaked gurnard. J Fish Biol 55:527–534

Divya PV, Nandakumar K (2006) Local drug delivery – Periocol® in periodontics. Trends Biomater Artif Organs 19:74–80

Eaglstein WH, Alvarez OM, Auletta M et al (1999) Acute excisional wounds treated with a tissue–engineered skin (Apligraf). Dermatol Surg 25:195–201

Eastoe JE (1957) The amino acid composition of fish collagen and gelatin. Biochem J 65(2):363–368

Ehrlich H (2010) Biological materials of marine origin. Invertebrates. Springer biological materials of marine origin by Ehrlich, Hermann Reproduced with permission of Springer in the format Book via Copyright Clearance Center

Ehrlich H (2012) Collagens of aquatic invertebrates: insights, trends and open questions. In: Proceedings of the 5th Freiberg Collagen Symposium, Freiberg, Germany, 4–5 September 2012

Ehrlich H, Deutzmann R, Brunner E et al (2010) Mineralization of the metre–long biosilica structures of glass sponges is templated on hydroxylated collagen. Nat Chem 2(12):1084–1088
Engel J (1997) Versatile collagens in invertebrates. Science 277:1785–1786
Exposito JY, Cluzel C, Garrone R et al (2002) Evolution of collagens. Anat Rec 268(3):302–316
Fernandes RM, Couto Neto RG, Paschoal CW et al (2008) Reprinted from Fernandes RM, Couto Neto RG, Paschoal CW et al (2008) Collagen films from swim bladders: preparation method and properties. Colloids Surf B Biointerfaces 62(1):17–21. Copyright (2008), with permission from Elsevier
Fujii R (1968) Fine structure of the collagenous lamella underlying the epidermis of the goby, *Chasmichthys gulosus*. Annot Zool Jpn 41:95–106
Garrone R (1978) Phylogenesis of connective tissue. In: Robert L (ed) Morphological aspects and biosynthesis of sponge intercellular matrix. S. Karger, Basel
Garrone R (1999) Evolution of metazoan collagens. Prog Mol Subcell Biol 21:119–139
Gilbert TW, Sellaro TL, Badylak SF (2006) Decellularization of tissues and organs. Biomaterials 27:3675–3683
Green EH, Tower RW (1901) The organic constituents of the scales of fish. Bull US Fish Comm 21:97–102. Copyright © 1901, National Climatic Data Center, Asheville, NC, USA. Reprinted with permission
Green EH, Tower RW (1902) The organic constituents of the scales of fish. Kessinger Publishing LLC, Whitefish, p 12
Gross J, Sokal Z, Rougvie M (1956) Structural and chemical studies of the connective tissue of marine sponges. J Histochem Cytochem 4:227–246
Guellec DL, Morvan-Dubois G, Sire J-Y (2004) Skin development in bony fish with particular emphasis on collagen deposition in the dermis of the zebrafish (*Danio rerio*). Int J Dev Biol 48:217–231. Reproduced with permission from The International Journal of Developmental Biology (Int. J. Dev. Biol.) (2004) 48:217–231. © UBC Press
Gustavson KH (1942) The directing influence of the organization of proteins upon their reactivity. Svensk kem Tidskr 54:74–83
Hayashi Y et al (2012) Reproduced from Hayashi Y, Yamada S, Yanagiguchi K, Koyama Z, Ikeda T (2012) Chitosan and fish collagen as biomaterials for regenerative medicine. Adv Food Nutr Res 65:107–120. Copyright (2012), with permission from Elsevier
Hickman et al (2000) Reprinted from Hickman D, Sims TJ, Miles CA et al (2000) Isinglass/collagen: denaturation and functionality. J Biotechnol 79(3):245–257. Copyright (2000), with permission from Elsevier
Higheberger HK (1961) Extraction of collagen. US Patent 2,979,438
Hofman KA, Newberry M (2011) Thermal transition properties of Hoki (*Macruronus novaezelandiae*) and Ling (*Genypterus blacodes*) skin collagens: implications for processing. Mar Drugs 9:1176–1186
Hollister SJ (2009) Scaffold design and manufacturing: from concept to clinic. Adv Mater 21:3330–3342
Hoyer et al (2012) Reprinted with permission from Hoyer B, Bernhardt A, Heinemann S et al (2012) Biomimetically mineralized salmon collagen scaffolds for application in bone tissue engineering. Biomacromolecules 13(4):1059–1066. Copyright (2012) American Chemical Society
Hutchinson JJ, Trueman CN (2006) Stable isotope analyses of collagen in fish scales: limitations set by scale architecture. J Fish Biol 69:1874–1880
Ikoma T, Kobayashi H, Tanaka J et al (2003) Physical properties of type I collagen extracted from fish scales of *Pagrus major* and *Oreochromis niloticas*. Int J Biol Macromol 32(3–5):199–204
Jiang ZN, Bo JQ, Zheng QX et al (2012) Extraction of collagen from fish scales with papain under ultrasonic pretreatment. Adv Mat Res 366:421–424
Jongjareonrak A, Benjakul S, Visessanguan W et al (2005) Isolation and characterisation of acid and pepsin–solubilised collagens from the skin of Brownstripe red snapper (*Lutjanus vitta*). Food Chem 93:475–484

Kalyani M (1997) Sugars linked with wish scale protein. Indian J Fish 26:228–229
Kawaguchi (1985) Reproduced from Kawaguchi T (1985) Chemical nature of collagen in the placoid-scale dentine of the blue shark, *Prionace glauca* L. Arch Oral Biol 30(5):385–390. Copyright (1985), with permission from Elsevier
Kawaguchi et al (2011) With kind permission from Springer Science+Business Media: Kawaguchi Y, Kondo E, Kitamura N, Arakaki K, Tanaka Y, Munekata M, Nagai N, Yasuda K (2011) *In vivo* effects of isolated implantation of salmon-derived crosslinked atelocollagen sponge into an osteochondral defect. J Mater Sci Mater Med 22(2):397–404. Copyright (2011) Springer Science+Business Media, LLC
Kawase et al (2010) With kind permission from Springer Science+Business Media: Kawase T, Okuda K, Kogami H et al (2010) Osteogenic activity of human periosteal sheets cultured on salmon collagen-coated ePTFE meshes. J Mater Sci Mater Med 21(2):731–739. Copyright © 2009, Springer Science+Business Media, LLC
Kimura S, Ohno Y (1987) Fish type I collagen: tissue–specific existence of two molecular forms, (a1)2a2 and ala2a3, in Alaska Pollack. Comp Biochem Physiol 88B:409–413
Kimura S, Kamimura T, Takema Y et al (1981) Lower vertebrate collagen, evidence for type I–like collagen in the skin of lamprey and shark. Biochim Biophys Acta 669:251–257
Kimura S, Ohno Y, Miyauchi Y et al (1987) Fish skin type I collagen: wide distribution of an a3 subunit in teleosts. Comp Biochem Physiol B 88:27–34
Kittiphattanabawon P, Benjakul S, Visessanguan W et al (2005) Characterisation of acid–soluble collagen from skin and bone of bigeye snapper (*Priacanthus tayenus*). Food Chem 89:363–372
Kittiphattanabawon P, Benjakul S, Visessanguan W et al (2010) Isolation and characterisation of collagen from the skin of brownbanded bamboo shark (*Chiloscyllium punctatum*). Food Chem 119:1519–1526
Lin CC, Ritch R, Lin SM et al (2010) A new fish scale–derived scaffold for corneal regeneration. Eur Cell Mater 19:50–57. Copyright (c) 2010, Lin et al. Reprinted with permission
Ludowieg JJ, Adams J, Wang AC et al (1973) The mammalian intervertebral disc. The collagen of whale fetal nucleus pulposus. Connect Tissue Res 2:21–29
Matsui R, Ishida M, Kimura S (1991) Characterization of an 3 chain from the skin type I collagen of chum salmon (*Oncorhynchus keta*). Comp Biochem Physiol B 99:171–174
Meunie FJ (1984) Spatial organization and mineralization of the basal plate of elasmoid scales in osteichthyans. Am Zool 24(4):953–964. By permission of Oxford University Press
Meunier FJ (1983) Les tissus osseux des Osteichlhyens. Structure, genese, croissance et evolution. Archives et documents, Micro–edition, Inst Ethnol SN.82–600–328
Mizuta S, Hwang J, Yoshinaka R (2002) Molecular species of collagen from wing muscle of skate (*Raja kenojei*). Food Chem 76:53–58
Mori H, Tone Y, Shimizu K, Zikihara K, Tokutomi S, Ida T, Ihara H, Hara M (2013) Studies on fish scale collagen of Pacific saury (*Cololabis saira*). Mater Sci Eng C 33:174–181. doi:10.1016/j.msec.2012.08.025
Mörner CT (1898) Die organische Grundsubstanz der Fischschuppen vom chemischen Gesichtspunkte aus betrachtet. Zeitschr f Physiol Chemie 24:125–137
Moura KO, Vieira EFS, Cestari AR (2012) Poly(glutaraldehyde)-stabilized fish scale fibrillar collagen–some features of a new material for heavy metal sorption. J Appl Polym Sci 124:3208–3221. © 2011 Wiley Periodicals, Inc. Reprinted with permission
Nadol JB, Gibbons JR, Porter KR (1969) A reinterpretation of the structure and development of the basement lamella: an ordered array of collagen in fish skin. Dev Biol 20:304–331
Nagai T (2004) Characterization of collagen from Japanese sea bass caudal fin as waste material. Eur Food Res Technol 218:424–427
Nagai T, Suzuki N (2000) Isolation of collagen from fish waste material – skin, bone and fins. Food Chem 68:277–281
Nagai T, Araki Y, Suzuki N (2002) Collagen of the skin of ocellate puffer fish (*Takifugu rubripes*). Food Chem 78:173–177

Nagai T, Izumi M, Ishii M (2004) Fish scale collagen: preparation and partial characterization. Int J Food Sci Technol 39:239–244
Nagai N, Mori K, Satoh Y et al (2007) *In vitro* growth and differentiated activities of human periodontal ligament fibroblasts cultured on salmon collagen gel. J Biomed Mater Res A 82(2):395–402
Nagai N, Mori K, Munekata M (2008a) Biological properties of crosslinked salmon collagen fibrillar gel as a scaffold for human umbilical vein endothelial cells. J Biomater Appl 23(3):275–287
Nagai N, Nakayama Y, Zhou YM et al (2008b) Development of salmon collagen vascular graft: mechanical and biological properties and preliminary implantation study. J Biomed Mater Res B Appl Biomater 87B(2):432–439
Nagai T, Suzuki N, Nagashima T (2008c) Collagen from common minke whale (*Balaenoptera acutorostrata*) unesu. Food Chem 111:296–301
Nagai N et al (2009) With kind permission from Springer Science+Business Media: Nagai N, Nakayama Y, Nishi S, Munekata M (2009) Development of novel covered stents using salmon collagen. J Artif Organs 12(1):61–66. Copyright © 2009, The Japanese Society for Artificial Organs
Nalinanon S, Benjakul S, Visessanguan W et al (2007) Use of pepsinfor collagen extraction from the skin of bigeye snapper (*Priacanthus tayenus*). Food Chem 104:593–601
Nalinanon S, Benjakul S, Visessanguan W et al (2008) Tuna pepsin: characteristics and its use for collagen extraction from the skin of threadfin bream (*Nemipterus* spp.). J Food Sci 73:C413–C419
Nalinanon S, Benjakul S, Kishimura H (2010) Collagens from the skin of arabesque greenling (*Pleurogrammus azonus*) solubilized with the aid of acetic acid and pepsin from albacore tuna (*Thunnus alalunga*) stomach. J Sci Food Agric 90(9):1492–1500
Nomura Y (2004) Properties and utilization of shark collagen. More efficient utilization of fish and fisheries products. Elsevier, New York
Nomura Y, Yamano M, Shirai K (1995) Renaturation of alpha 1 chains from shark skin collagen type I. J Food Sci 60:1233–1236
Nomura Y, Yamano M, Hayakawa C et al (1997) Structural property and in vitro self–assembly of shark type I collagen. Biosci Biotechnol Biochem 61:1919–1923
Nomura Y, Toki S, Ishii Y et al (2000a) The physicochemical property of shark type I collagen gel and membrane. J Agric Food Chem 48:2028–2032
Nomura Y, Toki S, Ishii Y et al (2000b) Improvement of the material property of shark type I collagen by comparison with pig type I collagen. J Agric Food Chem 48:6332–6633
Ogawa M, Moody MW, Portier RJ et al (2003) Biochemical properties of black drum and sheepshead seabream skin collagen. J Agric Food Chem 51(27):8088–8892
Parenteau-Bareil R, Gauvin R, Berthod F (2010) Collagen-based biomaterials for tissue engineering applications. Materials 3(3):1863–1887. © 2010 by the authors; licensee Molecular Diversity Preservation International, Basel, Switzerland. This article is an open-access article distributed under the terms and conditions of the Creative Commons Attribution license (http://creativecommons.org/licenses/by/3.0/)
Parravicini R, Cocconcelli F, Verona A et al (2012) Tuna cornea as biomaterial for cardiac applications. Tex Heart Inst J 39(2):179–183
Pati F, Datta P, Adhikari B et al (2012) Collagen scaffolds derived from fresh water fish origin and their biocompatibility. J Biomed Mater Res A 100(4):1068–1079. Copyright © 2012 Wiley Periodicals, Inc
Piez KA, Gross J (1960) The amino acid composition of some fish collagens: the relation between composition and structure. J Biol Chem 235:995–997
Rigby BJ (1968) Amino–acid composition and thermal stability of the skin collagen of the Antarctic ice–fish. Nature 219(5150):166–167
Rose C, Kumar M, Mandal AB (1988) A study of the hydration and thermodynamics of warm–water and cold–water fish collagens. Biochem J 249:127–133

Sadowska M, Kołodziejska I (2005) Optimisation of conditions for precipitation of collagen from solution using k–carrageenan. Studies on collagen from the skin of Baltic cod (*Gadus morhua*). Food Chem 91:45–49

Sadowska M, Kolodziejska I, Niecikowska C (2003) Isolation of collagen from the skin of Baltic cod (*Gadus morhua*). Food Chem 81:257–262

Sankar S, Sekar S, Mohan R et al (2008) Preparation and partial characterization of collagen sheet from fish (*Lates calcarifer*) scales. Int J Biol Macromol 42(1):6–9

Senaratne et al (2006) Reprinted from Senaratne LS, Park P-J, Kim S-K (2006) Isolation and characterization of collagen from brown backed toadfish (*Lagocephalus gloveri*) skin. Bioresource Technol 97(2):191–197. Copyright (2006), with permission from Elsevier

Seshaiya RV, Ambujabai P, Kalyani M (1963) Amino–acid composition of ichthylepidin of fish scales. JMU 32B:138

Shen X et al (2008) With kind permission from Springer Science+Business Media: Shen X, Nagai N, Murata M et al (2008) Development of salmon milt DNA/salmon collagen composite for wound dressing. J Mater Sci Mater Med 19(12):3473–3479. Copyright © 2008, Springer Science+Business Media, LLC

Simpson BK, Benjakul S, Nalinanon S et al (2012) Fish collagen. In: Food biochemistry and food processing, 2nd edn. Wiley. doi:10.1002/9781118308035.ch20

Singh VP, Nayak DG, Uppoor AS et al (2011) Clinical and radiographic evaluation of nano–crystalline hydroxyapatite bone graft (Sybograf®) in combination with bioresorbable collagen membrane (Periocol®) in periodontal intrabony defects. Dent Res J 9:60–67

Sire JY (1986) Ontogenic development of surface ornamentation in the scales of *Hemichromis bimaculatus* (Cichlidae). J Fish Biol 28:713–724

Skierka E, Sadowska M (2007) The influence of different acid and pepsin on the extractability of collagen from the skin of Baltic cod (*Gadus morhua*). Food Chem 105:1302–1306

Sugiura H, Yunoki S, Kondo E et al (2009) *In vivo* biological responses and bioresorption of tilapia scale collagen as a potential biomaterial. J Biomater Sci Polym Ed 20(10):1353–1368. Copyright © 2009 Taylor & Francis. Reprinted with permission

Tabarestani SH, Maghsoudlou Y, Motamedzadegan A et al (2012) Study on some properties of acid–soluble collagens isolated from fish skin and bones of rainbow trout (*Onchorhynchus mykiss*). Int Food Res J 19(1):251–257

Takahashi T, Yokoyama W (1954) Physico–chemical studies on the skin and leather of marine animals. XII. The content of hydroxyproline in the collagen of different fish skins. Bull Jpn Soc Sci Fish 20:525–529

Terada M, Izumi K, Ohnuki H et al (2012) Construction and characterization of a tissue-engineered oral mucosa equivalent based on a chitosan-fish scale collagen composite. J Biomed Mater Res 100B:1792–1802. Copyright © 2012 Wiley Periodicals, Inc

Tibbetts IR, Collette BB, Isaac R et al (2007) Functional and phylogenetic implications of the vesicular swimbladder of *Hemirhamphus* and *Oxyporhamphus convexus* (Beloniformes: Teleostei). Copeia 4:808–817

Toshiyuki I, Hisatoshi K, Junzo T (2003) Physical properties of type I collagen extracted from fish scales of *Pagrus major* and *Oreochromis niloticas*. Int J Biol Macromol 32:199–204

Venkatesh S, Dayananda C (2008) Properties, potentials, and prospects of antifreeze proteins. Crit Rev Biotechnol 28:57

Wang JH, Mizuta S, Yokoyama Y et al (2007a) Purification and characterization of molecular species of collagen in the skin of Skate (*Raja kenojei*). Food Chem 100:921–925

Wang L, An X, Xin Z et al (2007b) Isolation and characterization of collagen from the skin of Deep–Sea Redfish (*Sebastes mentella*). J Food Sci 72:E450–E455

Wang W, Itoh S, Aizawa T et al (2010) Development of an injectable chitosan/marine collagen composite gel. Biomed Mater 5(6):065009. © IOP Publishing. Reproduced with permission. All rights reserved

Wittenberg JB, Copeland DE, Haedrich FRL et al (1980) The swimbladder of deep–sea fish: the swimbladder wall is a lipid–rich barrier to oxygen diffusion. J Mar Biol Assoc UK 60:263–276

Woo JW, Yu SJ, Cho SM et al (2008) Extraction optimization and properties of collagen from yellowfin tuna (*Thunnus albacares*) dorsal skin. Food Hydrocoll 22:879–887

Wu CH, Chai HJ (2007) Collagen of fish scale and method of making thereof. United States patent application 20070231878

Xu Z, Ikoma T, Yoshioka T et al (2011) Effect of Glutaraldehyde on properties of membranes prepared from fish scale collagen. MRS proceedings 1418:2012. (doi: 10.1557/opl.2012.396), reproduced with permission. Copyright © 2012, Materials Research Society

Yan M, Li B, Zhao X et al (2008) Characterization of acid–soluble collagen from the skin of walleye pollack (*Theragra chalcogramma*). Food Chem 107:1581–1586

Yoshimura K, Terashima M, Hozan D et al (2000) Preparation and dynamic viscoelasticity characterization of alkali–solubilized collagen from shark skin. J Agric Food Chem 48:685–690

Youn, Shin (2009) Reprinted from Youn HS, Shin TJ (2009) Supramolecular assembly of collagen fibrils into collagen fiber in fish scales of red seabream, *Pagrus major*. J Struct Biol 168(2):332–336. Copyright (2009), with permission from Elsevier

Yunoki S, Suzuki T, Takai M (2003) Stabilization of low denaturation temperature collagen from fish by physical cross–linking methods. J Biosci Bioeng 96:575–577

Yunoki S, Nagai N, Suzuki T et al (2004) Novel biomaterial from reinforced salmon collagen gel prepared by fibril formation and cross–linking. J Biosci Bioeng 98(1):40–47

Zelechowska E, Sadowska M, Turk M (2010) Isolation and some properties of collagen from the backbone of Baltic cod (*Gadus morhua*). Food Hydrocoll 24:325–329

Zhang F, Wang A, Li Z et al (2011) Preparation and characterisation of collagen from freshwater fish Scales. Food Nutr Sci 2:818–823

Zylberberg L, Bereiter–Hahn J, Sire JY (1988) Cytoskeletal organization and collagen orientation in the fish scales. Cell Tissue Res 253:597–607

Chapter 9
Marine Gelatins

Abstract There are various reasons for considering the marine gelatin as an alternative to the terrestrian mammals' gelatin (bovine and pigs). Among them is the risk of transmission of infections illnesses, like spongiform encephalopathy, in bovine gelatin. From a social-cultural approach, some cultures reject animal gelatine due to their religious beliefs. One advantage of marine gelatin is that it has no risk associated with Bovine Spongiform Encephalopathy. In addition, it is considered as Kosher and Halal. This chapter is dedicated to methods of gelatin isolation from marine fish skins, scales, cartilage and bones. A variety of practical applications for fish gelatins in the form of gels, films and composite materials is discussed.

Since Papin (seventeenth century) gelatin was known as the substitute for meat in diet of poor people. In the time between 1803 and 1841 the nutritional value of this product was studied by different Academic Commissions (Viel and Fournier 2006). About 1870, gelatin becomes a component of photographic emulsions and remains to be an indispensable during long period of time. The replacement of the collodion wet process with a gelatin emulsion which could be dried, has been carried out by Dr. Maddox of England. Hilbert in 1906 patented the process entitled "*Manufacture of gelatin and glue from bones*" under US Patent Nr. 834806 (Hilbert 1906).

Today, pharmaceutical and edible gelatins differ from technical principally. Since 1906, extraction processes have been modified to a great extent with respect to extraction yields and physicochemical properties. Examples include using different organic acids for pre-treatment of the animal skins including fish, different salts for washing the skins, high-pressure treatment, and pepsin aided digestion (for review see Gomez-Guillen et al. (2009); Gomez-Guillen and Montero 2001a; as well as patents by Gomez-Guillen and Montero 2001b; Lefevre et al. 2002; Yang 2007). Commercial gelatin is classified into two types: Type A and Type B gelatin. Gelatin type A and gelatin type B are made by acidic and alkaline pretreatment, respectively, before heat processing. The difference between Gelatin A and B is that the alkaline pretreatment converts amide residues of glutamine and asparagines into glutamic and aspartic acid, which leads to a 25 % higher carboxylic acid content for gelatin B that for gelatin A. According to pretreatment approaches, alkaline treatment (type B gelatin) makes use of cattle hides and bones, however, acid pretreatment (type A gelatin) uses skins of pigs (Bae 2007).

H. Ehrlich, *Biological Materials of Marine Origin*, Biologically-Inspired Systems 4,
DOI 10.1007/978-94-007-5730-1_9

Here, I would like to insert some statistical information about gelatin as follow:

- "over 250,000 metric tons of gelatin are produced worldwide every year, 60 % of which is consumed by humans in a variety of products;
- the annual world output of gelatin has increased from year to year, with the highest source being pig-skin (46 %), followed by bovine hides (29.4 %), bones (23.1 %) and other sources (1.5 %);
- the fish gelatin production is very low, yielding only about 1 % of the annual world gelatin production," (Zhang et al. 2011; see also Han et al. 2009; Irwandi et al. 2009; Karim and Bhat 2009).

The increasing demand for non mammalian gelatin for kosher and halal food markets as well as outbreak of BSE, have revived industrial interest in gelatin of marine origin including fish (Jamilah and Harvinder 2002). As reviewed by Huda and Adzitey (2012), gelatin is broadly used in dairy products and confectionaries over a wide pH range due to its emulsifying properties as well as to stability, clarity, bland flavour, and, texture. Consequently, gelatin is suitable for use as a protein enrichment, adhesives, flocculating agent, salt reducer, and dietic food. Such forms of gelatin as tablets, capsules, and sponges, are well established in the pharmaceutical industry today (Herpandi et al. 2011). For gelatin manufacturers, yield from a particular raw material is also important. Recent experimental studies have shown that these quality parameters vary greatly depending on the biochemical characteristics of the raw materials, the manufacturing processes applied, and the experimental settings used for quality control tests (Johnston-Banks 1990; Norland 2003; Boran and Regenstein 2010).

Gelatin of fish waste origin including skin, scale and bones seems to be an attractive product due to its qualities and properties (Kittiphattanabawon et al. 2005; Huda and Adzitey 2012). For example, Guerard et al. (2001) reported that about 70 % of the original fish material after canned fish processing represent a mix of solid waste that include bone, skin, gills and several muscle remains. Additional examples can be found in literature:

"Potential sources of gelatin that have been underutilized include, for example, the skins from Alaska pollock (*Theragra chalcogramma*) and pink salmon (*Oncorhynchus gorbuscha*). These two species comprised approximately 73 % of the annual marine finfish catch of Alaska in 2000. It has been estimated that over a million tons of fish by-products are generated each year from the fishing industry in Alaska. Some of these by-products are converted into fish meal; however, much of this material is underutilized and dumped back into the ocean!" (Avena-Bustillos et al. 2011).

The main limiting factor for practical application of marine fish gelatins is their specific rheological and physic-chemical properties (Norland 1990; Leuenberger 1991; Gilsenan and Ross-Murphy 2000; Haug et al. 2004). Here, some examples:

- "cold-water fish gelatins, such as those extracted from pollock, cod, and salmon, have very low gelation and melting temperatures compared to mammalian and

warm-water fish gelatins. This is due to the cold-water fish having lower concentrations of proline and hydroxyproline than the other species;
- cold-water fish gelatin solutions behave as viscous liquids at room temperature, which could make them desirable for specific applications, such as ice cream, yogurt, dessert gels, confections, and imitation margarine;
- fish gelatin had less undesirable flavors and odors, as well as better release of aroma, than the same product made with pork gelatine with an equal Bloom value and higher melting point," (Avena-Bustillos et al. 2011; see also Avena-Bustillos et al. 2006; Choi and Regenstein 2000; Regenstein and Chaudry 2002; Solgaard et al. 2008).

Fish Skin Gelatin Due to low concentration of intra and inter-chain non-reducible crosslinks the collagen of the fish skins origin is highly soluble (Gimenez et al. 2005). Moreover, both technical (Gomez-Guillen et al. 2002; Fernandez-Dıaz et al. 2003; Songchotikunpan et al. 2008) and edible (Aberoumand 2010) gelatin can be produced from fish skins. Gimenez and co-workers, after analysis of papers by Gustavson 1956; Norland 1990 and Johnston-Banks 1990, described solubilisation of fish skin collagen as follow:

"Therefore, a mild acid pre-treatment is usually used for gelatin production. Such treatment leads to a type-A gelatin with an isoelectric point that can vary from 6.5 to 9. Increasing H^+ ions favours the access of water to collagen fibres. This water is held in by electrostatic forces between charged polar groups (electrostatic swelling) or by hydrogen bonding between uncharged polar groups and negative atoms (lyotropic hydration). The type and concentration of acid used strongly influences swelling properties and solubilisation of collagen. This leads to variations in molecular weight distribution in the resultant gelatins, depending on the persistence of some of the cross-links between collagen chains," (Giménez et al. 2005).

According to Asghar and Henrickson (1982), "the lyotropic effect of carboxylic acids on collagen seems to dominate the swelling capacity, rather than a specific ion effect. It is the non-ionized acid that acts as the swelling agent. This occurs by competition with the peptide group involved in intermolecular linking of the protein chain; and is due mainly to the hydrogen bonding power of the acid. Citric acid is widely used for the manufacture of foodgrade gelatin from fish skin because it does not introduce undesirable colour or odour to the gelatin. The overall properties of gelatins obtained following this procedure highly depend on the fish species used, and are largely attributed to differences in amino acid composition," (Giménez et al. 2005)

On example of megrim (*Lepidorhombus whiffiagonis*) skin, it was shown that also lactic acid can be used for collagen extraction (Gomez-Guillen et al. 2002; Gimenez et al. 2005). Different optimal conditions have been used to extract gelatin from fish skin: 0.115 M (84 min) acetic acid (Wang and Yang 2009), 0.115 M acetic acid plus 0.2 M (3 h) NaOH (Wang and Yang 2009), acetic acid plus ~0.1 M NaCl (Montero and Gomez-Guillen 2000), and acetic acid, NaOH and NaCl (Montero and Gomez-Guillen 2000). Comparative studies at nano scale into the effect of different pretreatment sets have only been done for acetic and acetic-NaOH (Wang and

Yang 2009). Up to 80.8 % gelatin recovery has been achieved in experiments with silver carp (*Hypophthalmichthys molitrix*) skin. The optimum extraction conditions were as follow: "50 °C for the extraction temperature, 0.1 N HCl for the acid concentration, 45 min for the acid pretreatment time, and finally 4:1 (v/w) for the water/skin ratio. The predicted responses for these extraction conditions were 630 g gel strength and 6.3 cP viscosity," (Boran and Regenstein 2009).

Gelatin was isolated from the skins of different fresh water species, including catfish (Yang et al. 2007; Yang and Wang 2009), silver carp (Boran and Regenstein 2009), grass carp (*Catenopharyngodon idella*) (Ladislaus et al. 2007); and marine fishes such as sharks (Hamada 1990), dogfish (*Squalus suckleyi*) (Geiger et al. 1962), yellowfin tuna (*Thunnus albacares*) (Andre et al. 2003; Cho et al. 2005; Gómez-Estaca et al. 2009b), megrim (*Lepidorhombus boscii*) (Montero and Gomez-Guillen 2000), cod (Gudmundsson and Hafsteinsson 1997), Atlantic salmon (Salmo salar) (Arnesen and Gildberg 2007), Chum salmon (*Oncorhynchus keta*) (Zhang et al. 2011), greater Lizardfish (*Saurida tumbil*) (Taheri et al. 2009), grouper (Rahman and Al-Mahrouqi 2009), pollock (Bower et al. 2010), bigeye snapper and brownstripe red snapper (Jongjareonrak et al. 2006), sin croaker (*Johnius dussumieri*) and shortfin scad (*Decapterus macrosoma*) (Cheow et al. 2007). Gelatins from the skin of four local marine fish, namely "kerapu" (*Epinephelus sexfasciatus*) (Fig. 9.1), "jenahak" (*Lutjianus argentimaculatus*), "kembung" (*Rastrelliger kanagurta*), and "kerisi" (*Pristipomodes typus*) have been successfully extracted by acid extraction (Irwandi et al. 2009).

Some of the processes for production of gelatins from fish skins are patented. For example, according to Grossman and Bergman (1992), a process for the production of gelatin from fish skins is comprised of the following steps:

(a) cleaning the skins in order to remove therefrom substantially all superfluous material;
(b) treating with dilute aqueous alkali;

Fig. 9.1 Gelatin extracted from marine fish *Epinephelus sexfasciatus* (Reprinted with permission from Irwandi et al. (2009). © 2009 IFRJ, Faculty of Food Science & Technology, UPM)

(c) washing with water until the washing water is substantially neutral;
(d) treating with dilute aqueous mineral acid;
(e) washing with water until the washing water is substantially neutral;
(f) treating with dilute aqueous citric acid and/or another suitable organic acid;
(g) washing with water until the washing water is substantially neutral;
(h) extracting with water at elevated temperatures not above about 55 °C the washed citric acid-treated skins. In practice, the present process employs much lower temperatures than known heretofore for the treatment steps, which results in a high quality product (e.g. absence of a fishy smell).

Fish Scale and Bone Gelatin Scales (Wangtueai and Noomhorm 2009) and bones (Alfaro et al. 2005) from many fish species proven to be a source for gelatin production (Cheow et al. 2007). Gelatin extraction procedure from fish scales is described in lizardfish (*Saurida* spp.) (Wangtueai and Noomhorm 2009) as follows:

"The thawed (kept in refrigerator at 9–10 °C for about 24 h) scales were treated with 2 volumes (v/w) of alkali solution (0.1–0.9 % NaOH) at room temperature (30 °C) for 1–5 h to remove the noncollagen protein and subcutaneous tissue after they were swollen. After the alkali treatment, the scales were neutralized by washing under running tap water until they had a pH of about 7. The scales were then subjected to a final wash with distilled water to remove any residual matter. The extractions were carried out in distilled water at control temperatures within the range of 70–90 °C for 1–5 h. The ratio used was 300 g (weight of wet scales) to 600 ml of distilled water. The coarse solids were filtered out with filter cloth, and this was followed by vacuum-filtering. The filtered solutions were evaporated under vacuum to 10 brix at 50 °C and dried in a vacuum dryer at 60 °C set pressure 0 mbar until brittle sheets were formed," (Wangtueai and Noomhorm 2009).

Some biochemical properties of scale and bone gelatin of marine fish species have been also investigated Ogawa et al (2004).

Fish Gelatin-Based Composites As reviewed by Bae et al (2009), researchers have investigated various approaches to modify fish gelatin in an effort to improve its functionality. The blending of fish gelatin with other biopolymers, such as j-carrageenan, chitosan, and pectin, is one possible way to improve the properties of fish gelatin. Furthermore, the addition of plasticizers, such as glycerol sorbitol, sucrose, polyethylene glycol, and salt agents (Sarabia et al. 2000; Koli et al. 2011) can improve the mechanical properties of fish gelatin films or gels. Additions of chemical cross-linking agents, such as glutaraldehyde, formaldehyde, and glyoxal or enzymes, such as microbial transglutaminase, have also been shown to improve the properties of fish gelatin. However, the chemical cross-linkers are toxic, which limits their use in food systems. Therefore, the use of enzymes as cross-linking agents could be a better alternative for food packaging (Bae et al. 2009).

Experience in manipulating the physical properties of fish gelatins has been gained during investigations of rheological (Gudmundsson 2002), mechanical (Chiou et al. 2006), cross-linking (Bode et al. 2011) and gelling. The properties of fish gelatins, as well as their interactions with other proteins (Badii and Howell

2005) and organic substances, guided the progress in developing of fish gelatin-based composites (Shakila et al. 2012). Here, are some examples.

Nanoclay composite film was produced using warm water fish gelatin as a base material and its physical, mechanical, and molecular weight change properties were observed after treatment with microbial transglutaminase (MTGase) (Bae et al. 2009). The viscosity of the MTGase-treated gelatin solution (2 % w/w) increased from 86.25 ± 1.77 (0 min) to 243 ± 12.37 cp (80 min). SDS–PAGE results indicated that the molecular weight of fish gelatin solutions increased after treatment with microbial transglutaminase. Tensile strength decreased from 61.60 ± 1.77 (0 min) to 56.42 ± 2.40 MPa (30 min), while E% increased from 13.94 ± 5.09 (0 min) to 15.78 ± 5.97 % (30 min) at 2 % (w/w) MTGase concentration. The oxygen permeability and water vapour permeability did not change as a function of treatment time at 2 % (w/w) MTGase concentration. The incorporation of nanoclay inhibited the increase of oxygen permeability. Film colour values (L, a, and b) did not change, but haze values increased from 5.24 ± 0.40 (0 min) to 6.44 ± 0.94 (50 min). XRD and TEM results suggested that the nanoclay was exfoliated in fish gelatin film (Bae et al. 2009).

Composite films were prepared from pectin and fish skin gelatin, as reported by Liu et al. (2007). The inclusion of protein promoted molecular interactions. This resulted in a well-organized homogeneous structure, as revealed by scanning electron microscopy and fracture-acoustic emission analysis. The resultant composite films showed an increase in stiffness and strength and a decrease in water solubility and water vapor transmission rate, in comparison with films cast from pectin alone. The composite films inherited the elastic nature of proteins, and were thus more flexible than the pure pectin films. Treating the composite films with glutaraldehyde/methanol induced chemical cross-linking with the proteins and reduced the interstitial spaces among the macromolecules and, consequently, improved their mechanical properties and water resistance. Treating the protein-free pectin films with glutaraldehyde/methanol also improved the Young's modulus and tensile strength, but showed little effect on the water resistance, as the treatment caused only dehydration of the pectin films and the dehydration is reversible. The composite films were biodegradable and possessed moderate mechanical properties and a low water vapor transmission rate (Liu et al. 2007).

The effectiveness of using hydroxylpropylmethylcellulose (HPMC) to enhance mechanical strength and thermal stability in fish skin gelatin was studied by Chen et al. (2009). The significant increase in absorbance observed after HPMC had been added to fish gelatin and then matured indicated successful formation of a composite gel. Increased gel strength and storage modulus (G') indicated the enhanced gelation ability of the matured composite gel, while increased melting temperature (T_m) and enthalpy (ΔH) indicated its improved thermal stability. Maturation-related rheological property improvements were more noticeable at 4 °C than 10 °C, but no apparent differences in Tm improvement were observed between 4 and 10 °C maturation. Nevertheless, the composite gel exhibited reversible cold and thermal gelation properties (Chen et al. 2009).

Due to the poor mechanical properties of the fish gelatin membranes, composite nanofibers made of fish gelatin and poly(L-lactide) (PLLA) were recently produced with a novel solution (An et al. 2010). The introduction of PLLA remarkably improved the mechanical properties of the gelatin membranes. With a combination of good biocompatibility and mechanical properties, fish gelatin/PLLA blending non-woven mats are considered to be very promising in the area of tissue regeneration.

Recently, Qazvini et al. (2012) reported on the development of "*self-healing fish gelatin/sodium montmorillonite biohybrid coacervates*". Complex coacervation driven by associative electrostatic interactions was studied by these authors in mixtures of exfoliated sodium-montmorillonite (Na^+-MMT) nanoplatelets and fish gelatin, at a specific mixing ratio and room temperature. Structural and viscoelastic properties of the coacervate phase were investigated as a function of pH by means of different complementary techniques. Independent of the technique used, the results consistently showed that there is an optimum pH value at which the coacervate phase shows the tightest structure with highest elasticity. The solid-like coacervates showed an obvious shear-thinning behavior and network fracture, but immediately recovered back into their original elastic character upon removal of the shear strain. The nonlinear mechanical response characterized by single step stress relaxation experiments revealed the same trend for the yield stress and isochronal shear modulus of the coacervates. As a function of pH, the modulus has a maximum at pH 3.0 and lower values at 2.5 and 3.5 pHs, followed by a very sharp drop at pH 4.0. Finally, small-angle X-ray scattering (SAXS) data confirmed that at pHs lower than 4.0 the coacervate phases were dense and structured with a characteristic length scale (ξ(SAXS)) of ~7–9 nm. Comparing the ξ(SAXS) with rheological characteristic length (ξ(rheol)) estimated from low-frequency linear viscoelastic data and network theory, it was concluded that both the strength of the electrostatic interactions and the conformation of the gelatin chains before and during of the coacervation process are responsible for the structure and rigidity of the coacervates (Qazvini et al. 2012).

9.1 Fish Gelatin-Based Films

Gelatin applications in foods are derived largely from its gelation and film-forming properties, which are a consequence of an extended, fibrous tertiary structure and a triple helical cross-linked quaternary structure (Simon-Lukasik and Ludescher 2004). In gelatin, the triple-helical fold can involve segments of several different chains. As a result of these interchain cross-links, a network is formed with interstitial space for water. As gelatin gels age, water is excluded and the protein matrix condenses into a rubbery film that vitrifies upon drying (Simon-Lukasik and Ludescher 2004). Recently, Carvalho et al. (2008) reported that adding a concentration step by evaporating at 60 °C before spray drying the gelatin from Atlantic halibut (*Hippoglossus hippoglossus*) skin resulted in sizeable differences in the physical properties of the corresponding films. Differences in the mechanical behaviour of these films were attributed to slight differences in the molecular weight distributions of the two types of gelatin, caused by protein heat degradation during the evaporation step. As a result, these authors concluded that gelatin having a predominance of lower-molecular weight fractions underwent greater plasticization by the added sorbitol molecules, culminating in greater breaking elongation and lower tensile strength of the resulting films.

Potential applications of edible films and coatings from biopolymers are to retard transport of gases (O_2 and CO_2), water vapour, and flavours for fruits and vegetables, confectioneries, frozen foods and meat products. Edible films can be formed by two main processes, i.e., casting and extrusion (Hernandez-Izquierdo and Krochta 2008). The film-formation process most often reported in the scientific literature is the casting method. Briefly, it involves dissolving the biopolymer and blending it with plasticizers and/or additives to obtain a film-forming solution, which is cast onto plates and then dried by driving off the solvent. The extrusion method relies on the thermoplastic behaviour of proteins at low moisture levels. Films can be produced by extrusion followed by heat-pressing at temperatures that are ordinarily higher than 80 °C. This process may affect film properties, but its use would enhance the commercial potential of films by affording a number of advantages over solution-casting, e.g., working in a continuous system with ready control of such process variables as temperature, moisture, size/shape, etc.

Fish gelatin films has been produced from fresh water fish like trout (Kim and Min 2012) and carp (Ninan et al. 2010) as well as from representatives of marine species like salmon (Díaz et al. 2011) and Allaska Pollack (Shiku et al. 2004). Studies on the production and characterization of films using fish gelatins are quite recent, and all fish gelatins have been observed to exhibit good film-forming properties; yielding transparent, nearly colourless, water soluble, and highly extensible films (Avena-Bustillos et al. 2006; Carvalho et al. 2008; Gomez-Guillen et al. 2007; Jongjareonrak et al. 2006, 2008; Zhang et al. 2007; Pranoto et al. 2007; Rattaya et al. 2009). Fish gelatin films have also been envisioned as coatings for food products (Berg et al. 1985).

There has been recently an increasing interest related to the filmogenic capacity and applicability as food packaging (Arvanitoyannis 2002). The type of packaging is an important factor to enhance the conservation and protection of perishable foods, specifically in those cases where oxidative and microbiological deterioration occurs. The majority of the packaging materials used to be of synthetic origin. Nowadays environmental motivations have resulted in an increasing effort to find biodegradable edible materials, with an emphasis upon recycling industrial wastes, or using renewable resources (Tharanathan 2003). This same sensitivity can be seen in the recent scientific literature focused on edible and/or biodegradable films with numerous references related to gelatine, either pure or mixed with other biopolymers (see for review Gomez-Guillen et al. 2007).

Gelation, oxygen permeability, and mechanical properties of mammalian and fish gelatin films have been investigated in details by Avena-Bustillos et al. (2011). Here, the results obtained from this comparative study are summarized. Mammalian gelatin solutions had the highest gel set temperatures, followed by warm-water fish and then cold-water fish gelatin solutions. These differences were related to concentrations of imino acids present in each gelatin, with mammalian gelatin having the highest and cold-water fish gelatin the lowest concentrations. Mammalian and warm-water fish gelatin films contained helical structures, whereas cold-water fish gelatin films were amorphous. This was due to the films being dried at room temperature (23 °C), which was below or near the gelation temperatures of mammalian and warm-water fish gelatin solutions and well above the gelation temperature of cold-water fish gelatin solutions. Tensile strength, percent elongation, and puncture deformation were highest in mammalian gelatin films, followed by warm-water fish gelatin film and then by cold-water fish gelatin films. Oxygen permeability values of cold-water fish gelatin films were significantly lower than those for mammalian gelatin films. These differences were most likely due to higher moisture sorption in mammalian gelatin films, leading to higher oxygen diffusivity (Avena-Bustillos et al. 2011).

The use of different cross-linking agents has been also reported for fish-gelatin films and their properties (Chiou et al. 2008, 2009). Thus, Yi et al. (2006) prepared films using a commercial high molecular weight cold-water fish gelatin plasticized with sorbitol by inducing enzymatic cross-linking with a microbial transglutaminase (MTGase). The tensile strength and oxygen permeability of the MTGase-modified films increased, while elongation decreased. The mechanical and barrier properties of the gelatin films were explainable in terms of the total free volume of the film matrix. Thanks to the triple-helix structures present in gelatin molecules, the gelatin matrix is usually compact, resulting in low oxygen permeability. The intra and intermolecular covalent bonds formed by MTGase could increase the free volume of the polymer matrix by hindering helical structure formation. The lower number of helical structures could decrease the flexibility of the gelatin matrix, while the higher degree of cross-linking could increase its strength.

Interestingly, ultraviolet-B radiation can induce cross-linking and improves physical properties of cold- and warm-water fish gelatin gels and films (Otoni et al. 2012). Cold- and warm-water fish gelatin granules were exposed to ultraviolet-B radiation for doses up to 29.7 J/cm^2. Solutions and films were prepared from the granules. SDS-PAGE and refractive index results indicated there was cross-linking of gelatin chains after exposure to radiation. It was observed that UV-B treated samples displayed higher gel strengths, with cold- and warm-water fish gelatin having gel strength increases from 1.39 to 2.11 N and from 7.15 to 8.34 N, respectively. In addition, both gelatin samples exhibited an increase in viscosity for higher UV doses. For gelatin films, the cold-water fish gelatin samples made from irradiated granules showed greater tensile strength. In comparison, the warm-water gelatin films made from irradiated granules had lower tensile strength, but better water vapor barrier properties. This might be due to the UV induced cross-linking in warm-water gelatin that disrupted helical structures (Otoni et al. 2012).

As represented above, a broad variety of fish gelatin-based films are examples of composite- or even multicomposite-containing (Shakila et al. 2012) biomaterials. For example, edible films based on fish-skin gelatin incorporated with chitosan and/or clove essential oil were investigated by Gómez-Estaca and co-workers (2009a, b) and their antimicrobial activity was tested on *Lactobacillus acidophilus, Pseudomonas fluorescens, Listeria innocua*, and *Escherichia coli*. The films incorporated with the clove essential oil were the most effective, although differences were observed depending on the biopolymeric matrix in which it was included.

When a clove added film was applied to the preservation of raw sliced salmon, a reduction of the growth of total bacteria was observed after 11 days of storage at 2 °C. This demonstrates that edible films based on fish gelatin can be used as an active packaging applied to fish products (Gómez-Estaca et al. 2009b).

In order to provide gelatin films with antioxidant capacity, two sulphur-free water-insoluble lignin powders (L1000 and L2400) were blended with a commercial fish-skin gelatin from warm water species at a rate of 85 % gelatin: 15 % lignin (w/w) (G–L1000 and G–L2400), using a mixture of glycerol and sorbitol as plasticizers (Núñez-Flores et al. 2013). The gelatin films lose their typical transparent and colourless appearance by blending with lignin. However, the resulting composite films gained in light barrier properties, which could be of interest in certain food applications for preventing ultraviolet-induced lipid oxidation. Lignin proved to be an efficient antioxidant at non-cytotoxic concentrations, however, no remarkable antimicrobial capacity was found.

Additionally, the development of fish gelatine-chitosan films modified with transglutaminase or 1-ethyl-3-(3-dimethylaminopropyl) carbodiimide (EDC) and plasticized with glycerol has been reported (Kolodziejska et al. 2006; Kolodziejska and Piotrowska 2007).

9.2 Shark Skin and Cartilage Gelatin

Shark is caught worldwide. From 1990 to 2003, the reported global catch of sharks increased by 22 %, 80 % of which was caught by 20 countries including Spain, France, and the United Kingdom. European Union (EU) countries caught nearly 115,000 metric tons of shark in 2003. Spain took the largest share at around 45 % of the EU total, followed by France (18 %), the United Kingdom (14 %) and Portugal (10.5 %) (Lack and Sant 2006). Most shark cartilage is utilized for the production of chondroitin sulphate and calcium. Gelatin can be prepared before and after extraction of chondroitin sulphate and calcium from shark cartilage (Kwak et al. 2009).

Shark skin collagen has such features as less imino acid residue content, lower denaturation temperature, stronger swelling ability, higher solubility in acid medium, and different fibril assembly from that of pig skin collagen. Moreover, the rheological properties of shark collagen solution are also different from those of land animal origin. As the feature of collagen must reflect its derivative gelatin, it is presumed that shark gelatin has also a different characteristic from that of land animals (Yoshimura et al. 2000). For example, sol-gel and gel-sol transition temperatures for shark gelatin were remarkably lower than those for pig gelatin (Yoshimura et al. 2000). Shark gelatin gel shows a narrower pH range in which a stable gel is formed compared with pig gelatin. Melting enthalpy of shark gelatin gel was greater than that of pig gelatin gel, and G of shark gelatin gel changed more extensively with rising temperature in comparison with pig gelatin gel. It is concluded that shark gelatin is different from pig gelatin not only in its gel characteristics, but also in its solution properties.

Shark skin gelatin properties also differ from those of bony fish. Gelatin film from shark skin has a lower water vapour permeability (WVP) and higher opacity at 280 nm (UV wavelength) than have gelatin films from other fish. This gelatin film can thus be applied to pharmaceutical products or foods rich in fat due to its excellent barrier properties against water vapour and UV. However, WVP of shark skin gelatin film was higher than that of other edible films (Limpisophon et al. 2009, 2010).

How is possible to isolate gelatin from the rigid shark skin? Recently, Kharyeki et al. (2011) proposed the following procedure. Frozen shark skin was thawed overnight at 4 °C, and then the residual meat on skin was removed manually. The skin was cut into the 2–3 cm^2 pieces and then washed with cold tap water (8 °C). About 100 g of skin was used for each treatment. To remove non-collagenous proteins, the prepared skin was treated with five volumes (v/w) of cold NaOH (0.01–1 N) at 4 °C. The samples were then washed with tap water until neutral or faintly basic pHs of wash water was obtained. The skins were then soaked in cold HCl (0.01–1 N) at 4 °C with a ratio of 1:5 (w/v). The samples were washed out with cold tap water. Each treatment was repeated three times, with a total time of 1 h. For water extraction, five volumes (w/v) of distilled water was added into the sample and then heated at 55 °C (±0.2) (3–8 h) in a water bath. After the extraction, gelatin solutions were filtered using two layer filter cloth to remove the skin residues.

Changes in functional properties of shark (*Isurus oxyrinchus*) cartilage gelatin produced by different drying methods have been reported by Kwak et al. (2009). Freeze-dried gelatin was found to have the strongest gel strength, while gelatins made at high temperatures formed weaker gels. The 135-kPa gel strength of freeze-dried gelatin was relatively high. While foam formation ability of the freeze-dried gelatin was the highest, its foam stability was the lowest. In addition, spray-dried gelatin had the best emulsion capacities. Dynamic viscoelastic properties of shark cartilage gelatins prepared by these drying methods were closely correlated with their gel strength. Elasticity modulus and loss modulus of the freeze-dried gelatin had higher values than those prepared by hot-air drying and spray drying; viscoelastic properties of the freeze-dried gelatin were maintained longer than those of other drying methods.

The method for extraction of gelatin from shark cartilage has been also proposed (see, for example, Cho et al. 2004). Thus, cleaned shark cartilage was soaked in eight volumes of (v/w) of sodium hydroxide solution (1–2 N) at 8 °C in a 200 rpm shaking incubator for 2–4 days, to remove the non-collagen protein and subcutaneous tissue after they were swollen. The alkali treated shark cartilage was washed, neutralized with 2 N HCl and rewashed. For hot-water extraction, seven volumes of (v/w) of distilled water were added. Gelatin was extracted at the pre-determined temperatures (40–80 °C) and times (1–5 h) in a water bath. The extracted solutions (pH 8) were centrifuged for 30 min at 900 g at 30 °C. The upper phase was vacuum-filtered with a filter paper, the filtered solution vacuum-concentrated to 10 brix at 60 °C, and finally dried for 24 h in a hot-air dryer.

Similar to other the fish gelatins discussed above, gelatins of shark origin can be produced in different physical forms and can be used as biocomposites. Incorporation of fatty acids (stearic and oleic) into edible films based on blue shark (*Prionace glauca*) skin gelatin was recently investigated to modify properties such as the water vapour barrier and flexibility due to their hydrophobicity and plasticizing effect, respectively (Limpisophon et al. 2010). Addition of stearic acid from 0 to 100 % of protein concentration in the film-forming solution considerably decreased water vapour permeability of gelatin– fatty acid emulsion films, compared to addition of oleic acid at the same fatty acid concentration. Increasing concentrations of both fatty acids decreased tensile strength, but increased elongation at break due to their plasticizing effect. At the same concentration, oleic acid gave a greater plasticizing effect than did stearic acid. On the other hand, transparency of the gelatin–stearic acid emulsion film was lower than that of the gelatin–oleic acid emulsion film. Faster stirring speed during homogenization improved properties of only the gelatin–stearic acid emulsion film.

Intriguingly, gelatin derived from shark skin was recently hydrolysed to obtain antifreeze peptides (Wang et al. 2012). Antifreeze proteins can inhibiting the growth of crystals, decreasing the injury of cells, and can retain the structure, texture and quality of productions. The purpose of the study reported was to obtain natural antifreeze peptides, and to investigate the hypothermia protection activity on bacteria. The most appropriate protease and hydrolysis time was selected with the index of the hypothermia protection activity on bacteria. The hydrolysate was subsequently added on to Sephadex G-50 gel filtration column and SP-Sephadex C-25 column to acquire high activity fractions. The fraction of cationic peptides termed P2 showed higher antifreeze activity. The hypothermia protection assay showed that the survival rate of *E. coli* was 80.8 % when the concentration of peptides complexes was at 500 μg/mL (Wang et al. 2012).

9.3 Conclusion

Gelatin is a multifunctional ingredient used in foods, pharmaceuticals, cosmetics, and photographic films as a gelling agent, stabilizer, thickener, emulsifier, and film former. As a thermoreversible hydrocolloid with a narrower gap between its melting and gelling temperatures, both of which are below human body temperature, gelatin provides unique advantages over carbohydrate-based gelling agents (Boran and Regenstein 2010). Gelatin is widely used as a medical biomaterial because it is readily available, cheap, biodegradable and demonstrates favorable biocompatibility (Elvin et al. 2010). In spite of the very intriguing title of the patent by Andre et al (2000) "*Use of collagen of aquatic origin for the production of supports for tissue engineering, and supports and biomaterials obtained*", there are unfortunately only few reports on biomedical application of fish gelatins. However, some of these biopolymers possess interesting properties. For example, Nagai et al (2008) investigated blood compatibility evaluation of elastic gelatin gel from salmon collagen. It was shown that the platelet adhesion rate was markedly lower on the elastic fish gelatin -gel compared to collagen-coated and fibrinogen-coated surfaces. Moreover this gel demonstrated good blood compatibility.

Zhang et al (2011) investigated oral administration of skin gelatin isolated from Chum salmon (*Oncorhynchus keta*) on wound healing in diabetic rats. Skin gelatin-treated diabetic rats showed a better wound closure, increased microvessel density, vascular endothelial growth factor, hydroxyproline and a reduced extent of inflammatory response. All parameters were significant ($P<0.05$) in comparison to vehicle-treated diabetic group. The authors proposed that oral administration of Chum Salmon skin gelatin might be a beneficial method for treating wound disorders associated with diabetes.

Altogether, fish gelatins are an example of supramolecular polymer networks (Seiffert and Sprakel 2012), which are three-dimensional structures of crosslinked macromolecules connected by transient, non-covalent bonds. They are a fascinating class of soft materials, exhibiting the interesting physio-chemical and materials properties discussed in this chapter.

References

Aberoumand A (2010) Edible gelatin from some fishes skins as affected by chemical treatments. World J Fish Mar Sci 2(1):59–61

Alfaro AT, Costa CS, Kuhn CR et al (2005) Process for preparation of gelatin from bones of king weakfish (*Macrodon ancylodon*) after acid pretreatment. Alimentaria 365:74–81

An K, Liu H, Guo S et al (2010) Preparation of fish gelatin and fish gelatin/poly(L-lactide) nanofibers by electrospinning. Int J Biol Macromol 47(3):380–388

Andre V, Abdul Malak N, Huc A (2000) Use of collagen of aquatic origin for the production of supports for tissue engineering, and supports and biomaterials obtained. US Patent 6,541,023

André F, Cavagna S, André C (2003) Gelatin prepared from tuna skin: a risk factor for fish allergy or sensitization? Int Arch Allergy Immunol 130:17–24

Arnesen JA, Gildberg A (2007) Extraction and characterisation of gelatine from Atlantic salmon (*Salmo salar*) skin. Bioresour Technol 98:53–57

Arvanitoyannis IS (2002) Formation and properties of collagen and gelatin films and coatings. In: Gennadios A (ed) Protein-based films and coatings. CRC Press, Boca Raton

Asghar A, Henrickson RL (1982) Chemical, biochemical, functional, and nutritional characteristics of collagen in food systems. In: Chischester CO, Mark EM, Stewart GF (eds) Advances in food research. Academic, London

Avena-Bustillos RJ, Olsen CW, Olson DA et al (2006) Water vapor permeability of mammalian and fish gelatin films. J Food Sci 71:E202–E207

Avena-Bustillos RJ, Chiou B, Olsen CW et al (2011) Gelation, oxygen permeability, and mechanical properties of mammalian and fish gelatin films. J Food Sci 76:E519–E524. Journal of Food Science © 2011 Institute of Food Technologists®. No claim to original US government works. Reprinted with permission

Badii F, Howell NK (2005) Fish gelatine: structure, gelling properties and interaction with egg albumen proteins. Food Hydrocoll 20:630–640

Bae H (2007) Fish gelatin-nanoclay composite film. Mechanical and physical properties, effect of enzyme cross-linking, and as a functional film layer. All dissertations. Paper 179. Copyright © 2007, Bae HJ. Reprinted with permission

Bae HJ, Darby DO, Kimmel RM et al (2009) Effects of transglutaminase-induced cross-linking on properties of fish gelatin–nanoclay composite film. Food Chem 114:180–189

Berg RA, Silver FH, Watt WR, Norland RE (1985) Fish gelatin in coating applications. In: Image Technol 1985 (38th Annual SPSE Conference), Atlantic City, pp. 106–109

Bode F, da Silva MA, Drake AF (2011) Enzymatically cross-linked tilapia gelatin hydrogels: physical, chemical, and hybrid networks. Biomacromolecules 12(10):3741–3752
Boran G, Regenstein JM (2009) Optimization of gelatin extraction from silver carp skin. J Food Sci 74:E432–E441. © 2009 Institute of Food Technologists®
Boran G, Regenstein JM (2010) Fish gelatin. Adv Food Nutr Rev 60:119–143
Bower CK, Avena-Bustillos RJ, Hietala KA et al (2010) Dehydration of pollock skin prior to gelatin production. J Food Sci 75(4):C317–C321
Carvalho RA, Sobral PJA, Thomazine M et al (2008) Development of edible films based on differently processed Atlantic halibut (*Hippoglossus hippoglossus*) skin gelatin. Food Hydrocoll 22(6):1117–1123
Chen HH, Lin CH, Kang HY (2009) Maturation effects in fishgelatin and HPMC composite gels. Food Hydrocoll 23:1756–1761
Cheow CS, Norizah MS, Kyaw ZY, Howell NK (2007) Preparation and characterisation of gelatins from the skins of sin croaker (*Johnius dussumieri*) and shortfin scad (*Decapterus macrosoma*). Food Chem 101:386–391
Chiou BS, Avena-Bustillos RJ, Shey J et al (2006) Rheological and mechanical properties of cross-linked fish gelatins. Polymer 47:6379–6386
Chiou BS, Avena-Bustillos RJ, Bechtel PJ et al (2008) Cold water fish gelatin films: effects of cross-linking on thermal, mechanical, barrier, and biodegradation properties. Eur Polym J 44:3748–3753
Chiou BS, Avena-Bustillos RJ, Bechtel PJ et al (2009) Effects of drying temperature on barrier and mechanical properties of cold-water fish gelatin films. J Food Eng 95:327–331
Cho SM, Kwak KS, Park DS et al (2004) Processing optimization and functional properties of gelatin from shark (*Isurus oxyrinchus*) cartilage. Food Hydrocoll 18:573–579
Cho SM, Gu YS, Kim SB (2005) Extracting optimization and physical properties of yellowfin tuna (*Thunnus albacares*) skin gelatin compared to mammalian gelatins. Food Hydrocoll 19:221–229
Choi SS, Regenstein JM (2000) Physicochemical and sensory characteristics of fish gelatin. J Food Sci 65:194–199
Díaz P, López D, Matiacevich S et al (2011) State diagram of salmon (*Salmo salar*) gelatin films. J Sci Food Agric 91(14):2558–2565
Elvin CM, Vuocolo T, Brownlee AG et al (2010) A highly elastic tissue sealant based on photopolymerised gelatin. Biomaterials 31(32):8323–8331
Fernandez-Dıaz MD, Monter P, Gomez-Guillen MC (2003) Effect of freezing fish skins on molecular and rheological properties of extracted gelatin. Food Hydrocoll 17:281–286
Geiger SE, Roberts E, Tomlinson N (1962) Dogfish gelatin. J Fish Res B Can 19(2):321–326
Gilsenan PM, Ross-Murphy SB (2000) Rheological characterisation of gelatins from mammalian and marine sources. Food Hydrocoll 14(3):191–195
Giménez B et al (2005) Reprinted from Giménez B, Turnay J, Lizarbe MA et al (2005) Use of lactic acid for extraction of fish skin gelatin. Food Hydrocoll 19(6):941–950. Copyright (2005), with permission from Elsevier
Gómez-Estaca J, López de Lacey A, Gómez-Guillén MC et al (2009a) Antimicrobial activity of composite edible films based on fish gelatin and chitosan incorporated with clove essential oil. J Aquat Food Prod Technol 18:46–52
Gómez-Estaca J, Montero P, Fernandez-Martin F et al (2009b) Physico-chemical and film-forming properties of bovine-hide and tuna-skin gelatin: a comparative study. J Food Eng 90(4):480–486
Gomez-Guillen MC, Montero P (2001a) Extraction of gelatin from megrim (*Lepidorhombus boscii*) skins with several organic acids. J Food Sci 66(2):213–216
Gomez-Guillen MC, Montero P (2001b) Method for the production of gelatin of marine origin and product thus obtained. International Patent PCT/S01/00275
Gomez-Guillen MC, Turnay J, Fernandez-Dıaz MD (2002) Structural and physical properties of gelatin extracted from different marine species: a comparative study. Food Hydrocoll 16:25–34

Gomez-Guillen MC, Ihl M, Bifani V et al (2007) Edible films made from tuna-fish gelatin with antioxidant extracts of two different murta ecotypes leaves (*Ugni molinae Turcz*). Food Hydrocoll 21:1133–1143

Gomez-Guillen MC, Perez-Mateos M, Gomez-Estaca J et al (2009) Fish gelatin: a renewable material for developing active biodegradable films. Trends Food Sci Technol 20:3–16

Grossman S, Bergman M (1992) Process for the production of gelatin from fish skins. US Patent 5,093,474

Gudmundsson M (2002) Rheological properties of fish gelatins. J Food Sci 67:2172–2176

Gudmundsson M, Hafsteinsson H (1997) Gelatin from cod skins as affected by chemical treatments. J Food Sci 62:37–39

Guerard F, Dufosse L, Broise DDL et al (2001) Enzymatic hydrolysis of proteins from yellowfin tuna *Thunnus albacares* wastes using Alcalase. J Mol Catal B: Enzym 11:1051–1059

Gustavson K (1956) The chemistry and reactivity of collagen. Academic, New York

Hamada M (1990) Effects of the preparation conditions on the physical properties of shark-skin gelatin gels. Nippon Suisan Gakkaishi 56:671–677

Han X, Xu Y, Wang J et al (2009) Effects of cod bone gelatin on bone metabolism and bone microarchitecture in ovariectomized rats. Bone 44:942–947

Haug IJ, Draget KI, Smidsrod A (2004) Physical and rheological properties of fish gelatine compared to mammalian gelatin. Food Hydrocoll 18(2):203–213

Hernandez-Izquierdo VM, Krochta JM (2008) Thermoplastic processing of proteins for film formation e a review. J Food Sci 73(2):30–39

Herpandi, Huda N, Adzitey F (2011) Fish bone and scale as a potential source of halal gelatin. J Fish Aquat Sci 6:379–389. Copyright © 2011 Academic Journals Inc. Reprinted with permission

Hilbert H (1906) Manufacture of gelatin and glue from bones. US Patent 834,806

Huda HN, Adzitey F (2012) Fish bone and scale as a potential source of halal gelatin. J Fish Aquat Sci 6(4):379–389

Irwandi J, Faridayanti S, Mohamed ESM et al (2009) Extraction and characterization of gelatin from different marine fish species in Malaysia. Int Food Res J 16:381–389

Jamilah B, Harvinder KG (2002) Properties of gelatins from skins of fish – black tilapia (*Oreochromis mossambicus*) and red tilapia (*Oreochromis nilotica*). J Food Chem 77:81–84

Johnston-Banks FA (1990) Gelatin. In: Harris P (ed) Food gels. Elsevier, London, pp 233–289

Jongjareonrak A, Benjakul S, Visessanguan W et al (2006) Skin gelatin from bigeye snapper and brownstripe red snapper: chemical compositions and effect of microbial transglutaminase on gel properties. Food Hydrocoll 20(8):1216–1222

Jongjareonrak A, Benjakul S, Visessanguan W et al (2008) Antioxidative activity and properties of fish skin gelatin films incorporated with BHT and a-tocopherol. Food Hydrocoll 22(3):449–458

Karim AA, Bhat R (2009) Fish gelatin: properties, challenges and prospects as an alternative to mammalian gelatins. Food Hydrocoll 23:563–576

Kharyeki ME, Rezaei M, Motamedzadegan A (2011) The effect of processing conditions on physico-chemical properties of whitecheek shark (*Carcharhinus dussumieri*) skin gelatin. Int Aquat Res 3:63–69

Kim D, Min SC (2012) Trout skin gelatin-based edible film development. J Food Sci 77(9):E240–E246

Kittiphattanabawon P, Benjakul S, Visessanguan W et al (2005) Characterization of acid soluble collagen from skin and bone of bigeye snapper (*Priacanthus tayenus*). Food Chem 89:363–372

Koli JM, Basu S, Nayak BB et al (2011) Improvement of gel strength and melting point of fish gelatin by addition of coenhancers using response surface methodology. J Food Sci 76(6):E503–E509

Kolodziejska I, Piotrowska B (2007) The water vapour permeability, mechanical properties and solubility of fish gelatine-chitosan films modified with transglutaminase or 1-ethyl-3-(3-dimethylaminopropyl) carbodiimide (EDC) and plasticized with glycerol. Food Chem 103(2):295–300

Kolodziejska I, Piotrowska B, Bulge M et al (2006) Effect of transglutaminase and 1-ethyl-3-(3-dimethylaminopropyl) carbodiimide on the solubility of fish gelatine-chitosan films. Carbohydr Polym 65(4):404–409
Kwak KS, Cho SM, Ji CI et al (2009) Changes in functional properties of shark (*Isurus oxyrinchus*) cartilage gelatin produced by different drying methods. Int J Food Sci Technol 44:1480–1484
Lack M, Sant G (2006) World shark catch, production and trade, 1990–2003. Australian Government, Department of the Environment and Heritage, and TRAFFIC Oceania
Ladislaus MK, Yan X, Yao W et al (2007) Optimization of gelatine extraction from grass carp (*Catenopharyngodon idella*) fish skin by response surface methodology. Bioresour Technol 98:3338–3343
Lefevre G, Biarrotte R, Takerkart G et al (2002) Process for the preparation of fish gelatin. US Patent 6,368,656
Leuenberger BH (1991) Investigation of viscosity and gelation properties of different mammalian and fish gelatins. Food Hydrocoll 5(4):353–361
Limpisophon K, Tanaka M, Weng WY et al (2009) Characterization of gelatin films prepared from under-utilized blue shark (*Prionace glauca*) skin. Food Hydrocoll 23:1993–2000
Limpisophon K, Tanaka M, Osako K (2010) Characterisation of gelatin–fatty acid emulsion films based on blue shark (*Prionace glauca*) skin gelatin. Food Chem 122:1095–1101
Liu L, Liu CK, Fishman ML et al (2007) Composite films from pectin and fish skin gelatin or soybean flour protein. J Agric Food Chem 55(6):2349–2355
Montero P, Gomez-Guillen MC (2000) Extracting conditions for megrim (*Lepidorhombus boscii*) skin collagen affect functional properties of the resulting gelatin. J Food Sci 65:434–438
Nagai N, Kubota R, Okahashi R et al (2008) Blood compatibility evaluation of elastic gelatin gel from salmon collagen. J Biosci Bioeng 106(4):412–415
Ninan G, Joseph J, Abubacker Z (2010) Physical, mechanical, and barrier properties of carp and mammalian skin gelatin films. J Food Sci 75(9):E620–E626
Norland RE (1990) Fish gelatin. In: Voigt MJ, Botta JR (eds) Advances in fisheries technology and biotechnology for increased profitability. Technomic Publisher Co. Inc., Lancaster
Norland (2003) Fish gelatin, technical aspects. Norland Products Inc., North Brunswick
Núñez-Flores R, Giménez B, Fernández-Martín F, López-Caballero ME, Montero MP, Gómez-Guillén MC (2013) Physical and functional characterization of active fish gelatin films incorporated with lignin. Food Hydrocoll 30(1):163–172
Ogawa M, Portier PJ, Moody MW (2004) Biochemical properties of bone and scale gelatin isolated from the subtropical fish black drum (*Pogomia cromis*) and sheepsheas seabream (*Archosargus probatocephalus*). Food Chem 88:495–501
Otoni CG, Avena-Bustillos RJ, Chiou BS et al (2012) Ultraviolet-B radiation induced cross-linking improves physical properties of cold- and warm-water fish gelatin gels and films. J Food Sci 77(9):E215–E223
Pranoto Y, Lee CM, Park HJ (2007) Characterizations of fish gelatin films added with gellan and k-carrageenan. LWT-Food Sci Technol 40:766–774
Qazvini NT, Bolisetty S, Adamcik J et al (2012) Self-healing fish gelatin/sodium montmorillonite biohybrid coacervates: structural and rheological characterization. Biomacromol 13(7):2136–2147
Rahman MS, Al-Mahrouqi AI (2009) Instrumental texture profile analysis of gelatin gel extracted from grouper skin and commercial (bovine and porcine) gelatin gels. Int J Food Sci Nutr 60(Suppl 7):229–242
Rattaya S, Benjakul S, Prodpran T et al (2009) Properties of fish skin gelatin film incorporated with seaweed extract. J Food Eng 95:151–157
Regenstein JM, Chaudry MM (2002) Kosher and Halal issues pertaining to edible films and coatings. In: Gennadios A (ed) Protein-based films and coatings. CRC Press LLC, Boca Raton
Sarabia AI, Gomez-Guillen MC, Montero P (2000) The effect of added salts on the viscoelastic properties of fish skin gelatin. J Food Chem 70:71–76

Seiffert S, Sprakel J (2012) Physical chemistry of supramolecular polymer networks. Chem Soc Rev 41(2):909–930
Shakila JR, Jeevithan E, Varatharajakumar A et al (2012) Comparison of the properties of multi-composite fish gelatin films with that of mammalian gelatin films. Food Chem 135(4):2260–2267
Shiku Y, Hamaguchi PY, Benjakul S et al (2004) Effect of surimi quality on properties of edible films based on Alaska Pollack. Food Chem 86:493–499
Simon-Lukasik KV, Ludescher RD (2004) Erythrosin B phosphorescence as a probe of oxygen diffusion in amorphous gelatin films. Food Hydrocoll 18:621–630
Solgaard G, Haug IJ, Draget KI (2008) Proteolytic degradation of cold-water fish gelatin solutions and gels. Int J Biol Macromol 43(2):192–197
Songchotikunpan P, Tattiyakul J, Supaphol P (2008) Extraction and electrospinning of gelatin from fish skin. Int J Biol Macromol 42:247–255
Taheri A, Abedian Kenari AM, Gildberg A et al (2009) Extraction and physicochemical characterization of greater Lizard fish (*Saurida tumbil*) skin and bone gelatin. J Food Sci 74(3):E160–E165
Tharanathan RN (2003) Biodegradable films and composite coatings: past, present and future. Trends Food Sci Technol 14(3):71–78
Viel C, Fournier J (2006) History of gelatin extraction process and debates of French academic commissions (XIXth century). Rev Hist Pharm (Paris) 54(349):7–28
Wang Y, Yang H (2009) Effects of concentration on nanostructural images and physical properties of gelatin from channel catfish skins. Food Hydrocoll 23:577–584
Wang SY, Zhao J, Xu ZB et al (2012) Preparation, partial isolation of antifreeze peptides from fish gelatin with hypothermia protection activity. Appl Mech Mater 140:411–415
Wangtueai and Noomhorm (2009) Reprinted from Wangtueai and Noomhorm (2009) Processing optimization and characterization of gelatin from lizardfish (*Saurida* spp.) scales. LWT- Food Sci Technol 42:825–834, Copyright (2009), with permission from Elsevier
Yang JH (2007) Fish gelatin hard capsule and its preparation method. US Patent 7,309,499
Yang H, Wang Y (2009) Effects of concentration on nanostructural images and physical properties of gelatin from channel catfish skins. Food Hydrocoll 23:577–584
Yang H, Wang Y, Regenstein JM, Rouse DB (2007) Nanostructural characterization of catfish skin gelatin using atomic force microscopy. J Food Sci 72(8):C430–C440
Yi JB, Kim YT, Bae HJ et al (2006) Influence of transglutaminase-induced cross-linking on properties of fish gelatin films. J Food Sci 71(9):376–383
Yoshimura K, Terashima M, Hozan D et al (2000) Physical properties of shark gelatin compared with pig gelatin. J Agric Food Chem 48:2023–2027
Zhang S, Wang Y, Herring JL, Oh J (2007) Characterization of edible film fabricated with channel catfish (*Ictalurus punctatus*) gelatin extract using selected pretreatment methods. J Food Sci 72(9):498–503
Zhang Z, Zhao M, Wang J et al (2011) Oral administration of skin gelatin isolated from Chum Salmon (*Oncorhynchus keta*) enhances wound healing in diabetic rats. Mar Drugs 9(5):696–711. Copyright © 2011 by the authors; licensee Molecular Diversity Preservation International, Basel, Switzerland. This article is an open-access article distributed under the terms and conditions of the Creative Commons Attribution license (http://creativecommons.org/licenses/by/3.0/)

Chapter 10
Marine Elastin

Abstract Elastic function for a lifetime in animals, including marine vertebrates is determined by the structural protein elastin. Extracellular matrix is the space where a monomer, tropoelastin, is rapidly transformed into its final polymeric form. Desmosine and isodesmosine serve as crosslinking molecules, binding the polymeric chains of amino acids into the 3D network of elastin. Elastin is located within such tissues and organs as skin, arteria, heart, notochord and swimbladder in marine fish. Interestingly, the lamprey possesses only elastin-like fibrillar proteins. The whale heart and aorta are the largest elastin-containing structures known in Nature.

According to definition, proposed by Miao et al. (2006), "elastin is the extracellular matrix protein responsible for properties of extensibility and elastic recoil in large blood vessels, lung, and skin of most vertebrates," (Miao et al. 2006). In 1963, Partridge and co-workers reported in *Nature* the first isolation of two new amino acids, named desmosine and isodesmosine, from the elastin of bovine ligamentum nuchae (Partridge et al. 1963). Further studies by Thomas et al. (1963) showed these compounds to be 1,3,4,5- and 1,2,3,5-tetra-substituted pyridinium salts, with each side chain having both a carboxyl-terminal and -amino terminal group. The side chain at position 1 was found to be a B-carbon straight chain. These amino acids seem to be essential components of elastin, and have been shown to be present in elastins obtained from various sources (Anwar 1966; Kielty et al. 2002). They serve as crosslinking molecules, binding the polymeric chains of amino acids into the 3D network of elastin (Piez 1968). Unique type of amino acid derived from the condensation of four lysine residues is known as desmosine (including the isomer isodesmosine) (Fig. 10.1) (see for review Anwar and Oda 1966; Partridge 1966; Petruska and Sandberg 1968; Franzblau and Lent 1968; Shimada et al. 1969). "This unique characteristic is considered useful in discriminating elastin breakdown-derived peptides from precursor elastin peptides. Based on this feature, desmosine has been extensively evaluated as a potentially attractive indicator of elevated lung elastic fibre turnover, and a marker for the effectiveness of agents with the potential to reduce elastin breakdown," (Luisetti et al. 2012; see for more information Turino et al. 2011).

Since the very first publications on elastin chemistry (Partridge et al. 1955), structure (Greenlee et al. 1966) and isolation in the chemically pure form (Steven et al. 1974; Starcher and Galione 1976; Soskel et al. 1987), numerous papers were

H. Ehrlich, *Biological Materials of Marine Origin*, Biologically-Inspired Systems 4,
DOI 10.1007/978-94-007-5730-1_10

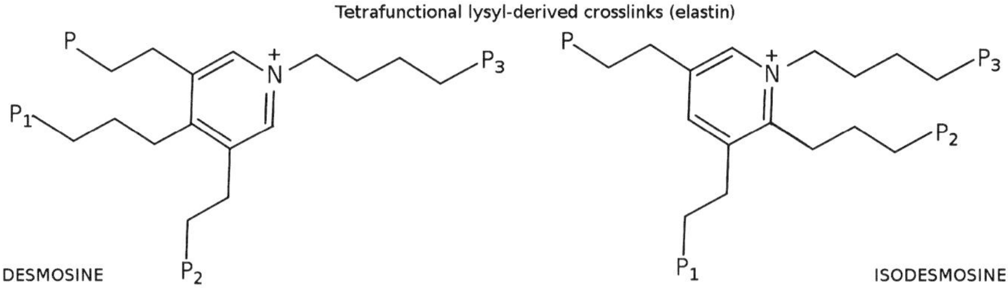

Tetrafunctional lysyl-derived crosslinks (elastin)

DESMOSINE ISODESMOSINE

Fig. 10.1 Elastin is the substrate for lysil oxidase which is responsible for oxidative deamination of certain lysine residues in this structural protein. Corresponding products of this reaction are represented here (Kielty et al. 2002)

related to physico-chemical (Mistrali et al. 1971; Gosline 1978; Lillie and Gosline 1996; Lillie et al. 1996) and mechanical (Gosline and French 1979; Aaron and Gosline 1981; Lillie and Gosline 1990; Lillie et al. 1994) properties, as well as the structural biology (Montes 1996) and genomics (He et al. 2007) of this bioelastomer. Special attention was paid to biomechanics of arterial elastin (see for details Gundiah et al. 2007, 2009). Tropoelastin is a 60–72 kDa protein of the extracellular matrix origin that is distributed in all vertebrates, with exception of cyclostomes (Wise and Weiss 2009). Molecular orientation of tropoelastin is determined by surface hydrophobicity (Le Brun et al. 2012).

Wise and Weiss (2009) characterized some properties of this protein as follow:

"Secreted tropoelastin is tethered to the cell surface, where it aggregates into organised spheres for cross-linking and incorporation into growing elastic fibres. Tropoelastin is characterised by alternating hydrophobic and hydrophilic domains, and is highly flexible. The conserved C-terminus is an area of the molecule of particular biological importance in that it is required for both incorporation into elastin, and for cellular interactions. Mature cross-linked tropoelastin gives elastin, which confers resilience and elasticity on a diverse range of tissues," (Wise and Weiss 2009; see also Sandberg et al. 1971).

According to the modern view (see for review Yeo et al. 2012), tropoelastin is secreted by elastogenic cells such as smooth muscle cells, endothelial cells, and fibroblasts. At the cell surface, the monomers cluster through hydrophobic domain interactions in an aqueous environment by the entropically driven process of coacervation. These tropoelastin assemblies remain attached through the C terminus to cell-surface integrin αvβ3 and glycosaminoglycans until deposition on microfibrillar scaffolds, which direct the shape and orientation of elastic fibers. Microfibrillar proteins recruit lysyl oxidase, which reacts with specific tropoelastin lysine residues to form cross-links. These cross-links occur at multiple sites in the molecule and are enriched in domains 19–25. Cross-linking imposes expansional constraints on elastin and renders elastic fibers resilient under repetitive mechanical stretching (Yeo et al. 2012).

A tropoelastin gene has been identified in amphibians as well as in avian and mammalian species (Chung et al. 2006). In contrast, teleosts species possess two tropoelastin genes with different tissue expression patterns. "General characteristics of tropoelastins, such as alternating arrangements of hydrophobic and crosslinking domains, are conserved across a wide phylogenetic range. However, the sequences of these domains are highly variable, particularly when amphibian and teleost tropoelastins are included," (Miao et al. 2009). It is suggested (Fritze et al. 2012) that human ageing is accompanied by a destruction of the elastic vascular structure: tropoelastin expression analysis shows that elastogenesis occurs throughout life with constantly decreasing levels.

How to replace damaged elastin-rich tissue using tropoelastin is one of the intriguing questions in biomedicine and biological materials science (Wise and Weiss 2009). The biomimetic potential of tropoelastin is characterized in the following form: "This modular, multifaceted molecule is being exploited to enhance the physical performance and biological presentation of engineered constructs to augment and repair human tissues. These tissues include skin and vasculature, and emphasize how growing knowledge of tropoelastin can be powerfully adapted to add value to pre-existing devices like stents and novel, multi-featured biological implants," (Mithieux et al. 2013).

Recently, attention is paid also to the tropoelastin-derived sequences which are crucial for understanding of the structural mechanism that underlies the elastomeric mechanical response of this specialized biological material. This statement can be confirmed by (Conticello and Carpenter Desai 2012):

"Tropoelastin and elastin-derived polypeptide sequences display a thermally reversible phase transition above a lower critical solution temperature, Tt, which coincides with the development of elastomeric restoring force in the material. The functionally critical properties of native elastins can be recapitulated in polypeptides that are composed of concatenated sequences of oligopeptide repeat motifs derived from tropoelastin; the most common of which are the pentapeptide sequences (Val-Pro-Gly-Xaa-Gly). Moreover, biosynthetic methods have been developed that enable the preparation of elastin-mimetic protein polymers that comprise complex sequences of defined macromolecular architecture (i.e., length, composition, and sequence); including multiblock copolymers. Thus, biosynthetic elastin-mimetic polypeptides represent the best-characterized biologically derived smart materials that have been prepared and analyzed to date," (Conticello and Carpenter Desai 2012).

Evolution of elastin is strongly related to the evolution of circulatory systems (see for review Sage and Gray 1976, 1979, 1980, 1981; Sage 1982, 1983; Chalmers et al. 1999; Faury 2001). In the classical paper entitled "*Evolution of elastin structure*" (Sage and Gray 1977), the authors tested the aortae of a number of vertebrates and invertebrates and reported as follow: "Comparison of purified elastins from several vertebrate groups reveals some striking differences in their amino acid compositions and properties, including the arrangement of the elastic fibers in the aorta. The patterns of variations in amino acid composition suggest a mode of evolution which is different from the slow accumulation of point mutations observed with globular proteins," (Sage and Gray 1977). Faury (2001) gives more detailed explanation:

> "Evolution of species has led to the appearance of circulatory systems. These include blood vessels and one or more pulsatile pumps, typically resulting in a low-pressurised open circulation in most invertebrates and a high-pressurised closed circulation in vertebrates. In both open and closed circulations, the large elastic arteries proximal to the heart damp out the pulsatile flow and blood pressure delivered by the heart, in order to limit distal shear stress and to allow regular irrigation of downstream organs.
>
> To achieve this goal, networks of resilient and stiff proteins adapted to each situation – i.e. low or high blood pressure -have been developed in the arterial wall to provide it with non-linear elasticity. In the low-pressurised circulation of some invertebrates, the mechanical properties of arteries can almost be entirely microfibril-based. In high-pressurised circulations, they are due to an interplay between a highly resilient protein, an elastomer in the octopus and elastin in most vertebrates, and the rather stiff protein collagen.
>
> In vertebrate development, elastin is incorporated in elastic fibres, on pre-deposited scaffold of microfibrils. The elastic fibres are then arranged in functional concentric elastic lamellae and, with the smooth muscle cells, lamellar units. The microfibrils may also play a direct functional role in all mature arteries of high- and low-pressurised circulations. Finally, blood pressure regularly increases with developmental stages it appears possible that the early deposition of microfibrils, which are highly-conserved in evolution, corresponds at least in part, to an early microfibril-driven elasticity in low-pressurised arteries, present across species. In vertebrates, when pressure developmentally rises above a threshold value, the vascular wall stress may turn on the expression of other resilient protein genes, including the elastin gene. Elastin would then be deposited on microfibrils, resulting in the elastic fibre network and elastic lamellae whose mechanical properties are adapted to allow for proper arterial work at higher pressures," (Faury 2001).

Therefore, below, I would like to discuss in particular data reported on the elastin properties of arteries in marine fish and mammals, including cetaceans. Intriguingly, elastin-based arteria of giant whale species are examples of the largest elastin-containing blood vessels known in Nature. In these large marine species, "this rubber-like protein forms a highly extensible tissue that has an elastic modulus of approximately 1 MPa, comparable with that of an ordinary rubber band," (Shadwick 1999; see also Aaron and Gosline 1980).

10.1 Elastin-Like Proteins in Lamprey

In contrast to gnathostomes with cartilage of fibrillar collagen origin, the same tissue in lampreys possesses elastin-like proteins as the dominant matrix components (Wright et al. 1988). So called *lamprin* has been identified as example of the elastin-like protein of lamprey trabecular cartilage (Robson et al. 1993; McBurney et al. 1996), although that of the pharyngeal cartilage has not been characterized (Robson et al. 1997). There are some elastin-like sequences reported for definitively non-collagenous lamprin (Robson et al. 2000). For example, the repetitive sequence GGLGY that was found in lamprin, appear also in other proteins, i.e. elastin, spidroin, spider minor ampullate silk proteins, in matrix proteins of the chorion, or the egg shell membrane of insects (Bochicchio et al. 2001).

"Lamprin genes from sea lamprey (*Petromyzon marinus*) contain either seven or eight exons, with exon 4 being alternatively spliced in all genes, resulting in a total of six different lamprin transcripts. All exon junctions are of class 1,1. An unusual feature of the lamprin gene structure is the distribution of the 3′ untranslated region sequence among multiple exons. Sequence and gene structure comparisons between lamprins, elastins, and insect structural proteins suggest that the regions of sequence similarity are the result of a process of convergent evolution," (Robson et al. 2000).

Here, the results of detailed investigations on elastin distribution in different cartilages of sea lamprey obtained by Wright and co-workers in 1988 using immunohistochemical and ultrastructural methods:

- "Chondrocytes are similar to those in hyaline cartilage.
- Lamprin fibrils and matrix granules, but no collagen fibrils, are found in a matrix arranged into pericellular, territorial, and interterritorial zones.
- Branchial, pericardial, and nasal cartilages differ from trabecular, annular, and piston cartilages in the organization of their matrix; as well as in the structural components of their matrix and perichondria.
- Immunoreactive elastin-like material is present within the perichondria and peripheral matrices of nasal, branchial, and pericardial cartilages in both larval and adult lampreys.
- Oxytalan, elaunin, and elastic-like fibers are dispersed between collagen fibers in the perichondrium. The matrix contains lamprin fibrils, matrix granules, and a band of amorphous material, which is reminiscent of elastin, in the periphery bordering the perichondrium.
- The presence of elastic-like fibers and elastin-like material within some lamprey cartilages implies that this protein may have evolved earlier in vertebrate history than has been previously suggested," (Wright et al. 1988).

Lysyl pyridinoline was found in high concentration in the elastin-like protein of lamprey branchial cartilage at a ratio of 7:1 to hydroxylysyl pyridinoline, the form that dominates in vertebrate collagens (Fernandes and Eyre 1999). Both forms of pyridinoline cross-link were absent from annular cartilage. Desmosine cross-links, which are characteristic of vertebrate elastin, were not detected in either form of lamprey cartilage. Pyridinoline cross-links are considered to be characteristic of collagen, so their presence in an elastin-like protein in a primitive cartilage poses evolutionary questions about the tissue, the protein, and the cross-linking mechanism. In contrast to ventral aorta in higher vertebrates, the wall of the same organ in lamprey is composed principally of microfibrils, and not of collagen and elastin. DeMont and Wright (1993) "measured the mechanical properties of the ventral aorta wall of the lamprey, and showed that the architecture provides a mechanical structure that does produce functional mechanical properties similar to aortae in higher vertebrates," (DeMont and Wright 1993).

It is proposed "that lampreys possess an ancestral type of cartilage that is similar to amphioxus gill cartilage. In this respect, gnathostome cartilage can be regarded as derived from the loss of an elastin-like protein as a cartilage component,"

(Ohtani et al. 2008). Two types of cartilages have been identified in the hagfish. The first one contains the *myxinin*, also the elastin-like protein as the main component. However, the second type of the hagfish cartilage superficially resembles notochord cells and is of collagenous origin (Wright et al. 1984).

10.2 Fish Elastin

Elastin was identified in both fresh water (Licht and Harris 1973; Serafini-Fracassini et al. 1978; Isokawa et al. 1988; Raso 1993) and marine (Sanchez-Quintana and Hurle 1987; Chow et al. 1989; Isokawa et al. 1990; Icardo and Colvee 2011; Braun et al. 2003) fish species. For example, tropoelastin genes in developing and adult zebrafish have been studied by Miao et al. (2006). These authors reported that: "Expression of genes was detected early in skeletal cartilage structures of the head, in the developing outflow tract of the heart, including the bulbus arteriosus and the ventral aorta, and in the wall of the swim bladder. While the temporal pattern of expression was similar for both genes, the upregulation of eln2 was much stronger than that of eln1. In general, both genes were expressed and their gene products deposited in most of the elastic tissues examined; with the notable exception of the bulbus arteriosus, in which eln2 expression and its gene product was predominant. This finding may represent a sub-specialization of eln2 to provide the unique architecture of elastin and the specific mechanical properties required by this organ," (Miao et al. 2006).

Thus, the places of elastin locations in fish organisms are:

- skin (Ferri 1980; Motta et al. 2012);
- arterial system (Isokawa et al. 1990; Bushnell et al. 1992);
- heart (Icardo and Colvee 2011);
- notochord (Sagstad et al. 2011) and
- swimbladder (Perrin et al. 1999; Robertson et al. 2007).

Skin Ferri in 1980 represented detailed study on elastic fibers, elaunin and oxytalan in the dermis of a teleost using optical and electron microscopy. The author reported that the dermo-epidermal junction is the location place of oxytalan that was identified in the form of fibers, which are bound by one of their extremities to the basal lamina. The other extremity is observed to be in continuity with elaunin fibers, which, however, cross the entire stratum compactum perpendicularly. These fibers are interconnected to other elaunin fibers, but in the deeper zone. Thick elastic fibers, forming an apparent continuous layer, have been identified between the stratum spongiosum and the stratum compactum (Ferri 1980). Elastin was also identified within skin of the blacktip shark (*Carcharhinus limbatus*) and of the shortfin mako (*Isurus oxyrinchus*). "The scales of the two species are anchored in the stratum laxum of the dermis. The attachment fibers of the scales in both species appear to be almost exclusively collagen, with elastin fibers visible in the stratum laxum of both species," as reported by Motta et al. (2012).

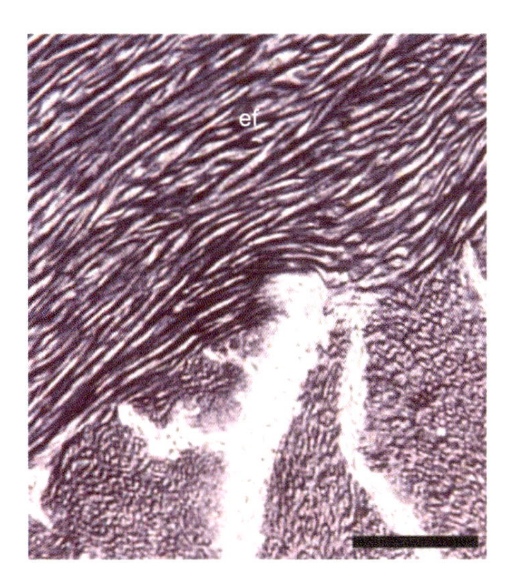

Fig. 10.2 Cross-section of the anterior bulbus of Yellowfin tuna. "The elastin is heavily stained and shows the circumferential alignment of the elastin fibres (top) in the media. The circumferential orientation of the elastin fibres is changed to longitudinal within the longitudinal elements. Scale bar, 100 μm" (Republished with permission of Company of Biologists Ltd., from Braun et al. (2003) Copyright © 2003 The Company of Biologists Ltd; permission conveyed through Copyright Clearance Center, Inc.)

Arterial System and Heart The pericardial cavity is the space where the *bulbus arteriosus* is located in teleost. The question concerning the arterial or cardiac nature of this organ is still open (Benjamin et al. 1983). The description of bulbus arteriosus is to be found in the work by Braun et al. (2003):

"Typically, it is swollen proximally, tapering distally into the ventral aorta. Within these parameters, a large variety of shapes occur, but the functional significance of the different shapes is unknown. The internal morphology of the bulbus also shows a large amount of variance between species. The lumen can be smooth, ridged or trabeculated. The trabeculae can be "spongy", or separated into discrete longitudinal and radial elements. The bulbar wall is composed of three layers: an intima composed of a single layer of endothelial cells, a thick media primarily composed of elastin (Fig. 10.2) and smooth muscle, and a collagenous adventitia surrounded by an outer layer of mesothelial cells," (Braun et al. 2003).

The methodological problem of how to differ between collagen and elastin fibers still exists. Recently Icardo (2013) investigated the localization of elastin and collagen in the teleost bulbus arteriosus histochemically as well as using transmission electron microscopy. Thus, "Martin's trichrome staining shows widespread distribution of collagen in the wall of the bulbus. However, Sirius red indicates that collagen is mostly restricted to the valves and to the subepicardial layer. This is confirmed by transmission electron microscopy. Trichrome staining gives false positives that may be related to the chemical characteristics of both matrix components and dyes. By contrast, Sirius red constitutes a highly reliable method to detect collagen distribution. On the other hand, orcein heavily stains the bulbus of all teleosts examined. This includes the bulbus of the Antarctic teleosts, which do not show structurally discernable elastin fibers. In these cases, orcein may be staining non-elastin components, or basic elastin components not assembled into larger units," (Icardo 2013; see also Icardo et al. 1999).

Also fish cardiac ventricle possesses elastin. For example the heart of icefish that can pump the very large volume of haemoglobin-less blood (Agnisola and Tota 1994).

Notochord Elastin genes were found to be specifically expressed in the notochord that "functions as the midline structural element of all vertebrate embryos, and allows movement and growth at early developmental stages," (Sagstad et al. 2011). The cells termed chordoblasts are responsible for elastin gene expression here.

Swimbladder The inner layer of the swimbladder in zebrafish wall contains elastin fibers. Corresponding confirmation has been reported by Perrin et al. (1999) as follow:

- "Elastase delivery to the swimbladder in vitro resulted in significant fragmentation of elastin in the anterior chamber.
- In situ hybridization of zebrafish embryos illustrated that elastin gene expression is evident in the developing gut tract prior to swimbladder morphogenesis.
- Northern blot analysis demonstrated that the major zebrafish elastin mRNA is 2.0 kb, which is significantly smaller than its higher vertebrate counterpart.
- Amino acid analysis of alkali-resistant protein from the anterior chamber of the adult zebrafish swimbladder showed a composition similar to higher vertebrate elastins, including significant amounts of desmosine crosslinks," (Perrin et al. 1999).

10.3 Cetacean Elastin

The protein elastin, a major component of the large, elastic arteries, provides the elasticity necessary for cyclic deformation of the arterial wall with minimal energy loss. Elastin provides elasticity and reversible stretch to accommodate pulsatile pressure waves from the heart. The arterial pressure pulse magnitude decreases with distance from the heart, due to the pulse-dampening function of the elastic arteries (Cheng and Wagenseil 2012). The amount of elastin also decreases with distance from the heart, as less pulse dampening is required in the distal vessels. The amount of collagen, which provides strength and limits stretch at high pressures, remains approximately constant with distance from the heart (Dobrin 1997). The elastic modulus of collagen is 100–1,000 times higher than elastin, so the decrease in the ratio of elastin to collagen provides a gradient of arterial stiffening down the vascular tree (Greenwald 2007). The large elastic or conducting arteries are those closest to the heart, and have relatively high elastin to collagen ratios. The role of elastin as an arterial pulse dampener, or windkessel (Dobrin 1997), is supported by evolutionary evidence.

Elastin is found exclusively in vertebrate animals, and its emergence coincides with the appearance of a closed circulatory system (Sage and Gray 1979). Turnover of elastin is slow, and assembly of elastic fibers is negligible in adult animals;

therefore, the elastic fibers created during development must last the entire lifetime of the animal (Cheng and Wagenseil 2012). According to Greenwald et al. (2001), unlike other proteins has a turnover rate close to zero, elastin undergoes fatigue failure in arteries due to cyclic stress. The protein becomes progressively fragmented with age, and is gradually replaced by much stiffer fibrous tissue. These changes are well advanced after one billion heartbeats, by which time most animals, whatever their size or heart-rate, are nearing the end of their lives. In man this number is achieved by the age of 30 years, which from an evolutionary point of view, is close to the end of useful life. The progressive failure of elastin with age and the consequent increase in arterial stiffness leads to an inexorable rise in blood pressure (Greenwald et al. 2001).

Now, I would like to represent the whale heart and arteries for better understanding of the role of elastin as a biological material with crucial importance for cetacean organisms. In their classical and well-illustrated paper entitled "*A Large Whale Heart Circulation*" Race and co-workers (1959), described the heart of adult sperm whale (*Physeter catadon*): The heart of the animal weighed 116 kg. "Crude ventricular volume and cardiac output was calculated to be 453 L per minute based on a rate of 10 beats per minute. *[…]* The aorta was found to be 20 cm in diameter and the wall to consist of very large interwoven bundles of elastin tissue and fibrous tissue, apparently devoid of muscle," (Race et al. 1959; see as example Fig. 10.3). There was no evidence of arteriosclerosis! (Race et al. 1959). It is known (Greenwald et al. 2001) that the pulse pressure generated by the heart is determined by the hydraulic input impedance of the circulation, which in turn is governed by the elastic properties and structural arrangement of elastin. Interestingly, the pulse pressure

Fig. 10.3 Fragment of the whale arteries with about 20 cm in diameter (Image courtesy of Andre Ehrlich)

depends on the elastic properties of the protein elastin and is independent of body mass (Greenwald et al. 2001).

Different species of marine mammals vary in diameter of their aortic arch that is enlarged relative to the descending aorta. Here, some data concerning specialization reported by Shadwick and Gosline in their impressive works:

- "The ratio of maximal diameter of the arch to that of the thoracic aorta is about 2.3 in the harbour seal (*Phoca vitulina*), 3.6 in the Weddell seal (*Leptonychotes weddelli*) and 3.2 in the fin whale (*Balaenoptera physalus*); compared with only 1.4 in the dog;
- in the harbour seal, more than 80 % of the volume change in the entire thoracic aorta that results from a pressure pulse occurs in the bulbous arch, and this is more than 90 % in the Weddell seal and fin whale. The enhanced capacitance of the arch in the harbour seal is primarily due to its larger diameter, as the relative wall thickness and elasticity of the arch and thoracic aorta are the same.
- the arch might be somewhat less stiff than the thoracic aorta. In addition to being greatly expanded, the aortic arch of the fin whale is also much more distensible than the relatively thin-walled and much stiffer descending aorta. At the estimated mean blood pressure, the elastic modulus of this vessel is 12 MPa, or 30 times that of the aortic arch," (Shadwick and Gosline 1995);
- "the aortic arch and the descending aorta in the fin whale (*Balaenoptera physalus*) are structurally and mechanically very different from comparable vessels in other mammals. Although the external diameter of the whale's descending thoracic aorta (approximately 12 cm) is similar to that predicted by scaling relationships for terrestrial mammals, the wall thickness:diameter ratio in the whale (0.015) is much smaller than the characteristic value for other mammals (0.05)," (Gosline and Shadwick 1996);
- "chemical composition studies indicate that the elastin:collagen ratio is high in the aortic arch (approximately 2:1), and that this ratio falls in the thoracic (approximately 1:2) and abdominal (approximately 1:3) aortas. However, the magnitude of the change in composition does not account for the dramatic difference in mechanical properties. This suggests that there are differences in the elastin and collagen fibre architecture of these vessels," (Gosline and Shadwick 1996);
- "the descending aorta contains dense bands of tendon-like, wavy collagen fibres that run in the plane of the arterial wall. They form a fibre-lattice that runs in parallel to the elastin lamellae and reinforces the wall, making it very stiff. The aortic arch contains a very different collagen fibre-lattice, in which fibres appear to have a component of orientation that runs through the thickness of the artery wall. This suggests that the collagen fibres may be arranged in series with elastin-containing elements. This difference in tissue architecture could account for both the lower stiffness and the extreme extensibility of the whale's aortic arch. Thus, both the structure and the mechanical behaviour of the lamellar units in the aortic arch and aorta of the whale have presumably been modified to produce the unusual mechanical and haemodynamic properties of the whale circulation," (Gosline and Shadwick 1996; see also Shadwick and Gosline 1994).

Mechanical design as well as the mechanical properties of whale arteries have been also investigated (Gosline and Shadwick 1996; Shadwick 1999). Special attention was paid to non-linear elasticity (Shadwick 1999).

In addition to its localization within circulatory system of cetaceans, elastin was found also as a structural component of other organs of these animals. For example, Prahl et al. (2009) reported on the presence of elastin in the nasal complex in the harbour porpoise (*Phocoena phocoena*). "A blade-like elastin body at the caudal wall of the epicranial respiratory tract may act as antagonist of the musculature that moves the blowhole ligament," (Prahl et al. 2009).

In the common dolphin (*Delphinus delphis*) elastin was found in the ovary (Takahashi et al. 2006).

10.4 Conclusion

Thus, according to the modern point of view "Elastin is a versatile elastic protein that dominates flexible tissues capable of recoil, and facilitates commensurate cell interactions in these tissues in all higher vertebrates. Elastin's persistence and insolubility hampered early efforts to construct versatile biomaterials. Subsequently the field has progressed substantially through the adapted use of solubilized elastin, elastin-based peptides and the increasing availability of recombinant forms of the natural soluble elastin precursor, tropoelastin," (Almine et al. 2010).

For example, the product with commercial name "BIO MARINE ELASTIN" is extracted by enzymatic hydrolysis of insoluble connective tissues of particular species of marine fish. Thanks to a special hydrolysis process, the long protein chains are cut and transformed into small peptides, polypeptides and amino acids that can penetrate into the skin (see for details www.cobiosa.com). The high insolubility of elastin has limited its use in biomedicine. Soluble derivatives of elastin including elastin peptides, digested elastins, and tropoelastin have much broader applications. Recombinantly produced tropoelastin, the soluble monomer of elastin, has been shown to exhibit many of the properties intrinsic to the mature biopolymer (Wise et al. 2009). As such, recombinant human tropoelastin provides a versatile building block for the manufacture of biomaterials with potential for diverse applications in elastic tissues. One of the major benefits of soluble elastins is that they can be engineered into a range of physical forms.

References

Aaron BB, Gosline JM (1980) Optical properties of single elastin fibres indicate random protein conformation. Nature 287:865–867

Aaron BB, Gosline JM (1981) Elastin as a random–network elastomer: a mechanical and optical analysis of single elastin fibres. Biopolymers 20:1247–1260

Agnisola C, Tota B (1994) Structure and function of the fish cardiac ventricle: flexibility and limitations. Cardioscience 5(3):145–153

Almine JF et al (2010) Reproduced from Almine JF, Bax DV, Mithieux SM et al (2010) Elastin-based materials. Chem Soc Rev 39:3371–3379. Copyright (2010) with permission of The Royal Society of Chemistry
Anwar RA (1966) Comparison of elastins from various sources. Can J Biochem 44:725–730
Anwar RA, Oda G (1966) The biosynthesis of desmosine and isodesmosine. J Biol Chem 241:4638–4641
Benjamin M, Norman D, Santer RM et al (1983) Histological, histochemical and ultrastructural studies on the bulbus arteriosus of the stickle–backs, *Gasterosteus aculeatus* and *Pungitius pungitius* (Pisces: Teleostei). J Zool (Lond) 200:325–346
Bochicchio B, Pepe A, Tamburro AM (2001) On (GGLGY) synthetic repeating sequences of lamprin and analogous sequences. Matrix Biol 20:243–250
Braun MH et al (2003) Republished with permission of The Company of Biologists Ltd, from Braun MH, Brill RW, Gosline JM et al (2003) Form and function of the bulbus arteriosus in yellowfin tuna (*Thunnus albacares*), bigeye tuna (*Thunnus obesus*) and blue marlin (*Makaira nigricans*): static properties. J Exp Biol 206:3311–3326. Copyright (2003) permission conveyed through Copyright Clearance Center, Inc
Bushnell PG, Jones DR, Farrell AP (1992) The arterial system. In: Hoar WS, Randall DJ, Farrell AP (eds) Fish physiology. Academic, New York
Conticello VP, Carpenter Desai HE (2012) Reprinted from Conticello VP, Carpenter Desai HE (2012) Elastins. In: Matyjaszewski K, Möller M (eds) Polymer science: a comprehensive reference, vol 9: polymers in biology and medicine, Elsevier, pp 71–103. Copyright © 2012, Elsevier B.V. All rights reserved. Reprinted with permission
Chalmers GWG, Gosline JM, Lillie MA (1999) The hydrophobicity of vertebrate elastins. J Exp Biol 202:301–314
Cheng JK, Wagenseil JE (2012) Extracellular matrix and the mechanics of large artery development. Biomech Model Mechanobiol. doi:10.1007/s10237-012-0405-8
Chow M, Boyd CD, Iruela–Arispe M et al (1989) Characterization of elastin protein and mRNA from salmonid fish (*Oncorhynchus kitsutch*). Comp Biochem Physiol B 93:835–845
Chung MI, Miao M, Stahl RJ et al (2006) Sequences and domain structures of mammalian, avian, amphibian and teleost tropoelastins: clues to the evolutionary history of elastins. Matrix Biol 25(8):492–504
DeMont ME, Wright GM (1993) With kind permission from Springer Science+Business Media: DeMont ME, Wright GM (1993) Elastic arteries in a primitive vertebrate: mechanics of the lamprey ventral aorta. Experientia 49(1):43–46. Copyright © Birkhäuser Verlag Basel, 1993
Dobrin PB (1997) Chapter 3: physiology and pathophysiology of blood vessels. In: Sidawy ANSB, DePalma RG (eds) The basic science of vascular disease. Futura Publishing, New York
Faury G (2001) Reprinted from Faury G (2001) Function-structure relationship of elastic arteries in evolution: from microfibrils to elastin and elastic fibres. Pathol Biol (Paris) 49(4):310–325, Copyright (2001), with permission from Elsevier
Fernandes RJ, Eyre DR (1999) The elastin–like protein matrix of lamprey branchial cartilage is cross–linked by lysyl pyridinoline. Biochem Biophys Res Commun 261(3):635–640
Ferri S (1980) Reprinted from Ferri S (1980) Elastic system fibers (oxytalan, elaunin and elastic fibers) in the skin of a freshwater teleost: optical and electron microscopy study. Arch Anat Microsc Morphol Exp 69(4):259–266. Copyright (1980) Masson, with permission from Elsevier
Franzblau C, Lent RW (1968) Studies on the chemistry of elastin. Brookhaven Symp Biol 21(2):358–377
Fritze O, Romero B, Schleicher M et al (2012) Age–related changes in the elastic tissue of the human aorta. J Vasc Res 49(1):77–86
Gosline JM (1978) The temperature–dependent swelling of elastin. Biopolymers 17:697–707
Gosline JM, French CJ (1979) Dynamic mechanical properties of elastin. Biopolymers 18:2091–2103
Gosline JM, Shadwick RE (1996) Republished with permission of The Company of Biologists Ltd, from Gosline JM, Shadwick RE (1996) The mechanical properties of fin whale arteries are explained by novel connective tissue designs. J Exp Biol 199(Pt 4):985–997. Copyright (1996); permission conveyed through Copyright Clearance Center, Inc

Greenlee TK, Ross R Jr, Hartman JL (1966) The fine structure of elastic fibers. J Cell Biol 30:59–65
Greenwald SE (2007) Ageing of the conduit arteries. J Pathol 211(2):157–172
Greenwald SE, Ryder GC, Martyn CN (2001) The aorta: built to last a lifetime? Engineering in Medicine and Biology Society. In: Proceedings of the 23rd annual international conference of the IEEE, vol 1. Istanbul, pp 180–183
Gundiah N, Ratcliffe MB, Pruitt LA (2007) Determination of strain energy function for arterial elastin: experiments using histology and mechanical tests. J Biomech 40(3):586–594
Gundiah N, Ratcliffe MB, Pruitt LA (2009) The biomechanics of arterial elastin. J Mech Behav Biomed Mater 2(3):288–296
He D, Chung M, Chan E et al (2007) Comparative genomics of elastin: sequence analysis of a highly repetitive protein. Matrix Biol 26(7):524–540
Icardo JM (2013) Reprinted from Icardo JM (2013) Collagen and elastin histochemistry of the teleost bulbus arteriosus: false positives. Acta Histochem 115(2):185–189. Copyright (2013), with permission from Elsevier
Icardo JM, Colvee E (2011) The atrioventricular region of the teleost heart. A distinct heart segment. Anat Rec (Hoboken) 294(2):236–242
Icardo JM, Colvee E, Cerra MC et al (1999) *Bulbus arteriosus* of the antarctic teleosts. II. The red–blooded *Trematomus bernacchii*. Anat Rec 256:116–126
Isokawa K, Takagi M, Toda Y (1988) Ultrastructural cytochemistry of trout arterial fibers as elastic components. Anat Rec 220:369–375
Isokawa K, Takagi M, Toda Y (1990) Ultrastructural and cytochemical study of elastic fibers in the ventral aorta of a teleost, *Anguilla japonica*. Anat Rec 226(1):18–26
Kielty CM, Sherratt MJ, Shuttleworth CA (2002) Elastic fibres. J Cell Sci 115:2817–2828
Le Brun AP, Chow J, Bax DV et al (2012) Molecular orientation of tropoelastin is determined by surface hydrophobicity. Biomacromolecules 13(2):379–386
Licht JH, Harris WS (1973) The structure, composition and elastic properties of the teleost bulbus arteriosus in the carp, *Cyprinus carpio*. Compar Biochem Physiol A Physiol 46:699–704
Lillie MA, Gosline JM (1990) The effects of hydration on the dynamic mechanical properties of elastin. Biopolymers 29:1147–1160
Lillie MA, Gosline JM (1996) Swelling and viscoelastic properties of osmotically stressed elastin. Biopolymers 39:641–652
Lillie MA, Chalmers GW, Gosline JM (1994) The effects of heating on the mechanical properties of arterial elastin. Connect Tissue Res 31:23–35
Lillie MA, Chalmers GW, Gosline JM (1996) Elastin dehydration through the liquid and vapour phase: a comparison of osmotic stress models. Biopolymers 39:627–639
Luisetti M, Stolk J, Iadarola P (2012) Desmosine, a biomarker for COPD: old and in the way. Eur Respir J 39(4):797–798. doi:10.1183/09031936.00172911. Reproduced with permission of the European Respiratory Society. Copyright remains with European Respiratory Society©
McBurney KM, Keeley FW, Kibenge FSB et al (1996) Spatial and temporal distribution of lamprin mRNA during chondrogenesis of trabecular cartilage in the sea lamprey. Anat Embryol 193:419–426
Miao M et al (2006) Reprinted from Miao M, Bruce AE, Bhanji T et al (2006) Differential expression of two tropoelastin genes in zebrafish. Matrix Biol 26(2):115–124. Copyright (2006), with permission from Elsevier
Miao M et al (2009) Reprinted from Miao M, Stahl RJ, Petersen LF et al (2009) Characterization of an unusual tropoelastin with truncated C-terminus in the frog. Matrix Biol 28(7):432–441, Copyright (2009), with permission from Elsevier
Mistrali F, Volpin D, Garibaldi GB et al (1971) Thermodynamics of elasticity in open systems. Elastin J Phys Chem 75:142–149
Mithieux SM et al (2013) Reprinted from Mithieux SM, Wise SG, Weiss AS (2013) Tropoelastin – a multifaceted naturally smart material. Adv Drug Deliv Rev 65(4):421–428. Copyright (2013), with permission from Elsevier
Montes GS (1996) Structural biology of the fibres of the collagenous and elastic systems. Cell Biol Int 20(1):15–27

Motta P et al (2012) Reprinted from Motta P, Habegger ML, Lang A et al (2012) Scale morphology and flexibility in the shortfin mako *Isurus oxyrinchus* and the blacktip shark *Carcharhinus limbatus*. J Morphol 273(10):1096–1110, with permission. Copyright © 2012 Wiley Periodicals, Inc

Ohtani K et al (2008) Reprinted from Ohtani K, Yao T, Kobayashi M et al (2008) Expression of Sox and fibrillar collagen genes in lamprey larval chondrogenesis with implications for the evolution of vertebrate cartilage. J Exp Zool B Mol Dev Evol 310(7):596–607, with permission. Copyright © 2008 Wiley-Liss, Inc., A Wiley Company

Partridge SM (1966) Biosynthesis and nature of elastin structures. Fed Proc 25(3):1023–1029

Partridge SM, Davis HF, Adair GS (1955) The chemistry of connective tissues. II. Soluble proteins derived from partial hydrolysis of elastin. Biochem J 61:11–17

Partridge SM, Elsden DF, Thomas J (1963) Constitution of the cross–linkages in elastin. Nature 197:1297–1298

Perrin S, Rich CB, Morris SM et al (1999) The zebrafish swimbladder: a simple model for lung elastin injury and repair. Connect Tissue Res 40(2):105–112. Copyright © 1999, Informa Healthcare. Reprinted with permission

Petruska JA, Sandberg LB (1968) The amino acid composition of elastin in its soluble and insoluble state. Biochem Biophys Res Commun 33(2):222–228

Piez KA (1968) Cross–linking of collagen and elastin. Annu Rev Biochem 37:547–570

Prahl S, Huggenberger S, Schliemann H (2009) Histological and ultrastructural aspects of the nasal complex in the harbour porpoise, Phocoena phocoena. J Morphol 270(11):1320–1337. Copyright © 2009 Wiley-Liss, Inc. Reprinted with permission by John Wiley and Sons

Race GJ, Edwards WLJ, Halden ER et al (1959) A large whale heart. Circulation 19:928–932. Copyright © 1959 American Heart Association, Inc. All rights reserved

Raso DS (1993) Functional morphology of laminin, collagen type IV, collagen bundles, elastin, proteoglycans in the bulbus arteriosus of the white bass, *Morone chrysops* (Rafinesque). Can J Zool 71:947–952

Robertson GN, McGee CAS, Dumbarton TC et al (2007) Development of the swimbladder and its innervation in the Zebrafish, Danio rerio. J Morphol 268:967–985

Robson P, Wright G, Sitarz E et al (1993) Characterization of lamprin, an unusual matrix protein from lamprey cartilage. Implications for evolution, structure, and assembly of elastin and other fibrillar proteins. J Biol Chem 268:1440–1447

Robson P, Wright GM, Youson JH et al (1997) A family of non–collagen–based cartilages in the skeleton of the sea lamprey, *Petromyzon marinus*. Comp Biochem Physiol 118B:71–78

Robson P et al (2000) Reprinted from Robson P, Wright GW, Youson JH et al (2000) The structure and organization of lamprin genes: multiple-copy genes with alternative splicing and convergent evolution with insect structural proteins. Mol Biol Evol 17(11):1739–1752, by permission of Oxford University Press

Sage H (1982) Structure–function relationships in the evolution of elastin. J Invest Dermatol 79:146s–153s

Sage H (1983) The evolution of elastin: correlation of functional properties with protein structure and phylogenetic distribution. Comp Biochem Physiol B 74(3):373–380

Sage H, Gray WR (1976) Evolution of elastin structure. In: Sandberg LB, Gray WR, Franzblau C (eds) Elastin and elastic tissue. Plenum Press, New York

Sage EH, Gray WR (1977) With kind permission from Springer Science+Business Media: Sage EH, Gray WR (1977) Evolution of elastin structure. Adv Exp Med Biol 79:291–312, Copyright © 1977, Springer

Sage H, Gray WR (1979) Studies on the evolution of elastin. I. Phylogenetic distribution. Comp Biochem Physiol 64B:313–327

Sage H, Gray WR (1980) Studies on the evolution of elastin. II. Histology. Comp Biochem Physiol 66B:13–22

Sage H, Gray WR (1981) Studies on the evolution of elastin. III. The ancestral protein. Comp Biochem Physiol 68B:473–480

Sagstad A et al (2011) With kind permission from Springer Science+Business Media: Sagstad A, Grotmol S, Kryvi H, Krossøy C, Totland GK, Malde K, Wang S, Hansen T, Wargelius A (2011) Identification of vimentin- and elastin-like transcripts specifically expressed in developing notochord of Atlantic salmon (*Salmo salar* L.). Cell Tissue Res 346(2):191–202. Copyright © 2011, Springer
Sanchez–Quintana D, Hurle JM (1987) Ventricular myocardial architecture in marine fishes. Anat Rec 217(3):263–273
Sandberg LB, Weissman N, Gray WR (1971) Structural features of tropoelastin related to the sites of cross–links in aortic elastin. Biochem 10:52–57
Serafini–Fracassini A, Field JM, Spina J et al (1978) The morphological organization and ultrastructure of elastin in the arterial wall of trout (*Salmo gairdneri*) and salmon (*Salmo salar*). J Ultrastruct Res 65:1–12
Shadwick RE (1999) Mechanical design in arteries. J Exp Biol 202:3305–3313. Printed in Great Britain © The Company of Biologists Limited 1999
Shadwick RE, Gosline JM (1994) Arterial mechanics of the fin whale suggest a unique hemodynamic design. Am J Physiol 267:R805–R818
Shadwick RE, Gosline JM (1995) Arterial windkessels in marine mammals. In: Ellington CP, Pedley TJ (eds) Biological fluid dynamics symposia of the Society for Experimental Biology 49. Company of Biologists Ltd, Cambridge, pp 243–252
Shimada W, Bowman A, Davis NR et al (1969) An approach to the study of the structure of desmosine and isodesmosine containing peptides isolated from the elastase digest of elastin. Biochem Biophys Res Commun 37:191–195
Soskel NT, Wolt TB, Sandberg LB (1987) Isolation and characterization of insoluble and soluble elastins. In: Cunningham LW (ed) Methods in enzymology, structural and contractile proteins, part D, extracellular matrix. Academic, Orlando
Starcher BC, Galione MJ (1976) Purification and composition of elastins from different animal species. Anal Biochem 74:441–447
Steven FS, Minns RJ, Thomas H (1974) The isolation of chemically pure elastins in a form suitable for mechanical testing. Connect Tissue Res 2:85–90
Takahashi Y, Ohwada S, Watanabe K et al (2006) Does elastin contribute to the persistence of corpora albicantia in the ovary of the common dolphin (*Delphinus delphis*). Mar Mamm Sci. doi:10.1111/j.1748-7692.2006.00050.x
Thomas J, Elsden DF, Partridge SM (1963) Partial structure of two major degradation products from the cross–linkages in elastin. Nature 200:651–652
Turino GM, Ma S, Lin YY et al (2011) Matrix elastin: a promising biomarker for chronic obstructive pulmonary disease. Am J Respir Crit Care Med 184(6):637–641
Wise, Weiss (2009) Reprinted from Wise and Weiss (2009) Tropoelastin. Int J Biochem Cell Biol 41(3):494–497. Copyright (2009), with permission from Elsevier
Wise SG, Mithieux SM, Weiss AS (2009) Engineered tropoelastin and elastin–based biomaterials. Adv Protein Chem Struct Biol 78:1–24
Wright GM, Keely FW, Youson JH et al (1984) Cartilage in the Atlantic hagfish, *Myxine glutinosa*. Am J Anat 169:407–424
Wright GM, Armstrong LA, Jacques AM et al (1988) Trabecular, nasal, branchial, and pericardial cartilages in the sea lamprey, *Petromyzon marinus*: fine structure and immunohistochemical detection of elastin. Am J Anat 182:1–15. Copyright © 1988 Wiley-Liss, Inc
Yeo GC, Baldock C, Tuukkanen A et al (2012) Tropoelastin bridge region positions the cell–interactive C terminus and contributes to elastic fiber assembly. Proc Natl Acad Sci U S A 109:2878–2883

Chapter 11
Marine Keratins

Abstract Keratins represent the largest subgroup among all intermediate filament proteins. These structural proteins are mechanically robust and chemically unreactive. This is due to tight packing of protein chains in the form of alpha-helices or β-sheets into supercoiled polypeptide chains crosslinked with disulphide bonds. This chapter is dedicated to keratins of marine fish, birds, reptilian and mammalian origin. Keratin-like biological materials from hagfish slime are also discussed.

As reported by Rouse and Van Dyke (2010), "the word 'keratin' first appears in the literature at around 1850 to describe the material that made up hard tissues such as animal horns and hooves (keratin comes from the Greek '*kera*' meaning horn). *[…]* The term '*keratin*' originally referred to the broad category of insoluble proteins that associate as intermediate filaments (IFs) *[see below]* and form the bulk of cytoplasmic epithelia and epidermal appendageal structures (i.e., hair, wool, horns, hooves and nails)," (Rouse and Van Dyke 2010). Principally, "keratins are expressed in all types of epithelial cells (simple, stratified, keratinized and cornified)," (Karantza 2011).

Historically, most attention was paid to mammalian keratins, including human. Here, I would like to take the liberty to summarize the history of mammalian keratins investigations, which was excellently reviewed by Mark Van Dyke (2012). Thus, "the earliest documented use of keratin in medicine comes from a Chinese herbalist named Li Shi-Zhen (Ben Cao Gang Mu. Materia Medica, a dictionary of Chinese herbs, written by Li Shi Zhen (1518–1593)). Over a 38-year period, he wrote a collection of 800 books known as the Ben Cao Gang Mu. These books were published in 1596, 3 years after his death. Among the more than 11,000 prescriptions described in these volumes, is a substance known as Xue Yu Tan, also known as Crinis Carbonisatus, that is made up of ground ash from pyrolized human hair. The stated indications for Xue Yu Tan were accelerated wound healing and blood clotting," (Rouse and Van Dyke 2010).

In the middle of nineteenth century, keratin intrigued scientists because of several specific properties. Most methods used for dissolving of other proteins at that time were ineffective in attempts to dissolve keratin. Studies on extraction procedures with respect to isolate keratins has been carried out from 1905 to 1935 using oxidative and reductive type of chemical reactions. Methods for breaking down the structures of hooves, horns, and hair had been developed by the late 1920s.

H. Ehrlich, *Biological Materials of Marine Origin*, Biologically-Inspired Systems 4,
DOI 10.1007/978-94-007-5730-1_11

Chronology of these experimental works is excellently represented in the work by Rouse and Van Dyke (2010) as follow:

"In 1934, a key research paper was published that described different types of keratins, distinguished primarily by having different molecular weights. This seminal paper demonstrated that there were many different keratin derivatives, and that each played a different role in the structure and function of the hair follicle. *[...]* In 1965, CSIRO scientist W. Gordon Crewther and his colleagues published the definitive text on the chemistry of keratins. This chapter in Advances in Protein Chemistry contained references to more than 640 published studies on keratins. *[...]* Advances in the extraction, purification and characterization of keratins, led to the exponential growth of keratin materials and their derivatives. In the 1970s, methods to form extracted keratins into powders, films, gels, coatings, fibers, and foams were developed and published by several research groups. All of these methods made use of the oxidative and reductive chemistries developed decades earlier, or variations thereof," (Rouse and Van Dyke 2010; see also Crewther et al. 1965; Goddard and Michaelis 1935).

Systematization of knowledge on structure, function and regulation of mammalian keratins synthesis led to their classification of these structural proteins into two distinct groups (see for review Block 1951; Langbein et al. 1999, 2001; Schweizer et al. 2006). According to Rouse and Van Dyke (2010), "'Hard' keratins form ordered arrays of IFs embedded in a matrix of cystine rich proteins and contribute to the tough structure of epidermal appendages. 'Soft' keratins preferentially form loosely-packed bundles of cytoplasmic IFs and endow mechanical resilience to epithelial cells. In 2006, Schweizer et al. developed a new consensus nomenclature for hard and soft keratins to accommodate the functional genes and pseudogenes for the full complement of human keratins. The structural subunits of both epithelial and hair keratins are two chains of differing molecular weight and composition (designated types I and II). They each contain non-helical end terminal domains and a highly-conserved, central alpha-helical domain.

The type I (acidic) and type II (neutral-basic) keratin chains interact to form heterodimers, which in turn further polymerize to form 10-nm intermediate filaments," (Rouse and Van Dyke 2010).

After 30 years keratin research, the total number of mammalian keratins has increased to 54, including 28 type I and 26 type II keratins (Karantza 2011). Chemistry of mammalian keratins (Crewther et al. 1965), their composition, structure and biosynthesis (Dobb and Rogers 1967; Fraser et al. 1972), biology and pathology (Moll et al. 2008), functions (Kirfel et al. 2003; Gu and Coulombe 2007; Magin et al. 2007), genetics (Hesse et al. 2001), mechanical (Yamada et al. 2003) and material properties (Hill et al. 2010; Rouse and Van Dyke 2010) are well studied. "Recently, keratins have been recognized as regulators of numerous cellular properties and functions. These include apico-basal polarization, motility, cell size, protein synthesis, membrane trafficking, and signalling. In cancer, keratins are extensively used as diagnostic tumor markers. Epithelial malignancies largely maintain the specific keratin patterns associated with their respective cells of origin, and, in many occasions, full-length or cleaved keratin expression (or lack thereof) in tumors and/

or peripheral blood carries prognostic significance for cancer patients. Quite intriguingly, several studies have provided evidence for active keratin involvement in cancer cell invasion and metastasis, as well as in treatment responsiveness, and have set the foundation for further exploration of the role of keratins as multifunctional regulators of epithelial tumorigenesis," (Karantza 2011).

Scientific data on isolation, purification, and characterization of keratin proteins of hair and wool fibers origin obtained during over the past century are used as the platform for the development of a keratin-based biomaterials today (Rouse and Van Dyke 2010). The first study using a keratin biomaterial was published in 1982. Noishiki and colleagues coated a heparinized keratin derivative onto a polymer stent and implanted it into a dog for more than 200 days without thrombosis (Noishiki et al. 1982). Keratin coatings and films, keratin-chitosan films, compression molded keratin films, keratin-hydroxyapatite sponges, modified keratin sponges containing growth factors, pure keratin scaffolds, and keratin-coated hydroxyapatite have all been developed and investigated during last 30 years (see for review Hill et al. 2010).

However, in contrast to keratins from human and terrestrial mammals (e.g. sheep) as well as feather keratins, little attention was paid to marine keratins of reptilians and fish origin. The most recent works on keratins are from marine turtles, and were published by Lorenzo Alibardi and co-workers. Fish keratins, including those from sharks (Schaffeld et al. 1998, 2001, 2004), were mostly investigated in terms of their molecular biology rather than for their bioinspirational potential (see for review Markl and Schechter 1998; Zimek et al. 2003; Schaffeld and Markl 2004; Krushna Padhi et al. 2006). Only fish with keratin-like fibrillar proteins, such as those from hagfish slime, have been intensively investigated in the Lab of Douglas Fudge as biological materials (represented below). The third group of marine keratins is related to cetacean keratins, mostly to baleen, which are also discussed in this chapter.

Marine Reptilian Keratins The heavily keratinised nature of the carapace and plastron of aquatic chelonians provides not only an impervious barrier to the environment, but a considerable degree of protection against attack. Such protection increases with age as the keratinised layer simultaneously thickens and hardens. The characteristic ridged pattern of the individual scales becomes more marked with age and, as in the newly emerged hatchling, is more pronounced over the carapace. Unlike the epidermis of many aquatic vertebrates, the chelonian shell is not covered with a layer of mucus (Fig. 11.1), which serves to reduce the frictional drag in water. It is customary, however, for the chelonian shell to acquire an overgrowth of algae. Although the ridges may have no significance in their own right, they may subserve a similar function by providing a suitable site at which mucous-secreting algae could anchor (Solomon et al. 1986).

Turtle's carapace is not only the one place of keratins location. For example, the peculiarly specialized esophagus of the Pacific ridley turtle (*Lepidochelys olivacea*) also possesses keratin. As described by Yoshie and Honma (1976), the esophagus is distinguished by a specialized remarkably keratinized squamous epithelium, and its

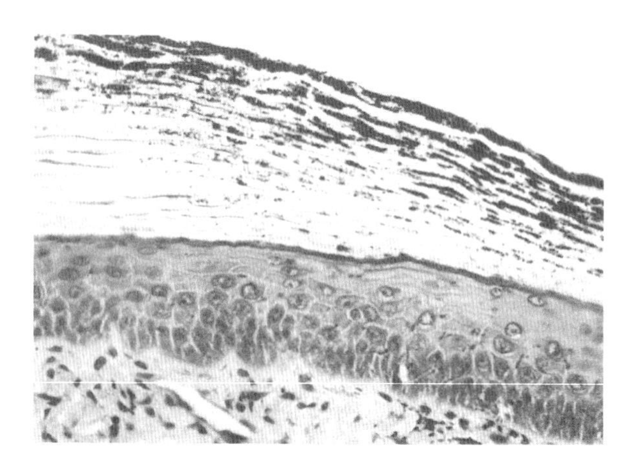

Fig. 11.1 Green turtle carapace. The other most layer is made of keratin. High magnification of growing region. ×300 (Reproduced with permission of Blackwell Publishing Ltd., from Solomon et al. (1986); permission conveyed through Copyright Clearance Center, Inc.)

major part is involved in synthesis of spine-like structures termed "pharyngeal teeth". Stratum corneum, stratum germinativum and stratum spinosum are three layers of the epithelium. The scaly keratin of possibly sloughing cells is the main component of the spine's surface. The more than 100 μm thick cornified layer is characteristic for the spine and can reach three to seven times the depth of corresponding layers in other parts. As natural alternative to marine turtle without toothed jaws, pharyngeal teeth are probably used for breaking food into small fragments (Yoshie and Honma 1976).

Comparative study on alpha- and beta- keratins in reptiles was initiated by Alexander (1970). From functional point of view, "α-keratins are crucial for water conservation (stratum corneum) in amniotes and for heat conservation (hair and fur) in mammals. Therefore these versatile biological materials likely played a crucial role in the successful invasion of terrestrial habitats by vertebrates," (Fudge et al. 2009b). Toni et al. (2007) characterized both types of these structural proteins as follow:

"The epidermis of reptiles *[…]* contains relatively large α-keratins of 40–68 kDa, and beta-keratins of smaller molecular weight (10–20 kDa) that possess different chemical and physical properties. α-keratins or cytokeratins constitute the intermediate-filament bundles and are present in most epidermal layers where they have a role in mechanical resistance, impart form to cells, or determine their changes in shape. Alpha-keratins are contained in keratinocytes of the alpha-layer of the epidermis, which mainly functions in permitting epidermal stretching and as a barrier against water-loss. Differently from alpha-keratins, beta-keratins are specific proteins utilized to make hard structures including scales, scutes, and claws (beak and feathers in birds). Beta-keratins are produced in the pre-corneous layer of the specialized epidermis of these skin appendages. The importance of beta-keratins in reptilian scales and their relatively small dimensions has led to recent studies of these proteins in detail; and some of their genes identified, sequenced, and cloned," (Toni et al. 2007).

Two types of keratinocyte differentiation has been described in the epidermis of turtles. So called softer keratinization was observed in the limbs, neck and tail, and harder keratinization – in the shell (see for review Dalla Valle et al. 2009a). Some amounts of ß-keratins are also synthetized in the soft epidermis of marine turtles (Alibardi et al. 2004; Alibardi and Toni 2006, 2007a). In various terrestrial tortoises, such structures as the precorneous layers of carapace and plastron epidermis, as well as the tough scales, are responsible for synthesis of the ß-keratins (Alibardi 2002, 2005, 2006). Different forms of ß-keratins in turtles are produced from low molecular weight precursors. Later these aggregate into larger forms during protein preparation (Alibardi and Toni 2006). Interestingly, β-keratins of crocodilians and birds origin are similar to those from turtle shell due to their glycine-proline-tyrosine rich content (Dalla Valle et al. 2009a; Fraser and Parry 2011).

Alibardi and Toni (2007b) also reported about the role of keratinocites in formation of a corneous cell envelope in crocodiles, as well as about characterization of keratins and associated proteins within crocodilian epidermis especially with respect to its corneification. These researchers showed that "some acidic alpha-keratins with a weight of 47–68 kDa are present. Smaller, 17–20 kDa beta-keratins with a prevalent basic pI (7.0–8.4) are also present. Acidic beta-keratins with a MW of 10–16 kDa are scarce, and may represent altered forms of the original basic proteins. Crocodilian beta-keratins are not recognized by a lizard beta-keratin antibody (A68B), and by a turtle beta-keratin antibody (A685). This result indicates that these antibodies recognize specific epitopes in different reptiles. Conversely, crocodilian beta-keratins cross-react with the Beta-universal antibody indicating they share a specific 20 amino acid epitope with avian beta-keratins. Although crocodilian beta-keratins are larger proteins than those present in birds, the obtained results indicate the presence of shared epitopes between avian and crocodilian beta-keratins. This is a good start for the future determination of the sequence of these proteins," (Alibardi and Toni 2007b). Unfortunately, there is a lack of information in detail on keratins from saltwater crocodile *C. porosus*, with exception of the paper by Alibardi and Toni (2007c).

Recently, Ye et al. (2010) showed that "beta-keratins in crocodiles reveal amino acid homology with avian keratins. Through the deduced amino acid sequence, these proteins are rich in glycine, proline and serine. The central region of the proteins is composed of two beta-folded regions, and show a high degree of identity with beta-keratins of aves and squamates. This central part is thought to be the site of polymerization to build the framework of beta-keratin filaments. It is believed that the beta-keratins in reptiles and birds share a common ancestry. Near the C-terminal, these beta-keratins contain a peptide rich in glycine-X and glycine-X-X. The distinctive feature of the region is some 12-amino acid repeats, which are similar to the 13-amino acid repeats in chick scale keratin, but are absent from avian feather keratin. From the phylogenetic analysis, the beta-keratins in crocodile have a closer relationship with avian keratins than the other keratins in reptiles," (Ye et al. 2010).

Principally, it is suggested that feather keratins diversified early from nonfeather keratins, deep in archosaur evolution (Dalla Valle et al. 2009b).

Marine Bird Keratins Marine birds are definitively related to marine vertebrates as described in chapters above, however, I take the liberty not to analyse the feather keratins here in detail due to the existing literature. Both the structural biology of feathers (Fraser and Parry 2008; Lingham-Soliar et al. 2010) and their mechanical properties (Bonser 1996, 2001; Bonser and Purslow 1995; Cameron et al. 2003; Bonser et al. 2004) are well described. However, I would like to note here that today much attention is paid to the photonic properties of bird feathers, because of the specific nanostructural organization of their keratins.

One of the intriguing topics concerning bird keratins is based on identification of the colour-producing intracellular β-keratin nanostructures observed in more than 20 different families of birds (Prum 2006). Many birds, including marine species, show broad variety of structural colours of their feathers. Note, that these colours are non-iridescent! Air nanostructures localized within the medullary cells of feather barbs and 3D sponge-like amorphous (or quasi-ordered) β-keratin are responsible for production of these colours (see for details Prum et al. 2009; Clarke et al. 2010; Stavenga et al. 2011; Shawkey et al. 2011; Maia et al. 2011, 2012; D'Alba et al. 2012). Two main classes of three-dimensional barb nanostructures are known, characterized by a tortuous network of air channels or a close packing of spheroidal air cavities (Saranathan et al. 2012). For example, the non-iridescent blue feather barbs of blue penguins (*Eudyptula minor*) contain a some kind of biophotonic nanostructure (D'Alba et al. 2011). "It is composed of parallel β-keratin nanofibres organized into densely packed bundles. Synchrotron small angle X-ray scattering and two-dimensional Fourier analysis of electron micrographs of the barb nanostructure revealed short-range order in the organization of fibres at the appropriate size scale needed to produce the observed colour by coherent scattering. These two-dimensional quasi-ordered penguin nanostructures are convergent with similar arrays of parallel collagen fibres in avian and mammalian skin, but constitute a novel morphology for feathers. The identification of a new class of β-keratin nano-structures adds significantly to the known mechanisms of colour production in birds, and suggests additional complexity in their self-assembly," (D'Alba et al. 2011).

Eliason and Shawkey (2012) investigated Dabbling ducks (Anatinae), which are "a speciose clade with substantial interspecific variation in the iridescent coloration of their wing patches (specula). In all species examined, speculum colour is produced by a photonic heterostructure. This consists of both a single thin-film of keratin and a two-dimensional hexagonal lattice of melanosomes in feather barbules. Although the range of possible variations of this heterostructure is theoretically broad, only relatively close-packed, energetically stable variants producing more saturated colours were observed, suggesting that ducks are either physically constrained to these configurations or are under selection for the colours that they produce," (Eliason and Shawkey 2012).

Recently, it was found that the sponge in the feather barbs of scarlet macaw (*Ara macao*) "is an amorphous diamond-structured photonic crystal with only short-range order. It possesses an isotropic photonic pseudogap that is ultimately responsible for the brilliant noniridescent coloration. The researchers further unravel an ingenious structural optimization for attaining maximum coloration, apparently resulting from natural evolution. Upon increasing the material refractive index above the level provided by nature, there is an interesting transition from a photonic pseudogap to a complete bandgap," (Yin et al. 2012).

The results discussed above additionally support the hypothesis that air nanostructures in feather barbs and colour-producing protein can be self-assembled. This is done by "arrested phase separation of polymerizing β-keratin from the cytoplasm of medullary cells. Such avian amorphous photonic nanostructures with isotropic optical properties may provide biomimetic inspiration for photonic technology" (Saranathan et al. 2012).

11.1 Intermediate Filaments

Epithelial cells are, probably, the place for location of keratins as one of the largest subgroup of intermediate filament (IF) proteins. IF of keratinous origin undergo complex regulation involving post-translational modifications and interactions with self and with various classes of associated proteins (Coulombe and Omary 2002). It was suggested that "the keratin-filament cycle of assembly and disassembly is a major mechanism of intermediate-filament network plasticity, allowing rapid adaptation to specific requirements, notably in migrating cells," (Kölsch et al. 2010).

Intermediate filaments (IFs), microfilaments (MFs) and microtubules (MTs) represent three main kinds of cytoskeletal filaments in eukaryotic cells (see for review Karantza 2011). Previously, IFs have been considered as "mechanical integrators of cellular space" (Lazarides 1980). Today, they are known to impart mechanical stability to cells and tissues (Kreplak and Fudge 2007). From structural view, there are some differences between IFs, MFs and MTs. For example,

- "MFs are composed of 6-nm intertwined actin chains and are responsible for resisting tension and maintaining cellular shape, forming cytoplasmic protuberances and participating in cell-cell and cell-matrix interactions;
- the IFs 10 nm in diameter are more stable than actin filaments, organize the internal three-dimensional cellular structure. Similarly to MFs, they also function in cell shape maintenance by bearing tension," (Karantza 2011; see also Herrmann et al. 2009; Pollard and Cooper 2009);
- "IFs are non-polar, and they do not exhibit large globular do mains. IF molecules associate via a coiled-coil interaction into dimers and higher oligomers;
- all IFs have a dimeric central rod domain, which is a coiled coil of two parallel α-helices flanked by head and tail domains of variable lengths. They are among the most chemically stable cellular structures, resisting high temperature, high salt and detergent solubilization;
- antiparallel molecular dimers (referred to as tetramers) polymerize in a staggered manner to make apolar protofilaments (3 nm in diameter). These in turn associate to protofibrils (4–5 nm in diameter), and then form the 10-nm Ifs;

- the MTs are hollow cylinders 23 nm in diameter, most commonly comprised of 13 protofilaments. These in turn are polymers of alpha- and beta-tubulin, and play key roles in intracellular transport and the formation of the mitotic spindle" (Norlén et al. 2007; see also Glotzer 2009; Karantza 2011; Steinert et al. 1993).

IFs differ from MTs and actin filaments due to that they are encoded by a large family of genes expressed in a tissue- and differentiation state-specific manner. Their rod domain amino-acid sequence is used as the key parameter for their classification. Thus, according to Human Intermediate Filament Mutation Database, www.interfil.org), there are 70 intermediate filament genes in six distinct classes in human alone. "Considering that many intermediate filament proteins can copolymerize with several different partners, the number of unique 10 nm intermediate filaments in humans is likely in the hundreds. The diversity of intermediate filaments in other species is not as well-described, but is undoubtedly impressive given the diversity in humans," (Fudge et al. 2009b). Here, some examples of the Ifs diversity represented in the paper by Herrmann et al. (2007) as follow:

"Types 1 and 2 IFs are found primarily in epithelial cells and include the acidic and basic keratins, respectively. Type 3 IFs include vimentin, desmin and glial fibrillary acidic protein; type 4 IFs assemble into neurofilaments; and type 5 IFs are the nuclear lamins," (Karantza 2011; see also Herrmann et al. 2007).

The assembly/disassembly cycle of keratin IF (Kölsch et al. 2010) allows rapid network remodeling within epithelial cells without network disruption. In brief, "the cycle begins with nucleation of keratin particles at the cell periphery, often in close proximity to lamellipodial focal adhesions (Fig. 11.2). This is followed by elongation of newly formed keratin particles during actin-dependent translocation toward the peripheral keratin network. After integration of precursor particles into the network, keratins IFs continue to move toward the nucleus and bundle. Some of them disassemble into soluble oligomers that rapidly diffuse throughout the cytoplasm and are available for another round of nucleation in the cell periphery. Others mature into a stable network that surrounds the nucleus, and is anchored to desmosomes and hemidesmosomes. Collectively, cycling allows the epithelial cytoskeleton to remain in motion without loss of structural integrity," (Windoffer et al. 2011).

How keratins provide rigidity and strength but at the same time remain dynamic and flexible is one of the key questions (Windoffer et al. 2011). Therefore, the biomechanical properties of intermediate filaments are intensively studied (see for review Ma et al. 1999; Bousquet et al. 2001; Kreplak et al. 2002; Russell et al. 2004; Kreplak and Fudge 2007; Wagner et al. 2007; Flitney et al. 2009; Leitner et al. 2012). The mechanical function of intermediate filaments is evident in many biological systems. For example, "the high storage modulus of the keratin IF network, coupled with its solid-like rheological behavior, supports the role of keratin IF as an intracellular structural scaffold that helps epithelial cells to withstand external mechanical forces" (Sivaramakrishnan et al. 2008).

Furthermore, in the work by Wagner and co-workers (2007), we can find following statement:

"Intermediate filaments, like their cytoskeletal counterpart F-actin, are examples of semi-flexible polymers. This is a class of polymers that are too soft to be truly

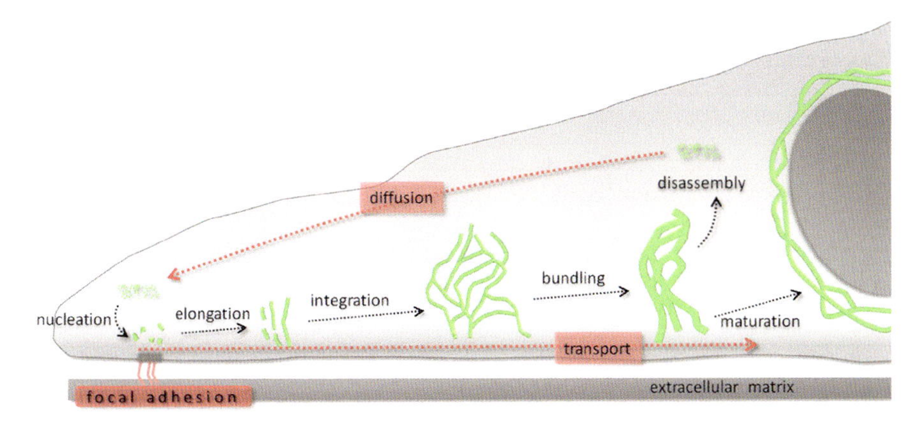

Fig. 11.2 The keratin cycle. According the description by Windoffer and co-workers, "soluble keratin oligomers assemble into particles in the cell periphery in proximity to focal adhesion sites (nucleation). These particles grow (elongation) and move toward the cell center in an actin dependent process (transport). Subsequently, elongated keratin IF particles are incorporated into the peripheral keratin IF network (integration). Filament bundling occurs during further centripetal translocation toward the nucleus (transport). Soluble oligomers dissociate (disassembly), diffuse throughout the cytoplasm (diffusion), and are reutilized for another cycle of keratin IF formation in the cell periphery. Alternatively, bundled filaments are stabilized (maturation), forming, e.g., the stable perinuclear cage (Republished with permission of ROCKEFELLER UNIVERSITY PRESS, from Windoffer et al. (2011); Copyright © 2011 Windoffer et al.; permission conveyed through Copyright Clearance Center, Inc.)

rodlike, but too stiff to be modeled as freely flexible chains; similar to those in rubberlike networks. More precisely, a filament is said to be semiflexible when its persistence length is of the same order as its contour length. Both in vivo and in vitro, IFs can be several microns long. When crosslinked into networks, the physically relevant length is the distance between network junction points, or crosslinks. This is generally on the order of a 100 to 1000 nm," (Wagner et al. 2007).

Aligned keratin IFs of a hard α-keratin fibre are embedded in an isotropic, high-sulphur protein matrix that can occupy up to 40 % of the fibre (Kreplak and Fudge 2007). According description made by Kreplak et al. (2004):

"Upon stretching a hard α-keratin fibre, the coiled coils, which have their axis parallel to the direction of applied stress, are transformed into a β-sheet structure with the b-strands running roughly parallel to the fibre axis. This is the so-called α-β transition which is the cornerstone of all molecular models of hard a-keratin mechanics for extensions above 5 %," (Kreplak et al. 2004; see also Crick 1952; Pauling and Corey 1951).

Also elastic properties of keratin-containing proteins as well as corresponding filamentous structures have been investigated. Young's modulus E of about 2GPa was measured for wet wool and hair, and seems to be characteristic for all single keratin filaments because of unifying principles in intermediate filament structure and assembly (Aebi et al. 1988). "Hence, IFs would be as rigid as F-actin (E 2.5 GPa) and MTs (E 1 GPa). In reality, IFs directly isolated from cultured cells and tissues or assembled from purified proteins in vitro appear much more flexible than

F-actin filaments and MTs when observed by electron microscopy. IFs are so much more flexible, in fact, that their highly curved appearance is used as a practical way of distinguishing them from other cytoskeletal filaments within whole tissue sections," (Kreplak and Fudge 2007).

There are numerous papers on keratins as intermediate filaments regarding to their isolation (Herrmann et al. 2004), atomic structure (Strelkov et al. 2001), molecular structure (Herrmann and Aebi 2004) and architecture (Strelkov et al. 2003), structural (Er Rafik et al. 2006) and genetic (Arora et al. 2008) polymorphism, self-organization properties (Lee and Coulombe 2009; Kim et al. 2010), genetics (Bawden et al. 2001), and function (Wang and Stamenovic 2000; Yoon et al. 2001; Coulombe et al. 2004), which have been published during last decade. I strongly recommend readers also to pay attention to the modern views on nanomechanical properties (Qin et al. 2009) as well as on computational and theoretical modeling of intermediate filament networks (Qin and Buehler 2012) recently proposed by Marcus Buehler and co-workers at MIT.

11.2 Hagfish Slime

Intriguingly, such animals as hagfish use intermediate filaments for defence (Downing et al. 1981a, b). In this case, IFs as strictly intracellular entities, possess novel property (Fudge et al. 2009b). Hagfishes (Fig. 11.3) are able to synthetize large quantities of IFs-containing slime when they are stressed or even provoked

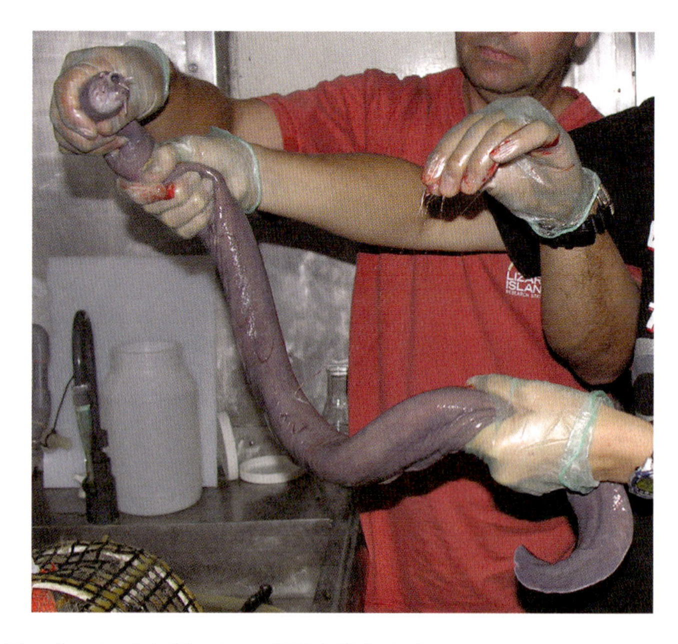

Fig. 11.3 The slimy hagfish (Courtesy: NOAA Fisheries)

(Strahan 1959, 1963; Martini 1998; Crystall 2000; Fudge 2001). Slime glands that line both sides of the hagfish's body are responsible for production and ejecting of the slime exudate (Downing et al. 1984; Spitzer et al. 1984, 1988; Spitzer and Koch 1998). In the classical paper by John Ferry published in 1941, he wrote about overproduction of colorless and transparent slime by a hagfish in sea water during a few seconds. The slime showed a tough, coherent, stringy consistency even being extremely diluted. The rigidity, exhibiting "elastic recoil", was reported as the special property of this highly viscous matter (Ferry 1941).

Hagfishes possess a large battery of slime glands (Blackstad 1963) and 90–200 associated slime pores running laterally along the full length of each side of their body (Zintzen et al. 2011). The histochemistry (Leppi 1967, 1968), ultrastructure (Fan 1965; Terakado et al. 1975), and function of these glands in different hagfish species are well investigated (see for details Newby 1946; Lametschwandtner et al. 1986; Luchtel et al. 1991; Spitzer and Koch 1998; Subramanian et al. 2008; Winegard and Fudge 2010).

The holocrine (slime glands) and merocrine (epidermal, through the skin) are two pathways of slime production in hagfish (Braun and Northcutt 1998). The hagfish epidermis is scaleless and consists of three cell types (small mucous cells, epidermal thread cells, and large mucous cells) (Fernholm 1981), which are located within collagenous fibrillar network that forms a flexible grayish "sheath". Two main types of gland cells, the gland mucous cells (GMCs) and gland thread cells (GTCs) (Fig. 11.4) can rupture and release their contents to the outside seawater immediately after the animal is disturbed. GTCs are responsible to release IFs, which are similar to skin keratins due to corresponding sequences. However, GMCs

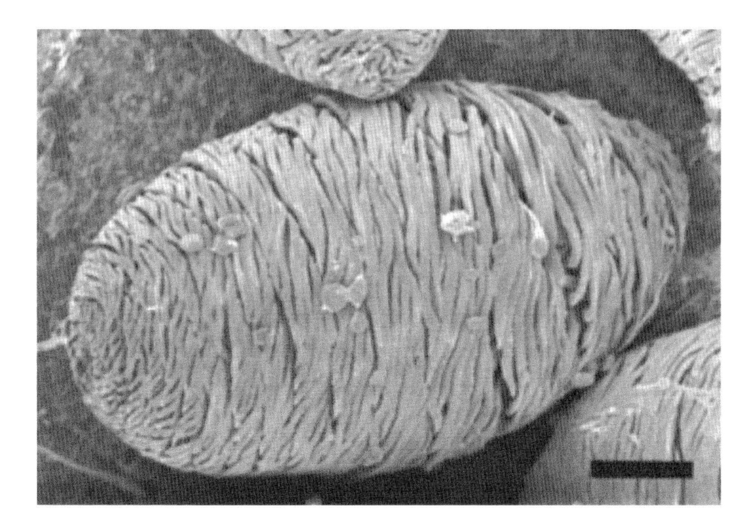

Fig. 11.4 SEM image of a hagfish gland thread cell without its plasma membrane, exposing the intricately coiled slime thread (Adapted from Kreplak and Fudge 2007). The slime fibre consists of a solid, nearly pure bundle of 103–104 IFs in cross-section that can be unravelled and used in vitro for investigation of mechanical properties. Scale bar 25 μm (Reprinted from Kreplak and Fudge (2007) with permission. Copyright © 2006 Wiley Periodicals, Inc.)

produce mucins, which interact with IFs, and show a high capacity for absorbing water (see for details Fudge et al. 2009a, b). According to the modern hypothesis (Lim et al. 2006; Zintzen et al. 2011), production of slime in modern hagfishes has numerous functions like immunological response to infectious pathogens as the epidermal slime of the more evolved fishes, or, decreasing competition for food by excluding other scavengers, frighten off different predators, and may also be a special biological approach that incapacitates prey by suffocating them (Lim et al. 2006; Zintzen et al. 2011).

Especially the presence of a fibrous component built from IFs within hagfish slime differ it from other, mostly mucin-based slimes (Fig. 11.5) (Koch et al. 1994, 1995). The diameter of IFs is about 10 nm. "These then bundle into a single protein polymer thread that takes up the vast majority of the cytoplasmic volume in mature cells. The thread is about 1 μm at its narrowest and about 3 μm at its widest, and when it is fully unravelled, it can be about 15 cm long," (Fudge et al. 2009a, b).

Koch and co-workers (1995) studied the Pacific hagfish (*Eptatretus stouti*) with respect to structural biology and biochemistry of its IFs. Here, the detailed description of the results obtained:

"It was determined that the deduced amino acid sequence of one thread IF chain (alpha, 66.6 kDa, native pl 7.5) of this species contained an atypical, threonine-rich central rod domain that had low identity (<30 %) with other vertebrate IF types. However, the N- and C-terminal domains exhibited several keratin-like features. From these and other unexpected characteristics, it was concluded that hagfish alpha is best categorized as a type II homologue of an epidermal keratin. The deduced sequence of a second thread IF subunit (gamma, 62.7 kDa, native pl 5.3) which is co-expressed and co-assembles in vitro with alpha in a 1:1 ratio was also

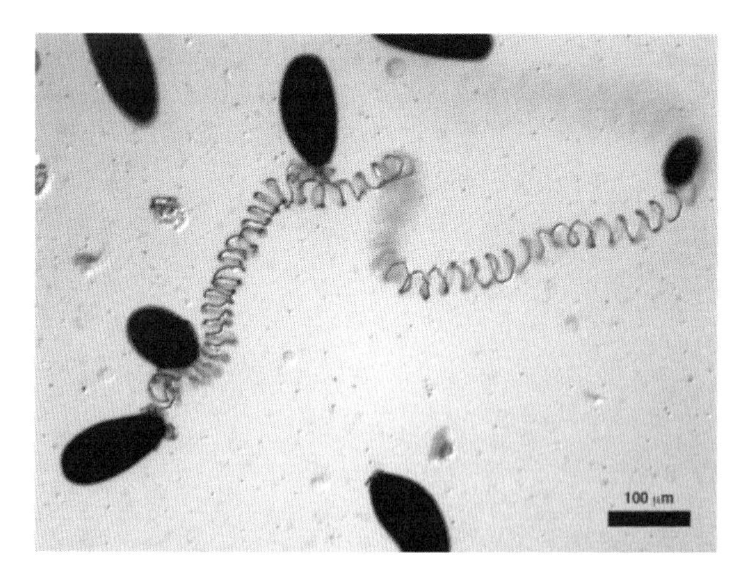

Fig. 11.5 TEM image of the Hagfish slime threads (Reprinted from Fudge et al. (2009b) by permission of Oxford University Press)

reported. As was found for alpha, the N- and C-terminal domains of gamma have keratin-like parameters, but the central rod has low identity to IFs of types I–V (<31 %), a cephalochordate IF (<29 %) and invertebrate IFs (<20 %) and no particular homology to type I or type II keratins. Central rod identity between gamma and alpha is also low (similar to 23 %), as is typical of comparisons between different rod types but atypical of similar rod types (>50 %). The central rods of both gamma and alpha lack the 42-residue insert of helix 1B present in lamins and invertebrate IFs, have unusually high threonine contents (gamma, 10 %; alpha, 13 %) compared to other IF types (2–5 %), contain a number of unexpected residues in consensus conserved sites, and employ a L12 segment of 21 residues rather than the 16 or 17 residues found in keratins. Theoretical analyses indicate that the hagfish molecules exist as coiled coil heterodimers (alpha/gamma) in which the chains are parallel, in axial register, and stabilized by significant numbers of ionic interactions. Fast Fourier-transform analyses revealed that the linear distribution period of similar to 9.55 for basic and acidic residues in other IF chains are not completely maintained, partly due to the high threonine content. The threonine residues occupy mainly outer sites b, c, f in the heptad substructure, possibly abetting parallel alignment of thousands of IFs within the thread, interactions with mucins at the thread periphery, and hierarchical IF chain assembly. It is suggested that the gamma and alpha chains from this most primitive extant vertebrate are type I and type II homologues of epidermal keratin chains, possibly related to early specialized keratins" (Koch et al. 1995).

On the micro level, the fully formed slime (Fig. 11.6) has been recently described as "a complex network capable of confining seawater to channels between the slime threads and ruptured mucins like a fine sieve. The interaction between the thread skeins and ruptured mucins is critical for the production of the mature slime," (Herr et al. 2010).

Fig. 11.6 TEM image of the Hagfish slime (Image courtesy of Dr. Gene Helfman)

The mechanical properties of the resulting IFs-based network determine the functional properties of hagfish slime. Recent works on the mechanical properties of fibres isolated from hagfish slime (Fudge et al. 2003, 2005; Lim et al. 2006; Hearle 2008) suggest that these unique fibrillar constructs may one day be replicated in a way that is economically viable and environmentally sustainable. Ewoldt et al. (2011) reported "the first experimental measurements of nonlinear rheological material properties of hagfish slime as a hydrated biopolymer/biofiber network, and developed a microstructural constitutive model to explain the observed nonlinear viscoelastic behavior. The linear elastic modulus of the network is observed to be G' about 0.02 Pa for timescales 0.1–10 s, making it one of the softest elastic biomaterials known," (Ewoldt et al. 2011). "Furthermore, the peculiar nature of hagfish slime inspires the possibility of a bio-inspired ultra-dilute elastic network which can be similarly deployed as a defensive mechanism for engineered systems," (Ewoldt et al. 2011).

Recently, hagfish slime threads have been proposed as "a biomimetic model for high performance protein fibres," (Fudge et al. 2010) in contrast to silk-like proteins. Douglas Fudge described the state of the art concerning this scientific direction as follow:

"The genes for these nanofibrous proteins, are far smaller than spidroin genes and not repetitive, making them more suitable for expression in bacteria. Another feature that makes them an attractive model is that they self-assemble from soluble precursors into 10 nm filaments in aqueous buffers. The key to the high strength and toughness of spider silk and hagfish threads is the β-sheet crystallites that simultaneously cross-link the protein molecules and arrange them into a structure in which 'sacrificial bonds' increase the energy required to break the material. The $\alpha \rightarrow \beta$ transition occurs far more readily in hagfish slime threads than it does in hard α-keratins, because slime threads lack elastomeric matrix proteins. In α-keratins resist these proteins swelling in water, and also provide a restoring force during and after longitudinal extension. Future work will focus on a more detailed characterization of the effects of cross-linking and annealing on fiber mechanics and structure, both in native slime threads and those that are produced from recombinant or isolated protein," (Fudge et al. 2010; see also Fudge and Gosline 2004; Koch et al. 1991, 1994, 1995).

11.3 Whale Baleen

Mysticeti and Odontoceti are the two suborders of the Cetacea order, which includes such marine mammals as dolphins, porpoises, and whales (Uhen 2010). Representatives of Mysticeti and Odontoceti whales differ from each other due to their feeding mechanism. The Mysticeti possess a special baleen-based filtration system, however the Odontoceti are equipped with teeth to tear their food (Werth 2001). "Cladistic analyses of living and extinct mysticetes support the hypothesis that mineralized teeth were lost in the common ancestor of crown Mysticeti,"

Fig. 11.7 Feeding gray whale. Baleen plates are well visible (Image courtesy of Yuri Yakovlev)

(Meredith et al. 2010). "Molecular sequences of three enamel-specific genes, ameloblastin (AMBN), enamelin (ENAM) and amelogenin (AMEL) contain stop codons and/or frameshift mutations in various mysticete species," (Meredith et al. 2010).

The upper jaw of the Mysticeti possess horn-like baleen that grows in the form of plates, which correspondingly grow too in parallel triangular sheets approximately 1 cm apart (Fig. 11.7).

As described by Lauffenburger (1993), "the plates grow continuously from the epidermal tissue of the gum and, like hair of other mammals, grow to maximum size sometime during adulthood. Seasonal variations in development result in visible growth ridges perpendicular to the length of each plate. When viewed from the exterior of the whale's mouth, the plates resemble rows of teeth on a comb," (Lauffenburger 1993). Baleen is not homologous to teeth in spite that it is secreted from gingival epithelia of the palate, and typically forms right and left racks of transversely oriented plates that extend into the oral cavity (Utrecht 1965). When the whale jaws are not completely closed, the frayed baleen functions as a sieve that entraps prey items but allows water to pass out of the mouth (Pivorunas 1979).

Depending on the species, a baleen plate can be 0.5–3.5 m long, and weigh up to 90 kg (Lauffenburger 1993). The color, quantity and size of baleen sheets are age and species dependent. In gray whales, a row of approximately 155 baleen plates grow from a foundation layer anchored to each side of the hard palate of the upper jaw (Pivorunas 1979). Bowhead whales are especially known for long baleen plates,

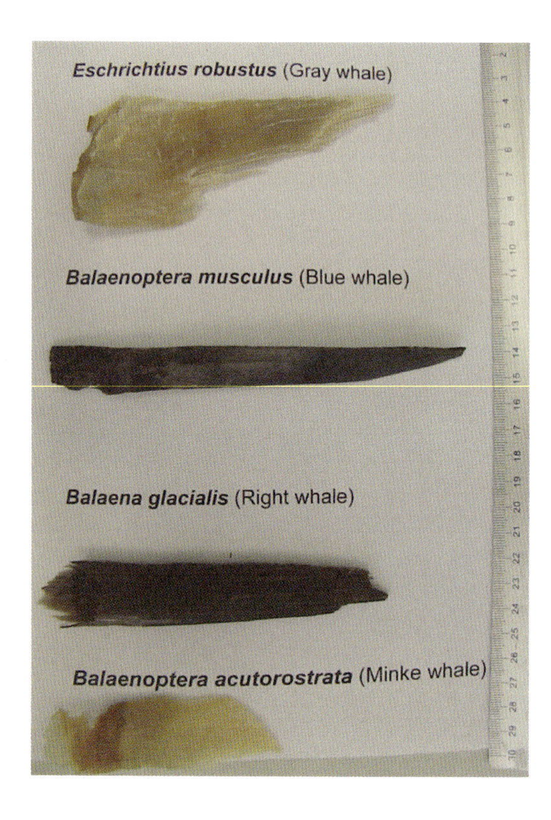

Fig. 11.8 Color diversity of whale's baleen (Image courtesy of Andre Ehrlich)

which can exceed 700 in number (Matthews 1968). The color of baleen can be, brown, gray, black, cream or even white. For example the gray whale baleen is coarse and bone colored (Fig. 11.8), however the right whale baleen is extremely fine and gray in color.

During the feeding, water with plankton is taken in through the open mouth and expelled out through the whale's filtering system based on baleen plates. Correspondingly, any prey taken in with the water is trapped on the baleen bristles, removed with the help of the tongue and then swallowed. Whales that feed on krill or other larger crustacea, as well as on fish, have very coarse bristles. Species that feed on small crustaceans possess baleen plates with very fine bristle (see for review Werth 2001; Brodie and Vikingsson 2009).

Here, some statistics:

- "extant species use this baleen filtering structure to consume as much as 600,000 kg of prey in a year," (Deméré et al. 2008; see also Gaskin 1982);
- "average daily krill ingestion of about 1300 and 835 L is predicted for the fin and sei whales, respectively. The whales seem to ingest about 30 % more than earlier reported of a prey, which has about 50 % of the salt concentration of seawater; thus maintaining the salt and water balance with a minimum of 1–2 % seawater ingestion" (Kjeld 2003).

Morphological descriptions of baleen have been reported since historical times including accounts by Zorgdrager (1720), Hunter (1787), Rosenthal (1829), Tullberg (1883), Flower (1883), Eschricht and Reinhardt (1866), Stevenson (1907), Ruud (1940) which provided considerable detail about baleen of different whale species. More recent descriptions of baleen have also focused on comparative size and numbers of baleen plates in Mysticeti (Cousteau and Paccalet 1988), on microscopic analysis of baleen as well as its cellular and subcellular morphology (Pfeiffer 1993; Fudge et al. 2009a), on growth (Sumich 2001), on measurement of carbon and nitrogen isotopes in baleen in order to assess life history feeding patterns of whales (Hobson and Schell 1998), on studies examining long-term changes of metal levels in whales (Hobson et al. 2004), on age estimation (Zeh et al. 2008), as well as on trace element chemistry of whale baleens in an attempt to reconstruct migration patterns (Freeburg et al. 2009). Studies on the isolation and sequencing of DNA from baleens have also been carried out (Kimura and Ozawa 1997; Rosenbaum et al. 1997). The use of whale baleen plates as a source of mitochondrial DNA (mtDNA) presents several advantages in comparison to whale bones. First, it is relatively easy to recover mtDNA from baleen because it is not locked up in a stony matrix, and there seems to be much more DNA in old baleen than in old bone. Second, baleen is usually morphologically diagnostic of the species, whereas single bone fragments often are not (Kimura and Ozawa 1997). Works on origin and evolution of baleen are still in progress (Meredith et al. 2010; Uhen 2010).

The structure and composition of baleen (St Aubin et al. 1984) as well as material properties (Szewciw et al. 2010) and some biomineralization aspects of baleen have been investigated (O'Connor 1987; Szewciw et al. 2010). Description of the plate of baleen was given by Lauffenburger (1993) as follow:

"A plate of baleen is composed of a sandwich-like, three-part structure: a central section of tubules in a cementing keratin matrix, flanked by a horny covering. Inside the whale's mouth, the horny covering and cementing matrix are worn away by the abrasive action of the tongue, revealing the central tubules or hairlike structures that form the mat or sieve during filtration. These tubes are highly calcified, containing crystals of hydroxyapatite, and they have central hollow cavities of varying sizes depending on the species. Calcification occurs in alternating rings around the outer edge of the tubes, an arrangement that strengthens the tubes without loss of flexibility. Baleen is the most highly calcified of the keratin materials and as a result is slightly more brittle," (Lauffenburger 1993; see also Halstead 1974; O'Connor 1987).

The biomineralogical characterization of baleen is rare. Here, only some data:

- "The sei whale (*Balaenoptera borealis*) has baleen that consists of 14.5 % hydroxyapatite (dry weight);
- Humpback (*Megaptera novaengliae*) and minke (*Balaenoptera acutorostrata*) baleen bristles contain intermediate (up to 4 %) and low (about 1 %) calcium salt concentrations, respectively," (Szewciw et al. 2010; see also Pautard 1963; St Aubin et al. 1984).

There is still lack on detailed information on material properties of whale baleen, with exception of the study reported by St Aubin et al. (1984) on the effects of petroleum on baleen plates. Recently, baleen has been reported as “an unusual mammalian keratin in that it never air dries,” (Szewciw et al. 2010). Material properties of this keratinous biological material in natural environments determine effective filter feeding mechanisms used by whales. Structural properties of baleen in different mysicetes including the influence of sediment abrasion are reported in several papers (see for details in Kasuya and Rice 1970; Rice and Wolman 1971; Nerini 1984).

Baleen, from human point of view, is elastic and robust wearing, features that make it useful for practical applications as a construction material. Historically, whaling industry was the source of baleen during centuries (Hunter 1787; Stevenson 1907; Lauffenburger 1993). “It found its height in the American market in the mid 1850s. It was used as an early form of plastic because of its ability to be bent and when cooled, retain its shape. Therefore it was used for corset stays, buggy whips, combs, brushes; and in native Alaskan culture, it was used extensively for making baskets,” (Lauffenburger 1993). In the nineteenth century, baleen was processed and sold in strips. At the peak of the baleen market, the amount of baleen harvested could pay for an entire whaling voyage (Matthews 1968).

Baleen preserves relatively well over time in some archeological sites and museums. Forensic science on keratin-based biomaterials is well established (Brenner et al. 1985; Kirkbride and Tungol 1999), however, there are no results regarding the forensic investigation on baleen identification.

11.4 Conclusion

The plasticity and network architecture of keratin IFs as well as their role in cell survival are in focus of detailed investigations today (Windoffer et al. 2011). The mechanical properties of epithelial cells are modulated by structural changes in keratin intermediate filament networks (Leitner et al. 2012). Therefore keratins in the form of IF, nanoorganized spongy structures (feather), or megascale constructs like whale baleen can be investigated as biological materials. They are very successful from an evolutionary point of view, especially in vertebrates. The question why chitin (the main structural biomaterial in all invertebrates) is changed to keratin in vertebrates is still open.

References

Aebi U, Haner M, Engel A (1988) Unifying principles in intermediate filament (IF) structure and assembly. Protoplasma 145:73–81

Alexander NJ (1970) Comparison of alpha and beta keratin in reptiles. Z Zellforsch Mikrosk Anat 110(2):153–165

Alibardi L (2002) Immunocytochemical observations on the cornification of soft and hard epidermis in the turtle *Chrysemys picta*. Zoology (Jena) 105:31–44

Alibardi L (2005) Proliferation in the epidermis of chelonians and growth of the horny scutes. J Morphol 265:52–69
Alibardi L (2006) Ultrastructural and immunohistochemical observations on the process of horny growth in chelonian shells. Acta Histochem 108:149–162
Alibardi L, Toni M (2006) Immunolocalization and characterization of beta–keratins in growing epidermis of chelonians. Tissue Cell 38:53–63
Alibardi L, Toni M (2007a) Immunological characterization of a newly developed antibody for localization of a beta–keratin in turtle epidermis. J Exp Zool 308B:200–208
Alibardi L, Toni M (2007b) Reprinted from Alibardi L, Toni M (2007b) Characterization of keratins and associated proteins involved in the corneification of crocodilian epidermis. Tissue Cell 39(5):311–323. Copyright © 2007, with permission from Elsevier
Alibardi L, Toni M (2007c) Beta–keratins of reptrilian scales share a central amino–acid sequence termed core–box. Res J Biol Sci 2(3):329–339
Alibardi L, Spisni E, Toni M (2004) Differentiation of the epidermis in turtle: an immunocytochemical, autoradiographic and electrophoretic analysis. Acta Histochem 106(5):379–395
Arora R, Bhatia S, Sehrawat A et al (2008) Genetic polymorphism of type 1 intermediate filament wool keratin gene in native Indian sheep breeds. Biochem Genet 46(9–10):549–556
Bawden CS, McLaughlan C, Nesci A et al (2001) A unique type I keratin intermediate filament gene family is abundantly expressed in the inner root sheaths of sheep and human hair follicles. J Invest Dermatol 116:157–166
Blackstad TW (1963) The skin and the slime gland. In: Brodal A, Fänge R (eds) The biology of myxine. Universitetsforlaget, Oslo
Block RJ (1951) Chemical classification of keratins. Ann N Y Acad Sci 53:608–612
Bonser RHC (1996) The mechanical properties of feather keratin. J Zool (Lond) 239:477–484
Bonser RHC (2001) The mechanical performance of medullary foam from feathers. J Mater Sci Lett 20:941–942
Bonser RHC, Purslow PP (1995) The Young's modulus of feather keratin. J Exp Biol 198:1029–1033
Bonser RHC, Saker L, Jeronimidis G (2004) Toughness anisotropy of feather keratin. J Mater Sci 39:2895–2896
Bousquet O, Ma L, Yamada S et al (2001) The nonhelical tail domain of keratin 14 promotes filament bundling and enhances the mechanical properties of keratin intermediate filaments in vitro. J Cell Biol 155:747–754
Braun C, Northcutt RG (1998) *Cutaneous exteroreceptors* and their innervation in hagfishes. In: Jorgensen L, Weber M (eds) The biology of hagfishes. Chapman & Hall, London
Brenner L, Squires PL, Garry M et al (1985) A measurement of human hair oxidation by Fourier transform infrared spectroscopy. J Forensic Sci 30:420–426
Brodie P, Vikingsson G (2009) On the feeding mechanisms of the Sei Whale (*Balaenoptera borealis*). J Northwest Atl Fish Sci 42:49–54
Cameron GJ, Wess TJ, Bonser RHC (2003) Young's modulus varies with differential orientation of keratin in feathers. J Struct Biol 143:118–123
Clarke JA, Ksepka DT, Salas–Gismondi R et al (2010) Fossil evidence for evolution of the shape and color of penguin feathers. Science 330:954–957
Coulombe PA, Omary MB (2002) 'Hard' and 'soft' principles defining the structure, function and regulation of keratin intermediate filaments. Curr Opin Cell Biol 14(1):110–122
Coulombe PA, Tong X, Mazzalupo S et al (2004) Great promises yet to be fulfilled: defining keratin intermediate filament function *in vivo*. Eur J Cell Biol 83(11–12):735–746
Cousteau JY, Paccalet Y (1988) Whales. H.N. Abrams, New York
Crewther WG, Fraser RDB, Lennox FG et al (1965) The chemistry of keratins. In: Anfinsen CB, Anson ML, Edsall JT, Richards FM (eds) Advances in protein chemistry. Academic, New York
Crick FHC (1952) Is alpha–keratin a coiled coil? Nature 170:882–883
Crystall B (2000) Monstrous mucus. New Sci 2229:38–41
D'Alba L, Saranathan V, Clarke JA et al (2011) Colour-producing β-keratin nanofibres in blue penguin (*Eudyptula minor*) feathers. Biol Lett 7(4):543–546, rsbl20101163, by permission of the Royal Society. Copyright © 2011, The Royal Society

D'Alba L, Kieffer L, Shawkey MD (2012) Relative contributions of pigments and biophotonic nanostructures to natural color production: a case study in budgerigar (*Melopsittacus undulatus*) feathers. J Exp Biol 215(Pt 8):1272–1277
Dalla Valle L, Nardi A, Gelmi C et al (2009a) Beta–keratins of the crocodilian epidermis: composition, structure, and phylogenetic relationships. J Exp Zool B Mol Dev Evol 312(1):42–57
Dalla Valle L, Nardi A, Toni M et al (2009b) Beta–keratins of turtle shell are glycine–proline–tyrosine rich proteins similar to those of crocodilians and birds. J Anat 214(2):284–300
Deméré TA, McGowen MR, Berta A et al (2008) Morphological and molecular evidence for a stepwise evolutionary transition from teeth to baleen in mysticete whales. Syst Biol 57(1):15–37, by permission of Oxford University Press
Dobb MG, Rogers GE (1967) Electron microscopy of fibrous keratins. Symposium on fibrous proteins 1:267–278
Downing SW, Salo WL, Spitzer RH, Koch EA (1981a) The hagfish slime gland: a model system for studying the biology of mucus. Science 214:1143–1145
Downing SW, Spitzer RH, Salo WL et al (1981b) Hagfish slime gland thread cells: organization, biochemical features, and length. Science 212:326–328
Downing SW, Spitzer RH, Koch EA et al (1984) The hagfish slime gland thread cell. I. A unique cellular system for the study of intermediate filaments and intermediate filament–microtubule interactions. J Cell Biol 98:653–669
Eliason CM, Shawkey MD (2012) A photonic heterostructure produces diverse iridescent colours in duck wing patches. J R Soc Interface 9(74):2279–2289, by permission of the Royal Society. Copyright © 2011, The Royal Society
Er Rafik M, Briki F, Burghammer M et al (2006) *In vivo* formation of the hard alpha–keratin intermediate filament along a hair follicle: evidence for structural polymorphism. J Struct Biol 154:79–88
Eschricht DF, Reinhardt J (1866) Recent memoirs on the cetacean. In: Flower WH (ed) On the greenland right whale. The Ray Soc, London
Ewoldt RH et al (2011) Reprinted from Ewoldt RH, Winegard TM, Fudge DS (2011) Non-linear viscoelasticity of hagfish slime. Int J Non-Linear Mech: Special issue on non-linear mechanics of biological structures 46(4):627–636. Copyright (2011), with permission from Elsevier
Fan WJW (1965) The fine structure of thread cell differentiation in the slime glands of the Pacific hagfish, *Polistrotrema stouti*. Anat Rec 151:348
Fernholm B (1981) Thread cells from the slime glands of hagfish (Myxinidae). Acta Zool 62:137–145
Ferry JD (1941) A fibrous protein from the slime of the hagfish. J Biol Chem 138:263–268. Copyright © 1941, by the American Society for Biochemistry and Molecular Biology
Flitney EW, Kuczmarski ER, Adam SA et al (2009) Insights into the mechanical properties of epithelial cells: the effects of shear stress on the assembly and remodeling of keratin intermediate filaments. FASEB J 23:2110–2119
Flower WH (1883) On whales, past and present, and their probable origin. Nature 28:199–202
Fraser RDB, Parry DAD (2008) Molecular packing in the feather keratin filament. J Struct Biol 162:1–13
Fraser RD, Parry DA (2011) The structural basis of the filament–matrix texture in the avian/reptilian group of hard β–keratins. J Struct Biol 173(2):391–405
Fraser RDB, MacRae TP, Rogers GE (1972) Keratins: their composition, structure and biosynthesis. Charles C Thomas, Springfield
Freeburg WE, Brault S, Mayo C et al (2009) Whale baleen trace element signatures: a predictor of environmental life history? American Geophysical Union, Fall Meeting 2009, abstract Nr. B32B–05
Fudge DS (2001) Hagfishes: champions of slime. Nat Aust 27:60–69
Fudge DS, Gosline JM (2004) Molecular design of the alpha–keratin composite: insights from a matrix–free model, hagfish slime threads. Proc Biol Sci 271:291–299
Fudge DS, Gardner KH, Forsyth VT et al (2003) The mechanical properties of hydrated intermediate filaments: insights from hagfish slime threads. Biophys J 85:2015–2027

Fudge DS, Levy N, Chiu S et al (2005) Composition, morphology and mechanics of hagfish slime. J Exp Biol 208:4613–4625
Fudge DS, Szewciw LJ, Schwalb AN (2009a) Morphology and development of blue whale baleen: an annotated translation of Tycho Tullberg's classic 1883 paper. Aquat Mamm 35:226–252
Fudge DS, Winegard T, Ewoldt RH et al (2009b) From ultra-soft slime to hard α-keratins: the many lives of intermediate filaments. Integr Comp Biol 49(1):32–39, by permission of Oxford University Press. Copyright © 2009, Oxford University Press. Reprinted
Fudge et al (2010) Hagfish slime threads as a biomimetic model for high performance protein fibres. Bioinspir Biomim 5:035002. doi:10.1088/1748-3182/5/3/035002. Copyright © 2014 IOP Publishing. Reproduced with permission. All rights reserved
Gaskin DE (1982) The ecology of whales and dolphins. Heinemann, Portsmouth
Glotzer M (2009) The 3Ms of central spindle assembly: microtubules, motors and MAPs. Nat Rev Mol Cell Biol 10:9–20
Goddard DR, Michaelis L (1935) Derivatives of keratin. J Biol Chem 112:361–371
Gu LH, Coulombe PA (2007) Keratin function in skin epithelia: a broadening palette with surprising shades. Curr Opin Cell Biol 19(1):13–23
Halstead LB (1974) Vertebrate hard tissues. Wykeham Publications, London
Hearle JWS (2008) An alternative model for the structural mechanics of hagfish slime threads. Int J Biol Macromol 42:420–428
Herr JE et al (2010) Republished with permission of The Company of Biologists Ltd., from Herr JE, Winegard TM, O'Donnell MJ et al (2010) Stabilization and swelling of hagfish slime mucin vesicles. J Exp Biol 213:1092–1099. doi:10.1242/jeb.038992. © 2010. Published by The Company of Biologists Ltd.; permission conveyed through Copyright Clearance Center, Inc.
Herrmann H, Aebi U (2004) Intermediate filaments: molecular structure, assembly mechanism, and integration into functionally distinct intracellular Scaffolds. Annu Rev Biochem 73:749–789
Herrmann H, Kreplak L, Aebi U (2004) Isolation, characterization, and in vitro assembly of intermediate filaments. Methods Cell Biol 78:3–24
Herrmann H, Bar H, Kreplak L et al (2007) Intermediate filaments: from cell architecture to nanomechanics. Nat Rev Mol Cell Biol 8:562–573
Herrmann H, Strelkov SV, Burkhard P et al (2009) Intermediate filaments: primary determinants of cell architecture and plasticity. J Clin Invest 119:1772–1783
Hesse M, Magin TM, Weber K (2001) Genes for intermediate filament proteins and the draft sequence of the human genome: novel keratin genes and a surprisingly high number of pseudogenes related to keratin genes 8 and 18. J Cell Sci 114:2569–2575
Hill P, Brantley H, Van Dyke M (2010) Some properties of keratin biomaterials: kerateines. Biomaterials 31:585–593
Hobson KA, Schell DM (1998) Stable carbon and nitrogen isotope patterns in baleen from eastern Arctic bowhead whales (*Balaena mysticetus*). Can J Fish Aquat Sci 55(12):2601–2607
Hobson KA, Riget FF, Outridge PM et al (2004) Baleen as a biomonitor of mercury content and dietary history of North Atlantic Minke Whales (*Balaenopetra acutorostrata*): combining elemental and stable isotope approaches. Sci Total Environ 331:69–82
Hunter J (1787) Observations on the structure and oeconomy of whales. Phil Trans R Soc 77:371–450
Karantza V (2011) Reprinted by permission from Macmillan Publishers Ltd: Oncogene, Karantza V (2011) Keratins in health and cancer: more than mere epithelial cell markers. Oncogene 30(2):127–138. Copyright © 2011, Rights Managed by Nature Publishing Group
Kasuya T, Rice DW (1970) Notes on baleen plates and on arrangement of parasitic barnacles of gray whale. Sci Rep Whales Res Inst 22:39–43
Kim JS, Lee CH, Coulombe PA (2010) Modeling the self–organization property of keratin intermediate filaments. Biophys J 99(9):2748–2756
Kimura T, Ozawa T (1997) Sample preparation and analysis of mitochondrial DNA from whale baleen plates. Mar Mamm Sci 13(3):495–498

Kirfel J, Magin TM, Reichelt J (2003) Keratins: a structural scaffold with emerging functions. Cell Mol Life Sci 60(1):56–71
Kirkbride KP, Tungol MW (1999) Infrared microspectroscopy of fibers. In: Robertson J, Grieve M (eds) Forensic examination of fibers, 2nd edn. Taylor & Francis Inc., Philadelphia, pp 179–222
Kjeld M (2003) Salt and water balance of modern baleen whales: rate of urine production and food intake. Can J Zool 81(4):606–616. © 2003 Canadian Science Publishing or its licensors. Reproduced with permission
Koch EA, Spitzer RH, Pithawalla RB et al (1991) Keratin–like components of gland thread cells modulate the properties of mucus from hagfish (*Eptatretus stouti*). Cell Tissue Res 264:79–86
Koch EA, Spitzer RH, Pithawalla RB et al (1994) An unusual intermediate filament subunit from the cytoskeletal biopolymer released extracellularly into seawater by the primitive hagfish (*Eptatretus stoutii*). J Cell Sci 107:3133–3144
Koch EA et al (1995) Reprinted from Koch EA, Spitzer RH, Pithawalla RB et al (1995) Hagfish biopolymer: a type I/type II homologue of epidermal keratin intermediate filaments. Int J Biol Macromol 17(5):283–292. Copyright © 1995, with permission from Elsevier
Kölsch A et al (2010) Republished with permission of COMPANY OF BIOLOGISTS, from Kölsch A, Windoffer R, Würflinger T, Aach T, Leube RE (2010) The keratin-filament cycle of assembly and disassembly. J Cell Sci 123(Pt 13):2266–2272. Copyright (2010); permission conveyed through Copyright Clearance Center, Inc.
Kreplak L, Fudge D (2007) Biomechanical properties of intermediate filaments: from tissues to single filaments and back. Bioessays 29:26–35. Copyright © 2006 Wiley Periodicals, Inc
Kreplak L, Franbourg A, Briki F et al (2002) A new deformation model of hard alpha–keratin fibres at the nanometer scale: implications for hard alpha–keratin intermediate filament mechanical properties. Biophys J 82:2265–2274
Kreplak L, Doucet J, Dumas P et al (2004) New aspects of the alphahelix to beta–sheet transition in stretched hard alpha–keratin fibres. Biophys J 87:640–647
Krushna Padhi B, Akimenko MA et al (2006) Independent expansion of the keratin gene family in teleostean fish and mammals: an insight from phylogenetic analysis and radiation hybrid mapping of keratin genes in zebrafish. Gene 368:37–45
Lametschwandtner A, Lametschwandtner U, Patzner RA (1986) The different vascular patterns of slime glands in the hagfishes *Myxine glutinosa* Linnaeus and *Eptatretus stouti* Lockington. A scanning electron microscope study of vascular corrosion casts. Acta Zool 67:243–248
Langbein L, Rogers MA, Winter H et al (1999) The catalog of human hair keratins. I. Expression of the nine type I members in the hair follicle. J Biol Chem 274:19874–19884
Langbein L, Rogers MA, Winter H et al (2001) The catalog of human hair keratins II. Expression of the six type II members in the hair follicle and the combined catalog of human type I and type II keratins. J Biol Chem 276:35125–35132
Lauffenburger JA (1993) Republished with permission of American Institute for Conservation of Historic and Artistic Works, from Lauffenburger JA (1993) Baleen in museum collections: its sources, uses, and identification. JAIC 32(3):213–230. Copyright © 1993; permission conveyed through Copyright Clearance Center, Inc.
Lazarides E (1980) Intermediate filaments as mechanical integrators of cellular space. Nature 283(5744):249–256
Lee CH, Coulombe PA (2009) Self–organization of keratin intermediate filaments into cross–linked networks. J Cell Biol 186:409–421
Leitner A, Paust T, Marti O et al (2012) Properties of intermediate filament networks assembled from keratin 8 and 18 in the presence of Mg^{2+}. Biophys J 103:195–201
Leppi TJ (1967) Histochemical studies on mucous cells in the skin and slime glands of hagfish. Anat Rec 157:278
Leppi TJ (1968) Morphochemical analysis of mucous cells in the skin and slime glands of hagfishes. Histochemie 16:68–78
Lim J, Fudge DS, Levy N, Gosline JM (2006) Hagfish slime ecomechanics: testing the gill–clogging hypothesis. J Exp Biol 209:702–710

Lingham–Soliar T, Bonser RHC, Wesley–Smith J (2010) Selective biodegradation of keratin matrix in feather rachis reveals classic bioengineering. Proc R Soc B 277:1161–1168

Luchtel DL, Martin AW, Deyrup–Olsen I (1991) Ultrastructure and permeability characteristics of the membranes of mucous granules of the hagfish. Tissue Cell 23:939–948

Ma L, Xu J, Coulombe PA et al (1999) Keratin filament suspensions show unique micromechanical properties. J Biol Chem 274:19145–19151

Magin TM, Vijayaraj P, Leube RE (2007) Structural and regulatory functions of keratins. Exp Cell Res 313:2021–2032

Maia R, D'Alba L, Shawkey MD (2011) What makes a feather shine? A nanostructural basis for glossy black colours in feathers. Proc Biol Sci 278(1714):1973–1980

Maia R, Macedo RH, Shawkey MD (2012) Nanostructural self–assembly of iridescent feather barbules through depletion attraction of melanosomes during keratinization. J R Soc Interface 9(69):734–743

Markl J, Schechter N (1998) Fish intermediate filament proteins in structure, function and evolution. Subcell Biochem 31:1–33

Martini FH (1998) The ecology of hagfishes. In: Jorgensen JM, Lomholt JP, Weber RE, Malte H (eds) The biology of hagfishes. Chapman & Hall, London

Matthews LH (1968) The whale. Crescent Books, New York

Meredith RW, Gatesy J, Cheng J et al (2010) Pseudogenization of the tooth gene enamelysin (MMP20) in the common ancestor of extant baleen whales. Proc R Soc B 278:993–1002. doi:10.1098/rspb.2010.1280, rspb20101280. By permission of the Royal Society. Copyright © 2011, The Royal Society

Moll R, Divo M, Langbein L (2008) The human keratins: biology and pathology. Histochem Cell Biol 129:705–733

Nerini M (1984) A review of gray whale feeding ecology. In: Jones ML, Swartz SL, Leatherwood S (eds) The gray whale *Eschrichtius robustus*. Academic, Orlando

Newby WW (1946) The slime glands and thread cells of the hagfish, *Polistrotrema stouti*. J Morphol 78:397–409

Noishiki Y, Ito H, Miyamoto T et al (1982) Application of denatured wool keratin derivatives to an antithrombogenic biomaterial: vascular graft coated with a heparinized keratin derivative. Kobunshi Ronbunshu 39(4):221–227

Norlén L et al (2007) Reprinted from Norlén L et al (2007) Structural analysis of vimentin and keratin intermediate filaments by cryo-electron tomography. Exp Cell Res 313(10):2217–2227. Copyright © 2007, with permission from Elsevier

O'Connor S (1987) The identification of osseous and keratinaceous materials at York, archaeological bone, antler and ivory. Occasional papers 5. United Kingdom Institute for Conservation, London

Pauling L, Corey RB (1951) The structure of hair, muscle, and related proteins. Proc Natl Acad Sci U S A 37:261–271

Pautard FG (1963) Mineralization of keratin and its comparison with the enamel matrix. Nature 199:531

Pfeiffer CJ (1993) Cellular structure of terminal baleen in various mysticete species. Aquat Mamm 18:67–73

Pivorunas A (1979) The feeding mechanisms of baleen whales. Am Sci 67:432–440

Pollard TD, Cooper JA (2009) Actin a central player in cell shape and movement. Science 326:1208–1212

Prum RO (2006) Anatomy, physics, and evolution of avian structural colors. In: Hill GE, McGraw KJ (eds) Bird coloration, vol 1, Mechanisms and measurements. Harvard University Press, Cambridge

Prum RO, Dufresne ER, Quinn T et al (2009) Development of colour–producing b–keratin nanostructures in avian feather barbs. J R Soc Interface 6:S253–S265

Qin Z, Buehler MJ (2012) Computational and theoretical modeling of intermediate filament networks: structure, mechanics and disease. Acta Mech Sin 28:941–950

Qin Z, Kreplak L, Buehler MJ (2009) Hierarchical structure controls nanomechanical properties of vimentin intermediate filaments. PLoS One 4(10):e7294

Rice DW, Wolman AA (1971) The life history and ecology of the gray whale (*Eschrichtius robustus*). Special publication (American Society of Mammalogists), no 3. American Society of Mammalogists, Stillwater, p 196

Rosenbaum HC, Egan MG, Clapham PJ et al (1997) An effective method for isolation of DNA from historical specimens of baleen. Mol Ecol 6:667–681

Rosenthal FC (1829) Ueber die barten des Schnabel–Walfisches (*Balaena rostrata*). Abhandl der Kon Akad der Wiss zu Berlin 127–132

Rouse JG, Van Dyke ME (2010) A review of keratin-based biomaterials for biomedical applications. Materials 3(2):999–1014. © 2010 by the authors; licensee Molecular Diversity Preservation International, Basel, Switzerland. This article is an open-access article distributed under the terms and conditions of the Creative Commons Attribution license. http://creativecommons.org/licenses/by/3.0/

Russell D, Andrews PD, James J et al (2004) Mechanical stress induces profound remodelling of keratin filaments and cell junctions in epidermolysis bullosa simplex keratinocytes. J Cell Sci 117:5233–5243

Ruud JT (1940) The surface structure of the baleen plates as a possible clue to age in whales. Hvalradets Skr 23:1–24

Saranathan V, Forster JD, Noh H et al (2012) Structure and optical function of amorphous photonic nanostructures from avian feather barbs: a comparative small angle X-ray scattering (SAXS) analysis of 230 bird species. J R Soc Interface 9(75):2563–2580, by permission of the Royal Society. Copyright © 2011, The Royal Society

Schaffeld M, Markl J (2004) Fish keratins. Methods Cell Biol 78:627–671

Schaffeld M, Löbbecke A, Lieb B, Markl J (1998) Tracing keratin evolution: catalog, expression patterns and primary structure of shark (*Scyliorhinus stellaris*) keratins. Eur J Cell Biol 77:69–80

Schaffeld M, Schultess J, Haberkamp M et al (2001) Intermediate filament protein evolution in fish: sequences from lamprey, shark, bichir, sturgeon and trout. Biol Cell 93:235

Schaffeld M, Höffling S, Markl J (2004) Sequence, evolution and tissue expression patterns of an epidermal type I keratin from the shark *Scyliorhinus stellaris*. Eur J Cell Biol 83:359–368

Schweizer J, Bowden PE, Coulombe PA et al (2006) New consensus nomenclature for mammalian keratins. JCB 174:169–174

Shawkey MD, Maia R, D'Alba L (2011) Proximate bases of silver color in anhinga (*Anhinga anhinga*) feathers. J Morphol 272(11):1399–1407

Sivaramakrishnan S, DeGiulio JV, Lorand L et al (2008) Micromechanical properties of keratin intermediate filament networks. PNAS 105(3):889–894. Copyright (2008) National Academy of Sciences, USA. Reprinted with permission

Solomon SE, Hendrickson JR, Hendrickson LP (1986) The structure of the carapace and plastron of juvenile turtles, *Chelonia mydas* (the green turtle) and *Caretta caretta* (the loggerhead turtle). J Anat 145:123–131

Spitzer RH, Koch EA (1998) Hagfish skin and slime glands. In: Jorgensen JM, Lomholt JP, Weber RE, Malte H (eds) The biology of hagfishes. Chapman & Hall, London, pp 109–132

Spitzer RH, Downing SW, Koch EA et al (1984) Hagfish slime gland thread cells. II. Isolation and characterization of intermediate filament components associated with the thread. J Cell Biol 98:670–677

Spitzer RH, Koch EA, Downing SW (1988) Maturation of hagfish gland thread cells: composition and characterization of intermediate filament polypeptides. Cell Motil Cytoskeleton 11:31–45

St Aubin DJ, Stinson RH, Geraci JR (1984) Aspects of the structure and composition of baleen, and some effects of exposure to petroleum hydrocarbons. Can J Zool 62:193–198

Stavenga DG, Tinbergen J, Leertouwer HL et al (2011) Kingfisher feathers—colouration by pigments, spongy nanostructures and thin films. J Exp Biol 214(Pt 23):3960–3967

Steinert PM, Marekov LN, Parry DA (1993) Conservation of the structure of keratin intermediate filaments: molecular mechanism by which different keratin molecules integrate into preexisting keratin intermediate filaments during differentiation. Biochemistry 32:10046–10056
Stevenson CH (1907) Whalebone: its production and utilization. U.S. Department of Commerce and Labor, Bureau of Fisheries document no. 626. U.S. Government Printing Office, Washington, DC. Reprinted in Bulletin from Johnny Cake Hill (Old Dartmouth Historical Society) 1965–66 (Winter):4–11
Strahan R (1959) Slime production in *Myxine glutinosa* Linnaeus. Copeia 2:165–166
Strahan R (1963) The behavior of myxinoids. Acta Zool 44:73–102
Strelkov SV, Herrmann H, Geisler N et al (2001) Divide–and–conquer crystallographic approach towards an atomic structure of intermediate filaments. J Mol Biol 306:773–781
Strelkov SV, Herrmann H, Aebi U (2003) Molecular architecture of intermediate filaments. Bioessays 25:243–251
Subramanian S, Ross NW, MacKinnon SL (2008) Comparison of the biochemical composition of normal epidermal mucus and extruded slime of hagfish (*Myxine glutinosa* L.). Fish Shellfish Immunol 25:625–632
Sumich JL (2001) Growth of baleen of a rehabilitating gray whale calf. Aquat Mamm 27:234–238
Szewciw LJ, de Kerckhove DG, Grime GW et al (2010) Calcification provides mechanical reinforcement to whale baleen α-keratin. Proc R Soc B 277(1694):2597–2605, by permission of the Royal Society
Terakado K, Ogawa M, Hashimoto Y, Matsuzaki H (1975) Ultrastructure of the thread cells in the slime gland of Japanese hagfishes, *Paramyxine atami* and *Eptatretus burger*. Cell Tissue Res 159:311–323
Toni M et al (2007) Reprinted with permission from Toni M et al (2007) Hard (beta-)keratins in the epidermis of reptiles: composition, sequence, and molecular organization. J Proteome Res 6(9):3377–3392. Copyright © 2007, American Chemical Society
Tullberg T (1883) Bau und entwicklung der barten bei balaenoptera sibbaldii. Nov Acta Reg Soc Sci Ups Ser III 11:1–36
Uhen MD (2010) The origin(s) of whales. Annu Rev Earth Planet Sci 38:189–219
Utrecht WLV (1965) On the growth of the baleen plate of the fin whale and the blue whale. Bijdr Dierk 35:3–38
Van Dyke ME (2012) Wound healing compositions containing keratin biomaterials. United States Patent 8258093
Wagner OI et al (2007) Reprinted from Wagner OI, Rammensee S, Korde N et al (2007) Softness, strength and self-repair in intermediate filament networks. Exp Cell Res 313(10):2228–2235. Copyright (2007), with permission from Elsevier
Wang N, Stamenovic D (2000) Contribution of intermediate filaments to cell stiffness, stiffening, and growth. Am J Physiol Cell Physiol 279:C188–C194
Werth AJ (2001) How do mysticetes remove prey trapped in baleen? Bull Mus Comp Zool 156:189–203
Windoffer R, Beil M, Magin TM et al (2011) Cytoskeleton in motion: the dynamics of keratin intermediate filaments in epithelia. JCB 194(5):669–678. © 2011 Windoffer et al. Reprinted with permission
Winegard TM, Fudge DS (2010) Deployment of hagfish slime thread skeins requires the transmission of mixing forces via mucin strands. J Exp Biol 213:1235–1240
Yamada S, Wirtz D, Coulombe PA (2003) The mechanical properties of simple epithelial keratins 8 and 18: discriminating between interfacial and bulk elasticities. J Struct Biol 143:45–55
Ye C et al (2010) With kind permission from Springer Science+Business Media: Ye C, Wu X, Yan P et al (2010) beta-Keratins in crocodiles reveal amino acid homology with avian keratins. Mol Biol Rep 37(3):1169–1174. Copyright © 2009, Springer Science+Business Media B.V
Yin H, Dong B, Liu X et al (2012) Amorphous diamond-structured photonic crystal in the feather barbs of the scarlet macaw. PNAS 109:10798–10801. Reprinted with permission

Yoon KH, Yoon M, Moir RD et al (2001) Insights into the dynamic properties of keratin intermediate filaments in living epithelial cells. J Cell Biol 153:503–516

Yoshie S, Honma Y (1976) Light and scanning electron microscopic studies on the esophageal spines in the Pacific ridley turtle, *Lepidochelys olivacea*. Arch Histol Jpn 38(5):339–346. Copyright © International Society of Histology and Cytology

Zeh JE, Rosa C, George JC et al (2008) Age estimation for young bowhead whales (*Balaena mysticetus*) using annual baleen growth increments. Can J Zool 86:525–538

Zimek A, Stick R, Weber K (2003) Genes coding for intermediate filament proteins: common features and unexpected differences in the genomes of humans and the teleost fish *Fugu rubripes*. J Cell Sci 116:2295–2302

Zintzen V, Roberts CD, Anderson MJ et al (2011) Hagfish predatory behaviour and slime defence mechanism. Sci Rep 1:131

Zorgrader CG (1720) Bloeyende Opkomst der Aloude en Hedendaaqsche Groenlandsche Visschery, plates, As Aio, Amsterdam, 81 p. The German translation printed in Leipzig, 1723

Chapter 12
Egg-Capsule Proteins of Selachians

Abstract Egg capsules of sharks, skates and chimaerods are made of very specific collagen and polyphenol-based fibrillar structures, which exhibit unusual preservation even when fossilized. The dogfish egg capsule is a composite material constructed largely from collagen fibrils with a highly ordered, kinked molecular arrangement set in a matrix containing spherical filler particles composed of hydrophobic protein. The capsule consists of collagenous three-dimensional very ordered open networks. This molecular construct is responsible not only for specific filtering properties, but also for strength of the unique structured capsule. Both capsular collagen and polyphenol-based proteins possess high biomimetic potential.

The egg capsules of sharks and rays have been described previously by a number of workers (Beard 1890; Widakowich 1906; Hussakof and Welker 1908; Clark 1922; Hobson 1930; Whitley 1938; Nalini 1940; Brown 1955; Krishnan 1959). Also attention has been paid to similar structures in Chimaeroids (Bessels 1869; Jaekel 1901; Gill 1905; Dean 1912; Brown 1946). The egg capsules of Elasmobranchii are widely recognised as important in species identification, and provide relevant information concerning their reproductive biology (Whitley 1939; Oddone et al. 2004).

According to the modern view, the "nidamental gland in contemporary elasmobranchs secretes a tertiary accessory envelope, called the egg capsule, around the fertilized ova as they pass through the oviduct and into the uterus," (Heiden et al. 2005). Except some rays, the encapsulation of fertilized eggs occurs in oviparous and nearly all viviparous species (Hamlett and Koob 1999). Due to peculiar thickness, the egg capsules of oviparous species are exceptionally durable. The biological sense of these structures is to protect embryos during protracted development in the marine environment (Hunt 1985; Knight et al. 1996). "*In viviparous* species however, the egg capsule is considerably thinner and, in sharks, often remains intact around developing embryos throughout the course of development in utero," (Heiden et al. 2005; see also Hamlett and Koob 1999). Egg capsules of selachians are diverse in their forms and can include tendrils, ribs, and ridges. The egg capsules of sharks are generally rectangular in shape with the corners prolonged into anterior and posterior pairs of horns. However, considerable variation exists in shape and size in the different species (Krishnan 1959) (Figs. 12.1 and 12.2).

As reviewed by Heiden et al. (2005), "many skate egg capsules have adhesive fibers that moor the capsules to the sea floor, or some other substrate. This protects

H. Ehrlich, *Biological Materials of Marine Origin*, Biologically-Inspired Systems 4,
DOI 10.1007/978-94-007-5730-1_12

Fig. 12.1 The egg capsules of sharks are diverse in shape, size and color (Image courtesy of Marcin Wysokowski)

Fig. 12.2 Reconstruction (*left*) and fossil specimen (*right*) of the 27 cm-long *Palaeoxyris friessi* n. sp. shark egg capsule (Adapted from Böttcher (2010) with permission)

the developing embryos from displacement, and to some extent from predation. The corners of the capsules produced by many oviparous dogfish (e.g., *Scyliorhinus canicula*) extend into dense tendrils that are much longer than the body of the capsule. These tendrils may be heavily spiralled or coiled, and are believed to help anchor the capsules (Knight et al. 1996). The capsules of *Heterodontus phillipi* and *Heterodontus galeatus* resemble a screw, and are literally screwed into narrow crevices in the substrate following oviposition," (Heiden et al. 2005). This both protects and supports the capsule (Smith 1942; McLaughlin and O'Gower 1971). Roland Brown characterized the egg capsules of living Chimaeroid species as consisting of two parts, dartlike to elliptic structures with characteristic morphological features. The part with leathery, centrally located embryo case possesses bilateral symmetry. The other part extends outward from the case in the horizontal plane, and is of membraneous nature (Brown 1946).

From biomaterials point of view, not only the form of egg capsules, but also their chemistry, composition, and mechanical properties are of special importance. Interestingly, besides teeth, scales and fin, spines are the only mineralized skeletal remains of fossil sharks. Their egg capsules are being increasingly reported from Paleozoic and Mesozoic brackish to fluvio-lacustrine deposits (Böttcher 2010; Fischer et al. 2010a, b). Studies on the chemical composition of these structures started at the beginning of the twentieth century. For example, the paper published in *Journal of Biological Chemistry* in 1908 was entitled "*Notes on the chemical nature of egg cases of two species of sharks*" (Hussakof and Welker 1908). Later egg capsule proteins were studied by Brown (1955) as well as by Cox et al. (1987). Today, very specific egg capsule collagens, as well as polyphenol-containing proteins, are the "prime examples" for the biomaterials community.

12.1 Collagen

Collagenous composite material of the egg capsule of selachian fish is known as "a *sophisticated collagenous material*" (Hepworth et al. 1994). Why? As reviewed by Knight and Vollrath (2001), "the Selachian egg case serves mainly as a shock absorber to protect the embryo. The wall of the case is a highly cross-linked collagenous composite material with a complex, multilamellate, highly hierarchical construction," (Knight and Vollrath 2001; see also Knight and Feng 1992, 1994a, b; Knight and Hunt 1974, 1976, 1986; Rusaouën et al. 1976; Hunt 1985; Knight et al. 1993; Gathercole et al. 1993; Knupp et al. 1996, 1998). Evidence for the collagenous nature of the dogfish egg case came from X-ray diffraction studies that indicated the existence of a 2.9 A° meridional arc typical of collagens, the thermal shrinkage characteristic curve (S-shaped with a mean half shrinkage temperature of 78 °C), and amino acid composition analysis showing that glycines accounted for about 16 % of the amino acid residues (Knight and Hunt 1974; Rusaouën et al. 1976).

In small-spotted catshark (*Scyliorhinus canicula*) the case wall is formed from about 20 lamellae. According detailed description made by Knight and Vollrath (2001), "each lamella is built up from a single layer of flattened collagenous ribbons. These overlap within the lamella, like tiles on a roof. The ribbons are held together by small quantities of amorphous matrix giving a 'sea and island' construction; in which the cross-sectional area of the islands is much greater than that of the sea. The ribbons are continuous throughout the length of the egg case. Each ribbon is formed from numerous transversely banded collagen fibrils arranged in a remarkably regular way that varies somewhat in different regions of the egg case.

Each collagen fibril is practically crystalline, being constructed from a tetragonal array of collagen molecules. These are regularly kinked approximately midway between their ends; with the kinked segments running at approximately 20° to the long axis of the fibril. The molecules are held together by extensive covalent cross-linking. This contributes to the material's mechanical and thermal stability, its extreme insolubility, and its resistance to enzymatic degradation," (Knight and Vollrath 2001; see also Knight et al. 1996).

Glycoxidative, dityrosyl, lysine-derived, and oxidative phenolic covalent cross-links (Hepworth et al. 1994) may contribute to the stability of this biological material. The role of disulphide bonds is also possible. The short length (approximately 40 nm) of the triple-helical collagen molecules with molecular weight of 34–35 kDa (Fig. 12.3) and an axial ratio (length to width) of about 20 make them ideally suited for nematic liquid crystallization (Knupp et al. 1998, 1999; Knight and Vollrath 2001).

The secretory cells of the D-zone of the nidamental gland of the dogfish are responsible for secretion and storage of the collagen of the egg capsules (Rusaouen-Innocent 1990). This protein assembles within the Golgi apparatus and migrates through the cytoplasm within storage granules. It "appears to pass through several morphologically distinct textures during storage, secretion and fibril formation which may represent different lyotropic liquid crystalline phases," as described in detail by Feng and Knight (1994b).

Interestingly, the collagen storage granules showed very specific structural behaviour in response to changes in pH for in vitro experiments (Feng and Knight 1994b). Thus, "from pH 2 to pH 4 most granules appeared completely amorphous; from pH 5 to pH 7 granules showed the following previously reported liquid crystalline textures: isotropic, lamellar, micellar, hexagonal columnar, transversely banded twisted nematic, and unhanded twisted nematic. At pH 8 granules showed both the hexagonal columnar phase (phase IV) and small quantities of the final fibrillar phase, together with a previously undescribed texture. The latter texture, which we refer to as phase VII, had a D period (17.5 nm) half that of the lamellar texture (phase II), and the final egg capsule fibrils (phase VI). From pH 9 to pH 11, only the final fibrillar texture (phase VI) and small quantities of the new texture (phase VII) were present," (Feng and Knight 1994b).

David Knight suggested the following causal sequence for collagen fibrillogenesis, orientation, storage and secretion, in this system:

"The collagen is maintained under acidic conditions during storage and secretion. Provided that the isoelectric point is fairly close to neutrality as it is in other collagens, the molecules will carry a net positive charge under these conditions. This will prevent them from approaching one another close enough to form intermolecular

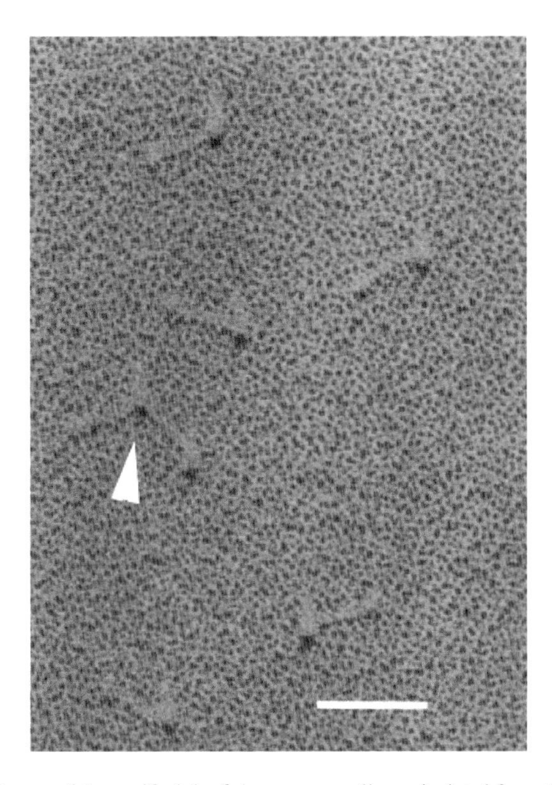

Fig. 12.3 TEM image of the purified dogfish egg case collagen isolated from the oviducal gland (scale bar 50 nm). According to original description, "the molecules are approximately 45 nm long, with a large noncollagenous domain at one end and a small one at the other. The molecules sometimes adhere to each other head-to-tail (*arrow*), as they do when incorporated into fibrils. The dimensions of the molecules make them ideal for nematic liquid crystallization," (Reprinted with permission from Knight and Vollrath 2001. Copyright (2001) American Chemical Society)

hydrogen bonds and covalent cross-links. This in turn will maintain a fluid state in which hydrophobic interactions between terminal peptides permit the formation of a sequence of liquid crystalline phases as water is removed. The fluidity of these phases allows the material to flow through the secretory apparatus and permits the remarkably precise orientation of the collagen molecules in the final material to be defined by a combination of shearing orientation and liquid crystallization. When the molecules are finally in position within the lumen of the nidamental gland, the pH is raised, allowing the helical portions of the molecules to come together by electrostatic interactions, in turn allowing the formation of intermolecular hydrogen bonds and then oxidative covalent cross-links," (Knight and Feng 1992, 1994a; Feng and Knight 1992,1994b).

Peptides from the dogfish egg capsule collagen have been partially sequenced, and show some homology with Type VI, Type X, though mainly Type IV, collagens (Luong et al. 1998; Knupp et al. 1998). However, the main principal difference of the dogfish egg case collagen is that this protein aggregates in the most ordered

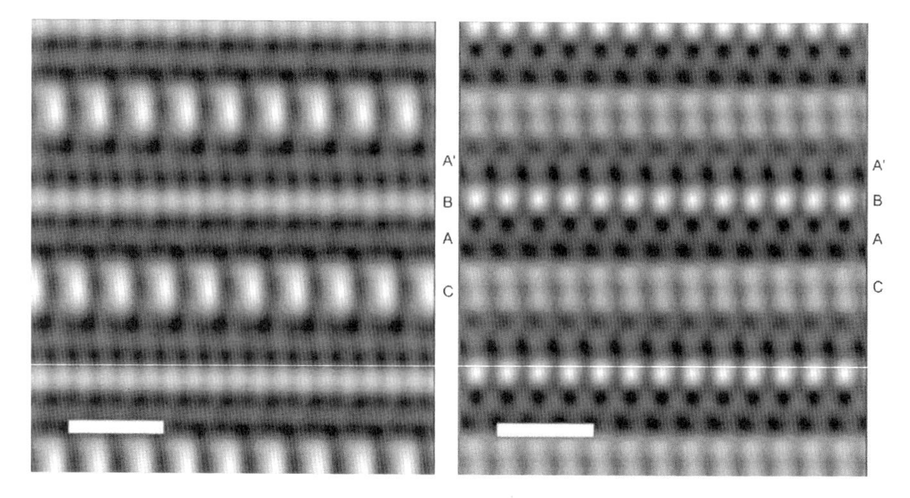

Fig. 12.4 Unique nanoarchitecture of the dogfish egg case collagen. The reconstruction obtained after filtering the corresponding Fourier transform images. Bar scale 20 nm (Reprinted from Knupp et al. 1998, Copyright (1998), with permission from Elsevier)

arrangement (Fig. 12.4). This unique molecular arrangement is responsible for strength as well as for specific filtering properties.

The wet longitudinal strips of the *S. canicula* egg case wall were object of investigations on their mechanical properties by Hepworth et al. (1994). These authors reported results as follow:

> "The strips are remarkably strong (tensile strength 11.9 MPa) and extensible (strain at fracture 0.39). The combination of high tensile strength and extensibility gives them a high toughness ($6x10^4$ J/m^2 or 6 kJ/kg for a specimen measuring 4x1x10 mm). This toughness is in the range of that of steels!
>
> An increase in axial periodicity in moderately strained material demonstrated by low-angle X-ray diffraction and TEM suggests that the kinks in the collagen molecules straighten out progressively as the material is strained and spring back when released. Recovery is substantially complete up to a global strain of 24 %. This suggests an entropic mechanism of energy storage," (Knight and Vollrath 2001).

It was reported that "as the material is progressively strained, the composite construction allows different layers of fibrils to rotate and fail successively depending on their initial orientation. Energy is thought to be dissipated by deformation of the matrix between the collagen fibrils as they rotate; and by the formation of numerous cracks as matrix and fibrils fail," (Knight and Vollrath 2001; see also Hepwoth et al. 1994). According to Knight and Vollrath (2001), "filler particles are present in high concentrations in the outer layers of the case. Ultrastructural evidence suggests these are bound into the meshwork of collagen fibrils and stretch when the egg case is strained, probably helping to store and/or dissipate energy," (Knight and Vollrath 2001).

Permeability, is the next interesting feature of the egg case due to their function as an ultrafilter, that is highly permeable to the products of nitrogenous excretion as well as to respiratory gases. For example, it was reported (Lombardi and Files 1993) that the egg capsule surrounding near-term *Mustelus canis* shark embryos was permeable to substances smaller than 5 kDa in size. Therefore it is not surprising that this transverse channels system with structural similarity to liquid crystalline construction have large biomimetic potential (Knight et al. 1996).

Collagen is also located within other structures observed on the surface of egg capsules, the tendrils. The tendrils of the egg capsule of the dogfish *S. canicula* appear to act as damped spring which becomes entangled with one another, attaching the capsule firmly to the seaweed *Halidrys siliquosa*. The tendrils have a unique structure (see for details Feng and Knight 1994a). Intriguingly, tendrils and capsule wall both appear to be constructed from similar components: transversely banded collagen fibrils and amorphous granules. Both are constructed from lamellae built up from laminae. However, the tendril has a spiral concentric construction apparently derived from rotation of the tendril within the tendril forming region; whereas the lamellae of the egg capsule form as a simple laminated extrusion. The tendril exhibits a marked primary twisting and secondary coiling; while the egg capsule of the dogfish is only slightly twisted. In the tendril, the fibrils are arranged at approximately right angles to one another in adjacent laminae. This is in contrast to the orthogonal arrangement with an angle of approximately 45° between fibrils in adjacent laminae in the capsule wall (Knight and Hunt 1976; Gathercole et al. 1993).

12.2 Polyphenol-Containing Egg Capsule Proteins

Formation of skate egg capsules begins in the nidamental gland with the secretion and assembly of capsular precursors (see for detail Koob and Cox 1988). These materials are white when assembled. They gradually develop color with time in utero, eventually producing the deep greenish brown characteristic of skate capsules at oviposition. In *Raja erinacea*, the tanning of capsules in utero is coincident with the introduction of catechols into the capsular matrix (Koob and Cox 1986). An enzymic activity able to oxidize catechols to quinones has been demonstrated histochemically in tanning capsules and nidamental glands from several oviparous elasmobranchs.

Brown (1955) reported that sections of newly formed *Raja* egg capsules turned brown upon incubation with tyrosine, and that this reaction could be blocked by potassium cyanide. She believed that these results demonstrated a polyphenol oxidase which would oxidize the polyphenol present in the capsule to quinone which, in turn, would tan the capsule (Koob and Cox 1988). A polyphenoloxidase was demonstrated histochemically in the shell glands of *S. canicula* by incubating sections of fixed glands with catechol (Threadgold 1957). Krishnan (1959) showed that both capsular material and sections of frozen glands from *Chiloscyiium griseum* oxidized catechol, and that the capsule had chemical properties like other quinone tanned matrices. He suggested that capsule formation involved a form of quinone autotanning catalyzed by a phenoloxidase (Koob and Cox 1988). Since works by Koob and Cox (1988) and Cox and Koob (1993), it was established that catechol oxidase plays a pivotal role during the formation of skate egg capsules by catalyzing the oxidation of capsular catechols to highly reactive quinones; forming dark pigments which tan the capsular matrix.

One of limiting factors that hindered detailed characterization of the proteins that comprise elasmobranch egg capsules is their strong resistance to acidic treatment. Only hot alkali and strong mineral acids can dissolve the tanned capsule (Hussakof and Welker 1911). Acid hydrolysis of the whole capsular material led to obtaining of the hydrolisate where elevated glycine, proline and hydroxyproline have been identified as dominant amino acids (Krishnan 1959; Gross et al. 1958; Knight and Hunt 1974; Cox et al. 1987). Thus, the collagenous origin of this biological material has been suggested. However, levels of glycine, proline and hydroxyproline in the capsules are generally lower than those of typical collagens. Identification of high levels of tyrosine in corresponding extracts of these capsules, suggesting the presence of another protein rich in tyrosine (Gross et al. 1958; Knight and Hunt 1974). Six proteins ranging in molecular mass from 100 to 15 kDa comprise the bulk of the skate egg capsule material (Koob and Cox 1993). In the little skate, *Leucoraja erinacea*, none of these proteins have an amino acid composition observed in interstitial collagens.

Nonetheless, glycine accounts for a relatively large proportion in each protein (50 % in the two low molecular mass proteins), and the larger molecular mass proteins contain substantial amounts of proline and hydroxyproline, suggesting that one or more of these proteins may contain collagenous domains. Of principal relevance to the polymerization mechanism described below, these proteins contain a relative abundance of tyrosine (25 % in the two low molecular mass proteins) (Koob 2002). Thus, skate egg capsules are polymerized by a quinone tanning process involving the introduction and oxidation of ortho-catechols following secretion and assembly of capsule precursors in the shell gland (Koob and Cox 1990). Catechol formation occurs via hydroxylation of tyrosine residues in the structural protein phase of the material. Approximately 50 % of the overall tyrosine residues are catalytically hydroxylated, forming peptide-bound 3,4-dihydroxyphenylalanine (an ortho-catechol).

As reported above, Catechol oxidase catalyzed oxidation of Dopa residues follows, producing Dopa quinone (Koob and Cox 1988). Dopa quinone then polymerizes the matrix through an as yet uncharacterized condensation reaction. Based on evidence from experiments with Dopa containing synthetic peptides, it seems likely that Dopa-quinones on neighboring protein monomers produce covalent cross-links via bisquinone formation (Koob 2002).

The resulting quinone-tanned material is mechanically and chemically stable. The polymerized material can be solubilized only with hydrolytic agents such as strong acids and bases (e.g. 6 N HCl at 100 °C and boiling 5 % KOH). It is refractory to proteinase attack, chaotropic agents, and organic solvents. As proposed by Thomas Koob, three aspects of this biological material elicit interest from a biomimetics perspective.

First, the polymerization that occurs after secretion and assembly of the precursors is very rapid. Capsule material in the little skate undergoes a transition from a chemically and mechanically friable substance to a leather-like material in less than 12 h, and this takes place entirely in the lumen of the reproductive tract. Second, the result of polymerization is a remarkably strong and durable protein composite, that in tensile tests exhibited a strength of 20 MPa and stiffness of 200 MPa. Third, and most relevant, is that polymerization occurs outside of the tissue of origin; and therefore mimicking the sclerotization process in the laboratory seemed feasible (Koob 2002).

A di-catechol from creosote bush (*Larrea tridentate*), so called nor-dihydroguaiaretic acid (NDGA), was proposed as a biomimetic analog to polyphenol-based biomaterial from skate egg capsules (Koob et al. 2001a, b; Koob and Hernandez 2002). NDGA has been characterized as follow:

"It caused a dose dependent increase in the material properties of reconstituted collagen fibers, achieving a 100-fold (!) increase in strength and stiffness over untreated fibers. The maximum tensile strength of the optimized NDGA treated fibers averaged 90 MPa; the elastic modulus of these fibers averaged 580 MPa. These properties were independent of strain rates ranging from 0.60 to 600 mm/min. Fatigue tests established that neither strength nor stiffness were affected after 80 k cycles at 5 % strain.

– treated fibers were not cytotoxic to tendon fibroblasts. Fibro- blasts attached and proliferated on NDGA treated collagen normally. NDGA-fibers did not elicit a foreign body response, nor did they stimulate an immune reaction over six weeks *in vivo*. The fibers survived 6 weeks with little evidence of fragmentation or degradation. The strength, stiffness and fatigue properties of the NDGA-treated fibers are comparable to those of tendon," (Koob 2002; see also Koob et al. 2001a, b; Koob and Hernandez 2003).

Due to tendon-like properties of NDGA treated collagen fibers, Koob and co-workers not exclude the development of neo-tendon by the colonizing fibroblasts *in vivo*.

12.3 Conclusion

I have no doubts that bioinspiration determined by hierarchically structured composite materials discussed in this chapter, is the driving force for development of novel specifically structured biocomposites. Novel and modern techniques shed new insights into their structural organization on a molecular level, as was shown for one example of egg capsule collagen. As characterized by Knupp and Squire (2005), "collagen Types IV, VI, VIII, X, and dogfish egg case collagen make linear and lateral associations to form open networks rather than fibers. The roles played by these network-forming collagens are diverse: they can act as support and anchorage for cells and tissues, serve as molecular filters, and even provide protective permeable barriers for developing embryos. Their functional properties are intimately linked to their molecular organization," (Knupp and Squire 2005). Approaches for stabilizing materials of collagenous origin with catechol containing monomers mimicking the biochemical situation within egg capsular layers, which were developed in the last decade, will definitively be continued in the future.

References

Beard J (1890) On the development of the common skate *Raia batis*. In: 8th annual report, Fisheries, Scotland
Bessels E (1869) Ueber fossile Selachier Eier. Ver vaterl Naturkunde Wurttemberg Jahresh 25:152–155
Böttcher R (2010) Description of the shark egg capsule *Palaeoxyris friessi* n. sp. From the Ladinian (*Middle Triassic*) of SW Germany and discussion of all known egg capsules from the Triassic of the Germanic Basin. Palaeodiversity 3:123–139
Brown RW (1946) Fossil egg capsules of chimaeroid fishes [From Germany, Wyoming, New Mexico, Montana, and Alaska]. J Paleontol 20(3):261–266
Brown CH (1955) Egg capsule proteins of selacians and trout. Quart J Microsc Sci 96:483–488
Clarck RS (1922) Rays and skates – egg capsules and young. J Mar Biol Assoc 12:543–577
Cox DL, Koob TJ (1993) Solid-state catechol oxidase in skate (*Raja erinacea*) egg capsule. Bull Mt Desert I Biol Lab 32:17–19
Cox DL, Mecham RP, Koob TJ (1987) Site specific variation in amino acid composition of the skate egg capsule *Raja erinucea*. J Exp Mar Biol Ecol 107:71–74
Dean B (1912) Orthogenesis in the egg capsules of Chimaera. Am Mus Nat Hist Bull 31:35–40
Feng D, Knight DP (1992) Secretion and stabilization of the layers of the egg capsule of the dog-fish *Scyliorhinus canicula*. Tissue Cell 24:778–790
Feng D, Knight DP (1994a) Structure and formation of the egg capsule tendrils in the dogfish *Scyliorhinus canicula*. Philos Trans Biol Sci 343:285–302
Feng D, Knight DP (1994b) Reprinted from Feng D, Knight DP (1994b) The effect of pH on fibrillogenesis of collagen in the egg capsule of the dogfish, *Scyliorhinus canicula*. Tissue Cell 26(5):649–659. Copyright (1994), with permission from Elsevier
Fischer J, Axsmith BJ, Ash SR (2010a) First unequivocal record of the hybodont shark egg capsule Palaeoxyris in the Mesozoic of North America. Neu Jhrb Geol Palaeontol Abh 255:327–344
Fischer J, Schneider JW, Ronchi A (2010b) New hybondontoid shark from the Permocarboniferous (Gzhelian–Asselian) of Guardia Pisano (Sardinia, Italy). Acta Palaeontol Pol 55:241–264
Gathercole LJ, Atkins EDT, Goldbeck-Wood EG et al (1993) Molecular bending and networks in a basement membrane like-collagen: packing in dogfish egg capsule collagen. Int J Biol Macromol 15:81–88
Gill T (1905) An interesting *Cretaceous chimaeroid* egg-case. Science 22:601–602
Gross J, Dumsha B, Glazer N (1958) Comparative biochemistry of collagen. Some amino acids and carbohydrates. Biochim Biophys Acta 30:293–297
Hamlett WC, Koob TJ (1999) Female reproductive system. In: Hamlett WC (ed) Sharks, skates, and rays. The biology of elasmobranch fishes. Johns Hopkins University Press, Baltimore
Heiden TCK, Haines AN, Manire C et al (2005) Structure and permeability of the egg capsule of the bonnethead shark, *Sphyrna tiburo*. J Exp Zool 303A:577–589. Copyright © 2005 Wiley-Liss, Inc. Reprinted with permission
Hepworth DG, Gathercole LJ, Knight DP et al (1994) Correlation of ultrastructure and tensile properties of a collagenous composite material, the egg capsule of the dogfish, *Scyliorhinus* spp, a sophisticated collagenous material. J Struct Biol 112:231–240
Hobson AD (1930) A note on the formation of the egg case of skate. J Mar Biol Assoc 16:577–581
Hunt S (1985) The selachian egg case collagen. In: Bairati A, Garrone R (eds) Biology of invertebrate and lower vertebrate collagens. Plenum Press, New York
Hussakof L, Welker WH (1908) Notes on the chemical nature of egg cases of two species of sharks. J Biol Chem 4:XLIV–XLV
Hussakof L, Welker WH (1911) Chemical notes on the egg capsules of two species of sharks. Biochem Bull 1:216–221
Jaekel O (1901) Ueber jurassiche Zahne und Eier von Chimariden. Neues Jahrb Beil 14:540–564

Knight DP, Feng D (1992) Formation of the dogfish egg capsule, a coextruded, multilayer laminate. Biomimetics 1:151–175

Knight DP, Feng D (1994a) Interaction of collagen with hydrophobic protein granules in the egg capsule of the dogfish *Scyliorhinus canicula*. Tissue Cell 26:155–167

Knight DP, Feng D (1994b) Some observations on the collagen fibrils of the egg capsule of the dogfish, *Scyliorhinus canicula*. Tissue Cell 26:385–401

Knight DP, Hunt S (1974) Fibril structure of collagen in egg capsules of dogfish. Nat Lond 249:379–380

Knight DP, Hunt S (1976) Fine structure of the dogfish egg case: a unique collagenous material. Tissue Cell 8:183–193

Knight DP, Hunt S (1986) A kinked molecular model for the collagen-containing fibrils in the egg case of the dogfish *Scyliorhinus caniculus*. Tissue Cell 18:201–208

Knight DP, Vollrath F (2001) Reprinted with permission from Knight DP, Vollrath F (2001) Comparison of the spinning of selachian egg case ply sheets and orb web spider dragline filaments. Biomacromolecules 2(2):323–34. Copyright © 2001, American Chemical Society

Knight DP, Feng D, Stewart M et al (1993) Changes in micromolecular organization in collagen assemblies during secretion in the nidamental gland and formation of the egg capsule wall in the dogfish *Scyliorhinus canicula*. Phil Trans R Soc Lond B 341:419–436

Knight DP, Hu XW, Gathercole LJ et al (1996) Molecular orientations in an extruded collagenous composite, the marginal rib of the egg capsule of the dogfish *Scyliorhinus canicula*: a novel lyotropic liquid crystalline arrangement and its origin in the spinnerets. Phil Trans R Soc Lond B 351:1205–1222

Knupp C, Squire JM (2005) Reprinted from Knupp C, Squire JM (2005) Molecular packing in network-forming collagens. Adv Protein Chem 70:375–403. Copyright 2005, with permission from Elsevier

Knupp C, Chew M, Morris E et al (1996) Three dimensional reconstruction of a collagen IV analogue in the dogfish egg case wall. J Struct Biol 117:209–221

Knupp C, Chew M, Squire J (1998) Collagen packing in the dogfish egg case wall. J Struct Biol 122:101–110

Knupp C, Luther PK, Morris EP et al (1999) Partially systematic molecular packing in the hexagonal columnar phase of dogfish egg case collagen. J Struct Biol 126:121–130

Koob TJ (2002) Reprinted from Koob TJ (2002) Biomimetic approaches to tendon repair. Comp Biochem Physiol A Mol Integr Physiol 133(4):1171–1192. Copyright © 2002, with permission from Elsevier

Koob TJ, Cox DL (1986) Studies on skate (*Raja erinacea*) egg capsule formation. II. Introduction of catechols occurs in utero. Bull Mt Desert I Biol Lab 26:109–112

Koob TJ, Cox DL (1988) Egg capsule catechol oxidase from the little skate *Raja erinacea* Mitchill, 1825. Biol Bull 175:202–221

Koob TJ, Cox DL (1990) Accumulation of calcium and magnesium in *Raja erinacea* egg capsule during formation and after oviposition. Bull Mt Desert I Biol Lab 28:26–27

Koob TJ, Cox DL (1993) Stabilization and sclerotization of *Raja erinacea* egg capsule proteins. Environ Biol Fishes 38:151–157

Koob TJ, Hernandez DH (2002) Material properties of polymerized NDGA-collagen fibers: development of biologically-based tendon constructs. Biomaterials 23:203–212

Koob TJ, Hernandez DJ (2003) Mechanical and thermal properties of novel polymerized NDGA-gelatin hydrogels. Biomaterials 24(7):1285–1292

Koob TJ, Willis TA, Hernandez DH (2001a) Biocompatibility of NDGA polymerized fibers. I. Evaluation of cytotoxicity with tendon fibroblasts *in vitro*. J Biomed Mater Res 56:31–39

Koob TJ, Willis TA, Hernandez DH (2001b) Biocompatibility of NDGA polymerized fibers. II. Attachment, replication and migration of tendon fibroblasts *in vitro*. J Biomed Mater Res 56:40–48

Krishnan G (1959) Histochemical studies on the nature and formation of egg capsules of the shark *Chiloscyllium griseum*. Biol Bull 117:298–307

Lombardi J, Files T (1993) Egg capsule structure and permeability in the viviparous shark, *Mustelus canis*. J Exp Zool 267:76–85
Luong TT, Boutillon MM, Garrone R et al (1998) Characterisation of selachian egg case collagen. Biochem Biophys Res Commun 250:657–663
McLaughlin RH, O'Gower AK (1971) Life history and underwater studies of a Heterodont shark. Ecol Monogr 41:271–289
Nalini KP (1940) Structure and function of the nidamental gland of *Chiloscyllium griseum*. Proc Md Acad Sci 12:189–214
Oddone MC, Marcal AS, Vooren CM (2004) Egg capsules of *Atlantoraja cyclophora* (Regan, 1903) and *A. plantana* (Günterm, 1880) (Pisces, Elasmobranchii, Rajidae). Zootaxa 426:1–4
Rusaouën M, Pujol JP, Bocquet J et al (1976) Evidence of collagen in the egg capsule of the dogfish, *Scyliorhinus canicula*. Comp Biochem Physiol B 53:539–543
Rusaouen-Innocent M (1990) A radioautographic study of collagen secretion in the dogfish nidamental gland. Tissue Cell 22:449–462
Smith BG (1942) The heterodontoid sharks: their natural history and the external development of *Heterodontus japonicus* based on notes and drawings by Bashford Dean. The Bashford Dean memorial volume: archaic fishes. Am Mus Nat Hist 8:649–770
Threadgold LT (1957) A histochemical study of the shell gland of *Scyliorhinus caniculus*. J Histochem Cytochem 5:159–166
Whitley GP (1938) The eggs of Australian sharks and rays. Aust Mus Mag 11(11):372–382
Whitley GP (1939) Taxonomic notes on sharks and rays. Aust Zool 9(3):227–262
Widakowich V (1906) Über Bau and Funktion des nidamental Organs von *Scyllium canicula*. Zeitschr f wiss Zool 80:1–21

Chapter 13
Marine Structural Proteins in Biomedicine and Tissue Engineering

Abstract Marine biopolymers like collagens, gelatins, keratins (keratin-like proteins), and elastins (elastin-like proteins) have been investigated during the last decade as potentially approvable, manufacturable, highly reproducible, approvable, and affordable biological materials. The development of biocompatible composites and vehicles of marine biopolymer origin for growth, retention, delivery, and differentiation of stem cells is of crucial importance for regenerative medicine.

All the marine structural proteins discussed have above possess high biomimetic potential. They either directly meet current challenges in many applied fields; or may be used as a source of bioinspiration for designing biomimetic materials and composites with hierarchical structure and appropriative mechanical properties. The application of new biomaterial technologies based on marine biopolymers offers the potential to direct the stem cell fate, targeting the delivery of cells and reducing immune rejection, thereby supporting the development of regenerative medicine (Armentano et al. 2013). The first goal is to develop porous biodegradable composite scaffolds, which can be designed with initial properties that reproduce the tension-compression nonlinearity, viscoelasticity and anisotropy of different tissues by introducing specific nanostructures. The second goal is more complex, as it deals with development of artificial organs. However, the first attempts have been carried out. Recently, dense and porous synthetic scaffolds were used as template for fish cells line, which, being attached, can proliferate and express fish skin components. This biomimetic approach has been used for development of a biosynthetic fish skin in vitro with future application in aquatic robots that can emulate living fish (Pouliot et al. 2004).

As reported above, marine collagens of vertebrate origin as well as gelatins as their derivatives are the most popular biological materials. These can be used as alternative to terrestrial mammal's collagens especially in tissue engineering. According to Glowacki and Mizuno (2008), "there are two major approaches to tissue engineering for regeneration of tissues and organs. One involves cell-free materials and/or factors and one involves delivering cells to contribute to the regeneration process. Of the many scaffold materials being investigated, collagen type I, with selective removal of its telopeptides, has been shown to have many advantageous features for both of these approaches," (Glowacki and Mizuno 2008).

H. Ehrlich, *Biological Materials of Marine Origin*, Biologically-Inspired Systems 4,
DOI 10.1007/978-94-007-5730-1_13

Recently, collagen-based porous, three-dimensional scaffolds have been widely used for regenerative medicine and especially for tissue engineering (Lu et al. 2010; Ko et al. 2010).

Fish collagen is a renewable source of material and has been successfully used for practical applications in biomedicine and tissue engineering in both pure (Kanayama et al. 2008) and crosslinked forms (Yunoki et al. 2007; Nagai et al. 2008). Numerous works were dedicated to shark collagens and their comparison with pig collagens. The results of these comparative studies prepared by Nomura and co-workers can be summarized as follow:

- "The physicochemical properties of shark type I collagen gel and membrane were not same as those of pig type I collagen. The denaturation temperature of shark collagen gel was about 15 °C lower.
- The breaking strength of shark collagen gel was greater, and shark collagen membrane had a greater mechanical strength and a higher water vapor sorption," (Nomura et al. 2000b).
- "Fibril reconstruction process that is, the nucleation and growth of mixed type I collagen fibril from sharks and pigs, progressed faster than that of the individual collagen species of shark or pig.
- The reconstructed mixed collagen fibril had a greater resistance to return to solution or melt into gelatin in comparison with the counterpart consisting solely of shark collagen.
- The denaturation temperature of the mixed collagen gel was about 10 °C higher than that of shark, and about 5 °C lower than that of pig. The breaking strength of the mixed collagen gel was tougher than that of pig, but weaker than that of shark," (Nomura et al. 2000a).

Nomura and Kitazume (2002) also investigated application of shark collagen as a substrate for zymography as well as a scaffold for cell culture. More detailed, "fibroblasts were cultured on a gel matrix of shark type I collagen at 30 °C. The collagen gel had contracted by 4 days of incubation. Individual fibroblasts were visible against the transparent background of the contracted collagen as long, lean star-shaped cells. It was showed that the matrix metalloproteinases (MMPs) from fibroblasts secreted from the medium more easily digested shark gelatin than pig gelatin. MMP-2, -9, and that of potential form were recognizable in the zymographic gel of shark gelatin," (Nomura and Kitazume 2002).

Fish scales collagen (FSC) was used to manufacture the scaffold for corneal tissue engineering (Krishnan et al. 2012). "The *ex vivo* cultured limbal stem cells over a biocompatible scaffold are used in the management of limbal stem cell deficiency as an ideal replacement for human amniotic membrane (HAM). It is important that the mechanical and physical strengths of FSC were comparable to that of HAM's. Under microscopic observation, epithelial migration was noted at the end of 48 h from limbal explants plated on FSC; and after 72 h on HAM. By the end of the 15th day, 90–100 % confluent growth resembling the morphological features of limbal epithelium was seen. FSCs were optically clear with sufficient strength, and gave encouraging results in culture studies; the same may be tried as a potential candidate for corneal transplantation after *in vivo* studies," (Krishnan et al. 2012).

Collagens of fish origin also possess very specific properties that arise from their chemical compositions. Recently, Raabe et al. (2010) "investigated the effect of transforming growth factor beta 1 (TGF-β1), in comparison to hydrolyzed fish collagen, in terms of the chondrogenic differentiation potential of Adipose-derived stromal cells (ADSCs). Adipose-derived stromal cells are multipotent cells which, in the presence of appropriate stimuli, can differentiate into various lineages such as the osteogenic, adipogenic and chondrogenic lines. In this study, chondrogenesis was as effectively induced by hydrolyzed fish collagen as it was successfully induced by TGF-β1. These findings demonstrated that hydrolyzed fish collagen alone has the potential to induce and maintain ADSCs-derived chondrogenesis. These results support the application of ADSCs in equine veterinary tissue engineering, especially for cartilage repair," (Raabe et al. 2010).

In contrast to collagens and gelatins of marine origin keratins, with the exception of keratin-like filaments from hagfish slime, cannot be used as bulk biomaterials. However, their potential for bioinspiration is definitively important. Recently, Douglas Fudge and co-workers published their work entitled "Hagfish slime threads as a biomimetic model for high performance protein fibres" (Fudge et al. 2010). There we can find the motivations underpinning their research:

"Textile manufacturing is one of the largest industries in the world, and synthetic fibres represent two-thirds of the global textile market. Synthetic fibres are manufactured from petroleum-based feedstocks, which are becoming increasingly expensive as demand for finite petroleum reserves continues to rise. For the last three decades, spider silks have been held up as a model that could inspire the production of protein fibres exhibiting high performance and ecological sustainability. Unfortunately, artificial spider silks have yet to fulfil this promise. Previous work on the biomechanics of protein fibres from the slime of hagfishes *[see Sect. 11.2]* suggests that these fibres might be a superior biomimetic model to spider silks. Based on the fact that the proteins within these 'slime threads' adopt conformations that are similar to those in spider silks when they are stretched, it was hypothesized that draw processing of slime threads should yield fibres that are comparable to spider dragline silk in their mechanical performance," (Fudge et al. 2010). These authors showed that "draw-processed slime threads are indeed exceptionally strong and tough. Additionally, post-drawing steps such as annealing, dehydration and covalent cross-linking can dramatically improve the long-term dimensional stability of the threads. The data presented in this paper suggest that hagfish slime threads are a model that should be pursued in the quest to produce fibres that are ecologically sustainable and economically viable," (Fudge et al. 2010).

"Assuming that it is possible to express large quantities of intermediate filament proteins in cheap expression systems like bacteria, spinning those proteins into fibres with useful material properties will also be a challenge, as it has been for those working on artificial spider silks. The fact that intermediate filament proteins self-assemble into 10 nm wide nanofilaments in aqueous solutions may be an important asset of the intermediate filament model. In addition, the secondary structure of intermediate filament proteins is dominated by an α-helical rod domain, which very readily forms coiled coils with the proper polymerization partner. The introspective nature of hydrogen bonding in α-helices makes these proteins less likely to aggregate with other proteins than the far more gregarious spidroins, which have a strong tendency to form β-sheets and β-sheet crystals. The spinning process for artificial slime threads therefore only needs to align the α-helices and arrange the proteins into a coherent fibre. Once these conditions are met, subjecting the proteins to modest stresses that exceed the yield stress (about 3 MPa) during draw processing should open up the α-helices. This may allow for the formation of β-sheets among adjacent proteins and the formation of a strong, tough protein fibre," (Fudge et al. 2010; see also Fudge et al. 2003).

Marine elastins are also discussed above. Although today fish elastins are normally used as source for obtaining their hydrolysates for the cosmetic industry (see for example Nishida et al. 2012), their biomimetic potential is absolutely indisputable. Principles of marine elastin structural organization, and their physic-chemical and material properties, determine current approaches where elastin-like biocomposites are designed and used in tissue engineering (Wise et al. 2009; Kothapalli and Ramamurthi 2009; Nivison-Smith et al. 2010; Girotti et al. 2011; Prieto et al. 2011; Grover et al. 2012).

As discussed Waterhouse and co-workers (2011):

"In addition to its critical role in maintaining vessel integrity and elastic properties under pulsatile flow, elastin plays an important role in signalling and regulating luminal endothelial and smooth muscle cells in the arterial wall. Despite its well-established significance in the vasculature and its growing use as a biomaterial in tissue engineering, the hemocompatibility of elastin is often overlooked. Past studies pointing to the potential of arterial elastin and decellularized elastin as nonthrombogenic materials have begun to be realized, with elastin scaffolds and coatings displaying increased hemocompatibility." (Waterhouse et al. 2011). Recently, interesting results were reported by Shiratsuchi et al. (2010):

"it was showed that elastin peptides prepared from marine fish elastin inhibit collagen-induced platelet aggregation and stimulate migration and proliferation of human skin fibroblasts. These results suggest that piscine elastin peptides can be applied as useful biomaterials in which elasticity, antithrombotic property, and the enhancement of cell migration and proliferation are required," (Shiratsuchi et al. 2010).

13.1 Conclusion

In conclusion, I also take the liberty to concentrate your attention on polysaccharides of marine vertebrate origin; by which I mean proteoglycans and glycosaminoglycans. Proteoglycans (Tingbø et al. 2012) as well as glycosaminoglycans (GAGs) like chondroitin sulfate (Hashiguchi et al. 2011), dermatan sulfate (Chatziioannidis et al. 1999; Sakai et al. 2003; Dhahri et al. 2010), aggrecan (Kakizaki et al. 2011) and hyaluronan (Pfeiler et al. 2002) has been also isolated from marine vertebrates, mostly from fish.

Hyaluronic acid (HA), or hyaluronan is a linear naturally occurring polysaccharide formed from repeating disaccharide units of D-glucuronate and N-acetyl-D-glucosamine (see for review Kogan et al. 2007). Here, the characterization of this biopolymer given by Volpi et al. (2009):

"Despite its relatively simple structure, HA is an extraordinarily versatile glycosaminoglycan currently receiving attention across a wide front of research areas. It has a very high molar mass, usually in the order of millions of Daltons, and possesses interesting visco-elastic properties based on its polymeric and polyelectrolyte characteristics. HA is omnipresent in the human body and in other vertebrates. It occurs in almost all biological fluids and tissues, although the highest amounts of HA are found in the extracellular matrix of soft connective tissues. HA is involved in several key processes, including cell signaling, wound repair and regeneration, morphogenesis, matrix organization and pathobiology. Clinically, it is used as a diagnostic marker for many disease states including cancer, rheumatoid arthritis, liver pathologies, and as an early marker for impending rejection following organ transplantation. It is also used for supplementation of impaired synovial fluid in arthritic patients and for patients following cataract surgery, and as a filler in cosmetic and soft tissue surgery," (Volpi et al. 2009). Additional applications include use "as a device in several surgical procedures, particularly as an anti-adhesive following abdominal procedures, and also in tissue engineering," (Volpi et al. 2009; see also Prestwich 2011).

Hyaluronan has been identified as the principal glycosaminoglycan in the highly hydrated, extracellular body matrix of the larval stage (leptocephalus) of seven species of true eels (Teleostei: Elopomorpha: Anguilliformes) and the ladyfish *Elops saurus* (Elopiformes), and was found as a minor GAG component in the bonefish *Albula* sp. (Albuliformes) (Pfeiler et al. 2002). The authors reported that "in addition to its presumed role in maintaining the structural integrity and hydration of the gelatinous body of the leptocephalus, HA is postulated to function as a storage polysaccharide in those species in which it is the predominant GAG," (Pfeiler et al. 2002).

Both heparin cofactor II and antithrombin are possible mediators of synthesis of a dermatan sulfate, an example of highly sulfated polysaccharide, which have been isolated from the ray *Raja radula* skin. It was showed that dermatan sulfate exhibited a high anticoagulant activity (Ben Mansour et al. 2009a, b, 2010). Similar activity of a dermatan sulfate from the skin of the shark *Scyliorhinus canicula* was also reported by Dhahri et al. (2010). The substance of the shark origin showed anticoagulant effect that was higher than that isolated from pork. Fish dermatan sulfate was not effective in platelet aggregation and, therefore, may be useful in anticoagulant therapy (Dhahri et al. 2010).

References

Armentano I, Fortunati E, Mattioli S, Rescignano N, Kenny JM (2013) Biodegradable composite scaffolds: a strategy to modulate stem cell behaviour. Recent Pat Drug Deliv Formul 7(1):9–17

Ben Mansour M, Dhahri M, Vénisse L et al (2009a) Mechanism of thrombin inhibition by heparin cofactor II and antithrombin in the presence of the ray (*Raja radula*) skin dermatan sulfate. Thromb Res 123(6):902–908

Ben Mansour M, Majdoub H, Bataille I et al (2009b) Polysaccharides from the skin of the ray *Raja radula*. Partial characterization and anticoagulant activity. Thromb Res 123(4):671–678

Ben Mansour M, Dhahri M, Hassine M et al (2010) Highly sulfated dermatan sulfate from the skin of the ray *Raja montagui*: anticoagulant activity and mechanism of action. Comp Biochem Physiol B Biochem Mol Biol 156(3):206–215

Chatziioannidis CC, Karamanos NK, Anagnostides ST (1999) Purification and characterisation of a minor low–sulphated dermatan sulphate–proteoglycan from ray skin. Biochimie 81(3):187–196

Dhahri M, Mansour MB, Bertholon I et al (2010) Anticoagulant activity of a dermatan sulfate from the skin of the shark *Scyliorhinus canicula*. Blood Coagul Fibrinolysis 21(6):547–557

Fudge DS, Gardner KH, Forsyth VT et al (2003) The mechanical properties of hydrated intermediate filaments: insights from hagfish slime threads. Biophys J 85:2015–2027

Fudge DS, Hillis S, Levy N et al (2010) Hagfish slime threads as a biomimetic model for high performance protein fibres. Bioinspir Biomim 5:035002. doi:10.1088/1748-3182/5/3/035002. Copyright © 2014 IOP Publishing. Reprinted with permission

Girotti A, Fernández–Colino A, López IM et al (2011) Elastin–like recombinamers: biosynthetic strategies and biotechnological applications. Biotechnol J 6(10):1174–1186

Glowacki J, Mizuno S (2008) Collagen scaffolds for tissue engineering. Biopolymers 89(5):338–344. Copyright © 2008 Wiley Periodicals, Inc. Reprinted with permission

Grover CN, Cameron RE, Best SM (2012) Investigating the morphological, mechanical and degradation properties of scaffolds comprising collagen, gelatin and elastin for use in soft tissue engineering. J Mech Behav Biomed Mater 10:62–74

Hashiguchi T, Kobayashi T, Fongmoon D et al (2011) Demonstration of the hepatocyte growth factor signaling pathway in the *in vitro* neuritogenic activity of chondroitin sulfate from ray fish cartilage. Biochim Biophys Acta 1810(4):406–413

Kakizaki I, Tatara Y, Majima M et al (2011) Identification of proteoglycan from salmon nasal cartilage. Arch Biochem Biophys 506(1):58–65

Kanayama T, Nagai N, Mori K (2008) Application of elastic salmon collagen gel to uniaxial stretching culture of human umbilical vein endothelial cells. J Biosci Bioeng 105(5):554–557

Ko YG, Grice S, Kawazoe N et al (2010) Preparation of collagen–glycosaminoglycan sponges with open surface porous structures using ice particulate template method. Macromol Biosci 10(8):860–871

Kogan G, Soltés L, Stern R et al (2007) Hyaluronic acid: a natural biopolymer with a broad range of biomedical and industrial applications. Biotechnol Lett 29(1):17–25

Kothapalli CR, Ramamurthi A (2009) Biomimetic regeneration of elastin matrices using hyaluronan and copper ion cues. Tissue Eng Part A 15(1):103–113

Krishnan S, Sekar S, Katheem MF et al (2012) Fish scale collagen–a novel material for corneal tissue engineering. Artif Organs 36(9):829–835. © 2012, Copyright the Authors. Artificial Organs © 2012, International Center for Artificial Organs and Transplantation and Wiley Periodicals, Inc. Reprinted with permission

Lu H, Ko YG, Kawazoe N (2010) Cartilage tissue engineering using funnel–like collagen sponges prepared with embossing ice particulate templates. Biomaterials 31(22):5825–5835

Nagai N, Mori K, Munekata M (2008) Biological properties of crosslinked salmon collagen fibrillar gel as a scaffold for human umbilical vein endothelial cells. J Biomater Appl 23(3):275–287

Nishida K, Tateishi C, Tsuruta D et al (2012) Contact urticaria caused by a fish–derived elastin–containing cosmetic cream. Contact Dermatitis 67:171–172

Nivison–Smith L, Rnjak J, Weiss AS (2010) Synthetic human elastin microfibers: stable cross–linked tropoelastin and cell interactive constructs for tissue engineering applications. Acta Biomater 6(2):354–359

Nomura Y, Kitazume N (2002) Use of shark collagen for cell culture and zymography. Biosci Biotechnol Biochem 66(12):2673–2676. Copyright © 2002 Taylor & Francis

Nomura Y, Toki S, Ishii Y (2000a) Improvement of the material property of shark type I collagen by composing with pig type I collagen. J Agric Food Chem 48(12):6332–6336

Nomura Y, Toki S, Ishii Y et al (2000b) The physicochemical property of shark type I collagen gel and membrane. J Agric Food Chem 48(6):2028–2032

Pfeiler et al (2002) Reprinted from Pfeiler E, Toyoda H, Williams MD, Nieman RA (2002) Identification, structural analysis and function of hyaluronan in developing fish larvae (leptocephali). Comp Biochem Physiol B Biochem Mol Biol 132(2):443–451. Copyright (Year), with permission from Elsevier

Pouliot R, Azhari R, Qanadilo HF et al (2004) Tissue engineering of fish skin: behavior of fish cells on poly(ethylene glycol terephthalate)/poly(butylene terephthalate) copolymers in relation to the composition of the polymer substrate as an initial step in constructing a robotic/living tissue hybrid. Tissue Eng 10(1–2):7–21

Prestwich GD (2011) Hyaluronic acid–based clinical biomaterials derived for cell and molecule delivery in regenerative medicine. J Control Release 155(2):193–199

Prieto S, Shkilnyy A, Rumplasch C et al (2011) Biomimetic calcium phosphate mineralization with multifunctional elastin–like recombinamers. Biomacromolecules 12(5):1480–1486

Raabe et al (2010) With kind permission from Springer Science+Business Media: Raabe O, Reich C, Wenisch S et al (2010) Hydrolyzed fish collagen induced chondrogenic differentiation of equine adipose tissue-derived stromal cells. Histochem Cell Biol 134(6):545–554. Copyright © 2010, Springer

Sakai S, Kim WS, Lee IS et al (2003) Purification and characterization of dermatan sulfate from the skin of the eel, *Anguilla japonica*. Carbohydr Res 338(3):263–269

Shiratsuchi E, Ura M, Nakaba M et al (2010) Elastin peptides prepared from piscine and mammalian elastic tissues inhibit collagen-induced platelet aggregation and stimulate migration and proliferation of human skin fibroblasts. J Pept Sci 16:652–658. doi:10.1002/psc.1277. Copyright © 2010 European Peptide Society and John Wiley & Sons, Ltd. Reprinted with permission

Tingbø MG, Pedersen ME, Kolset SO et al (2012) Lumican is a major small leucine–rich proteoglycan (SLRP) in Atlantic cod (*Gadus morhua* L.) skeletal muscle. Glycoconj J 29(1):13–23

Volpi N, Schiller J, Stern R et al (2009) Role, metabolism, chemical modifications and applications of hyaluronan. Curr Med Chem 16(14):1718–1745. Copyright © 2009, Bentham Science Publisher. Reprinted by permission of Eureka Science Ltd

Waterhouse A, Wise SG, Ng MK, Weiss AS (2011) Elastin as a nonthrombogenic biomaterial. Tissue Eng Part B Rev 17(2):93–99. The publisher for this copyrighted material is Mary Ann Liebert, Inc. publishers. Reprinted with permission

Wise SG, Mithieux SM, Weiss AS (2009) Engineered tropoelastin and elastin–based biomaterials. Adv Protein Chem Struct Biol 78:1–24

Yunoki S, Mori K, Suzuki T (2007) Novel elastic material from collagen for tissue engineering. J Mater Sci Mater Med 18(7):1369–1375

Chapter 14
Epilogue

My efforts in the present work were to show the structural and chemical diversity in biological materials derived from marine fish, reptiles, birds and mammals as well as their material properties. The scientific history of the discovery of these materials spans the last 150 years, and our own recent results may stimulate other researchers to rise to new challenges. In the interest of, my book is dedicated to biological materials isolated, observed, or described in marine vertebrate organisms selected by the author. Rather than a comprehensive work, this book is intended to provide an overview of a few organisms that demonstrate the high potential of future discovery in this field.

In these concluding remarks, I want to focus the attention of readers on some open questions, as well as on additional topics of interest. These include fish gills, amphibian skin and sucking disk of sharksuckers.

Fish Gills Unique biomaterials-containing structures not discussed in this work are fish gills. The function of gills is well described in the paper by Campbell et al. (2008).

"A fish continuously pumps water through its mouth and over gill arches, using coordinated movements of the jaws and operculum (gill cover) for this ventilation (Fig. 14.1). A swimming fish can simply open its mouth and let water flow past its gills. Each gill arch has two rows of gill filaments, composed of flattened plates called lamellae. Blood flowing through capillaries within the lamellae picks up oxygen from the water. Notice that the counter current flow of water and blood maintains a concentration gradient down which O_2 diffuses from the water into the blood over the entire length of a capillary," Campbell et al. (2008).

Two parallel sheets of epithelia separated by a narrow space for blood circulation and for the exchange of respiratory gases are the main structural components of gill lamella (Olson 2002; Wilson and Laurent 2002; Evans et al. 2005). Correspondingly, surface area is increased due to this specific lamellar structure. Also, so called collagen columns (Hughes and Grimstone 1965; Bettex-Galland and Hughes 1973; Wright 1973), which presumably function to prevent ballooning of the lamellae, are of crucial importance (Fig. 14.2). Normally, collagen columns are surrounded by

H. Ehrlich, *Biological Materials of Marine Origin*, Biologically-Inspired Systems 4,
DOI 10.1007/978-94-007-5730-1_14

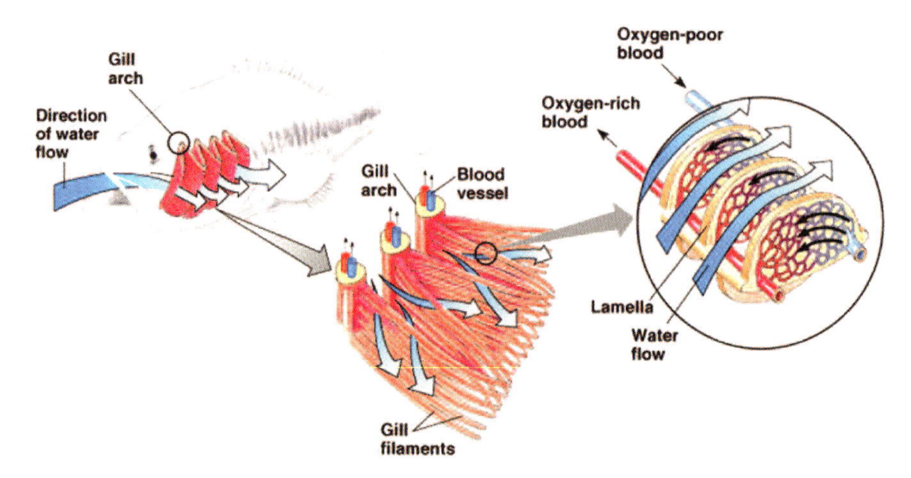

Fig. 14.1 Schematic of the teleost fish gill (CAMPBELL, NEIL A.; REECE, JANE B., BIOLOGY, 8th, ©2008. Printed and Electronically reproduced by permission of Pearson Education, Inc., Upper Saddle River, New Jersey). See text for details

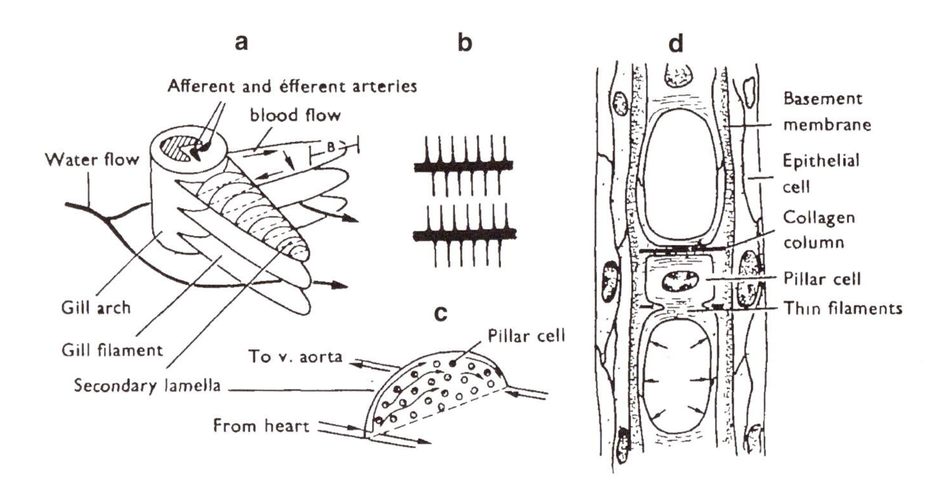

Fig. 14.2 Diagrams to show the structure of fish gills. (**a**) Single gill arch with two rows of filaments. Secondary lamellae are drawn on the upper surface of one filament. Directions of water and blood flow are indicated. (**b**) Longitudinal section of two filaments showing secondary lamellae projecting above and below each filament. (**c**) A single secondary lamella to show blood flow between pillar cells. (**d**) Section across a secondary lamella showing epithelial, basement membrane, and pillar cell flange layers separating blood from the water. Pillar cells with fine filaments are shown, and a column connects the two basement membrane layers (Republished with permission of The Company of Biologists Ltd, from Bettex-Galland and Hughes 1973; permission conveyed through Copyright Clearance Center, Inc.)

plasma membranes composed of pillar cells. The function of these endothelial cells is to isolate columns from the circulation (see for details Hughes and Grimstone 1965; Wright 1973; Olson 2002; Wilson and Laurent 2002; Evans et al. 2005). Thus two parallel sheets of respiratory epithelium are connected due to pillar cells with spool-shaped, cylindrical morphology. "Usually five to eight collagen columns are enfolded by the plasma membrane of a pillar cell. In the peripheral cytoplasm, pillar cells have numerous myofilaments that run parallel to the collagen columns," (Kato et al. 2007; see also Bettex-Galland and Hughes 1973).

Thus, in the case of fish gills we can observe very a unusual role for collagen in the form of columns that must be definitively investigated in the future.

Recently, visiting the Zoological Museum at Christian-Albrecht University in Kiel (Germany), I found a well prepared skeleton of whale shark (Fig. 14.3) with amazing gill rakers. Gill rakers attached to the branchial arches have been hypothesized to be a component of all filtration mechanisms in fish. The functions of the whale shark's gills are:

- to filter plankton using sieves-like microstructured rakers;
- to extracting oxygen from large volumes of seawater.

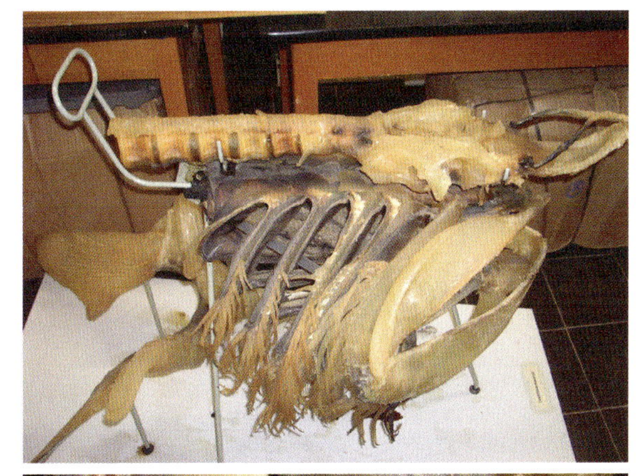

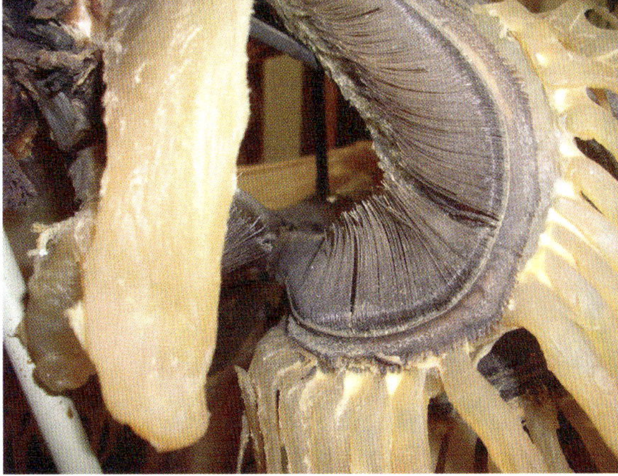

Fig. 14.3 Gill rakers of whale shark are morphologically very similar to whale baleen (Image courtesy of Andre Ehrlich)

However, my interest is based on the high structural similarity between these gill rackers and those of whale baleen. Intriguingly, our first measurements on mechanical properties of both biological materials showed very similar results. Now, we are starting experimental work to investigate the chemical composition of gill rackers from whale sharks. It would be very surprising to obtain analytical results confirming the keratinous nature of these gill rackers, as sharks and whales are related to fish and mammals, respectively. This challenging and exciting task stimulates our attempts to obtain new knowledge.

Amphibian Skin Desalination or other water purification by use of membranes are topics of much current interest. The skin of amphibians seems to be the source of bioinspiration for this application. Both phenomena, which player crucial role in amphibian evolution, like gas exchange and free water movement across skin are based on specific permeability of their skin. "Amphibians generally have very high rates of water turnover unless, like fossorial forms, they exploit environmental situations where water fluxes across their highly permeable skins are low. Many amphibians have remarkable tolerances to osmotic imbalances, and are able to rectify these rapidly when water becomes available," (Shoemaker and Nagy 1977; see also Neill 1958). Salinity potentially influences the abundance and distribution of amphibians by imposing osmotic stress (Smith et al. 2007). The issue of salinity effects on amphibian biology and ecology has recently drawn more attention, for at least two reasons (see for review Wu and Kam 2009): first, several studies have revealed that amphibians breed in brackish water more commonly than originally thought, and that they exhibit interesting physiological and ecological adjustments to increased salinity stress; second, amphibians are important indicator species for freshwater ecosystems that face increased salinization due to natural causes, or to anthropogenic ones such as salt pollution resulting from road de-icing salt runoff.

In contrast to amphibians, reptiles, due to their low skin permeability, uricotelism, and salt glands, are well adapted to remain in water. Amphibians adapted to saline media store solutes and remain at least slightly hyperosmotic to their environment (see for review Schoemaker and Nagy 1977).

As discussed by Natchev and co-authors:

"Due to their highly vascularised and permeable skins, amphibians have a low capacity to handle increased salinity in the waters they inhabit. Nevertheless, a few species are able to live in brackish and even hypersaline waters. Reports on adult amphibians, as well as their tadpoles, living under saline conditions are numerous; but only a few amphibians can survive prolonged exposure to sea water. Among anuran species, adaptations to elevated salinity differ considerably. One ranid species – the crab eating frog, *Fejervarya cancrivora* (Gravenhorst 1829) – is considered to possess the highest salinity tolerance among amphibians. In this species, 50 % of the larvae survived in up to 80 % sea water – equivalent to 0.5 %M Cl," (Natchev et al. 2011; see also Beebee 1985; Dunson 1977; Karraker 2007; Sillero and Ribeiro 2010; Spotila and Berman 1976; Wu and Kam 2009).

Aside from *F. cancrivora*, only two other anurans, the bufonid *Pseudepidalea viridis* and the pipid *Xenopus laevis*, are capable of tolerating higher than brackish salinities (Balinsky et al. 1972; Hillmann et al. 2009).

The Crab-eating Frog is broadly distributed in India and Asia including mangroves, marshes, coastal scrub, and disturbed forests where it can tolerate both brackish and sea water. Phylogenetic analysis supported the monophyly of the *Fejervarya* species, and the genome organization of *F. cancrivora* mitochondrial DNA differs from that of neobatrachian frogs and typical vertebrates (Ren et al. 2009).

"The exceptional ability of *F. cancrivora* to survive in saline media is probably related to the exceptionally high rates of urea production exhibited in this species. Experimentation with unfed amphibians may lead to underestimation of salinity tolerance, because these fasting animals must deaminate a significant fraction of their protein to produce and maintain high concentrations of urea," (Schoemaker and Nagy 1977). As the most salt-tolerant of all frogs (Hillmann et al. 2009) it has been found under laboratory conditions to tolerate salinities of up to 29 ppt (Gordon et al. 1961). Adults of this species survive in high salinity by maintaining a hyperosmotic plasma that uses the increased concentrations of sodium, chloride, and urea (Gordon et al. 1961; Schmidt-Nielsen and Lee 1962; Gordon and Tucker 1968). Thus, the plasma of *Fejervarya* species helps them to survive in salt aquatic niches by topping up the ionic concentration with non-ionic solubilised urea.

Adult frogs have a different, and rather more unusual, method of osmotic regulation. Instead of being osmoregulators and maintaining an imbalance between the osmotic concentration of their internal fluids and that of the exterior, they are partial osmo-conformers. Internal osmotic concentration is matched with that of the exterior, at least in a hyperosmotic medium. This is brought about not by manipulating ion levels, but by 'topping up' the ionic concentration of the plasma with the non-ionic solute urea $CO(NH_2)_2$. Other amphibia show slightly raised urea levels under conditions of water shortage, but employing high urea levels to maintain osmotic balance with the environment is a habit shared only with elasmobranch fish (sharks, skates and rays) (Gordon et al. 1961). Urea is the standard nitrogenous excretory product of ordinary adult amphibia, and is normally voided at the earliest convenient opportunity. Indeed, given the solubility of urea, it is not an easy substance for a largely aquatic animal to retain, and it is not understood how *Rana cancrivora* (*Fejervarya raja*) achieves this. Not only is it soluble and difficult to retain, urea is toxic: the concentrations of urea that occur in the frog's plasma at exterior salinities of 80 % sea water, should denature enzymes and affect the binding of oxygen by haemoglobin. Somehow *R. cancrivora* copes with these hazards (Hogarth 1999).

The crab-eating frog remains somewhat hypertonic to the external medium, even in the most concentrated solutions that have been tested. This situation is maintained by the formation of urine which is more dilute than the plasma. Since the skin is permeable to water, the inevitable result is an osmotic inflow of water from the medium, and this inflow permits the formation of urine without any necessity for the frog to drink the external medium. There may be several advantages in this situation (Schmidt-Nielsen and Lee 1962). One is that the gastro-intestinal tract is not loaded with an excessive intake of magnesium and sulphate, which constitute roughly one-tenth of the salts in sea water. Another advantage in the formation of dilute urine is that this urine, if retained in the bladder, can serve as a water reservoir. Osmoregulation of the crab-eating frog in sea water resembles that of elasmobranchs, except in that there is no evidence of active tubular reabsorption of urea in the frog (Smith 1936).

The phenomenon of salt water resistance observed for this animal is also very intriguing from biomaterial science point of view, especially because of the skin properties. Recent studies on crab-eating frogs in Indonesia confirms suggestions that these saltwater amphibians possess special skin resistance to water loss that allows them to reduce desiccation during the critical period of acclimation from land to salt water aerials (Wygoda et al. 2011). Note that semi-aquatic species typically exhibit no skin resistance to water loss. In 2010 Wygoda and co-workers discovered crab-eating frogs on Hoga Island, Indonesia. Here, some results as follow:

"The discovery presented an ideal opportunity to investigate water conservation in perhaps the world's most unique amphibian. Using gravimetric wind tunnel methods, the frog's ability to resist water loss across the skin was tested. High resistance values are common for arboreal and desert fossorial frogs, which may have values up to 300 $s*cm^{-1}$. However, semi-aquatic species, like the crab-eating frog, typically exhibit no skin resistance, so that water evaporates at rates equal to free water surface (i.e. 0 sec/cm). Surprisingly, it was found a skin resistance value of 0.27 with a standard error of $\pm$ 0.06 $s*cm^{-1}$. While this value seems small, the relationship between vapour density, water loss, and skin resistance means that modest increases in resistance may dramatically reduce overall skin evaporation rates," (Bennett et al. 2011).

The adaptive advantage of evaporative water loss reduction in *F. cancrivora* may be related to its amphibious behaviour and the time course of the mechanism by which it develops the ability to survive in waters of high salinity. Crab-eating Frogs often move between fresh water and saltwater habitats (Wells 2007), but their compensatory establishment of elevated plasma solute levels is not accomplished immediately upon entry into high salinity water. Indeed, Dicker and Elliott (1970) found that freshwater-acclimated frogs placed in solutions of >270 mOsm initially lost water osmotically that they did not recover for several days. Extrapolating results of Dicker and Elliott (1970) to full-strength seawater (ca. 2,000 mOsm) shows that freshwater-acclimated frogs moving into the sea would initially lose water at a rate of about 41 mg cm^{-2} h^{-1}. A 50-g frog in seawater would fare somewhat better, reaching the same dehydration level in ca. 4 h. Clearly, initial ventures into the marine environment must be short-lived for these frogs, especially for smaller juveniles, and it is during time on land that reduced evaporative water loss would be most beneficial. A low-magnitude cutaneous resistance, for example, could prolong survival time on land by several hours. The same 2.35-g frog in air, with a cutaneous resistance of 0.27 s cm^{-1} and in the water-conserving posture, could increase its time out of water by 1.5 h over a frog with no cutaneous resistance. The larger 50-g frog under similar conditions could in-crease time on land by nearly 7 h. Reduced evaporative water loss probably continues to provide an advantage to frogs even after they fully acclimate to marine conditions, in that they must regularly venture back onto land to restore urea that is continuously lost to the sea faster than it can be generated metabolically (Gordon and Tucker 1968).

The mechanism responsible for reduced evaporative water loss in *F. cancrivora* is unknown. The only mechanism known to provide for reduced evaporative water loss in frogs in a given environment other than through cranial co-ossification or cocoon formation (see for review Wygoda et al. 2011) is by lipid secretion onto the skin surface from lipid glands or lipid-secreting mucous glands (reviewed by Lillywhite 2006). The histology of *F. cancrivora* skin has been examined and was found to be unique among amphibians in having (in addition to mucous glands) two previously undescribed glands (vacuolated and mixed) and no granular glands (Seki et al. 1995). Because Seki et al. (1995) did not stain their skin samples for lipids, it is not known whether any of these gland types are lipid sources. Further examination of *F. cancrivora* skin might help elucidate the mechanism that provides this species with its low level of resistance to evaporative water loss.

Sucking Disk The fish species termed *Echeneis naucrates* and known as the sharksucker was originally described by Linnaeus in 1758. The genus name, *Echeneis*, is derived from the Greek '*echein*' meaning to hold and 'nays' meaning ship; suckling fish, or remora (see for review Richards 2006). In ancient time, Greek sailors believed in mysterious magical powers of sharksuckers which could slow down or even stop their ships.

Transportation of one organism by another, phenomenon known as phoresy, is related to characteristic behavioural features of remoras, which usually can be observed on large fish species like swordfish or tuna, or even, in the case of small specimens, traveling in the mouths or gills of large manta rays, sailfish, and ocean sunfish.

After the young remoras are about 1 cm long, the sucking disc became visible and is fully formed when the fish reaches about 3 cm. In adult forms, the sucking disc is well visible due to a series of paired, transversely oriented, "*pectinated lamellae*" which forms the most conspicuous structure on the surface of dorsoventrally flattened head. The pair of pectinated lamellae together with the interneural ray and intercalary bone represent the skeleton of the sucking disc (Bonnell 1962). "The main part of each pectinated lamella is formed by bilateral extensions of the base of the fin spine just above its proximal tip, each of which develops a row of spinous projections, or spinules, along its posterior margin. The number of rows and the number of spinules increase with size, and they become autogenous from the body of the lamellae," (Britz and Johnson 2012) (Fig. 14.4). The spino-occipital and anterior spinal nerves are responsible for innervation of the muscles and the skin of the disc that is originally belonged to the anterior region of the trunk (Houy 1910). "A series of lateral line nerve innervated sense organs at the rim of the disk's suction cup are similar in structure to Meissner's corpuscles in mammals, sensitive to touch, and play an important role during the attachment of the remora. The remarkable behaviour of remoras to attach to large animals in the water has led to a similarly fascinating, but true, use of remoras for catching fish and turtles. This is only possible because of the great force that is needed to dislodge a remora once it has attached" (Britz and Johnson 2012; see also Fulcher and Motta 2006).

Fig. 14.4 Sharksuckers (*above*) possess unique sucking discs (*below*) (Image courtesy of George Burgess)

What about the mechanism of disc action? This is described by Fulcher and Motta (2006) as follow:

"Disk muscles erect or depress the numerous paired laminae, or toothed plates, which bear two to four rows of posteriorly directed spinules. The erect laminae create a sub-ambient chamber, allowing these fish to adhere to other fish and inanimate objects. Resting sub-ambient suction pressure differentials were recorded, as were the greatest sub-ambient pressure differentials as the fish were pulled posteriorly to simulate drag induced by a swimming host. The resting pressure differential averaged –0.5 kPa, with no significant difference between Plexiglas® and shark skin surfaces. With a force applied to their caudal peduncle, the echeneids generated suction pressure differentials averaging –92.7 kPa within the disk cavity while attached to Plexiglas. On shark skin, the use of spinules increased friction and reduced the maximum sub-ambient suction pressure differential to –46.6 kPa; considerably more force (17.4 N) was required to dislodge the echeneids from the shark skin than from the smooth Plexiglas (11.2 N)," (Fulcher and Motta 2006).

Here, I absolutely agree with Fulcher and Motta that "future research should investigate the relationships between sub-ambient suction pressure differential, disk size, and host specificity, as well as the possibility of greater reliance on spinules in species that adhere to rough-scaled hosts. Furthermore, future research should examine the anterior blind chamber and its purported role as a pneumatic pump, and extend our findings by investigating the relationship between drag and suction pressure differentials when remoras are exposed to water of varying velocities in flume chambers," (Fulcher and Motta 2006). It would be also be highly useful to obtain knowledge about the nanostructural organization of the disc, and the molecular interrelations on this level.

This book covers many different fields, and I hope it provides reader with a current survey of biomimetic applications for a diverse range of marine vertebrates. While certain subjects I have covered here obviously warrant a more in-depth review, I hope this work server to stimulate interest and further investigations into these diverse interdisciplinary topics.

References

Balinsky JB, Dicker SE, Elliot AB (1972) The effect of long term adaptation to different levels of salinity on urea synthesis and tissue amino acid concentrations in *Rana cancrivora*. Comp Physiol Biochem 43B:71–82

Beebee TJC (1985) Salt tolerance of Natterjack Toad (*Bufo calamita*) eggs and larvae from coastal and inland populations in Britain. Herpetol J 1:14–16

Bennett W, Dabruzzi T, Wygoda M (2011) Saltwater frogs exhibit water conservation to exploit their environment. Published on-line http://www.biodiversityscience.com/2011/04/27/saltwater-frogs-water-conservation/. Accessed 15 May 2014. © 2014 Biodiversity Science. All Rights Reserved

Bettex-Galland M, Hughes GM (1973) Contractile filamentous material in the pillar cells of fish gills. J Cell Sci 13:359–370

Bonnell B (1962) Structure of the sucker of Echeneis. Nature 196:1114–1115

Britz R, Johnson GD (2012) Ontogeny and homology of the skeletal elements that form the sucking disc of remoras (Teleostei, Echeneoidei, Echeneidae). J Morphol 273:1353–1366. Copyright © 2012 Wiley Periodicals, Inc. Reprinted with permission

Campbell NA, Reece JB, Urry LA et al (2008) Circulation and gas exchange, Chapter 42. In: Biology, 8th edn. Pearson Benjamin Cummings, San Francisco. ©2008. Printed and Electronically reproduced by permission of Pearson Education, Inc., Upper Saddle River, NJ

Dicker SE, Elliott AB (1970) Water uptake by the crab-eating frog *Rana cancrivora*, as affected by osmotic gradients and by neurohypophysial hormones. J Physiol 207:119–132

Dunson WA (1977) Tolerance to high temperature and salinity by tadpoles of the Philippine frog *Rana cancrivora*. Copeia 1977:375–378

Evans DH, Piermarini PM, Choe KP (2005) The multifunctional fish gill: dominant site of gas exchange, osmoregulation, acid-base regulation, and excretion of nitrogenous waste. Physiol Rev 85:97–177

Fulcher BA, Motta PJ (2006) Suction disk performance of echeneid fishes. Can J Zool 84:42–50. Copyright © 2008 Canadian Science Publishing or its licensors. Reproduced with permission

Gordon MS, Tucker VA (1968) Further observations on the physiology of salinity adaptation in the crab-eating frog (*Rana cancrivora*). J Exp Biol 49:185–193

Gordon MS, Schmidt-Nielsen K, Kelly HM (1961) Osmotic regulation in the crab-eating frog (*Rana cancrivora*). J Exp Biol 38(3):659–678

Hillmann SS, Withers PC, Drewes RC, Hillyard SD (2009) Ecological and environmental physiology of amphibians. Oxford University Press, New York

Hogarth PJ (1999) The biology of mangroves. Oxford University Press, Oxford, 228 p

Houy R (1910) Beiträge zur Kenntnis der Haftscheibe von Echeneis. Zool Jb Anat Ontog Tiere 29(for 1909):101–138

Hughes GM, Grimstone AV (1965) The fine structure of the secondary lamellae of the gills of *Gadus pollachius*. Q J Microsc Sci 106:343–353

Karraker N (2007) Are embryonic and larval green frogs (*Rana clamitans*) insensitive to road deicing salt? Herpetol Conserv Biol 2:35–41

Kato A, Nakamura K, Kudo H et al (2007) Characterization of the column and autocellular junctions that define the vasculature of gill lamellae. J Histochem Cytochem 55(9):941–953. Copyright © 2007, The Histochemical Society. Reprinted by Permission of SAGE Publications.)

Lillywhite HB (2006) Water relations of tetrapod integument. J Exp Biol 209:202–226

Natchev N, Tzankov N, Gemel R (2011) Green frog invasion in the Black Sea: habitat ecology of the *Pelophylax esculentus* complex (Anura, Amphibia) population in the region of Shablenska Tuzla lagoon in Bulgaria. Herpetol Notes 4:347–351. Copyright © 2013, Societas Europaea Herpetologica. Reprinted with permission

Neill WT (1958) The occurrence of amphibians and reptiles in saltwater areas, and a bibliography. Bull Mar Sci Gulf Caribb 8:1–9

Olson KR (2002) Vascular anatomy of the fish gill. J Exp Zool 293:214–231

Ren Z, Zhu B, Ma E et al (2009) Complete nucleotide sequence and gene arrangement of the mitochondrial genome of the crab-eating frog *Fejervarya cancrivora* and evolutionary implications. Gene 441:148–155

Richards WJ (2006) Echeneidae: Remoras. In: Richards WJ (ed) Early stages of Atlantic fishes, vol 2, An identification guide for the Western Central North Atlantic. Taylor & Francis, Boca Raton, pp 1433–1438

Schmidt-Nielsen K, Lee P (1962) Kidney function in the crab-eating frog (*Rana cancrivora*). J Exp Biol 39:167–177

Seki T, Kikuyama S, Yanaihara N (1995) Morphology of the skin glands of the crab-eating frog (*Rana cancrivora*). Zool Sci 12:523–626

Shoemaker V II, Nagy KA (1977) Reproduced with permission of Annual Review of Physiology: Shoemaker V II, Nagy KA (1977) Osmoregulation in amphibians and reptiles. Annu Rev Physiol 39:449–471, by Annual Reviews. http://www.annualreviews.org

Sillero N, Ribeiro R (2010) Reproduction of *Pelophylax perezi* in brackish water in Porto (Portugal). Herpetol Notes 3:337–340

Smith HW (1936) The retention and physiological role of urea in the elasmobranchii. Biol Rev 11:49–82

Smith MJ, Schreiber ESG, Scroggie MP et al (2007) Associations between anuran tadpoles and salinity in a landscape mosaic of wetlands impacted by secondary salinisation. Freshw Biol 52:75–84. doi:10.1111/j.1365-2427.2006.01672.x

Spotila JR, Berman EN (1976) Determination of skin resistance and the role of the skin in controlling water loss in amphibians and reptiles. Comp Biochem Physiol Part A 55:407–411

Wells KD (2007) The ecology and behavior of amphibians. University of Chicago Press, Chicago

Wilson JM, Laurent P (2002) Fish gill morphology: inside out. J Exp Zool 293:192–213

Wright DE (1973) The structure of the gills of the elasmobranch, *Scyliorhinus canicula* (L.). Z Zellforsch Mikrosk Anat 144:489–509

Wu CS, Kam Y-C (2009) Effects of salinity on the survival, growth, development, and metamorphosis of *Fejervarya limnocharis* tadpoles living in brackish water. Zoolog Sci 26:476–482

Wygoda ML, Dabruzzi TF, Bennett WA (2011) Cutaneous resistance to evaporative water loss in the crab-eating frog (*Fejervarya cancrivora*). J Herpetol 45(4):417–420

Index

H. Ehrlich, *Biological Materials of Marine Origin*, Biologically-Inspired Systems 4,
DOI 10.1007/978-94-007-5730-1